Springer Tracts in Modern Physics

Volume 283

Springer Tracts in Modern Physics provides comprehensive and critical reviews of topics of current interest in physics. The following fields are emphasized:

- Elementary Particle Physics
- Condensed Matter Physics
- Light Matter Interaction
- Atomic and Molecular Physics
- Complex Systems
- Fundamental Astrophysics

Suitable reviews of other fields can also be accepted. The Editors encourage prospective authors to correspond with them in advance of submitting a manuscript. For reviews of topics belonging to the above mentioned fields, they should address the responsible Editor as listed in “Contact the Editors”.

More information about this series at http://www.springer.com/series/426

Miron Amusia · Vasily Shaginyan

Strongly Correlated Fermi Systems

A New State of Matter

Springer

Miron Amusia
Racah Institute of Physics
The Hebrew University
Jerusalem, Israel

A. F. Ioffe Physical-Technical Institute
St. Petersburg, Russian Federation

Vasily Shaginyan
Petersburg Nuclear Physics Institute
of National Research Centre
"Kurchatov Institute"
Gatchina, Russia

Clark Atlanta University
Atlanta, USA

ISSN 0081-3869 ISSN 1615-0430 (electronic)
Springer Tracts in Modern Physics
ISBN 978-3-030-50358-1 ISBN 978-3-030-50359-8 (eBook)
https://doi.org/10.1007/978-3-030-50359-8

This Springer imprint is published by the registered company Springer Nature Switzerland AG
The registered company address is: Gewerbestrasse 11, 6330 Cham, Switzerland

Preface

Strongly correlated Fermi systems are numerous and represented by high number of heavy-fermion systems such as heavy-fermion (HF) metals, high-temperature superconductors, quantum spin liquids (QSL) confined in frustrated insulators, quasicrystals, two-dimensional (2D), and one-dimensional (1D) quantum liquids like 2D ^{3}He, and electrons in metal-oxide-semiconductor field-effect transistor (MOSFETs). For brevity, we shall call these strongly correlated Fermi systems heavy-fermion compounds. Numerous experimental facts collected on HF compounds reveal their thermodynamic, transport, relaxation, etc. properties, thus representing all fields of the condensed matter physics. These facts allow us to conclude that the physics of HF compounds represents a new edition of the condensed matter physics, for the observed behavior is quite different from that of conventional metals and insulators. We note that the book "Theory of Heavy-Fermion Compounds" (Springer Series in Solid-State Sciences, Volume 182) offers the theory of the fermion condensation (FC) that explains high number of the corresponding important experimental facts. However, the theory of the fermion condensation is still under construction; therefore, the problem of elaborating the new edition of the condensed matter physics is both arduous and of great importance.

In this book, combining analytical considerations with arguments based entirely on experimental grounds, we show that experimental facts collected on very different strongly correlated Fermi systems demonstrate a remarkable commonality among them, as expressed in universal scaling behavior of their thermodynamic, transport, and relaxation properties, independent of the great diversity in their individual microstructure and microdynamics, see Chap. 7. The universal behavior exhibited by this class of systems, generically known as HF systems or compounds, being analogous to that commonality expressed in gaseous, liquid, and solid states of matter, leads us to consider such HF systems as manifestations of a new state of matter arising from the presence of flat bands in their excitation spectra. Such flat bands come from the formation of FC due to a specific quantum phase transition known as the topological fermion-condensation quantum phase transition (FCQPT) that transforms the Fermi surface into the Fermi volume, and makes a flat band, see Chaps. 3, 4 and

Sect. 4.1.3. We show that in very different substances and under very different external conditions the universal FCQPT, at microscopic level, takes place determining the macroscopic properties and behavior of numerous substances, the number of which increases as an avalanche. It is quite natural to consider all these substances as the new state of matter, since their behavior near the phase transition acquires important similarities making them universal. The idea of such a phase transition, making flat bands, started long ago, in 1990 [V. A. Khodel, V. R. Shaginyan, JETP Lett. **51**, 553 (1990)], at first as a curious mathematical option, and now it is a rapidly expanding, vibrant field with uncountable applications.

To make the book understandable as much as possible, and for the reader convenience we, when considering the formation of the new state of matter, give the necessary elements of the theory within a particular chapter, illuminating our consideration by the corresponding experimental facts. Because of big diversity of the considered HF compounds, we hope that such a presentation allows the reader to learn the particular physical process without a laborious recursion to the corresponding chapters of the book.

A fruitful concept of "classical" condensed matter physics is the Landau paradigm of quasiparticles. This concept permits to represent any object as a certain ground state and its elementary excitations in the form of weakly interacting quasiparticles. Modern experimental findings in HF compounds put such fundamental concepts of "classical" condensed matter physics, as quasiparticle, under scrutiny. As a result, there has been a growing body of theoretical and experimental studies showing that the conventional picture of quasiparticles is not always correct for system of strongly interacting fermions. As examples of systems exhibiting significant deviations from the above quasiparticle picture serve chemical compounds with heavy fermions (HF compounds), whose experimental behavior is strongly different from that predicted by the Landau Fermi liquid theory based on the concept of quasiparticles whose basic characteristic, namely, the effective mass m^* weakly depends on external parameters such as temperature T, magnetic field B, pressure P, etc. The whole body of these "strange" experimental facts is now commonly attributed to as non-Fermi liquid (NFL) behavior. Since the quasiparticle paradigm is inherent in other branches of condensed matter physics, the same NFL behavior is exhibited at low temperatures by 2D electron gas and even ^{3}He, where neutral atoms are fermions with spin 1/2. We believe that to meet the above challenges of adequate description of these systems new views on old ideas are necessary. Our proposal is to modify the notion of Landau quasiparticle based within the framework of the FC theory. Namely, our monograph explains how the Landau quasiparticle paradigm can be modified to describe universally the new state of matter formed by the wide class of HF compounds. We present a theoretical approach, that is, the theory of FC, which generates modified and extended quasiparticles paradigm, and permits to naturally describe the basic properties and the scaling behavior of the broad variety abovementioned substances. We present a comprehensive analysis of existing experimental results for all listed systems with the NFL behavior like HF metals, high-temperature superconductors, quantum spin liquids, quasicrystals, and 2D and 1D HF compounds. The common feature of these systems is that their physical properties can hardly be understood within the framework of the Landau Fermi liquid theory, and their behavior is so unusual that the traditional Landau quasiparticles paradigm fails to describe it. Nonetheless, these substances are unexpectedly identical despite their diversity, and these very different HF compounds exhibit uniform scaling behavior that allows us to consider them as a manifestation of the new state of matter. The important unique peculiarity of our monograph is that we compare a great deal of our theoretical consideration based on the FC theory with numerous experimental facts. As a result, the monograph contains about 150 figures which facilitate understanding both the presented theory and the experimental facts collected on a very broad variety of HF compounds. It also explains striking NFL anomalies of the above class of substances that are not observed in the common solid state physics. For example, we analyze such anomalies as asymmetric tunneling conductivity $\sigma_d(V)$ with respect to change of the bias voltage V to $-V$ so that $\sigma_d(V) \neq \sigma_d(-V)$. This asymmetry points to a violation of both the time and charge symmetries.

After reading the book, one can see that HF compounds with quite different microscopic nature exhibit the same NFL behavior, while the data collected on very different strongly correlated fermionic systems manifest the universal behavior so that these substances are unexpectedly identical despite their diversity. For the reader convenience, the analysis is carried out in the broad context of explanation of salient and unusual experimental results. The numerous calculations of the thermodynamic, relaxation, and transport properties, being in good agreement with experimental facts, offer the reader solid grounds to learn the FC theory implications and applications. Finally, the reader will learn that the FC theory develops unexpectedly simple, yet complete and uniform description of the NFL behavior of many different classes of substances. The reader will learn that all these classes of substances or HF compounds exhibit the uniform behavior typical for a special state of matter like gas, liquid, or solid state.

We also consider the specific properties of HF compounds that make them cardinally different from those considered by the conventional solid state physics. For example, we consider a quantum spin liquid that determines the properties of frustrated insulators represented by quantum magnets and 2D ^{3}He; the main properties of quasicrystals are also explained. These substances exhibit the thermodynamic properties of HF metals. While contrast to metals, quantum magnets are insulators. We also demonstrate that the fermion condensation gives ground to the quasi-classical physics in HF compounds. This permits us to gain more insights into the puzzling NFL physics of HF compounds and explain challenging experimental facts related to Planckian metals. We analyze the asymmetry of the conductivity $\sigma_d(V)$ revealed by the methods of scanning tunneling microscopy and point-contact spectroscopy and explain its behavior under the application of magnetic field. The violation of the Wiedemann-Franz law in HF metals is also considered. Special consideration is given to the unusual behavior of high-T_c

superconductors that cannot be explained within the framework of the classical Bardeen-Cooper-Schrieffer theory.

This monograph is written on the base of our theoretical investigations as well as on the theoretical and experimental studies of other researchers. The interested reader can find our original papers, which we used to write the book, at free access offered by the arXiv.org: https://arxiv.org/search/?query=shaginyan&searchtype=all.

The book is intended for undergraduate, graduate students, and researchers in the condensed matter physics and the other fields of physics. Also, the material of the book has widely been presented in the form of lectures at Clark Atlanta University (USA), Hebrew University of Jerusalem (Israel), St. Petersburg University (Russia), Syktyvkar University (Russia), and Opole University (Poland).

We are indebted to many of our colleagues for long-lasting collaboration of subject of our present monograph. To name a few, they are S. A. Artamonov, A. Bulgac, J. W. Clark, V. T. Dolgopolov, J. Dukelsky, V. I. Isakov, G. S. Japaridze, V. A. Khodel, A. Z. Msezane, Yu. G. Pogorelov, P. Schuck, A. A. Shashkin, V. A. Stephanovich, M. V. Zverev, and G. E. Volovik. We acknowledge with pleasure that are collaborating with most of the listed people in the frame of virtual International Laboratory FERMION CONDENSATION http://fc.komisc.ru/index.php.

Jerusalem, Israel/St. Petersburg, Russia — Miron Amusia
Gatchina, Russia/Atlanta, USA — Vasily Shaginyan

Acknowledgements We acknowledge the hospitality of Petersburg Nuclear Physics Institute of National Research Centre "Kurchatov Institute", Hebrew University, and Clark Atlanta University, who strongly support our work on this book. This work was partly supported by U.S. DOE, Division of Chemical Sciences, Office of Energy Research, AFOSR and by Center for Theoretical Studies of Physical Systems at Clark Atlanta University.

Contents

Acronyms

2D	Two-dimensional
$\mathbf{A_q}$	The hyperfine coupling constant of the muon or nucleus with the spin excitation at wave vector **q**
A_{HT}, A_{LT}	High- and low-temperature resistance coefficients
a_B	Bohr radius
$A_w(B,T)$, A_w	The constants of temperature T and magnetic field B dependent parts of thermal resistance
a	Lattice constant
AC, ac	Alternating current
AF, AFM	Antiferromagnet, antiferromagnetic
B, H	Magnetic field strength
B_{c0}, H_{c0}	The critical field at which the magnetic field-induced QCP takes place
B_{c2}, H_{c2}	The upper critical fields for superconductors
B_{eff}	Effective magnetic field
B_c	Critical magnetic field
B_{inf}	The magnetic field taken at the inflection point
$C(T)$	Heat capacity
C, C_{mag}	Specific heat of quantum spin liquid
$c_{i\sigma}$, $c_{i\sigma}^{\dagger}$	Creation and annihilation operators for electron on site i with spin index $\sigma = \uparrow, \downarrow$
$C_n(T)$, $C_s(T)$	Heat capacity in the normal and superconducting state
c_s, c_t	Sound and zero-sound velocities, respectively

c	The speed of light
$D(k, \omega)$	Boson propagator
DC	Direct current
DFT	Density Functional Theory
DMFT	Dynamic Mean-Field Theory
DOS	Density of states
DP	Dulong-Petit
$E_{DMFT}[\rho(r), G]$	Total energy functional in DMFT theory
$E_{gs}(\kappa(\mathbf{p}), \mathbf{n}(\mathbf{p}))$	Ground state energy
$E_0(A)$	Ground state energy of atomic nucleus as a function of mass number
e	The charge of electron
$F(\mathbf{p}, \mathbf{p_1})$, $F_{\sigma,\sigma_1}(\mathbf{p}, \mathbf{p}_1, n)$, $F(\mathbf{p}, \mathbf{p}_1, n)$, $f(\mathbf{p}, \mathbf{p_1})$, $U(\lvert\mathbf{p}-\mathbf{p}'\rvert)$	Landau interaction (amplitude)
f_1	First harmonic of Landau interaction (amplitude)
F^s_L, F^a_L	Dimensionless spin-symmetric and spin-antisymmetric parts of Landau interaction (amplitude)
$\bar{f}_i$	Occupation number in two-dimensional electron system
$s-$, $p-$, $d-$ and $f-$ electron shells	Electron shells with orbital quantum number from 0 to 3, respectively
$F(n(\mathbf{p}, T))$	Free energy as a functional of electron occupation number $n(\mathbf{p}, T)$
$F^+(\mathbf{p}, \omega)$, F^+, $G(\mathbf{p}, \omega)$, G	Gorkov Green's functions of a superconductor
FC	Fermi condensate or Fermi condensation
FCQPT	Fermion-condensation quantum phase transition
FL	Fermi liquid
FM	Ferromagnet, ferromagnetic
G_A, G_R	The advanced and retarded quasiparticle Green's functions, respectively
$\hbar = h/2\pi$	Planck constant
HF	Heavy fermions
^{3}He, ^{4}He	Helium isotopes with 3 and 4 nucleons
$I_{n_\mathbf{p}}$	Collision integral as a function of quasiparticles occupation number $n_\mathbf{p}$
I	The electric current
I^s, I^{as}	Symmetric and asymmetric parts of the electric current
k_B	Boltzmann constant

K	Compressibility and also Kadowaki-Woods ratio
$L(T)$, L_0	Lorentz number at $T \neq 0$ and $T = 0$ respectively
λ	The London penetration depth
LFL	Landau Fermi Liquid
LMR	Longitudinal magnetoresistance
LSDA	Local spin density approximation
$\hat{\mathbf{M}}$, $\underline{\mathbf{M}}$	Average magnetization vector
$\mathscr{M}$	Magnetization modulus
M, $M(T, B)$	Magnetization as a function of temperature T and magnetic field B
M^*, M^*_{mag}, M^*_{FC}, m^*	Quasiparticle effective mass
m, m_e, M	Free electron mass
M_{nor}	Normalized magnetization
MOSFET	Metal-Oxide-Semiconductor Field-Effect Transistor
MR, $R^{\rho}_M(B, T)$	Magnetoresistance
n	Charge carrier concentration
n_i	Density of electrons at the site i
n_s	Density of electrons in two-dimensional electron system
$n(\mathbf{p}, T)$	Quasiparticle occupation number (or distribution function) as a function of their momentum $\mathbf{p}$ and temperature T
$N(\xi, T)$	The density of states
$N_S(E)$	The density of states of a superconductor
$N(E_F)$ or $N(0)$	The density of states at the Fermi surface
NFL	Non-Fermi Liquid
NMR	Nuclear magnetic resonance
$\mathbf{p_i}$, $\mathbf{p_f}$	The borders of Fermi volume, where fermion condensate is localized
$\mathbf{p}^{\uparrow}$, $\mathbf{p}^{\downarrow}$	Momenta of Fermi surfaces with spin up and spin down
p_F	Fermi momentum
P	Pressure
PM	Paramagnet, paramagnetic
QC, QCs	Quasicrystals
QCL	Quantum-critical line
QCP	Quantum-critical point
QPT	Quantum Phase Transition
QSL	Quantum spin liquid

$R_{nl}(r)$	Radial part of wave functions, determined by two quantum numbers n (main) and l (orbital)
$R_H(B)$	Hall coefficient
r_s	Wigner-Seitz radius
$S(n(\mathbf{p}, T))$, $S(T)$	The entropy as a function of quasiparticles occupation number $n_{\mathbf{p}}$ and temperature T
S_{AF}, $S_{NFL}(T)$	Entropy in the antiferromagnetic and non-Fermi liquid phases, respectively; latter is a function of temperature T
S_0	The residual entropy of Fermi condensate at $T = 0$
$S(T)$	Seebeck coefficient
SC	Superconducting
SCDFT	Superconducting Density Functional Theory
SCQSL	Strongly correlated quantum spin liquid
SDA	Spin density approximation
S_T	Thermoelectric power
SUSY	Supersymmetry
$\mathscr{T}(j, \omega = 0)$	Vertex part of a diagram
t_{ij}	The hopping matrix element for electrons between sites i and j (e.g., in Hubbard model)
T_c	Critical temperature
T_D, Ω_D, Ω_t	Debye temperature
$T[\rho, G]$	Kinetic energy term in DMFT theory
T	Tesla
$T^*, T^*(B)$	Crossover/transition temperature
T_p^*	The temperature at which the pseudogap is closed
T_f	Temperature above which the effect of FC is insignificant
T_{inf}	The temperature taken at the inflection point
TMR	Transversal magnetoresistance
T_{NL}	The Néel temperature
TPT	Topological phase transition
TT	Topological transitions
$1/T_1$	The spin-lattice relaxation rate
$1/T_1^{\mu}$, $1/T_{1N}^{\mu}$	Muon (ordinary and normalized, respectively) spin-lattice relaxation rates

U, U_i	Local Coulomb repulsion between two electrons occupying the same site i (e.g., in Hubbard model)
vHs	van Hove singularity
V_{KS}	Kohn-Sham potential
$V_{xc}[\rho, G]$	Exchange-correlation potential in DMFT theory
V	Applied voltage
V	Volume
v_t	The velocity of transverse zero sound
$w(B,T)$, w	Thermal resistivity
w_0	Residual thermal resistivity
WF	Wiedemann-Franz
x_{FC}	Critical number density in *FCQPT* point
x	Electron (hole) number density, doping
$z(\mathbf{p})$, z	Quasiparticle weight, renormalization factor, so-called z-factor
$\alpha_{FC}(T)$	Thermal expansion coefficient for the system with fermion condensate
α_R, $\alpha_R(T,B)$	Exponent in power-law of resistance temperature dependence
$\alpha(T)$	Thermal expansion coefficient
α, β, g	Coupling constant, Landau interaction constant
α_{cr}, b_c, g_c	Critical value of Landau interaction constant
α	Critical exponent
$\chi_{AC}(B,T)$	AC magnetic susceptibility
χ	Magnetic susceptibility, spin susceptibility
$\chi_0(\mathbf{q},\omega)$	The linear response function for the noninteracting Fermi liquid
$\chi(\mathbf{q},\omega,g)$	The response function of the system
$\chi(\mathbf{q},\omega \to 0)$	Low-frequency dynamical magnetic susceptibility
$\chi'(\mathbf{p},\omega,T)$, $\chi''(\mathbf{p},\omega,T)$	The real and imaginary parts of the spin susceptibility, respectively
$\delta\mathbf{p} = \mathbf{p_f} - \mathbf{p_i}$	Momentum interval occupied by fermion condensate
$\delta(x)$	Dirac delta function
$\Delta(\mathbf{p})$, $\Delta(\phi)$	Superconducting gap as a function of momentum and angle, respectively
$\Delta\rho_L(B,T)$	Classical contribution to magnetoresistance

$\Delta\sigma_d(V)$	The asymmetric part of the differential conductivity
ε_F, E_F	Fermi energy
ε_i	Single-electron energy
$\varepsilon(\mathbf{p},T)$, $\varepsilon(\mathbf{p},T,n)$	Quasiparticle single-particle energy as a function of momentum **p**, temperature T and occupation number n
γ_0	Sommerfeld coefficient
$\Gamma(T)$	Grüneisen ratio
γ, $\gamma(T)$	The quasiparticle damping or width of quasiparticle state
Γ	The scattering amplitude
Γ^{ω}	The ω-limit of the scattering amplitude
Γ^{k}	The k-limit of the scattering amplitude
θ	The step function
$\kappa(\mathbf{r},\mathbf{r}_1)$, $\kappa(\mathbf{p})$	Anomalous density or the order parameter
$\kappa(B,T)$, κ	Thermal (heat) conductivity as a function of magnetic field B and temperature T
$I(B,T) = (\kappa(B,T) - \kappa(0,T))/\kappa(0,T)$	The reduced heat conductivity
κ_d	Dielectric constant
$I_N(B,T)$	The normalized reduced heat conductivity
λ_0	Coupling constant of superconducting (pairing) interaction
$\lambda_0 V(\mathbf{p}_1,\mathbf{p}_2)$	Pairing interaction
μ, $\mu(T)$, μ_σ,$\mu(A)$	Chemical potential, also as a function of temperature T, spin σ , and mass number A
μ_B	Bohr magneton
∇	Nabla operator
v_F	Fermi velocity
Ω_F	The volume of Fermi sphere
Ω_{FC}	The volume occupied by fermion condensate
Ω_0	The characteristic frequency of zero sound
$\hbar\omega_c = \hbar eB/m^*c$	Cyclotron splitting
Φ_{lm}	Angular part of wave functions
Φ, $\Phi[n(r,\kappa(r,r_1))]$	Thermodynamic potential as a function of quasiparticle occupation number n
$\Pi(\mathbf{j})$	Polarization operator
π	Pi-number

$\Psi_\sigma(\mathbf{r})$	Field operator for annihilation of an electron with spin $\sigma = \uparrow, \downarrow$ at the position $\mathbf{r}$
$\psi_i(\mathbf{r})$	Single-electron wave function
ρ_∞	Critical density
$\rho(T)$	Electric resistance
ρ_0	Residual (electric) resistance
ρ_0^{LFL}, ρ_0^{NFL}, ρ_0^{nem}	Residual resistance in the Landau Fermi liquid state, non-Fermi liquid state, and the nematic phase
$\rho(\mathbf{r})$, ρ_x	Electronic charge density
ρ	The number density of a system
ρ_{inf}	The magnetoresistance taken at the inflection point
σ_{tr}	Transport cross section
$\sigma_d(V)$	Differential tunnel conductivity
$\Sigma(\mathbf{p}, \varepsilon)$	Self-energy part
$\sigma(T)$, σ	The electrical conductivity
τ_q	The lifetime of quasiparticles
τ	The collision time
$\theta(p)$	Theta function or Heaviside function
$\phi_\lambda(\mathbf{r})$	Wave functions

Chapter 1
Introduction

Abstract In this chapter, we outline the difference between HF compounds and common solids, which is strongly and weakly correlated systems, and show that the former cannot be described adequately in the framework of ubiquitous methods like the band theory of solids and the Landau Fermi liquid (LFL) theory. We familiarize the reader with theoretical approaches to strongly correlated systems and to the Landau theory of Fermi liquid in particular. We show that the LFL theory is insufficient to describe so-called non-Fermi liquid properties of HF compounds, for these compounds located near a topological fermion-condensation quantum phase transition. To drive the system in question to the phase transition, the correlation is to be strong rather than weak. The aims and goals of the book are also discussed.

1.1 General Considerations

When we are dealing with an ensemble of particles, which do not interact with each other, the problem of describing their properties and behavior can be completely solved analytically. The situation is drastically different, when even a weak interparticle interaction is "turned on": In this case, the problem ceases to be single-particle and many-body effects enter the scene. Note that even a small interparticle interaction can lead to profound effects, e.g., the well-known Efimov effect and its generalization or the superconductivity or superfluidity of weakly interacting Fermi gas. As modern physics is dealing primarily with ensembles of interacting particles, one can say that its central problem is the many-body theory. Indeed, all our surroundings, from cosmic bodies to tiny objects like molecules, are made of many constituents.

To form a body of any size, the constituent particles have to interact with each other. Such an interaction is achieved by the exchange of these particles by quanta of interaction field—photons, gluons, and mesons. However, if the speed of interacting particles is much less than the speed of light, the interaction can be limited to either some potentials, or, for charged objects and not too small distances, to pure Coulomb interaction.

M. Amusia and V. Shaginyan, *Strongly Correlated Fermi Systems*,
Springer Tracts in Modern Physics 283,
https://doi.org/10.1007/978-3-030-50359-8_1

The materials that are well understood usually contain the ensemble of free or almost free weakly interacting (or weakly correlated) particles. To understand the latter materials, band theory, which represent electrons in the form of extended plane waves, is a good starting point. That theory helps capture the delocalized nature of electrons in metals. It is valid provided that the Coulomb interaction energy of electrons is much smaller than their kinetic energy. However, there are important physical systems for which interactions between the electrons are not weak, and these interactions play a major role in determining the properties of such systems. They are usually called strongly correlated systems. Examples here can be metals with open $d-$ and $f-$ electron shells, where electrons occupy narrow bands. There, electrons experience strong Coulomb repulsion because of their spatial confinement in those bands. The effect of correlations on the physical properties of such objects is often profound. The interplay of the d and f electrons internal degrees of freedom like spin, charge, and orbital moment can exhibit a lot of exotic phenomena at low temperatures. That interplay makes strongly correlated electron systems extremely sensitive to small variations of external parameters, such as temperature, pressure, or magnetic field.

1.2 Strong and Weak Interparticle Interactions

In materials called HF compounds, mobile electrons (or quasiparticles) at low temperatures behave as if their masses were a hundred times the mass of an electron in a silicon or simple metal. Such systems exhibit a great variety of interesting phenomena like anomalies in electric and thermal conductivity, quantum phase transitions between magnetically ordered state and superconductivity, emergence and dissociation of local magnetic moments, etc. This rich variety of the phenomena makes their experimental and theoretical studies all the more difficult. Quantum mechanical study of strongly correlated fermionic systems is performed using two approaches: ab initio numerical electronic structure simulations and model theoretical treatment. In the first case, real materials can be described with all details of their chemical composition and crystal structure fully taken into account. In model approaches, such subtleties are usually neglected in favor of more rigorous mathematical treatment of the problem. The result yields the dependencies (in graphical and sometimes in analytical form) of system physical properties upon external parameters like temperature, pressure, and/or external fields. But nowadays even in the latest theoretical treatment, the solutions of derived equations could be obtained by means of numerical calculations only.

In this context, we mention so-called optical lattices, consisting of boson atoms ensemble in a periodic potential. At low temperatures, such Bose particles can condense, i.e., most of the particles can be found in a state of zero momentum, which helps to minimize their kinetic energy. Below, we will see that similar phenomenon can occur in systems with strongly correlated fermions, forming fermion condensate (FC) and giving rise to completely new physics, which proves to be the same for many seemingly different systems [1]. We will see also that the notion of topological fermion-condensation quantum phase transition (FCQPT), forming the fermion condensation, and related to flat bands permits to elegantly explain the whole bunch of puzzling experimental facts in systems with strongly correlated fermions [1]. At this moment, another general approach suitable for understanding strongly correlated systems does not exist. For example, certain aspects of systems exhibiting quantum Hall effect or some magnetic materials are already understood. But the other systems are understood very poorly. For instance, the nature of high-temperature superconductivity remains mysterious despite decades of intensive work. Most importantly, we do not have a unified view on the fermionic systems for which interparticle interactions not only cannot be neglected but play a decisive role in the formation of their observable properties.

1.3 Theoretical Approaches to Strongly Correlated Systems

A natural question appears: How to solve a many-body problem? In other words, how to describe a system of large number of interacting constituents knowing how to deal with single free particle motion and its interaction with the other particle and environment? This problem is not new and by far not specific to quantum systems. Indeed, this problem existed already at the time when Newton created his mechanics and discovered the law of gravitation. However, in the case of relatively weak gravitational interaction, only two-body dynamics is essential so that the rest of particle ensemble could be treated as small perturbation. For instance, the Moon and all other planets of Solar system affect the Earth rotation around the Sun as a small, although noticeable perturbation. Only recently it became clear that in some cases even weak gravitational interaction can lead to prominent many-body effects, so-called resonance phenomena.

In quantum mechanics, the many-body problem became essential almost immediately after discovery of the Schrödinger equation [2] in 1926. The motivation was the fact that the potential, in which the given quantum particle (say, electron) moves, is created by the rest of the particles, among which this electron resides. At that time, it had already been recognized that the electron-electron interaction in atoms plays less important role than their interaction with the atomic nucleus. This hierarchy is similar to that in Solar system in regard to inter-planet and Sun-planet interactions. Already in 1928 Hartree suggested the concept of self-consistent field [3, 4] that permitted to consider many-electron atoms treating their electrons as moving in a field essentially modified as compared to the pure Coulomb field of an atomic nucleus. 2 years later, Fock took into account the fact that all electrons are identical Fermi particles so that their wave function has to be antisymmetric under permutation of any two electrons [5, 6]. As a result, famous Hartree-Fock equation has been formulated. This equation is in use until now, being applied not only to single atoms, but to multi-atomic objects like molecules, clusters, fullerenes, and even bulk solids. The application of the Hartree-Fock equation is not limited to the systems with Coulomb interparticle interaction. The same equation can be successfully applied to nucleus as a system consisting of protons and neutrons. Actually, Hartree-Fock equation considers only non-perturbative part of interparticle interaction; it can also be applied to systems of Bose particles where it is called Gross-Pitaevskiy equation.

The remaining part of interparticle interaction that is neglected in the Hartree-Fock framework is usually called *residual interaction*. This part of interaction leads to so-called *correlation effects*, or *interparticle correlations*. Obviously, the Hartree-Fock equation is able to describe many-body system satisfactorily only if correlations are negligible. However, as it is known now, there exist a number of many-body systems, where correlations lead to decisively important *collective phenomena*.

We note here that Hartree-Fock approach along with little earlier suggested Thomas-Fermi approach [7, 8] were the first representative of so-called first principle (ab initio) methods that were used as a first step to incorporate the mentioned above collective phenomena in the physical properties of solids and strongly correlated systems in particular. The aim of such methods is to calculate the macroscopic, observable properties of a body having in employing its microscopic characteristics like its crystalline symmetry and interatomic interaction potentials. At the first glance, if all microscopic details were known, the problem could be solved exactly utilizing (many-particle) Schrödinger equation. Unfortunately, this is not the case for majority of real systems, since the required calculations are too difficult and cumbersome. Only molecules consisting of small number of atoms as well as single atoms can be more or less satisfactorily treated by this straightforward method. In this approach, their wave functions are represented as linear combinations of corresponding Slater determinants.

Further development of ab initio methods was related to the breakthrough, made by Hohenberg-Kohn theorem, which puts a foundation to the density functional theory (DFT) [9–12]. The Hohenberg-Kohn theorem states that a system's wave function is defined uniquely by its charge density. The idea of DFT is a direct generalization of Hartree-Fock self-consistent field approach. Namely, in DFT, the differential equations called Kohn-Sham equation for single-electron wave functions are solved with certain effective potential, which, in turn, is determined self-consistently by that equation's solutions. This Kohn-Sham equation is derived by variation of the energy functional $E[\rho]$, depending on the local electronic charge density $\rho = \rho(\mathbf{r})$. This functional, expressing the total energy of many-body system, has the form [11]

$$E[\rho] = T[\rho] + \int \rho(\mathbf{r}) V_{\text{ext}}(\mathbf{r}) \mathrm{d}^3 r + \int \mathrm{d}^3 r \, \mathrm{d}^3 r' \frac{\rho(\mathbf{r})\rho(\mathbf{r}')}{|\mathbf{r} - \mathbf{r}'|} + E_{\text{xc}}[\rho], \qquad (1.1)$$

where $T[\rho]$ is kinetic energy, $V_{\text{ext}}(\mathbf{r})$ is the external potential acting on the electrons, third term is the Hartree contribution to the Coulomb interaction between the charges, and the last term is the so-called exchange and correlation energy contributions $E_{\text{xc}}[\rho]$. The exchange contribution takes into account the partial suppression of Coulomb repulsion between electrons in the same spin state due to Pauli principle. The DFT method would be effective if one could construct the explicit form of the term E_{xc} in (1.1). However, this is not the case, as for that one needs to solve the system of functional equations exactly [13].

To handle this problem, the DFT self-consistent approach along with approximations (like, for instance, local density approximation (LDA) and its spin counterpart local spin density approximation (LSDA), see [14] for details, for exchange-correlation energy E_{xc} had been utilized in DFT. Also, as particular representation of wave functions corresponding to $\rho(\mathbf{r})$ cannot be defined uniquely in DFT, this quantity is expressed usually via single-electron wave functions $\psi_i(\mathbf{r})$

$$\rho(\mathbf{r}) = \sum_{i=1}^{N} |\psi_i(\mathbf{r})|^2, \tag{1.2}$$

where N is the number of electrons in a system.

Minimizing the functional (1.1) over the functions $\psi_i(\mathbf{r})$ (with respect to the normalization conditions for ψ_i) generates Kohn-Sham equations:

$$\left[-\frac{\nabla^2}{2m} + V_{\text{KS}}[\rho(\mathbf{r})]\right] \psi_i = \varepsilon_i \psi_i, \tag{1.3}$$

where ε_i are the single-electron eigenvalues, Kohn-Sham potential V_{KS} represents a static mean field (signifying the explicit neglecting of correlation effects when all electrons feel the same potential) of the electrons and has to be determined in a self-consistent way

$$V_{\text{KS}}[\rho(\mathbf{r})] = V_{\text{ext}}(\mathbf{r}) + V_{\text{xc}}[\rho(\mathbf{r})] + 2\int \frac{\rho(\mathbf{r}')\mathrm{d}^3r'}{|\mathbf{r}-\mathbf{r}'|},\ V_{\text{xc}}[\rho(\mathbf{r})] = \frac{\delta E_{\text{xc}}[\rho(\mathbf{r})]}{\delta\rho(\mathbf{r})}, \tag{1.4}$$

where $\delta/\delta\rho$ means the variational derivative. Equation (1.4) allows one to calculate electronic charge density and total energy for the ground state of a system. The DFT with LDA and/or LSDA approximations for $E_{\text{xc}}[\rho]$ had been successfully applied to describe the electronic properties of atoms, molecules, and solids, where correlation effects are not too strong. In the systems with strong correlations like transition metal oxides, the DFT gave incorrect metallic ground state in the absence of long-range magnetic order.

As in 1960s the computational power was too feeble to calculate real things by ab initio methods, the so-called model Hamiltonian approach comes into play. Namely, the full many-body Hamiltonian is simplified to account for only a few relevant degrees of freedom—typically, the valence electron orbitals near the Fermi level. One of the simplest models of correlated electrons is the Hubbard Hamiltonian [15], which takes into account the interplay between electron hopping and local on-site repulsion:

$$\mathscr{H} = \sum_{i,j,\sigma} t_{ij} c_{i\sigma}^{\dagger} c_{j\sigma} + U \sum_i n_{i\uparrow} n_{i\downarrow}, \tag{1.5}$$

where $c_{i\sigma}$ and $c_{i\sigma}^{\dagger}$ are creation and annihilation operators for electron on site i with spin index $\sigma = \uparrow, \downarrow$, $n_i = c_{i\sigma}^{\dagger} c_{i\sigma}$ is the density of electrons at site i, U is a local Coulomb interaction between two electrons occupying the same site i. The matrix element t_{ij} describes hopping of electrons with between sites i and j. It is usually supposed that hopping is not zero for nearest neighbors (with number z_n) only. Such simplified treatment of electron-electron Coulomb repulsion turns out to be very convenient to study the itinerant electron systems, especially their magnetic states. A wide variety of analytical and numerical methods have been utilized to study strongly correlated electron systems in the framework of Hubbard's model. But despite that, the Hubbard model (1.5) could be solved only for limited number of cases.

It had been observed in 1989 [16] that the consideration of the electron correlations on the lattices with large (actually infinite) number of nearest neighbors z_n simplifies the problem essentially so that it can be solved exactly for any strength of Coulomb repulsion. The simplification at $z_n \to \infty$ occurs due to the fact that in this case one can neglect the spatial fluctuations leaving only time-dependent on-site fluctuations. This fact laid the foundation of dynamical mean-field theory (DMFT), which mapped lattice models (like Hubbard one [17, 18]) to effective impurity problem with correlated electrons feeling mean field which is dynamic, i.e., time- or energy-dependent. Solution of the latter effective impurity problem can be further used to construct the self-energy for lattice Green's function G that, in turn, gives new approximation for above dynamic mean field. The DMFT solutions become exact as the number of neighbors increases. Similar to DFT, the DMFT also relies on energy functional $E_{\text{DMFT}}[\rho(\mathbf{r}), G]$, which in this case depends on the above local Green's function G:

$$E_{\text{DMFT}}[\rho(\mathbf{r}), G] = T[\rho(\mathbf{r}), G] + \int \rho(\mathbf{r}) V_{\text{ext}}(\mathbf{r}) \mathrm{d}^3 r + \\ + \int \mathrm{d}^3 r \, \mathrm{d}^3 r' \frac{\rho(\mathbf{r})\rho(\mathbf{r}')}{|\mathbf{r} - \mathbf{r}'|} + E_{\text{xc}}[\rho(\mathbf{r}), G], \tag{1.6}$$

where $G = G_{i\sigma}(t - t') = - < c_{i\sigma}(t) c_{i\sigma}^{\dagger}(t') >$ specifies the probability to create an electron with spin index σ at a site i at time t' and annihilate it at the same site at a time $t > t'$. In practical calculations, this function is found self-consistently from the above effective single-impurity problem, where the rest of the electrons are considered as the bath for given impurity one. The theory is called "dynamic" as Green's function depends explicitly on time or frequency arguments. Also, the approximations for $E_{\text{xc}}[\rho(\mathbf{r}), G]$ are provided.

Although dynamical mean-field methods clearly represent a new advance in many-body physics, they are still unable to capture many effects in strongly correlated systems, e.g., the divergent behavior of the effective mass. In spite of this, the generalization of DFMT to account for real materials is an active area of research. The above consideration supports the view that only ab initio methods in principle cannot give sufficient insights into the physics of strongly correlated systems because of their computational complexity, still beyond the reach of modern computer's possibilities. More insight has been achieved on the way of model assumptions and simplification of initial many-body Hamiltonian, leaving in it only the most essential terms, which, on the other hand, could be assessed both analytically and numerically. Nonetheless, these methods meet with enormous difficulties while being applied to consideration of excitation spectra of many-body systems, while these spectra define the thermodynamic, transport, and relaxation properties.

The complimentary approach that determined an essential breakthrough in the understanding of many-body systems was the introduction of so-called *quasiparticles*. Namely, the real interacting (with potential V) particle ensemble is substituted by quasiparticles that interact via some *effective* interaction V_{eff} and represent the excitation on top of some basic vacuum ground state.

Then, the natural question appears on how to construct the quasiparticles and their effective interaction. At the first glance, this is a simple problem, if the interaction is weak. But what is the meaning of weakness if, roughly speaking, the effect of interaction includes in some cases large numbers generated by the corresponding phase transition. So, the weakness can be overcome by the large number, leading to macroscopic correlated coherent phenomena like superfluidity and/or superconductivity.

The first attempt in this direction was undertaken by Landau in 1956 [19], when he assumed the existence of quasiparticles, and afterward, when he, together with his colleagues, gave a proof of the developed approach on the grounds of the general many-body theory [20]. He suggested and proved that for systems where the interaction is relatively strong, quasiparticles can be introduced as well- defined objects, provided that the Pomeranchuk stability conditions are not violated [21], see for details Chaps. 2 and 4. The ground state energy E in that case is assumed to be a functional of the quasiparticle distribution function $n(\mathbf{p})$, $E[n(\mathbf{p})]$, with the single-particle spectrum (or the dispersion law) $\varepsilon(\mathbf{p}) = \delta E[n(\mathbf{p})]/\delta n(\mathbf{p})$, see for details Chap. 3. The Landau picture of quasiparticles as long-living excitations weakly interacting with each other works very well for many objects. This system of quasiparticles forms the Landau quasiparticle paradigm. As an example, Landau treated the liquid ^{3}He, which is a liquid, whose constituents are Fermi particles—two-electron atoms with ^{3}He nuclei. He demonstrated that at low temperatures the excitations of such a system could be described as those of a quasiparticle gas. Nevertheless, in a number of recently studied strongly correlated systems, the above quasiparticles neither interact weakly nor remain well defined due to the violation of the stability conditions [21]. As we shall see in Chap. 2, this paradigm is to be substituted by an extended one.

It is worth noting that in case of Landau Fermi liquid at temperature $T = 0$ in spite of strong interparticle interaction the distribution of quasiparticles $n(\mathbf{p})$ as a function of their momentum p can be presented as a Fermi step of noninteracting fermions [20]:

$$n(\mathbf{p}) = \begin{cases} 1, & p \leq p_F \\ 0, & p > p_F, \end{cases} \tag{1.7}$$

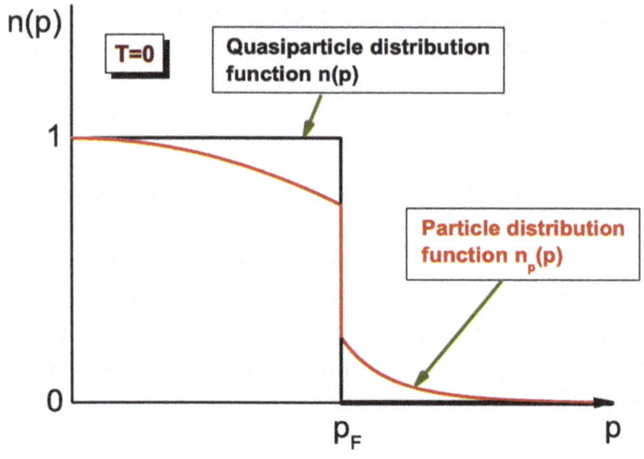

Fig. 1.1 Quasiparticle $n(\mathbf{p})$ and particle $n_p(\mathbf{p})$ distribution functions at $T = 0$ are shown by the arrows. $n(\mathbf{p})$ is given by (1.7) and $n_p(\mathbf{p})$ by (1.9) with the jump at $p = p_F$

where p_F is the so-called Fermi momentum of the system. It is essential that the densities (i.e., total numbers per unit volume) of particles and quasiparticles coincide, e.g., in three-dimensional (3D) case they both equal to

$$\rho = \frac{p_F^3}{3\pi^2}. \tag{1.8}$$

Throughout, *if otherwise is not stated*, we adopt the system of units, where $\hbar = c = 1$, where c is the speed of light.

As to the momentum dependence of the *particle* distribution function $n_p(\mathbf{p})$, it has, according to Migdal [22], a tail at $p > p_F$, in contrast to the distribution function of quasiparticles. This tail decreases as p increases. This decrease is fast, but not exponential. In the same paper [22], it had been demonstrated that at $p = p_F$ the function $n_p(\mathbf{p})$ has a step:

$$0 < n_p(p_F) = z < 1. \tag{1.9}$$

It is seen from (1.7), (1.9), and Fig. 1.1 that both the quasiparticles and the particles distribution functions have jumps at the Fermi momentum p_F.

The single-particle spectrum $\varepsilon(\mathbf{p})$ for Landau quasiparticle at $(p - p_F)/p_F \ll 1$ is similar to that of an excitation of a gas of noninteracting Fermi particles at low temperatures T

$$\varepsilon(p) - \mu = \frac{p_F(p - p_F)}{M^*}, \tag{1.10}$$

where μ is the chemical potential and M^* is the effective mass that, from case to case, differs from mass M in different systems. Here, M is the bare mass of the particles constituting the many-body system under consideration. In the many-body theory, one usually investigates systems consisting of a single-particle type, having the same mass, spin, and charge. Generalization to multi-component systems is very often not a simple and straightforward task.

A quasiparticle excitation can decay by creating two and more quasiparticles. If this decay proceeds sufficiently fast, the concept of quasiparticle as well-defined system's excitation faces difficulties. More precisely, before the decay process quasiparticle lives for time (so-called *quasiparticle lifetime*), which is reciprocal to its width γ. It had been also demonstrated by Migdal [22] that the quasiparticle width can be estimated by the following relation:

$$\gamma \sim \frac{(p - p_F)^2}{2M^*}. \tag{1.11}$$

Note that $\gamma \ll \varepsilon(p) - \mu = p_F(p - p_F)/M^*$ at $(p - p_F)/p_F \ll 1$, and therefore quasiparticle can be considered to be reasonably stable, if they are close enough to the Fermi surface at any interparticle interaction strength. The simplest excitation of a Fermi system, which interacts strongly with an external field, can be viewed as an absorption of its quanta with low excitation energy ω and small (as compared to p_F) momentum k by a combination of a quasiparticle of energy $\varepsilon(p)$ (with $p > p_F$) and quasihole $\varepsilon(p')$ (with $p' \leq p_F$). Note that the energy and momentum conservation is valid:

$$\omega(k) = \varepsilon(p) - \varepsilon(p'), \ \mathbf{k} = \mathbf{p} - \mathbf{p}'. \tag{1.12}$$

The presence of effective interaction, transforming one electron-hole pair into the second, third, and so on, leads to collective response of the Fermi system on the external field. Due to the presence of V_{eff}, so-called coherent collective excitations in the Fermi system are formed. One of examples is a sound-like excitation with $\omega = c_t k$ termed *zero sound* with propagation velocity $c_t \geq p_F/M^*$ [20]. General expressions to determine the dependence of excitation energy ω upon momentum k had been derived [20, 23]. It is essential to have in mind that there exists a profound analogy between quasihole in condensed matter and antiparticle in elementary particles theory.

The concept of relatively weakly interacting quasiparticles permits to express the low-temperature thermodynamic and electrodynamic properties of the Fermi system under consideration via several numbers characterizing the quasiparticles and their effective interaction. For instance, the specific heat C turned out to be simply proportional to M^*, which in turn is proportional to the system's temperature T

$$C \propto M^* T. \tag{1.13}$$

At some parameters of the effective interaction, a Fermi liquid can undergo *phase transitions* to other states like ferromagnetic and antiferromagnetic. This corresponds to solutions of Landau equation determining collective excitations that have zero or pure imaginary energy for nonzero momentum. Pomeranchuk derived the conditions [21] yielding the value and sign of V_{eff}, which is necessary for the susceptibility toward a fictitious external field to become infinite. This means that the system is unstable with respect to transition to certain fundamentally different (than the initial one) ground state, which is already stable for above V_{eff}. The phase transitions are of particular interest, since they describe possible system evolution under the action of the above mentioned weak perturbations. The stability conditions itself can only signal on possible instability of initially chosen system's ground state. But they are not enough for finding the state of a system formed after instability has been developed.

It was common wisdom for many years that if we augment the above Pomeranchuk conditions by the one, determining the transition to the superconducting state, we are able to describe all possible phase transitions of a Fermi system. However, for ^{3}He the experimental situation is not that clear. The investigation of the heat capacity C at temperature dropping from tenths of Kelvins down to millikelvin starts to contradict the linear law (1.13). The data, instead of temperature independence, demonstrated a more complex law like $M^* = a + b \ln T$ with very strange at that time possibility of diverging in the limit $T = 0$ [24]. Starting from millikelvin, new phases, including a superfluid one, were discovered in liquid ^{3}He [25]. Arguments appear that for any system a superfluidity is inevitable for sufficiently low T and therefore strictly speaking the above-presented quasiparticle concept in liquid ^{3}He is not valid at low temperatures. Nonetheless, the Landau Fermi liquid approach along with quasiparticles paradigm has been successfully applied to studies of electrons in solids and in metals in particular [20, 23]. It appeared that vast majority of properties of electron liquid in metals can be well understood in the framework of the temperature-independent Landau-type quasiparticles so that this approach becomes almost universal in quantum liquids description in condensed matter. However, further development of the condensed matter physics shows convincingly that more and more properties of solids either cannot be satisfactorily described by the ordinary Fermi liquid picture and/or require a complete revision of the quasiparticle paradigm. It turned out that latter revision is highly nontrivial task. It appeared, however, that additionally to Fermi liquid instabilities (related to Pomeranchuk's conditions and to transition to superfluid (superconducting) state), another instability was overlooked. Indeed, it has never been discussed what happens to the system if some interparticle interaction generates the divergence of the effective mass $M^* \to \infty$ rather than the susceptibility. In this case, as we will see in Chap. 2, all quasiparticle excitation energies equal to zero so that the system resembles a condensate of Fermi particles or Fermi condensation/condensate (FC) [13, 26–29]. Numerous studies have confirmed the possibility of the FC state that exists for certain interparticle interaction potentials and demonstrates its unusual properties. The possibility of the fermion-condensation quantum phase transition (FCQPT), preceding FC, does not abandon the concept of quasiparticles. On the contrary, it demonstrates that during this phase transition the quasiparticle effective mass M^* become dependent on external parameters like temperature, pressure, and/or external field. This situation opposes the Landau Fermi liquid picture, where the effective mass weakly depends on the external parameters, being approximately a constant, determined by particle density and their interaction. The above consideration makes it clear that FCQPT has a quantum nature, determining a non-Fermi liquid (NFL) behavior or quantum criticality in strongly correlated systems. Moreover, FCQPT is a unique topological phase transition that separates LFL from a liquid of a new type with specific topological charge [30] and forms a new state of matter represented by strongly correlated Fermi systems, see Chaps. 4 and 7.

In the beginning, the idea of FC has not been considered by the community too seriously. It was interpreted as a mathematical trick rather than an approach to describe real phenomena. It turns out, however, that the above idea is capable to deliver the adequate description of a huge body of otherwise unexplained experimental data. It has been demonstrated several times that classical Landau theory of Fermi liquid is insufficient and too narrow to encompass many experimentally important objects. A variety of these effects appeared in a number of newly discovered systems, including recently experimentally discovered flat bands in graphene and those that can be explained only in the framework of the FC theory (see, e.g., Chaps. 18, 19 and 21 and [1, 27–29]). Abundance of data stimulates to present these results as a book, what makes data better accessible not only to experts, but also to graduate students who will, hopefully, use this approach in their future research. This approach can be considered as complimentary, but also in many cases as a unique tool to treat and describe real systems, exhibiting quantum criticality.

The physics of quantum matter occupies a substantial part of modern physics. In contrast to the general belief, quantum matter can show up even at room temperature, as it takes place for electrons in metals, which behave in accordance with the above the famous Landau Fermi liquid (LFL) theory based on the quasiparticle paradigm [19, 20, 25]. We consider the quantum criticality in HF compounds begotten by quantum phase transitions with their quantum-critical points located at $T = 0$. The quantum criticality describes the new state of matter which exhibits a universal behavior and can hardly be understood within the framework of the LFL theory. Thus, quantum criticality in materials of significant theoretical and practical interest requires a new theoretical input. Furthermore, there are indications that the relevant new physics demands a departure from the quasiparticle paradigm of LFL theory. One could also expect that we have to confine our consideration to ultralow temperatures. This is not the case since HF compounds, for instance, reveal significant deviation from LFL properties, termed as NFL behavior, for the temperatures as high as tens of Kelvins. On the other hand, as we shall see, HF compounds at their quantum criticality can exhibit a quasi-classical behavior.

From now on, we call HF compounds "strongly correlated Fermi systems" as well. Strongly correlated Fermi systems, represented by HF compounds, high-T_c superconductors, strongly correlated insulators with spin liquid, quasicrystals, and two-dimensional (2D) Fermi liquids are among the most intriguing and best experimentally studied fundamental systems in physics. However, until very recently, these lacked theoretical description [27–29]. The properties of these materials differ dramatically from those of ordinary Fermi systems [27–29, 31–41]. For instance, in the case of HF metals, the strong electron-electron correlations lead to a renormalization of M^*, which may exceed the ordinary, so-called "bare" electron mass by several orders of magnitude or even become infinitely large. The effective mass strongly depends on temperature, pressure, or applied magnetic field. HF metals exhibit NFL behavior and unusual power laws of the temperature dependence of the thermodynamic properties at low temperatures. To describe this NFL behavior, the ideas based on the concept of quantum, thermal fluctuations, and Kondo lattice in a quantum-critical point (QCP) have been utilized, see, e.g., [31, 33, 42–45]. These ideas, however, could not provide a universal description of NFL properties. This generated a real crisis in physics of HF systems and to overcome this crisis. A new quantum phase transition, responsible for the observed behavior was suggested [27–29, 36–40, 42, 46]. Below, we will show that the theory of the fermion condensation permits to resolve many puzzles of the condensed matter physics of strongly correlated Fermi systems, which represent the new state of matter.

1.4 Quantum Phase Transitions and NFL Behavior of HF Compounds

The unusual properties and NFL behavior observed in high-T_c superconductors, HF metals, 2D Fermi systems and other HF compounds, such as novel insulators and quasicrystals, are assumed to be determined by various magnetic quantum phase transitions [27–29, 31–38, 40–43]. Since a quantum phase transition occurs at $T = 0$, the control parameters are all but temperature, i.e., composition, electron (hole) number density x, pressure, magnetic field strength B, etc. A quantum phase transition occurs at a quantum- critical point, which separates the ordered phase (emerging as a result of quantum phase transition) from disordered one. It is usually assumed that magnetic (e.g., ferromagnetic and antiferromagnetic) quantum phase transitions with corresponding critical fluctuations are responsible for the NFL behavior. The critical point of such a phase transition can be shifted to $T = 0$ by varying the above-mentioned control parameters.

Universal behavior can be expected only if the system under consideration is sufficiently close to a quantum-critical point, e.g., when the correlation length is much larger than the microscopic length scale, so that critical quantum and thermal fluctuations determine the anomalous contribution to the thermodynamic functions of a substance. Quantum phase transitions of this type are so widespread [32–34, 38–42] that we call them ordinary quantum phase transitions [47]. In this case, the physics is determined by thermal and quantum fluctuations of the critical state, while quasiparticle excitations are destroyed by these fluctuations. Conventional arguments that quasiparticles in strongly correlated Fermi liquids "get heavy and die" at a quantum-critical point commonly employ the well-known formula based on the assumptions that the z-factor (the quasiparticle weight in the single-particle state, see above) vanishes at the points of second-order phase transitions [46]. However, it has been shown that this scenario is problematic, see Chap. 4, Sect. 4.1.4 [48, 49].

The order parameter fluctuations, developing an infinite correlation range, and the vanishing of quasiparticle excitations are considered to be the main reason for the NFL behavior of heavy-fermion metals, 2D fermion systems, and high-T_c superconductors [33, 34, 38, 41, 42, 50]. However, this approach faces certain difficulties. Critical behavior in HF metals is observed experimentally at high temperatures compared to the effective Fermi temperature T_k. For instance, the thermal expansion coefficient $\alpha(T)$, which is a linear function of temperature for normal LFL, $\alpha(T) \propto T$, demonstrates the experimental $\sqrt{T}$ temperature dependence in $CeNi_2Ge_2$ as the temperature decreases from 6 K to at least 50 mK (i.e., varies by two orders of magnitude) [43]. Such behavior can hardly be explained within the framework of the above critical fluctuation theory. Obviously, such a situation is realizable only at a low-temperature regime, $T \to 0$, when the critical fluctuations make the leading contribution to the entropy and when the correlation length is much longer than the microscopic length scale. At a certain temperature T_k, this macroscopically large correlation length must be destroyed by ordinary thermal fluctuations, and the corresponding universal behavior must disappear.

Another difficulty is in explaining the restoration of the LFL behavior under the application of magnetic field B, as observed in HF metals and in high-T_c superconductors [31, 43, 51]. At $T \to 0$ for the LFL state, the following relations are valid for the electric resistivity $\rho(T) = \rho_0 + AT^2$, the heat capacity $C(T) = \gamma_0 T$, and the magnetic susceptibility $\chi = \text{const}$. It turns out that the coefficients in the above laws depend on the magnetic field strength B. Namely, $A = A(B)$, the Sommerfeld coefficient $\gamma_0(B) \propto M^*$, and the magnetic susceptibility $\chi = \chi(B)$. These quantities depend on B in such a way that $A(B) \propto \gamma_0^2(B)$ and $A(B) \propto \chi^2(B)$, which implies that the Kadowaki-Woods relation $K = A(B)/\gamma_0^2(B)$ [52] is B-independent and is preserved [43]. Such universal behavior, quite natural when quasiparticles with the effective mass M^* play the main role, can hardly be explained within the framework that assumes the absence of quasiparticles or Kondo lattice. We emphasize here that quasiparticles are absent in ordinary quantum phase transitions in the vicinity of QCP. Indeed, there is no reason to expect that γ_0, χ, and A are affected by the fluctuations in such a correlated fashion that preserves the Kadowaki-Woods ratio.

For instance, the Kadowaki-Woods relation does not agree with the spin-density-wave scenario [43] and with the results of research in quantum criticality based on the renormalization-group approach [53]. Moreover, measurements of charge and heat transfer have shown that the Wiedemann-Franz law holds in some high-T_c superconductors [51, 54] and HF metals [55–58]. All this suggests that quasiparticles do exist in such metals, and this conclusion is also corroborated by photoemission spectroscopy results [59, 60]. The inability to explain peculiarities of behavior of HF metals mentioned above in the framework of theories based on ordinary quantum phase transitions implies that another important concept introduced by Landau, the order parameter, also ceases to operate (see, e.g., [38, 40, 42, 46]). Thus, we are left without the most fundamental principles of many-body quantum physics [19, 20, 25], and many interesting phenomena associated with the NFL behavior of strongly correlated Fermi systems remain unexplained.

NFL behavior manifests itself in the power-law behavior of the physical quantities of strongly correlated Fermi systems located close to their QCPs, with exponents different from those of ordinary Fermi liquids [61, 62]. It is a common belief that the main output of theory is the explanation of these exponents which are at least dependent on the magnetic character of QCP and dimensionality of the system. On the other hand, the NFL behavior cannot be captured by these exponents as shown

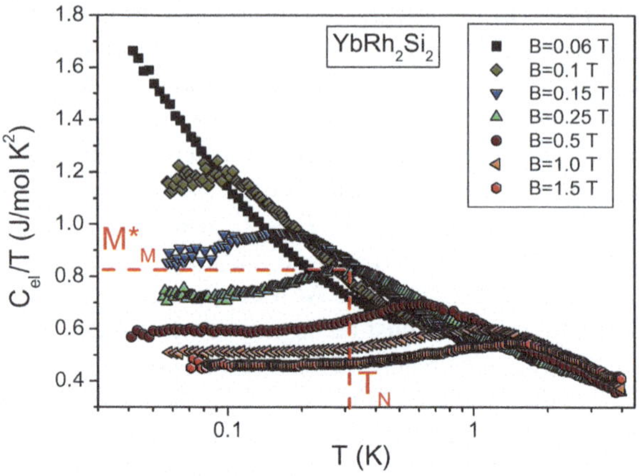

Fig. 1.2 Electronic specific heat of $YbRh_2Si_2$, C/T, versus temperature T as a function of magnetic field B [61] shown in the legend. $M^*_M \propto C_{el}/T$ and T_N are shown as the example for $B = 0.25$ T

in Fig. 1.2. Indeed, the specific heat C/T exhibits a behavior that is to be described as a function of both temperature T and magnetic B field rather than by a single exponent. One can see that at low temperatures C/T demonstrates the LFL behavior which is changed by the transition regime at which C/T reaches its maximum and finally C/T transforms into NFL behavior as a function of T at fixed B. It is clearly shown in Fig. 1.2 that, in particular, in the transition regime, these exponents may have little physical significance.

To show that the behavior of $C/T \propto M^*$ reported in Fig. 1.2 is of general character, we show that in the QCP vicinity it is helpful to use "internal" scales to measure the effective mass $M^* \propto C/T$ and temperature T [27–29, 63, 64]. As it is shown in Fig. 1.2, a maximum structure in $C/T \propto M^*_M$ at temperature T_M appears under the application of magnetic field B and T_M shifts to higher T as B is increased. The value of the Sommerfeld coefficient $C/T = \gamma_0$ is saturated toward lower temperatures decreasing at elevated magnetic field. To obtain the normalized effective mass M^*_N, we use M^*_M (maximal value of the effective mass) and T_M as above "internal" scales: The maximum value of C/T has been used to normalize C/T, and T was normalized by T_M. In Fig. 1.3, the obtained $M^*_N = M^*/M^*_M$ as a function of normalized temperature $T_N = T/T_M$ is shown by symbols. Note that we have excluded the experimental data for magnetic field $B = 0.06$ T. In that case, as will be shown, $T_M \to 0$ and the corresponding T_M and M^*_M are unavailable. It is seen that the LFL and NFL states are separated by the transition (or crossover) regime (or region) where M^*_N reaches its maximal value. Figure 1.3 reveals the scaling behavior of the normalized experimental curves—the curves at different magnetic fields B merge into a single one in terms of the normalized variable $y = T/T_M$. As it is shown in Fig. 1.3, the normalized effective mass $M^*_N(y)$ extracted from the measurements is not a constant, as would be for LFL, and shows the scaling behavior over three decades in normalized temperature y. It is shown in Figs. 1.2 and 1.3 that the NFL behavior and the associated scaling extend at least to temperatures up to few Kelvins. Scenario, where order parameter fluctuations with infinite (or sufficiently large) cor-

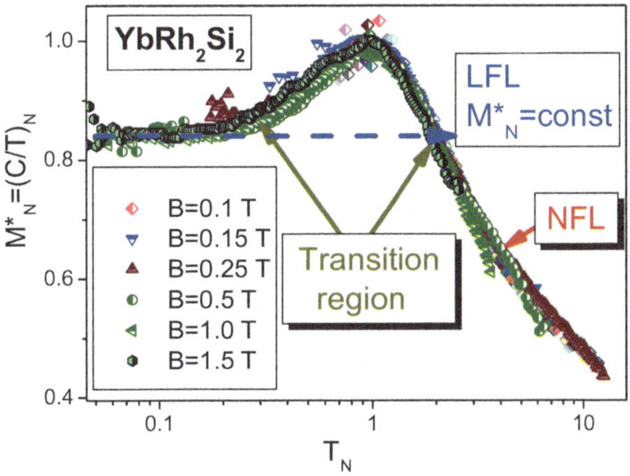

Fig. 1.3 Scaling behavior of the archetypical HF metal $YbRh_2Si_2$. The normalized effective mass M^*_N versus normalized temperature T_N. M^*_N is extracted from the measurements of the specific heat C/T on $YbRh_2Si_2$ in magnetic fields B [61] listed in the legend. Constant effective mass M^*_N inherent in normal Landau Fermi liquids is depicted by the solid line. The transition region separates the LFL and NFL behaviors

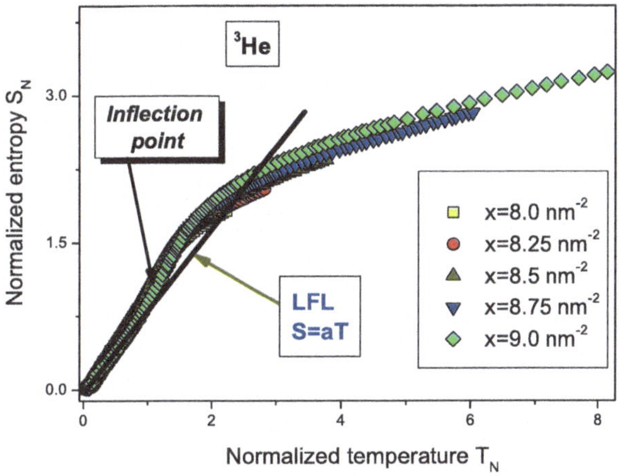

Fig. 1.4 The normalized entropy S_N versus normalized temperature T_N. S_N is extracted from the measurements of the entropy S on ^{3}He at different number densities x [65] shown in the legend. The behavior of the entropy $S \propto T$ inherent in normal Landau Fermi liquids is represented by the solid line

relation length and time develop the NFL behavior, can hardly occur at such high temperatures. We now briefly discuss how the scaling behavior of the normalized entropy is reported in Fig. 1.4, revealing the quantum criticality observed in 2D ^{3}He [65]. This quantum criticality is extremely significant as it allows us to detect the scaling behavior in the 2D system formed by ^{3}He atoms which are essentially different from electrons. As we shall see, the dependence of some observable like the entropy, obviously, do not have "peculiar points" like maxima. The normalization is to be performed in the other points like the inflection point at $T = T_{\rm inf}$ shown in Fig. 1.4 by the arrow; for details, see Chap. 21 and Sect. 21.4. It is shown in Fig. 1.4 that the normalized experimental curves $S(T)$ taken at different values of the number densities x merge into single curve $S_N(T_N) = S(T/T_{\rm inf})/S(T_{\rm inf})$. The observed behavior of S_N strongly deviates from that of the LFL one and cannot be described by a function $S_n(T_N) \simeq T_N^{\beta}$.

Thus, we conclude that the proper explanation of scaling behavior of both $M^*_N(y)$ and $S_N(y)$ shown in Figs. 1.3 and 1.4 is a challenge for theories of critical behavior of HF metals. While the existing theories are primarily dealing with calculations of so-called critical exponents β that characterize $M^*_N(y)$ and S_N at $y \gg 1$, they overlook the regime, signifying the transition from LFL to NFL behavior, and are unable to explain both the LFL and scaling behavior, emerging under the application of magnetic field. As we mentioned above, this transition regime is indeed related to the quantum criticality of systems located near FCQPT. Another part of the problem is the remarkably large temperature ranges over which the NFL behavior is observed. Thus, we conclude that the influence of the critical point extends over a wide range in $T > 0$. This is the regime of quantum criticality, which is crucial for interpreting a wide variety of experiments. As we will see below, the above large temperature ranges are precursors of the quantum-critical point related to FCQPT and the emergence of new quasiparticles. The latter fact, in turn, generates the scaling behavior of the normalized effective mass that allows to explain the thermodynamic, transport, and relaxation properties of HF compounds at the LFL, transition, and NFL regimes.

Taking into account the simple behavior shown in Figs. 1.3 and 1.4, a natural question appears: What theoretical concepts can replace the Fermi liquid paradigm with the notion of the effective mass in cases where the LFL theory breaks down? So far, such a concept within the framework of ordinary quantum phase transitions approach is not available [27–29, 33]. Therefore, here in this book, we focus on the FCQPT concept that preserves the notion of quasiparticles and is intimately related to the unlimited growth of the effective mass M^*. We shall show that this approach is capable to reveal the scaling behavior of the effective mass and to deliver an adequate theoretical explanation of a vast majority of experimental results collected on strongly correlated Fermi systems like HF metals, quasicrystals, quantum spin liquids, 2D Fermi systems, etc. As we shall see, all these HF compounds exhibit a universal scaling behavior at their quantum criticality, and constitute a new state of matter. Thus, whichever mechanism drives the system to FCQPT, the system demonstrates the universal behavior, while there are lot of such mechanisms or tuning parameters like the pressure, number density, magnetic field, chemical doping, frustration, etc.

In contrast to the Landau paradigm based on the assumption that M^* is a constant as shown by the solid line in Fig. 1.3, in FCQPT approach the effective mass M^* of new quasiparticles depends strongly on T, x, B, etc. Therefore, to explain numerous experimental data, the extended quasiparticles paradigm is to be introduced. The main point here is that the new well-defined quasiparticles (with M^* depending on external parameters) determine, as before, the thermodynamic, relaxation, and transport properties of strongly correlated Fermi systems in wide temperature range. The FCQPT approach had been already successfully applied to describe the thermodynamic properties of such different strongly correlated systems as ^{3}He on the one hand and complicated HF compounds and insulators with spin liquid on the other hand [27–29, 48, 66–69].

1.5 Main Goals of the Book

Based on the theory of HF compounds [1, 29], we present in this monograph the new state of matter formed by strongly correlated Fermi systems, that is, HF metals, high-T_c superconductors, substances with quantum spin liquids, quasicrystals, and 2D systems like ^{3}He. This extremely wide diversity of strongly correlated fermion systems introduces the new state of matter by demonstrating the universal behavior of the thermodynamic, transport, and relaxation properties. Such a universal behavior is not seen in any other state of matter, for HF compound exhibit the uniform quantum criticality, defined by the topological FCQPT [1, 13, 26–29]. For the reader convenience, we briefly discuss the construction of the fermion-condensation theory that is capable to explain the vast majority of experimental facts in strongly correlated Fermi systems, while detailed explanations of the construction are presented by [1, 13, 29]. Our analysis goes in the context of salient experimental results. Our calculations of the non-Fermi liquid behavior, the scales and thermodynamic, relaxation, and transport properties are in good agreement with experimental facts. We focus on the scaling behavior of the thermodynamic, transport, and relaxation properties that can be revealed from both experimental facts and theoretical analysis. We do not discuss, however, the specific features of strongly correlated systems in full, starting from their microscopic analysis; instead, we focus on the universal behavior of such systems that can be revealed by the introducing internal scales, as it is explained in Sect. 1.4 and Chap. 7. We also do not discuss the physics of Fermi systems that are not related to condensed matter. Namely, these are neutron stars, atomic clusters and nuclei, quark plasma, and ultra-cold gases in traps, where, according to our views, the fermion condensate induced by FCQPT can also exist [1, 70–75]. Ultra-cold gases in traps are interesting because their easy tuning allows selecting the values of the parameters required for observations of quantum-critical point and FC. We also do not discuss in detail the specific microscopic mechanisms of quantum criticality related to the emergence of FCQPT in a specific compound. Such mechanisms rely heavily on crystalline structure, frustration, and related flat bands of a specific substance, and development of these mechanisms seems not feasible in a near future. In contrast, we consider general mechanisms leading to the formation of flat bands and analyze the common properties of strongly correlated Fermi systems in close connection with accessible experimental facts. For example, the mechanism of quantum criticality, observed in f-electron materials can take place in the systems when the centers of merged single-particle levels "get stuck" at the Fermi surface. One observes that this could provide a simple mechanism for pinning narrow bands in solids to the Fermi surface [75]. Also, we consider high-T_c superconductors within a coarse-grained model based on the FCQPT theory just to illuminate their generic relationships with HF metals. When studying quasicrystals, quantum spin liquids, and 2D systems, we propose the corresponding mechanisms as well. To stress the ubiquitous features of FCQPT, in Chap. 18, Sect. 18.3 we consider its possible role in the emergence of the Universe.

Experimental studies of the properties of quantum phase transitions and their quantum- critical points are very important for understanding the physical nature of strongly correlated Fermi systems. The experimental data, gathered on different strongly correlated Fermi systems, complement each other. In the case of high-T_c superconductors, only few experiments dealing with their QCPs have been conducted. This is because the corresponding QCPs are in the superconductivity range at low temperatures, and the physical properties of the respective quantum phase transition are altered by the superconductivity. As a result, high magnetic fields are needed to destroy the superconducting state. Such experiments can be conducted for HF metals. Experimental research provides data for HF metals behavior, shedding light on the nature of their critical points and phase transitions [43, 51, 54, 56]. Hence, a complete understanding of unusual physical properties related to NFL behavior can be achieved on the base of simultaneous and comparative experimental studies of high-T_c superconductors, HF metals, and other correlated Fermi systems [57, 59, 60].

Since we are concentrated on the properties that are non-sensitive to the detailed crystalline structure of the system, we can avoid difficulties associated with the anisotropy generated by the crystal lattice of solids, its special features, defects, etc. We study the universal behavior of high-T_c superconductors, HF metals, quasicrystals, quantum spin liquids, and 1D and 2D Fermi systems at low temperatures using the model of a homogeneous HF liquid [27–29, 63, 64]. The model is quite meaningful as we consider the low-temperature scaling behavior of these compounds. This behavior is closely related to the scaling of quantities like effective mass, heat capacity, thermal expansion coefficient, etc. The aforementioned scaling is determined, primarily, by long-wavelength properties of the corresponding compound. In other words, they are dealing with transfers of momenta with wave vectors that are small compared to those of the reciprocal lattice constant. The high momentum contributions can therefore be ignored substituting the lattice by the jelly model.

We analyze the universal properties of HF compounds systems using the FCQPT theory [1, 13, 26–29, 76], because the behavior of HF metals already suggests that their unusual properties can be associated with the QPT related to the infinite increase in the effective mass at the critical point. Moreover, we shall see that both the scaling behavior and the quantum criticality displayed in Figs. 1.3 and 1.4 can be quite naturally understood within the framework of the above FCQPT, which gives explanations of the NFL behavior observed in a wide range of HF compounds. To our best knowledge, by now the fermion-condensation theory is the only theory that allows one to explain the wide range of experimental facts collected on HF compounds [1, 29].

References

1. M.Y. Amusia, K.G. Popov, V.R. Shaginyan, W.A. Stephanowich, *Theory of Heavy-Fermion Compounds*. Springer Series in Solid-State Sciences, vol. 182 (Springer, Berlin, 2015)
2. E. Schrödinger, Naturwissenschaften **14**, 664 (1926)
3. D.R. Hartree, Math. Proc. Camb. Philos. Soc. **24**, 89 (1928)
4. D.R. Hartree, Math. Proc. Camb. Philos. Soc. **24**, 111 (1928)
5. V.A. Fock, Z. Phys. **61**, 126 (1930)
6. V.A. Fock, Z. Phys. **62**, 795 (1930)
7. L.N. Thomas, Proc. Camb. Philos. Soc. **23**, 542 (1927)
8. E. Fermi, Rend. Accad. Lincei **6**, 602 (1927)
9. P. Hohenberg, W. Kohn, Phys. Rev. **136**, B864 (1964)
10. W. Kohn, L.J. Sham, Phys. Rev. **140**, A1133 (1965)
11. H. Eschrig, *Fundamentals of Density Functional Theory* (Teubner, Stuttgart, 1996)
12. W. Kohn, Rev. Mod. Phys. **71**, 1253 (1999)
13. V.A. Khodel, V.R. Shaginyan, V.V. Khodel, Phys. Rep. **249**, 1 (1994)
14. R. Jones, O. Gunnarson, Rev. Mod. Phys. **61**, 689 (1989)
15. J. Hubbard, Proc. R. Soc. Lond. **276**, 238 (1963)
16. W. Metzner, D. Vollhardt, Phys. Rev. Lett. **62**, 324 (1989)
17. A. Georges, G. Kotliar, Phys. Rev. B **45**, 6479 (1992)
18. M. Jarrell, Phys. Rev. Lett. **69**, 168 (1992)
19. L.D. Landau, Zh. Eksp. Teor. Fiz. **30**, 1058 (1956)
20. E.M. Lifshitz, L.P. Pitaevskii, *Statisticheskaya Fizika (Statistical Physics), Part 2*. Course of Theoretical Physics (Nauka, Moscow, 1978)
21. I.Y. Pomeranchuk, Zh. Eksp. Teor. Fiz. **35**, 524 (1958)
22. A.B. Migdal, Sov. Phys. JETP **5**, 333 (1957)
23. A.A. Abrikosov, L.P. Gorkov, I.E. Dzyaloshinski, *Methods of Quantum Field Theory in Statistical Physics* (Dover, New York, 1975)
24. M.Y. Amusia, V.E. Starodubsky, Phys. Lett. A **35**, 115 (1971)
25. D. Pines, P. Noziéres, *Theory of Quantum Liquids* (Benjamin, New York, 1966)
26. V.A. Khodel, V.R. Shaginyan, JETP Lett. **51**, 553 (1990)
27. V.R. Shaginyan, M.Y. Amusia, K.G. Popov, Phys. Usp. **50**, 563 (2007)
28. V.R. Shaginyan, M.Y. Amusia, K.G. Popov, Usp. Fiz. Nauk **177**, 586 (2007)
29. V.R. Shaginyan, M.Y. Amusia, A.Z. Msezane, K.G. Popov, Phys. Rep. **492**, 31 (2010), arXiv:1006.2658
30. G.E. Volovik, JETP Lett. **53**, 222 (1991)
31. G.R. Stewart, Rev. Mod. Phys. **73**, 797 (2001)
32. C.M. Varma, Z. Nussionov, W. van Saarloos, Phys. Rep. **361**, 267 (2002)
33. H.v. Löhneysen, A. Rosch, M. Vojta, P. Wölfle, Rev. Mod. Phys. **79**, 1015 (2007)
34. M. Vojta, Rep. Prog. Phys. **66**, 2069 (2003)
35. V.I. Belyavsky, Y.V. Kopaev, Phys. Usp. **49**, 441 (2006)
36. J. Custers, P. Gegenwart, H. Wilhelm, K. Neumaier, Y. Tokiwa, O. Trovarelli, C. Geibel, F. Steglich, C. Pépin, P. Coleman, Nature **424**, 524 (2003)
37. T. Senthil, M.P.A. Fisher, Phys. Rev. B **62**, 7850 (2000)
38. T. Senthil, M. Vojta, S. Sachdev, Phys. Rev. B **69**, 035111 (2004)
39. T. Senthil, S. Sachdev, M. Vojta, Phys. B **359–361**, 9 (2005)
40. T. Senthil, A. Vishwanath, L. Balents, S. Sachdev, M.P.A. Fisher, Science **303**, 1490 (2004)
41. P. Coleman, in *Lectures on the Physics of Highly Correlated Electron Systems VI*, ed. by F. Mancini (American Institute of Physics, New York, 2002), p. 79
42. P. Coleman, A.J. Schofield, Nature **433**, 226 (2005)
43. R. Kúchler, N. Oeschler, P. Gegenwart, T. Cichorek, K. Neumaier, O. Tegus, C. Geibel, J.A. Mydosh, F. Steglich, L. Zhu, Q. Si, Phys. Rev. Lett. **91**, 066405 (2003)
44. N.E. Hussey, Nat. Phys. **3**, 445 (2007)

45. P. Gegenwart, T. Westerkamp, C. Krellner, Y. Tokiwa, S. Paschen, C. Geibel, F. Steglich, E. Abrahams, Q. Si, Science **315**, 969 (2007)
46. P. Coleman, C. Pépin, Q. Si, R. Ramazashvili, J. Phys. Condens. Matter **13**, R723 (2001)
47. V.R. Shaginyan, J.G. Han, J. Lee, Phys. Lett. A **329**, 108 (2004)
48. V.A. Khodel, J.W. Clark, M.V. Zverev, Phys. Rev. B **78**, 075120 (2008)
49. V.A. Khodel, JETP Lett. **86**, 721 (2007)
50. P. Coleman, **1**, 1 (2010), http://arxiv.org/abs/cond-mat/:1001.0185v1
51. C. Proust, E. Boaknin, R.W. Hill, L. Taillefer, A.P. Mackenzie, Phys. Rev. Lett. **89**, 147003 (2002)
52. K. Kadowaki, S.B. Woods, Solid State Commun. **58**, 507 (1986)
53. A.J. Millis, A.J. Schofield, G.G. Lonzarich, S.A. Grigera, Phys. Rev. Lett. **88**, 217204 (2002)
54. R. Bel, K. Behnia, Y. Nakajima, K. Izawa, Y. Matsuda, H. Shishido, R. Settai, Y. Onuki, Phys. Rev. Lett. **92**, 217002 (2004)
55. J. Paglione, M. Tanatar, D. Hawthorn, E. Boaknin, R.W. Hill, F. Ronning, M. Sutherland, L. Taillefer, C. Petrovic, P. Canfield, Phys. Rev. Lett. **91**, 246405 (2003)
56. J. Paglione, M.A. Tanatar, D.G. Hawthorn, F. Ronning, R.W. Hill, M. Sutherland, L. Taillefer, C. Petrovic, Phys. Rev. Lett. **97**, 106606 (2006)
57. F. Ronning, R.W. Hill, M. Sutherland, D.G. Hawthorn, M.A. Tanatar, J. Paglione, L. Taillefer, M.J. Graf, R.S. Perry, Y. Maeno, A.P. Mackenzie, Phys. Rev. Lett. **97**, 067005 (2006)
58. F. Ronning, C. Capan, E.D. Bauer, J.D. Thompson, J.L. Sarrao, R. Movshovich, Phys. Rev. B **73**, 064519 (2006)
59. J.D. Koralek, J.F. Douglas, N.C. Plumb, Z. Sun, A.V. Fedorov, M.M. Murnane, H.C. Kapteyn, S.T. Cundiff, Y. Aiura, K. Oka, H. Eisaki, D.S. Dessau, Phys. Rev. Lett. **96**, 017005 (2006)
60. S. Fujimori, A. Fujimori, K. Shimada, T. Narimura, K. Kobayashi, H. Namatame, M. Taniguchi, H. Harima, H. Shishido, S. Ikeda, D. Aoki, Y. Tokiwa, Y. Haga, Y. Onuki, Phys. Rev. B **73**, 224517 (2006)
61. N. Oeschler, S. Hartmann, A. Pikul, C. Krellner, C. Geibel, F. Steglich, Phys. B **403**, 1254 (2008)
62. P. Gegenwart, T. Westerkamp, C. Krellner, M. Brando, Y. Tokiwa, C. Geibel, F. Steglich, Phys. B **403**, 1184 (2008)
63. V.R. Shaginyan, M.Y. Amusia, K.G. Popov, Phys. Lett. A **373**, 2281 (2009)
64. V.R. Shaginyan, M.Y. Amusia, K.G. Popov, S.A. Artamonov, JETP Lett. **90**, 47 (2009)
65. M. Neumann, J. Nyéki, B. Cowan, J. Saunders, Science **317**, 1356 (2007)
66. V.R. Shaginyan, A.Z. Msezane, K.G. Popov, V.A. Stephanovich, Phys. Rev. Lett. **100**, 096406 (2008)
67. V.R. Shaginyan, A.Z. Msezane, K.G. Popov, Phys. Rev. B **84**, 060401(R) (2011)
68. V.R. Shaginyan, A.Z. Msezane, K.G. Popov, G.S. Japaridze, V.A. Stephanovich, Europhys. Lett. **97**, 56001 (2012)
69. V.R. Shaginyan, A.Z. Msezane, K.G. Popov, V.A. Khodel, Phys. Lett. A **376**, 2622 (2012)
70. V.A. Khodel, J.W. Clark, M.V. Zverev, Phys. Rev. Lett. **87**, 031103 (2001)
71. M.Y. Amusia, A.Z. Msezane, V.R. Shaginyan, Phys. At. Nucl. **66**, 1850 (2003)
72. V.A. Khodel, J.W. Clark, L. Hoachen, M.A. Zverev, Pisma v ZHETF **84**, 703 (2006)
73. G.E. Volovik, Acta Phys. Slov. **56**, 49 (2006)
74. G.E. Volovik, in *Quantum Analogues: From Phase Transitions to Black Holes and Cosmology*. Springer Lecture Notes in Physics, vol. 718, ed. by W.G. Unruh, R. Schutzhold (Springer, Orlando, 2007), p. 31
75. V.A. Khodel, J.W. Clark, H. Li, M.V. Zverev, Phys. Rev. Lett. **98**, 216404 (2007)
76. V.R. Shaginyan, JETP Lett. **79**, 286 (2004)

Chapter 2
Landau Fermi Liquid Theory

Abstract This chapter is devoted to consideration of the Landau theory of Fermi liquid. It has a long history and remarkable results in describing numerous properties of the electron liquid in ordinary metals and Fermi liquids of the ^{3}He type. The theory is based on the assumption that elementary excitations determine the physics at low temperatures, resembling that of weakly interacting Fermi gas. These excitations behave as quasiparticles with a certain effective mass. The effective mass M^* exhibits a simple universal behavior, for it depends weakly on the temperature, pressure, and magnetic field strength and is a parameter of the theory. Microscopically deriving the equation determining the effective mass, we go beyond Landau approach and analyze more complicated universal behavior of M^* at the fermion-condensation quantum phase transition, see Chaps. 3 and 7.

2.1 Quasiparticle Paradigm

One of the most complex problems of modern condensed matter physics is the problem of the structure and properties of Fermi systems with strong interparticle interaction. The first consistent theory of above systems had been proposed by Landau. The main idea of Landau theory of Fermi liquids, later called "normal", was to describe the strongly interacting multi-fermion system in terms of some ground state and its elementary excitations, which, being quantized, generate so-called Landau quasiparticles. Contrary to ordinary particles in a Fermi liquid, which are strongly interacting fermions, the quasiparticles are fictitious particles which are indeed the quanta of above elementary excitations. In such a picture, the real physical interaction between fermions (which in general case is unknown) is substituted by some effective phenomenological interaction between quasiparticles, formulated in terms of so-called Landau's interaction (or amplitudes). This paradigm permits to reduce the problem formally to the weakly interacting Fermi gas and introduces the effective interaction between quasiparticles [1–3]. The Landau approach can be regarded

M. Amusia and V. Shaginyan, *Strongly Correlated Fermi Systems*,
Springer Tracts in Modern Physics 283,
https://doi.org/10.1007/978-3-030-50359-8_2

as an effective low-energy theory with the high-energy degrees of freedom being eliminated by introducing the above quasiparticle interaction, which substitute the strong interaction between real particles. The invariability of the ground state of the Landau Fermi liquid (LFL) is determined by the Pomeranchuk stability conditions: A system becomes unstable when at least one Landau interaction becomes negative and reaches its critical value [2–4]. We note that the new phase, in which stability is restored, can also be described, in principle, by the LFL theory.

To make the presentation self-contained, here we briefly recapitulate the main ideas of the LFL theory [1–3, 5]. The theory is based on the quasiparticle paradigm, which states that quasiparticles are elementary weakly excited states of Fermi liquids and are therefore specific excitations determining their low-temperature thermodynamic and transport properties. In the case of the electron liquid, the quasiparticles are characterized by the electron quantum numbers and the effective mass M^*. As it is shown in Chap. 3, the ground state energy of the system is a functional of the quasiparticle occupation numbers (or the quasiparticle distribution function) $n(\mathbf{p}, T)$, and the same is true for the free energy $F(n(\mathbf{p}, T))$, the entropy $S(n(\mathbf{p}, T))$, and other thermodynamic functions. In the case of solid when the fermion system interacts with the lattice, the ground state $E(n(\mathbf{p}))$ is the functional of $n(\mathbf{p})$ only at low temperatures when one can omit contributions coming from the lattice, for example, such as phonons.

We can find the distribution function from the minimum condition for the free energy $F = E - TS$ (hereafter we use the units, where $k_B = \hbar = 1$)

$$\frac{\delta(F - \mu x)}{\delta n(\mathbf{p}, T)} = \varepsilon(\mathbf{p}, T) - \mu(T) - T \ln \frac{1 - n(\mathbf{p}, T)}{n(\mathbf{p}, T)} = 0. \tag{2.1}$$

Here, μ is the chemical potential fixing the number density x:

$$x = \int n(\mathbf{p}, T) \frac{d\mathbf{p}}{(2\pi)^3} \tag{2.2}$$

and

$$\varepsilon(\mathbf{p}, T) = \frac{\delta E(n(\mathbf{p}, T))}{\delta n(\mathbf{p}, T)} \tag{2.3}$$

is the quasiparticle energy. The quasiparticle energy, just as the energy E, is a functional of $n(\mathbf{p}, T)$: $\varepsilon = \varepsilon(\mathbf{p}, T, n)$. The entropy $S(n(\mathbf{p}, T))$ related to quasiparticles is given by the well-known expression [1–3]

$$\begin{aligned} S(n(\mathbf{p}, T)) = -2 \int \Big[& n(\mathbf{p}, T) \ln(n(\mathbf{p}, T)) + (1 - n(\mathbf{p}, T)) \\ & \times \ln(1 - n(\mathbf{p}, T)) \Big] \frac{d\mathbf{p}}{(2\pi)^3}, \end{aligned} \tag{2.4}$$

which follows from combinatorial reasoning. Equation (2.1) is usually written in the standard form of the Fermi-Dirac distribution,

$$n(\mathbf{p},T)=\left\{1+\exp\left[\frac{(\varepsilon(\mathbf{p},T)-\mu)}{T}\right]\right\}^{-1}. \tag{2.5}$$

At $T\to 0$, (2.1) and (2.5) reduce to $n(p,T\to 0)\to\theta(p_F-p)$ if the derivative $\partial\varepsilon(p\simeq p_F)/\partial p$ is finite and positive. Here, p_F is the Fermi momentum and $\theta(p_F-p)$ is the unit step function. The single-particle energy can be approximated as $\varepsilon(p\simeq p_F)-\mu\simeq p_F(p-p_F)/M^*_L$, where M^*_L is the effective mass of the Landau quasiparticle,

$$\frac{1}{M^*_L}=\frac{1}{p}\frac{d\varepsilon(p,T=0)}{dp}|_{p=p_F}. \tag{2.6}$$

In turn, the effective mass M^*_L is related to the bare electron mass m by the well-known Landau equation [1–3, 5]

$$\frac{1}{M^*_L}=\frac{1}{m}+\sum_{\sigma_1}\int\frac{\mathbf{p}_F\mathbf{p}_1}{p_F^3}F_{\sigma,\sigma_1}(\mathbf{p_F},\mathbf{p}_1)\times\frac{\partial n_{\sigma_1}(\mathbf{p}_1,T)}{\partial p_1}\frac{d\mathbf{p}_1}{(2\pi)^3}, \tag{2.7}$$

where $F_{\sigma,\sigma_1}(\mathbf{p_F},\mathbf{p}_1)$ is the Landau interaction, which depends on the momenta $\mathbf{p_F}$ and $\mathbf{p}$ and spin indices σ,σ_1. For simplicity, we omit the spin indices in the effective mass as M^*_L is almost completely spin-independent in the case of a homogeneous liquid and weak magnetic fields. We note that (2.7) is exact, as it is shown in Chap. 3. The Landau interaction F is given by

$$F_{\sigma,\sigma_1}(\mathbf{p},\mathbf{p}_1,n)=\frac{\delta^2E(n)}{\delta n_\sigma(\mathbf{p})\delta n_{\sigma_1}(\mathbf{p}_1)}. \tag{2.8}$$

2.2 Pomeranchuk Stability Conditions

The stability of the ground state of LFL is determined by the Pomeranchuk stability conditions: The considered system becomes unstable when at least one harmonic of Landau interaction becomes negative and reaches its critical value [2–5]

$$F_L^{a,s}=-(2L+1). \tag{2.9}$$

Here, F^a_L and F^s_L are the dimensionless spin-symmetric and spin-antisymmetric Landau interactions, and L is the angular momentum related to the corresponding Legendre polynomials P_L,

$$F_{\sigma,\sigma_1}(\mathbf{p},\mathbf{p}_1) = \frac{1}{N}\sum_{L=0}^{\infty} P_L(\Theta)\left[F^a_{L\ \sigma,\sigma_1} + F^s_L\right]. \tag{2.10}$$

Here, Θ is the angle between momenta $\mathbf{p}$, and the dimensionless Landau interaction reads

$$F_{\sigma,\sigma_1}(\mathbf{p},\mathbf{p}_1) = N F_{\sigma,\sigma_1}(\mathbf{p},\mathbf{p}_1,n), \tag{2.11}$$

where the density of states $N = M^*_L p_F/(2\pi^2)$ and the Landau interaction $F_{\sigma,\sigma_1}(\mathbf{p},\mathbf{p}_1,n)$ is given by (2.8). It follows from (2.7) that

$$\frac{M^*_L}{m} = 1 + \frac{F^s_1}{3}. \tag{2.12}$$

In accordance with the Pomeranchuk stability conditions, it is seen from (2.12) that $F^s_1 > -3$; otherwise, the effective mass becomes negative leading to unstable state when it is energetically favorable to excite quasiparticles near the Fermi surface. Below we again omit the spin indices σ for simplicity.

2.3 Thermodynamic and Transport Properties

To deal with the transport properties of Fermi systems, one needs a transport equation describing slowly varying disturbances of the quasiparticle distribution function $n_{\mathbf{p}}(\mathbf{r},t)$ which depends on position $\mathbf{r}$ and time t. As long as the transferred energy ω and momentum q of the external field quanta are much smaller than the energy and momentum of the quasiparticles, $qp_F/(TM^*_L) \ll 1$, the quasiparticle distribution function $n(\mathbf{q},\omega)$ satisfies the transport equation [1–3, 5]

$$\frac{\partial n_{\mathbf{p}}}{\partial t} + \nabla_{\mathbf{p}}\varepsilon_{\mathbf{p}}\nabla_{\mathbf{r}}n_{\mathbf{p}} - \nabla_{\mathbf{r}}\varepsilon_{\mathbf{p}}\nabla_{\mathbf{p}}n_{\mathbf{p}} = I[n_{\mathbf{p}}]. \tag{2.13}$$

The left-hand side of (2.13) describes the dissipationless dynamic of quasiparticles in phase space. The quasiparticle energy $\varepsilon_{\mathbf{p}}(\mathbf{r},t)$ now depends on its position and time, and the collision integral $I[n_{\mathbf{p}}]$ measures the rate of the distribution function variation due to collisions. The transport equation (2.13) allows one to calculate all the transport properties of a Fermi system. It should be emphasized here the role of Umklapp processes, in particular, these processes lead to a nonzero contribution of FC to ρ_0 [6–8], and is associated with the presence of the crystal lattice, more precisely, with the Umklapp processes, violating momentum conservation [9]. We assume the presence of Umklapp processes in all cases when the violation of momentum conservation is of importance.

It is a common belief that the equations of this subsection are phenomenological and inapplicable to describe Fermi systems where the effective mass M^* depends strongly on temperature T, external magnetic fields B, pressure P, etc. At the same time, many experimental data collected for HF metals reveal that the quasiparticle effective mass strongly depends on T, B, and doping (or the number density) x. Moreover, in those cases, the effective mass M^* can reach very high values or even diverge. As we have seen in Sect. 1.1, such a behavior is so unusual that the traditional Landau quasiparticle paradigm fails to describe it. Therefore, to reconcile the Landau quasiparticle picture with the experimental situation in the real substances, the extended quasiparticle paradigm is to be introduced with still well-defined quasiparticles determining as before the thermodynamic and transport properties of strongly correlated Fermi systems. In this paradigm, the effective mass M^* becomes a function of T, x, and B so that just this dependence leads to the experimentally observed NFL behavior [8, 10–16]. As we shall see in Sect. 2.3.1 and Chap. 3, the extended quasiparticle paradigm permits to derive (2.7) microscopically. Thus, (2.7) is exact rather than phenomenological. On the other hand, there is nothing wrong with any phenomenological theory; the most fundamental equations (e.g., the Schrödinger equation) are phenomenological, for nobody can derive them from some basic principles, see, e.g., [17].

2.3.1 Equation for the Effective Mass

To derive the equation determining the effective mass, we consider the model of a homogeneous HF liquid and employ the density functional theory for superconductors (SCDFT) [18] which allows us to consider E as a functional of the occupations numbers $n(\mathbf{p})$ [13, 19–21]. As a result, the ground state energy E of the normal state becomes the functional of the occupation numbers and the function of the number density x, $E = E(n(\mathbf{p}), x)$, while (2.3) gives the single-particle spectrum. Upon differentiating both sides of (2.3) with respect to $\mathbf{p}$ and after some algebra involving integration by parts, we obtain [12, 13, 19, 20]

$$\frac{\partial \varepsilon(\mathbf{p})}{\partial \mathbf{p}} = \frac{\mathbf{p}}{m} + \int F(\mathbf{p}, \mathbf{p}_1, n) \frac{\partial n(\mathbf{p}_1)}{\partial \mathbf{p}_1} \frac{d\mathbf{p}_1}{(2\pi)^3}. \tag{2.14}$$

To calculate the derivative $\partial\varepsilon(\mathbf{p})/\partial\mathbf{p}$, we employ the functional representation

$$\begin{aligned} E(n) = &\int \frac{p^2}{2m} n(\mathbf{p}) \frac{d\mathbf{p}}{(2\pi)^3} \\ &+ \frac{1}{2} \int F(\mathbf{p}, \mathbf{p}_1, n)|_{n=0}\, n(\mathbf{p}) n(\mathbf{p}_1) \frac{d\mathbf{p} d\mathbf{p}_1}{(2\pi)^6} + \cdots \end{aligned} \tag{2.15}$$

It is seen directly from (2.14) that the effective mass is given by the well-known Landau equation

$$\frac{1}{M^*} = \frac{1}{m} + \int \frac{\mathbf{p}_F \mathbf{p}_1}{p_F^3} F(\mathbf{p}_F, \mathbf{p}_1, n) \frac{\partial n(p_1)}{\partial p_1} \frac{d\mathbf{p}_1}{(2\pi)^3}. \tag{2.16}$$

Here we suppress the spin indices for simplicity. To calculate M^* as a function of T, we construct the free energy $F = E - TS$, where the entropy S is given by (2.4). Minimizing F with respect to $n(\mathbf{p})$, we arrive at the Fermi-Dirac distribution, (2.5). The above derivation shows that (2.14) and (2.16) are exact and allow to calculate the behavior of both $\partial\varepsilon(\mathbf{p})/\partial\mathbf{p}$ and M^* which now is a function of temperature T, external magnetic field B, number density x, and pressure P rather than a constant. As we shall see, it is just this feature of M^* that forms both the scaling and the NFL behavior observed in measurements on HF metals.

It is assumed in LFL theory that M^*_L is positive, finite, and independent of external factors like field and/or temperature. As a result, the temperature-dependent corrections to M^*_L, the quasiparticle energy $\varepsilon(\mathbf{p})$ and other quantities begin with the term proportional to T^2 in 3D systems. In this case, the effective mass is given by (2.16), and the specific heat C is as follows [1] :

$$C = \frac{2\pi^2 NT}{3} = \gamma_0 T = T\frac{\partial S}{\partial T}, \tag{2.17}$$

and the spin susceptibility

$$\chi = \frac{3\gamma_0\mu_B^2}{\pi^2(1+F_0^a)}, \tag{2.18}$$

where μ_B is the Bohr magneton and $\gamma_0 \propto M^*_L$. In the LFL case, upon using the transport equation (2.13), one finds for the electrical resistivity at low T [5]

$$\rho(T) = \rho_0 + AT^{\alpha_R}, \tag{2.19}$$

where ρ_0 is the residual resistivity, the exponent $\alpha_R = 2$, and $A \propto (M^*)^2$ is the coefficient determining the charge transport. This coefficient is proportional to the quasiparticle-quasiparticle scattering cross section. Equation (2.19) symbolizes and defines the LFL behavior observed in normal metals.

Due to the normalization condition used in (2.11), (2.12) defines the implicit dependence of the effective mass M^* on the Landau interaction. Taking into account (2.12) and introducing the density of states of a free Fermi gas, $N_0 = mp_F/(2\pi^2)$, we obtain the effective mass M^* as an explicit function of the Landau interaction and the bare mass m [22–24]

$$\frac{M^*}{m} = \frac{1}{1 - F^1/3}, \tag{2.20}$$

where $F^1 = N_0 f^1$, and $f^1(p_F, p_F)$ is the p-wave component of the Landau interaction. Since $x = p_F^3/3\pi^2$ in the LFL theory, the Landau interaction can be written as $F^1(p_F, p_F) = F^1(x)$. Provided that at a certain critical point $x_{\rm FC}$, the denominator $(1 - F^1(x)/3)$ tends to zero, i.e., $(1 - F^1(x)/3) \propto (x - x_{\rm FC}) + a(x - x_{\rm FC})^2 + \cdots \to 0$, we find that [25, 26]

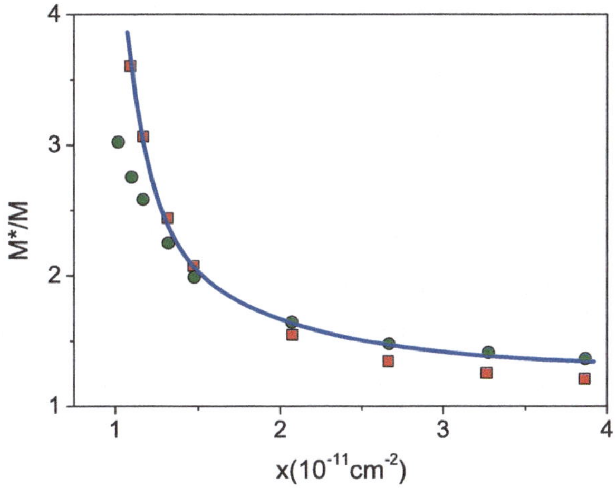

Fig. 2.1 The ratio M^*/M in a silicon MOSFET as a function of the electron number density x, with M being the bare mass of electron. The squares mark the experimental data on the Shubnikov-de Haas oscillations. The data obtained by applying a parallel magnetic field are marked by circles [27–29]. The solid line represents the function (6.52)

$$\frac{M^*(x)}{m} \simeq a_1 + \frac{a_2}{x - x_{FC}} \propto \frac{1}{r}, \tag{2.21}$$

where a_1 and a_2 are constants and $r = (x - x_{FC})/x_{FC}$ is the "distance" from QCP x_{FC} where $M^*(x \to x_{FC}) \to \infty$. We note that the divergence of the effective mass given by (2.20) and (2.21) do preserve the Pomeranchuk stability conditions, see (2.9), for the divergence takes place at $F^1 \geq 3$. As it was discussed above, (2.20) and (2.21) represent the explicit formula for the effective mass, while (2.12) represents an implicit formula for the effective mass.

The behavior of $M^*(x)$ described by formula (2.21) is in good agreement with the results of experiments [27–31] and calculations [32, 33]. In the case of electron systems, (2.21) holds for $x > x_{FC}$, while for 2D ^{3}He we have $x < x_{FC}$ so that always $r > 0$, see Sect. 4.1.5 and Fig. 4.2 for details. Such behavior of the effective mass is observed in HF metals, which have a fairly flat and narrow conduction band corresponding to a large effective mass, with a strong correlation and the effective Fermi temperature $T_k \sim p_F^2/M^*(x)$ of the order of several dozen kelvins or even lower [34].

The effective mass as a function of the electron density x in a silicon MOSFET (Metal-Oxide-Semiconductor Field-Effect Transistor), approximated by (2.21), is shown in Fig. 2.1. The constants a_1, a_2, and x_{FC} are taken as fitting parameters. We see that (2.21) provides a good description of the experimental results.

The divergence of the effective mass $M^*(x)$ discovered in measurements involving 2D ^{3}He [30, 31] is illustrated in Fig. 2.2. Figures 2.1 and 2.2 show that the description provided by (2.21) does not depend on the nature of constituting fermions and is in good agreement with the experimental data.

To illustrate the ability of the extended quasiparticle paradigm to capture the observed scaling behavior of M^*, it is instructive to explore it briefly here, while more detailed consideration is reserved for Chap. 7. For that, we write the quasiparticle

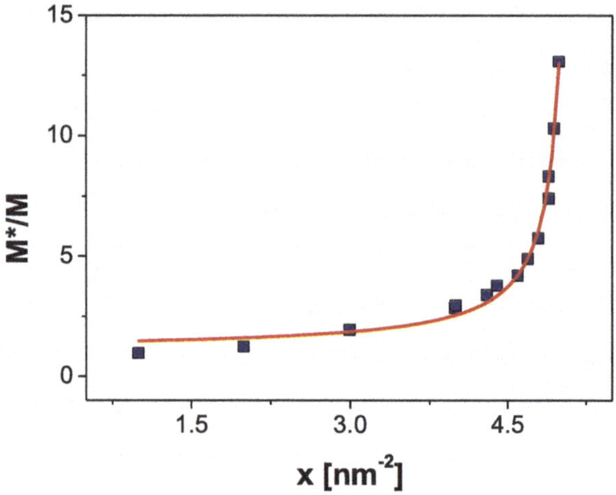

Fig. 2.2 The ratio M^*/M in 2D ^{3}He as a function of the density x of the liquid, obtained from heat capacity and magnetization measurements. Here M is the bare mass of ^{3}He. The experimental data are marked by black squares [30, 31], and the solid line represents the function given by (2.21), where $a_1 = 1.09$, $a_2 = 1.68\,\mathrm{nm}^{-2}$, and $x_{\rm FC} = 5.11\,\mathrm{nm}^{-2}$

distribution function as $n_1(\mathbf{p}) = n(\mathbf{p}, T) - n(\mathbf{p})$, with $n(\mathbf{p})$ being the step function, and (2.16) then becomes

$$\frac{1}{M^*(T)} = \frac{1}{M^*} + \int \frac{\mathbf{p}_F\mathbf{p}_1}{p_F^3} F(\mathbf{p}_{\rm F}, \mathbf{p}_1) \frac{\partial n_1(p_1, T)}{\partial p_1} \frac{d\mathbf{p}_1}{(2\pi)^3}. \tag{2.22}$$

At QCP $x \to x_{\rm FC}$, the effective mass $M^*(x)$ diverges and (2.22) becomes homogeneous determining M^* as a function of temperature while the system exhibits the NFL behavior. If the system is located before QCP, M^* is finite so that at low temperatures the integral on the right-hand side of (2.22) represents a small correction to $1/M^*$ and the system demonstrates the LFL behavior shown in Figs. 1.2 and 1.3. The LFL behavior assumes that the effective mass is independent of temperature, $M^*(T) \simeq$ const, as shown by the horizontal line in Fig. 1.3. Obviously, the LFL behavior takes place only if the second term on the right-hand side of (2.22) is small in comparison with the first one. In the case of common metals, the effective mass $M^* \sim m$ and the Fermi energy is $E_F \sim 1$ eV. Then, the integral I on the right-hand side of (2.22) satisfies the condition $M^*I \ll 1$ even at room temperatures. As a result, $M^*(T) \simeq M^*(T = 0)$. As soon as the effective mass $M^*(x)$ diverges, as it is shown in Fig. 2.2 the condition $M^*I \ll 1$ ceases to be valid at some temperature $T \sim T_M$. Thus, at temperatures $T \sim T_M$ the system enters the transition regime: M^* grows, reaching its maximum M^*_M at $T = T_M$, with subsequent diminishing. As shown in Fig. 1.3, near temperatures $T \geq T_M$ the last "traces" of LFL regime disappear, the second term in (2.22) starts to dominate, and again this equation becomes homogeneous so that the NFL behavior is restored, manifesting itself in decreasing M^* as a function of T. As we shall see below, the above solution of (2.22) generates the scaling behavior of the effective mass, resulting in the experimentally observed universal (i.e., independent on microscopic structure of specific substance) properties of HF compounds at their quantum criticality.

References

1. L.D. Landau, Zh. Eksp. Teor. Fiz. **30**, 1058 (1956)
2. E.M. Lifshitz, L.P. Pitaevskii, *Statisticheskaya Fizika (Statistical Physics), Part 2*. Course of Theoretical Physics (Nauka, Moscow, 1978)
3. E.M. Lifshitz, L.P. Pitaevskii, *Statisticheskaya Fizika (Statistical Physics), Part 2*. Course of Theoretical Physics (Pergamon Press, Oxford, 1980)
4. I.Y. Pomeranchuk, Zh. Eksp. Teor. Fiz. **35**, 524 (1958)
5. D. Pines, P. Noziéres, *Theory of Quantum Liquids* (Benjamin, New York, 1966)
6. V.R. Shaginyan, A.Z. Msezane, K.G. Popov, G.S. Japaridze, V.A. Stephanovich, Europhys. Lett. **97**, 56001 (2012)
7. V.R. Shaginyan, A.Z. Msezane, K.G. Popov, V.A. Khodel, Phys. Lett. A **376**, 2622 (2012)
8. M.Y. Amusia, K.G. Popov, V.R. Shaginyan, W.A. Stephanowich, *Theory of Heavy-Fermion Compounds*. Springer Series in Solid-State Sciences, vol. 182 (Springer, Berlin, 2015)
9. E.M. Lifshitz, L.P. Pitaevskii, *Physical Kinetics* (Pergamon Press, Oxford, 1981)
10. V.R. Shaginyan, M.Y. Amusia, K.G. Popov, Phys. Usp. **50**, 563 (2007)
11. V.R. Shaginyan, M.Y. Amusia, A.Z. Msezane, K.G. Popov, Phys. Rep. **492**, 31 (2010), arXiv:1006.2658
12. V.A. Khodel, J.W. Clark, M.V. Zverev, Phys. Rev. B **78**, 075120 (2008)
13. V.R. Shaginyan, M.Y. Amusia, K.G. Popov, Phys. Lett. A **373**, 2281 (2009)
14. J. Dukelsky, V. Khodel, P. Schuck, V. Shaginyan, Z. Phys. **102**, 245 (1997)
15. V.A. Khodel, V.R. Shaginyan, in *Condensed Matter Theories*, vol. 12, ed. by J. Clark, V. Plant (Nova Science Publishers Inc., New York, 1997), p. 221
16. J.W. Clark, V.A. Khodel, M.V. Zverev, Phys. Rev. B **71**, 012401 (2005)
17. R.B. Laughlin, D. Pines, Proc. Natl. Acad. Sci. USA **97**, 28 (2000)
18. L.N. Oliveira, E.K.U. Gross, W. Kohn, Phys. Rev. Lett. **60**, 2430 (1988)
19. V.R. Shaginyan, Phys. Lett. A **249**, 237 (1998)
20. M.Y. Amusia, V.R. Shaginyan, Phys. Lett. A **269**, 337 (2000)
21. V.R. Shaginyan, JETP Lett. **68**, 527 (1998)
22. D. Vollhardt, Rev. Mod. Phys. **56**, 99 (1984)
23. M. Pfitzner, P. Wölfle, Phys. Rev. B **33**, 2003 (1986)
24. D. Vollhardt, P. Wölfle, P.W. Anderson, Phys. Rev. B **35**, 6703 (1987)
25. V.M. Yakovenko, V.A. Khodel, JETP Lett. **78**, 850 (2003)
26. V.R. Shaginyan, JETP Lett. **77**, 99 (2003)
27. A.A. Shashkin, S.V. Kravchenko, V.T. Dolgopolov, T.M. Klapwijk, Phys. Rev. B **66**, 073303 (2002)
28. A.A. Shashkin, M. Rahimi, S. Anissimova, S.V. Kravchenko, V.T. Dolgopolov, T.M. Klapwijk, Phys. Rev. Lett. **91**, 046403 (2003)
29. S.V. Kravchenko, M.P. Sarachik, Rep. Prog. Phys. **67**, 1 (2004)
30. A. Casey, H. Patel, J. Nyeki, B.P. Cowan, J. Saunders, Low Temp. Phys. **113**, 293 (1998)
31. A. Casey, H. Patel, J. Cowan, B.P. Saunders, Phys. Rev. Lett. **90**, 115301 (2003)
32. Y. Zhang, V.M. Yakovenko, S.D. Sarma, Phys. Rev. B **71**, 115105 (2005)
33. Y. Zhang, S.D. Sarma, Phys. Rev. B **70**, 035104 (2004)
34. G.R. Stewart, Rev. Mod. Phys. **73**, 797 (2001)

Chapter 3
Density Functional Theory of Fermion Condensation

Abstract In this chapter, a density functional formulation of the fermion condensation is presented. It is demonstrated that the standard Kohn-Sham scheme is not valid beyond the topological FCQPT since the functional of the free energy F starts to depend on the quasiparticle density matrix. In the case of infinite homogeneous system, the ground state energy E becomes a functional of the quasiparticle occupation numbers $n(\mathbf{p})$, $E[n(\mathbf{p})]$. Thus, our consideration demonstrates that both the Landau Fermi liquid theory and the fermion-condensation theory are microscopic theories rather than phenomenological ones. We also present both the functional equation that defines the functional and a procedure to solve the equation. Our consideration also furnishes an opportunity to calculate the real single-particle excitation spectra of superconducting systems within the density functional theory. We show that the standard superconducting state, taking place in conventional metals, is strongly modified by the FC state, as it is shown in detail in Chap. 23.

3.1 Introduction

An essential progress has been achieved in describing multielectron systems, which is connected mainly with the construction of the density functional theory (DFT) [1–4]. As it is mentioned in Chap. 1, this DFT in principle permits to calculate precisely the characteristics of the ground state of the system, such as the ground state energy and the single-particle potential. However, out of its scope are those dynamic properties of the system which are closely related to the behavior of the system in time-dependent external fields. This shortcoming of DFT was eliminated in [5], where the density functional method was generalized to time-dependent densities. After this, DFT acquires the ability to consider formally static and dynamic characteristics of a multielectron system. However, this version of DFT is also unable to derive the density functional itself. So, for its application, some approximations

M. Amusia and V. Shaginyan, *Strongly Correlated Fermi Systems*,
Springer Tracts in Modern Physics 283,
https://doi.org/10.1007/978-3-030-50359-8_3

of this functional must be constructed. Most frequently used is the local density approximation (LDA), in which one has [1–4],

$$E_{xc}[\rho] = \int \varepsilon_{xc}(\rho(\mathbf{r}))\rho(\mathbf{r})\,d^3\mathbf{r}.$$

Here E_{xc} is the exchange-correlation part of the density functional, $\varepsilon_{xc}(\rho(\mathbf{r},t))$ is the exchange-correlation energy of homogeneous electron gas of the density $\rho(\mathbf{r},t)$, related to a single particle. The exchange-correlation potential is of the form

$$v_{xc}(\mathbf{r}) = \frac{\delta E_{xc}}{\delta\rho(\mathbf{r})}|_{\rho=\rho_0}, \tag{3.1}$$

with ρ_0 being the ground state one-particle density. The exchange-correlation kernel is given by

$$f_{xc}(\mathbf{r},t;\mathbf{r}',t') = \delta(t-t')\delta(\mathbf{r}-\mathbf{r}')\frac{d^2}{d\rho^2}[\rho(\mathbf{r})\varepsilon_{xc}(\rho(\mathbf{r}))]|_{\rho=\rho_0}. \tag{3.2}$$

However, in this case, the potential $v_{xc}(\mathbf{r})$ (3.2) has a wrong asymptotic behavior at large distances r from a system, since the exchange-correlation kernel $f_{xc}(\mathbf{r},t;\ \mathbf{r}',t')$ is of zero range [1–4]. These limitations are quite essential, because, for instance, the correct asymptotic $v_{xc}(\mathbf{r})$ at long distances r is of decisive importance for the correct description of autoionization resonances, photoionization, Van der Waals forces, etc. The finite range dependence of the effective interaction is also essential in considering processes related to transferring considerable momentum such as the electron scattering on atoms and the photo effect [1–4].

On the other hand, given the exchange-correlation action functional $A_{xc}[\rho]$, one could construct the single-particle exchange-correlation potential v_{xc} and f_{xc},

$$v_{xc}(\mathbf{r}_1,t_1) = \frac{\delta A_{xc}}{\delta\rho(\mathbf{r}_1,t_1)};\quad f_{xc}(\mathbf{r}_1,t_1,\mathbf{r}_2,t_2) = \frac{\delta v_{xc}(\mathbf{r}_1,t_1)}{\delta\rho(\mathbf{r}_2,t_2)}, \tag{3.3}$$

and employ (3.3) instead of incorrect (3.1) and (3.2). However, it is seen from (3.3) that f_{xc}, being symmetric under exchange $(\mathbf{r}_1,t_1)$ and $(\mathbf{r}_2,t_2)$, cannot be a casual function. As a result, v_{xc} cannot also be a casual function, as it should be. Thus, one has to conclude that (3.3) is not valid.

We suggest an approach to DFT, taking as a basis an exact functional equation for the effective interaction. As a result, we obtain equations for the effective interaction, or f_{xc}, v_{xc} and the linear response function. Taking these equations as a basis, we explore the relations between the action functional, the linear response function, and the single-particle potential. Then, a density functional theory of the fermion condensation (FC) is developed. The FC is a new phase transition, associated with rearrangement of the single-particle degrees of freedom, and analogous to Bose condensation but for fermions. The main feature of this transition is the appearance of a plateau in the single-particle excitation spectrum or a flat band, at the Fermi level [6–8]. As consequence of this, the standard Kohn-Sham scheme is no longer valid, while a system with the fermion condensate represents an example of strongly correlated liquid. Our consideration also furnishes an opportunity to calculate the single-particle excitation spectra of superfluid systems [9, 10].

3.2 Functional Equation for the Effective Interaction

The key point of our consideration is a functional equation for the effective interaction R, see [9, 11] and references therein. Here, we outline only the main points of this derivation, grounded on the conventional formalism of the quantum mechanical action integral A.

The action integral A of a system is given by

$$A[\rho] = \int dt \langle\varphi(t)\left| i\frac{\partial}{\partial t} - \widehat{H}_\lambda(t)\right|\varphi(t)\rangle, \tag{3.4}$$

with $\varphi(t)$ being time-dependent wave function. It is a unique functional of the density and must have a stationary point at the correct time-dependent density $\rho_0(\mathbf{r}, t)$ [5],

$$\left.\frac{\delta A}{\delta\rho}\right|_{\rho=\rho_0} = 0. \tag{3.5}$$

The Hamiltonian in (3.4) contains the kinetic energy of the electrons $\widehat{T}$, their mutual Coulomb repulsion $\widehat{H}_{\rm int}$, and an external field $\widehat{U}$

$$\widehat{H}_\lambda(t) = \widehat{T} + \widehat{H}_{\rm int} + \widehat{U}, \tag{3.6}$$

with

$$\widehat{T} = \int \widehat{\psi}^+(\mathbf{r})\left(-\frac{\nabla^2}{2}\right)\widehat{\psi}(\mathbf{r})d^3\mathbf{r}, \tag{3.7}$$

$$\widehat{H}_{\rm int} = \frac{1}{2}g\int \widehat{\psi}^+(\mathbf{r}_1)\widehat{\psi}^+(\mathbf{r}_2)v(\mathbf{r}_1-\mathbf{r}_2)\widehat{\psi}(\mathbf{r}_2)\widehat{\psi}(\mathbf{r}_1)d^3\mathbf{r}_1 d^3\mathbf{r}_2, \tag{3.8}$$

$$\widehat{U} = \int \lambda v_{\rm ext}(\mathbf{r}, t)\widehat{\psi}^+(\mathbf{r})\widehat{\psi}(\mathbf{r})d^3\mathbf{r} + \int U_{\rm ext}(\mathbf{r})\widehat{\psi}^+(\mathbf{r})\widehat{\psi}(\mathbf{r})d^3\mathbf{r}. \tag{3.9}$$

Here $v(\mathbf{r}_1 - \mathbf{r}_2)$ is the Coulomb interaction, $\lambda v_{\rm ext}(\mathbf{r}, t)$ is an auxiliary time-dependent external field with λ being sufficiently weak coupling constant, while $U_{\rm ext}$ is an external field, formed, for instance, by a static contribution of the electron-nucleus interaction. For the sake of simplicity, we omit here the spin variables.

The functional A can be written as [5]

$$A[\rho] = A_1[\rho] + \int \lambda v_{\text{ext}}(\mathbf{r})\rho(\mathbf{r})d^3\mathbf{r},$$

with A_1 being a universal functional. The total derivative of A with respect to the coupling constant g is given by [11]

$$\frac{dA}{dg} = \frac{\partial A}{\partial g} + \left.\frac{\delta A}{\delta \rho}\right|_{\rho_0} \frac{\partial \rho}{\partial g} = \frac{\partial A}{\partial g} = \frac{\partial A_1}{\partial g}. \tag{3.10}$$

The term $(\delta A/\delta \rho)$ in (3.10) vanishes due to (3.5). We can define the universal functional

$$H_{\text{int}}(g', [\rho]) \equiv \int \langle \varphi(t)|\widehat{H}_{\text{int}}|\varphi(t)\rangle dt \tag{3.11}$$

$$= \frac{1}{2} g \int \rho(\mathbf{r}_1, t_1)\rho(\mathbf{r}_2, t_2)v(\mathbf{r}_1 - \mathbf{r}_2)\delta(t_1 - t_2)dt_1dt_2d^3\mathbf{r}_1d^3\mathbf{r}_2 + A_{\text{xc}}.$$

Making use of (3.10) one immediately obtains

$$\frac{dA_1}{dg} = \frac{\partial A_1}{\partial g} = \frac{1}{g} H_{\text{int}}(g, [\rho]), \tag{3.12}$$

or

$$A_1 = A_0 + \int_0^g H_{\text{int}}(g', [\rho]) \frac{dg'}{g'}, \tag{3.13}$$

where A_0 is the action integral of a system of independent particles. The universal functional is given by the well-known relation (e.g., [11])

$$\begin{aligned} H_{\text{int}}(g, [\rho]) = &\frac{1}{2} g \int \rho(\mathbf{r}_1, t_1)\rho(\mathbf{r}_2, t_2)v(\mathbf{r}_1 - \mathbf{r}_2)\delta(t_1 - t_2)dt_1dt_2d^3\mathbf{r}_1d^3\mathbf{r}_2 \\ &+ \frac{1}{2} \int [\chi(\mathbf{r}_1, \mathbf{r}_2, t_1, t_2) - \delta(\mathbf{r}_1 - \mathbf{r}_2)\rho(\mathbf{r}_1, t)] \\ &\times \delta(t_1 - t_2)gv(\mathbf{r}_1 - \mathbf{r}_2)\, dt_1dt_2d^3\mathbf{r}_1d^3\mathbf{r}_2. \end{aligned} \tag{3.14}$$

Here $\chi(\mathbf{r}_1, \mathbf{r}_2, t_1, t_2)$ is the linear response function

$$\chi(\mathbf{r}_1, \mathbf{r}_2, t_1, t_2) = \frac{\delta \rho(\mathbf{r}_1, t_1)}{\delta \lambda v_{\text{ext}}(\mathbf{r}_2, t_2)}. \tag{3.15}$$

It is not very difficult to demonstrate that

$$\frac{\delta^2 A_1}{\delta \rho(\mathbf{r}_1, t_1)\delta \rho(\mathbf{r}_2, t_2)} = -\chi^{-1}(\mathbf{r}_1, \mathbf{r}_2, t_1, t_2). \tag{3.16}$$

Indeed,

$$A([\rho]) \;=\; A_1([\rho]) + \int \lambda v_{\text{ext}}(\mathbf{r},t)\rho(\mathbf{r},t)\mathrm{d}^3\mathbf{r}\mathrm{d}t. \tag{3.17}$$

Here A_1 is the action integral when $\lambda = 0$. Noticing that

$$\left.\frac{\delta A}{\delta\rho}\right|_{\rho=\rho_0+\delta\rho} \;=\; 0,$$

and taking into account the relations

$$\left.\left.\frac{\delta A_1}{\delta\rho}\right|_{\rho_0} + \left.\frac{\delta^2 A_1}{\delta\rho^2}\right|_{\rho_0}\delta\rho + \lambda v_{\text{ext}}\right|_{\lambda\to 0} = 0,$$

$$\left.\frac{\delta A_1}{\delta\rho}\right|_{\rho_0} \;=\; 0,$$

one obtains (3.16). Here $\delta\rho$ is the density fluctuations induced by the external field. In the same way,

$$\frac{\delta^2 A_0([\rho])}{\delta\rho^2} = \left.\frac{\delta^2 A_1(g,[\rho])}{\delta\rho^2}\right|_{g=0} = \chi_0^{-1}. \tag{3.18}$$

In the same way, recalling the fundamental one-to-one correspondence between time-dependent densities and time-dependent potentials [5], one gets

$$\frac{\delta^2 A_1}{\delta\lambda v_{\text{ext}}(\mathbf{r}_1,t_1)\delta\lambda v_{\text{ext}}(\mathbf{r}_2,t_2)}|_{\rho_0} = \chi(\mathbf{r}_1,\mathbf{r}_2,t_1,t_2). \tag{3.19}$$

$$\frac{\delta^2 A_0}{\delta\lambda v_{\text{ext}}(\mathbf{r}_1,t_1)\delta\lambda v_{\text{ext}}(\mathbf{r}_2,t_2)}|_{\rho_0} = \chi_0(\mathbf{r}_1,\mathbf{r}_2,t_1,t_2). \tag{3.20}$$

Let us introduce the definition for R

$$\frac{\delta^2 A_1}{\delta\rho^2}(g,[\rho]) - \frac{\delta^2 A_1}{\delta\rho^2}(0,[\rho]) = -\chi^{-1} + \chi_0^{-1} \equiv R. \tag{3.21}$$

Then, one gets from (3.11), (3.16) and (3.18)

$$R(\mathbf{r}_1,t_1,\mathbf{r}_2,t_2) - gv(\mathbf{r}_1-\mathbf{r}_2) = f_{\text{xc}}(\mathbf{r}_1,t_1,\mathbf{r}_2,t_2) = \frac{\delta^2 A_{\text{xc}}[\rho]}{\delta\rho(\mathbf{r}_1,t_1)\delta\rho(\mathbf{r}_2,t_2)},$$

$$R(\mathbf{r}_1,\mathbf{r}_2,t_1,t_2) = gv(\mathbf{r}_1-\mathbf{r}_2) + \frac{\delta^2}{\delta\rho(r_1,t_1)\delta\rho(r_2,t_2)}\int_0^g H_{\text{int}}(g',[\rho])\frac{\mathrm{d}g'}{g'}. \tag{3.22}$$

On using (3.21), the linear response function, given by (3.16), can be expressed as

$$\chi(\mathbf{r}_1, \mathbf{r}_2, t_1, t_1) = \chi_0(\mathbf{r}_1, \mathbf{r}_2, t_1, t_2) \tag{3.23}$$

$$+\int \chi_0(\mathbf{r}_1, \mathbf{r}_1{}', t_1, t_1')R(\mathbf{r}_1{}', \mathbf{r}_2{}', t_1', t_2')\chi(\mathbf{r}_2{}', \mathbf{r}_2, t_2', t_2)\mathrm{d}^3\mathbf{r}_1{}'\mathrm{d}\mathbf{r}_2{}'\mathrm{d}t_1'\mathrm{d}t_2'.$$

In the stationary case, $\lambda v_{\mathrm{ext}} = 0$, and after the Fourier transformation with respect to $(t_1 - t_2)$, one obtains the equation for $R(\mathbf{r}_1, \mathbf{r}_2, \omega)$

$$R(\mathbf{r}_1, \mathbf{r}_2, \omega) = gv(\mathbf{r}_1 - \mathbf{r}_2) \tag{3.24}$$

$$-\frac{1}{2}\frac{\delta^2}{\delta\rho(\mathbf{r}_1, \omega)\delta\rho(\mathbf{r}_2, -\omega)}\int \chi(\mathbf{r}_1{}', \mathbf{r}_2{}', iw)v(\mathbf{r}_1{}' - \mathbf{r}_2{}')\mathrm{d}^3\mathbf{r}_1{}'\mathrm{d}^3\mathbf{r}_2{}'\frac{\mathrm{d}w}{2\pi}\mathrm{d}g',$$

while the Fourier transform of (3.23) for $\chi(\mathbf{r}_1, \mathbf{r}_2, t_1, t_2)$ with respect to $(t_1 - t_2)$ reads

$$\chi(\mathbf{r}_1, \mathbf{r}_2, \omega) = \chi_0(\mathbf{r}_1, \mathbf{r}_2, \omega) + \int \chi_0(\mathbf{r}_1, \mathbf{r}'_1, \omega)\, R(\mathbf{r}'_1, \mathbf{r}'_2, \omega)\chi(\mathbf{r}'_2, \mathbf{r}_2, \omega)\mathrm{d}^3\mathbf{r}'_1\mathrm{d}^3\mathbf{r}'_2. \tag{3.25}$$

In the case of homogeneous system, the Fourier transform with respect to the spatial coordinates $(\mathbf{r}_1 - \mathbf{r}_2)$ and time $(t_1 - t_2)$ can be carried out and (3.16) recasts into the form

$$\frac{\delta^2 A}{\delta\rho(\mathbf{q}, \omega)\delta\rho(-\mathbf{q}, -\omega)} = -\chi^{-1}(\mathbf{q}, \omega). \tag{3.26}$$

It is seen from (3.16) that the linear response function is symmetric under exchange of $(\mathbf{r}_1, t_1)$ and $(\mathbf{r}_2, t_2)$. Thus, we conclude that one has to use the noncausal linear response function instead of the causal linear response function. This question will be more closely investigated in Sect. 3.3.

Let us consider a system of N interacting particles in its ground state. According to the prescriptions of the density functional method instead of this a system of N fictitious particles may be considered which are moving in an effective field $U(\mathbf{r})$, determined by equation [11]

$$U(\mathbf{r}) = \frac{\delta}{\delta\rho(\mathbf{r})}(E_0 - T_0). \tag{3.27}$$

Here T_0 is the kinetic energy of noninteracting fictitious particles, while E_0 is the total energy of the system. We remark that T_0 and E_0 are functionals of the density ρ. In what follows we will not emphasize the difference between these particles and the fictitious ones if this will not lead to confusion.

Equation (3.27) may be presented as

$$U(\mathbf{r}) = \int g\, v(\mathbf{r}-\mathbf{r}')\rho(\mathbf{r}')\,\mathrm{d}^3\mathbf{r}' + \frac{\delta}{\delta\rho(\mathbf{r})} E_{\mathrm{xc}} = \frac{\delta}{\delta\rho(\mathbf{r},t)} (A_1 - A_0)\Big|_{\rho=\rho_0}. \tag{3.28}$$

The functional derivative in (3.27) is calculated at $\rho = \rho_0(\mathbf{r})$, where $\rho_0(\mathbf{r})$ is the ground state one-particle density. This condition is valid also for other functional derivatives calculated below. If the ground state is considered (the functional derivative is calculated at $\rho = \rho_0(\mathbf{r})$), being the unique functional of the density, the effective potential must be independent of time. The effective interaction R is determined by the equation

$$R(\mathbf{r}_1,\mathbf{r}_2,t_1,t_2) = \frac{\delta}{\delta\rho(\mathbf{r}_2,t_2)} U(\mathbf{r}_1).$$

Of course, here $R(\mathbf{r}_1,\mathbf{r}_2,t_1,t_2)$ depends only upon the time difference $(t_1 - t_2) = t$:

$$R(\mathbf{r}_1,\mathbf{r}_2,t_1-t_2) = gv(\mathbf{r}_1-\mathbf{r}_2) + \frac{\delta}{\delta\rho(\mathbf{r}_2,t)} U_{\mathrm{xc}}(\mathbf{r}_1). \tag{3.29}$$

Here $U_{\mathrm{xc}}(\mathbf{r})$ is the exchange-correlation part of $U(\mathbf{r})$. The functional E_{xc} may be expressed via the linear response function [11],

$$E_{\mathrm{xc}} = -\frac{1}{2}\int [\chi(\mathbf{r}_1,\mathbf{r}_2,iw) + 2\pi\rho(\mathbf{r}_1)\delta(w)\delta(\mathbf{r}_1-\mathbf{r}_2)]v(\mathbf{r}_1-\mathbf{r}_2)\,\mathrm{d}g'\,\frac{\mathrm{d}w}{2\pi}\mathrm{d}^3\mathbf{r}_1\mathrm{d}^3\mathbf{r}_2. \tag{3.30}$$

Here, the integration over the coupling constant goes from 0 to its real value g. Let us now complement (3.28), (3.29), and (3.30) by (3.25) connecting the function $\chi(\mathbf{r}_1,\mathbf{r}_2,\omega)$ and the effective interaction $R(\mathbf{r}_1,\mathbf{r}_2,\omega)$ which is the Fourier image of the time-dependent function $R(\mathbf{r}_1,\mathbf{r}_2,t_1-t_2)$ given by (3.29),

$$\chi(\mathbf{r}_1,\mathbf{r}_2,\omega) = \chi_0(\mathbf{r}_1,\mathbf{r}_2,\omega) + \int \chi_0(\mathbf{r}_1,\mathbf{r}'_1,\omega)\,R(\mathbf{r}'_1,\mathbf{r}'_2,\omega)\chi(\mathbf{r}'_2,\mathbf{r}_2,\omega)\mathrm{d}^3\mathbf{r}'_1\mathrm{d}^3\mathbf{r}'_2.$$

Here $\chi_0(\mathbf{r}_1,\mathbf{r}_2,\omega)$ is the linear response of fictitious Kohn-Sham particles, moving in an effective single-particle field $U(\mathbf{r})$. We note that exchange-correlation energy E_{xc} reduces to the exchange energy E_x if $\chi(\mathbf{r}_1,\mathbf{r}_2,\omega)$ is replaced by $\chi_0(\mathbf{r}_1,\mathbf{r}_2,\omega)$ in (3.30). In this case, the integration over the coupling constant is trivial. The function $\chi_0(\mathbf{r}_1,\mathbf{r}_2,\omega)$ is determined by the expression (e.g., [11])

$$\chi_0(\mathbf{r}_1,\mathbf{r}_2,\omega) = \sum_i n_i[\varphi_i^*(\mathbf{r}_1)\varphi_i(\mathbf{r}_2)G(\mathbf{r}_1,\mathbf{r}_2,\varepsilon_i+\omega) + \varphi_i(\mathbf{r}_1)\varphi_i^*(\mathbf{r}_2)G^*(\mathbf{r}_1,\mathbf{r}_2,\varepsilon_i-\omega)]. \tag{3.31}$$

Here n_i is occupation numbers, $G(\mathbf{r}_1,\mathbf{r}_2,\omega)$ is the single-particle Green's function determined by the equation

$$G(\mathbf{r}_1, \mathbf{r}_2, \omega) = \sum_i \varphi_i^*(\mathbf{r}_1)\varphi_i(\mathbf{r}_2)\left(\frac{1-n_i}{\omega-\varepsilon_i+i\eta} + \frac{n_i}{\omega-\varepsilon_i-i\eta}\right),$$

where $\varphi_i(\mathbf{r})$ and ε_i are the single-particle Kohn-Sham wave functions and energies, respectively, η is an infinitesimally small positive number. These functions and energies are determined by the equation

$$\left(-\frac{\nabla^2}{2M} + U(\mathbf{r})\right)\varphi_i(\mathbf{r}) = \varepsilon_i\,\varphi_i(\mathbf{r}), \tag{3.32}$$

while the single-particle density $\rho(\mathbf{r})$ is given by

$$\rho(\mathbf{r}) = \sum_i n_i|\varphi_i(\mathbf{r})|^2. \tag{3.33}$$

Substituting (3.30) into (3.28) and using (3.29), we get

$$\begin{aligned} R(\mathbf{r}_1, \mathbf{r}_2, t_1 - t_2) &= g\,v(\mathbf{r}_1 - \mathbf{r}_2) - \frac{1}{2}\frac{\delta^2}{\delta\rho(\mathbf{r}_1, t_1)\delta\rho(\mathbf{r}_2, t_2)} \\ &\times \int [\chi(\mathbf{r}'_1, \mathbf{r}'_2, iw) + 2\pi\rho(\mathbf{r}'_1)\delta(\mathbf{r}'_1 - \mathbf{r}'_2)\delta(w)] \\ &\times v(\mathbf{r}'_1 - \mathbf{r}'_2)\,\mathrm{d}g'\,\frac{\mathrm{d}\omega}{2\pi}\,\mathrm{d}^3\mathbf{r}'_1\,\mathrm{d}^3\mathbf{r}'_2. \end{aligned} \tag{3.34}$$

Thus (3.25), (3.28), (3.30), and (3.34) form the system of equations which permits to calculate the functions $R(\mathbf{r}_1, \mathbf{r}_2, \omega)$, $U(\mathbf{r})$, $\chi(\mathbf{r}_1, \mathbf{r}_2, \omega)$ and exchange-correlation functional E_{xc} of a many-electron system, interacting with each other by the pair potential $gv(\mathbf{r}_1 - \mathbf{r}_2) = g/|\mathbf{r}_1 - \mathbf{r}_2|$. An approximate procedure—so-called local approximation—for the evaluation of the functional derivatives has been developed. As a result, we have constructed the linear response function and calculated the correlation energy and the excitation spectra of the electron systems. We were also able to obtain the approximate analytic expression $R_{HF}(q, \rho)$ for the effective interaction in an electron gas R. The approximation R_{HF},

$$R_{HF}(q, \rho, g) = \frac{4\pi g}{q^2} + R_F(q, \rho, g),$$

$$R_F(q, \rho, g) = -\frac{g\pi}{p_F^2}\left[\frac{q^2}{12p_F^2}\ln\left|1 - \frac{4p_F^2}{q^2}\right| - \frac{2p_F}{3q}\ln\left|\frac{2p_F - q}{2p_F + q}\right| + \frac{1}{3}\right],$$

describes the correlation energy in a wide range of density variation much better than the well-known RPA [9–11]. Here $p_F = (3\pi^2\rho)^{1/3}$, with ρ being the density of an electron system.

Table 3.1 The correlation energy per electron in eV of an electron gas of the density r_s. The Monte Carlo results $\varepsilon_{\rm MC}$ [12] are compared with the RPA calculations, denoted by $\varepsilon_{\rm RPA}$, and our calculations [11]. ε_c denotes the results of exact calculations, and ε_{c1} denotes the results when the effective interaction R was approximated by $R_{\rm HF}$

r_s	$\varepsilon_{\rm MC}$	ε_c	ε_{c1}	$\varepsilon_{\rm RPA}$
1	−1.62	−1.62	−1.62	−2.14
3	−1.01	−1.01	−1.02	−1.44
5	−0.77	−0.77	−0.80	−1.16
10	−0.51	−0.51	−0.56	−0.84
20	−0.31	−0.31	−0.38	−0.58
50	−0.16	−0.17	−0.22	−0.35

Here $r_s = (3/4\pi\rho a_B)^{1/3}$, and a_B is the Bohr radius. One can see from Table 3.1 and $R_{\rm HF}$ permits to describe the correlation energy in a rather broad interval of the density variation.

3.3 Relations Between the Action Integral, Single-Particle Potential, and Effective Interaction

Now let us consider some difficulties in connecting the quantum mechanical action integral $A[\rho]$, the single-particle potential $v_s[\rho]$, and the linear response function $\chi[\rho]$ [9, 10]. The roots of the difficulties are in the fact that the linear response function χ, as well as the effective interaction R (see (3.16), (3.22)), has to be symmetrical functions with respect to its arguments, while, on the other hand, these functions, being a casual functions of time, cannot be symmetric under permutation of t_1 and t_2. To clarify this problem, we calculate the l.h.s. of (3.19) directly, using Green's function technique [11]. Considering $\lambda v_{\rm ext}$ as a small perturbation, one gets

$$\frac{\delta^2 A_1}{\delta\lambda v_{\rm ext}(\mathbf{r}_1,t_1)\delta\lambda v_{\rm ext}(\mathbf{r}_2,t_2)}\,|_{\rho=\rho_0} \tag{3.35}$$

$$= -i(<|\,T\rho(\mathbf{r}_1,t_1)\rho(\mathbf{r}_2,t_2)\,|> - <|\,\rho(\mathbf{r}_1,t_1)\,|><|\,\rho(\mathbf{r}_2,t_2)\,|>)$$

$$= \chi(\mathbf{r}_1,t_1,\mathbf{r}_2,t_2).$$

Here, $<\cdots>$ denotes the average taken over the ground state, while T means time ordering operator. One can now conclude that the second functional derivative of the action integral A_1 appeared in (3.19) is defined by the noncausal linear response function, which, as it should be, is a symmetrical function of its variables. Upon the Fourier transformation with respect $(t_2 - t_1)$, it takes the form

$$\chi(\mathbf{r}_1,\mathbf{r}_2,\omega)=\sum_{k\neq 0}\left[\frac{<0\mid\hat{\rho}^\dagger(\mathbf{r}_1)\mid k><k\mid\hat{\rho}(\mathbf{r}_2)\mid 0>}{\omega-(E_k-E_0)+i\delta}\right. \tag{3.36}$$

$$\left.-\frac{<0\mid\hat{\rho}^\dagger(\mathbf{r}_2)\mid k><k\mid\hat{\rho}(\mathbf{r}_1)\mid 0>}{\omega+(E_k-E_0)-i\delta}\right].$$

Here $\hat{\rho}(\mathbf{r})$ is the electron density operator, and $\mid 0>$ and $\mid n>$ are the many-particle system's exact ground and excited states, respectively. E_0 and E_k are energies of these states. It is seen from (3.36) that χ has poles in II and IV quadrants of the complex ω plane. And it includes exactly the same information related to the properties of a system as the casual linear response function, while both these functions coincide on the imaginary axis of the plane, being related one to another by the analytical continuation [13]. Just in the same way, the response functions of higher orders can be constructed. Thus, having in hand the action $A_1[\rho]$, one can explore both the ground state and excited state of a system.

Now consider the effective interaction R given by (3.21). In that case one should define the inverse operators χ^{-1} and χ_0^{-1}. This definition is obvious since on the imaginary axis of the ω plane the linear response function is an analytical function, not vanishing at any point of this axis [14]. Thus, the inverse linear response function can be obtained by analytically continuing from the imaginary axis into the other region of the ω plane. As seen from (3.21) and (3.22), the effective interaction $R^{(0)}\equiv R$ is given by the second derivative of the action A_1. In the same way, calculating the functional derivatives of higher orders, one can obtain the effective interaction $R^{(l)}$, $l\geq 1$,

$$R^{(l)}(\mathbf{r}_1,t_1,\mathbf{r}_2,t_2,\ldots,\mathbf{r}_{l+2},t_{l+2})=\frac{\delta^{l+2}A_1[\rho]}{\delta\rho(\mathbf{r}_1,t_1)\delta\rho(\mathbf{r}_2,t_2)\ldots\delta\rho(\mathbf{r}_{l+2},t_{l+2})}\,|_{\lambda v_{\rm ext}=0}\,. \tag{3.37}$$

Then, taking the Fourier transform of the effective interaction with respect to time, performing the analytical continuation to guarantee the causality, and upon carrying out the inverse Fourier transform, one can get the casual effective interaction $\Re^{(l)}$. These $\Re^{(l)}$ represent the terms of series, which define the casual effective interaction $\Re[\rho]$,

$$\Re[\rho](\mathbf{r}_1,t_1,\mathbf{r}_2,t_2) \tag{3.38}$$

$$=\sum_{l\geq 0}\int\frac{1}{l!}\Re^{(l)}(\mathbf{r}_1,t_1,\mathbf{r}_2,t_2,\ldots,\mathbf{r}_{l+2},t_{l+2})\delta\rho(\mathbf{r}_3,t_3)\ldots\delta\rho(\mathbf{r}_{l+2},t_{l+2})$$

$$d^3\mathbf{r}_3dt_3\ldots d\mathbf{r}_{l+2}dt_{l+2}.$$

Now, having at hand $\Re$, one can construct the single-particle potential $v_s[\rho](\mathbf{r},t)$ by integrating an equation determines the potential [11],

$$\frac{\delta v_s[\rho](\mathbf{r}_1,t_1)}{\delta\rho(\mathbf{r}_2,t_2)}=\frac{v_{\mathrm{xc}}[\rho](\mathbf{r}_1,t_1)}{\delta\rho(\mathbf{r}_2,t_2)}+gv(\mathbf{r}_1-\mathbf{r}_2)=\Re[\rho](\mathbf{r}_1,t_1,\mathbf{r}_2,t_2). \tag{3.39}$$

Equation (3.39) can be integrated as simply as one can integrate the Tailor expansion of a function,

$$v_s[\rho](\mathbf{r}_1,t_1)=\int\Re^{(0)}(\mathbf{r}_1,t_1,\mathbf{r}_2,t_2)\delta\rho(\mathbf{r}_2,t_2)\mathrm{d}^3\mathbf{r}_2\mathrm{d}t_2 \tag{3.40}$$

$$+\sum_{l\geq 1}\int\frac{1}{(l+1)!}\Re^{(l)}(\mathbf{r}_1,t_1,\mathbf{r}_2,t_2,\ldots,\mathbf{r}_{l+2},t_{l+2})\delta\rho(\mathbf{r}_2,t_2)\ldots\delta\rho(\mathbf{r}_{l+2},t_{l+2})$$

$$\mathrm{d}^3\mathbf{r}_2\mathrm{d}t_2\ldots\mathrm{d}^3\mathbf{r}_{l+2}\mathrm{d}t_{l+2}.$$

The physical meaning of (3.40) is quite transparent: as it is seen from (3.38) and (3.39), the functions $\Re^{(l)}$ are directly obtained by calculating the corresponding functional derivatives of v_s with respect to the density. This fact shows that $v_s[\rho]$ contains a great deal of information, and reciprocally, to construct v_s one needs the same information. We remark that v_{xc} (3.39) cannot be presented as a functional derivative, and therefore, it is not directly linked to $A[\rho]_{\mathrm{xc}}$, while, in accordance to (3.21), f_{xc} can be presented as the second functional derivative of A_{xc}.

3.4 DFT and Fermion Condensation

Let us outline the key points of the FC theory [6–8, 11]; for detailed consideration, see Chap. 4. FC is related to a new class of solutions of the Fermi-liquid-theory equation [15]

$$\frac{\delta(F-\mu N)}{\delta n(p,T)}=\varepsilon(p,T)-\mu(T)-T\ln\frac{1-n(p,T)}{n(p,T)}=0, \tag{3.41}$$

for the quasiparticle distribution function $n(p,T)$, depending on the momentum p and temperature T. Here, F is the free energy, μ is the chemical potential, while $\varepsilon(p,T)=\delta E_0/\delta n(p,T)$ is the quasiparticle energy, being a functional of $n(p,T)$ just like the energy E_0 and the other thermodynamic functions. Equation (3.41) is usually rewritten in the form of the Fermi-Dirac distribution

$$n(p,T)=\left\{1+\exp\left[\frac{(\varepsilon(p,T)-\mu)}{T}\right]\right\}^{-1}. \tag{3.42}$$

In homogeneous matter, the standard solution $n_F(p,T=0)=\theta(p_F-p)$, with p_F being the Fermi momentum, is obtained assuming $\mathrm{d}\varepsilon(p,T=0)/\mathrm{d}p$ to be positive and finite near the Fermi level. T-dependent corrections to the effective mass M^*, quasiparticle energy, and the other quantities start with T^2-terms [15]. New FC solutions of (3.42) possess at low T the shape of the spectrum $\varepsilon(p, T)$ linear in T [16]:

$$\varepsilon(p, T) - \mu(T) = T\nu_0(p), \tag{3.43}$$

within the interval $p_i < p < p_f$ occupied by the fermion condensate. Inserting (3.43) into (3.42), one finds its distribution $n_0(p)$ at $T = 0$

$$n_0(p) = [1 + e^{\nu_0(p)}]^{-1}, \quad p_i < p < p_f, \tag{3.44}$$

which drastically differs from the Fermi step. At $T \to 0$, the function $n_0(p)$, being continuous within the FC interval, admits a finite limit for the logarithm in (3.41), yielding at $T = 0$ [6]

$$\varepsilon(p) = \frac{\delta E_0}{\delta n(p)} = \mu, \; p_i < p < p_f. \tag{3.45}$$

When fermion condensation is just starting, the momenta obey the equality $p_i = p_F = p_f$. This fact means, as follows from (3.45), that $M^* \to \infty$, for the flat band $\varepsilon(p) = \mu$ takes place at the only point $p = p_F$. Then, at finite temperatures and when FC has taken place, the effective mass, as follows from (3.43), $M^* \sim 1/T$. Within the FC interval, the solution $n_0(p)$ deviates from the Fermi step function $n_F(p)$ in such a way that the energy $\varepsilon(p)$ stays constant: $\varepsilon(p) = \mu$, while outside of this region, $n_0(p)$ coincides with $n_F(p)$. We see that the occupation numbers $n_0(p)$ become variational parameters: the solution $n_0(p)$ emerges if the energy E_0 is lowered by alteration of the standard occupation numbers. We recall that the idea of a multi-connected Fermi sphere, resulting in lowering of the energy, was suggested [17]. FC can be considered as a generalization of this idea. The quasiparticle formalism is applicable to this problem if the damping of the condensate states is small compared to their energy. This condition holds for superfluid systems and for normal ones provided the ratio of the fermion-condensate density ρ_c to the density ρ of the system is small.

As we have mentioned above, calculations within the density functional theory (DFT) failed to reproduce the smooth segments in the single-particle excitation spectra, while these smooth segments are the "calling card" of FC. Thus, a density functional formulation of FC is of crucial importance, since it will allow to understand the roots of these difficulties. As we shall see, the local single-particle potential of DFT [2] beyond the FC phase transition becomes a nonlocal one. One of the remarkable peculiarities of the FC phase transition is related to spontaneous breaking of gauge symmetry: the superconductivity order parameter $\kappa(p) =< a_{\sigma,\mathbf{p}} a_{-\sigma,-\mathbf{p}} >$ has a nonzero value $\kappa_0(p) = \sqrt{n_0(p)[1 - n_0(p)]}$ even if the gap Δ vanishes [7, 18]. On one hand, this circumstance allows us to apply methods of the DFT of superconductivity [5] to consider FC. On the other hand, this consideration also gains new insights into the density functional theory of superconductivity. As a result, one gets the possibility to calculate the single-particle excitation spectrum [18].

3.5 DFT, The Fermion Condensation, and Superconductivity

In the DFT of superconductivity, there exists a unique functional $F(T)$ of two densities, namely, the normal density of an electron system ρ and the anomalous density κ [5]

$$\rho(\mathbf{r}) = \sum_\sigma < \psi_\sigma^+(\mathbf{r})\psi_\sigma(\mathbf{r}) >; \quad \kappa(\mathbf{r}_1, \mathbf{r}_2) =< \psi_\uparrow(\mathbf{r}_1)\psi_\downarrow(\mathbf{r}_2) > . \tag{3.46}$$

The functional $F[\rho, \kappa]$ is given by

$$\begin{aligned} F[\rho, \kappa] &= T_s[\rho, \kappa] - TS_s[\rho, \kappa] + \int (v_{\text{ext}}(\mathbf{r}) - \mu)\rho(\mathbf{r})\mathrm{d}^3\mathbf{r} + \int \frac{\rho(\mathbf{r}_1)\rho(\mathbf{r}_2)}{|\mathbf{r}_1 - \mathbf{r}_2|}\mathrm{d}^3\mathbf{r}_1\mathrm{d}^3\mathbf{r}_2 \\ &- \int \kappa^*(\mathbf{r}_1, \mathbf{r}_2)V(\mathbf{r}_1, \mathbf{r}_2, \mathbf{r}_3, \mathbf{r}_4)\kappa(\mathbf{r}_3, \mathbf{r}_4)\mathrm{d}^3\mathbf{r}_1\mathrm{d}^3\mathbf{r}_2\mathrm{d}^3\mathbf{r}_3\mathrm{d}^3\mathbf{r}_4 + F_{\text{xc}}[\rho, \kappa] \\ &= E[\rho, \kappa] - TS_s[\rho, \kappa] + \int (v_{\text{ext}}(\mathbf{r}) - \mu)\rho(\mathbf{r})\mathrm{d}^3\mathbf{r}. \end{aligned} \tag{3.47}$$

Here $T_s[\rho, \kappa]$ and $S_s[\rho, \kappa]$ stand for the kinetic energy and the entropy of a noninteracting system, while $F_{\text{xc}}[\rho(\mathbf{r}), \kappa(\mathbf{r}_1, \mathbf{r}_2)]$ is the exchange-correlation free energy functional, V is a pairing interaction, and v_{ext} is an external potential. The fourth and fifth terms on the r.h.s. of (3.47) are the Hartree term due to the Coulomb forces and the pairing interaction, respectively. We suppose V to be sufficiently weak like the model BCS interaction. The last equality in (3.47) can be considered as the definition of E. For the densities (3.46), one can employ a quite general form [5]

$$\rho(\mathbf{r}_1) = \sum_{\sigma,n} [|\phi_{\sigma,n}(\mathbf{r}_1)|^2|v_{\sigma,n}|^2(1 - f_{\sigma,n}) + |\phi_{\sigma,-n}(\mathbf{r}_1)|^2|u_{\sigma,n}|^2 f_{\sigma,n}], \tag{3.48}$$

and

$$\kappa(\mathbf{r}_1, \mathbf{r}_2) = \frac{1}{2}\sum_n [\phi^*_{\sigma,n}(\mathbf{r}_1)\phi_{-\sigma,-n}(\mathbf{r}_2) + \phi^*_{\sigma,n}(\mathbf{r}_2)\phi_{-\sigma,-n}(\mathbf{r}_1)]v^*_{\sigma,n}u_{\sigma,n}(1 - 2f_{\sigma,n}), \tag{3.49}$$

with the coefficients $v_{\sigma,n}$ and $u_{\sigma,n}$, obeying the normalization conditions,

$$|v_{\sigma,n}|^2 + |u_{\sigma,n}|^2 = 1. \tag{3.50}$$

Here n denotes the quantum numbers such as the momentum p in the case of homogeneous matter or the crystal momentum and the band index in the solid state. We introduce a *one-quasiparticle* density matrix $\rho_1(\mathbf{r}_1, \mathbf{r}_2)$,

$$\rho_1(\mathbf{r}_1,\mathbf{r}_2)=\sum_{\sigma,n}[\phi^*_{\sigma,n}(\mathbf{r}_1)\phi_{\sigma,n}(\mathbf{r}_2)|v_{\sigma,n}|^2(1-f_{\sigma,n})+\phi^*_{\sigma,-n}(\mathbf{r}_1)\phi_{\sigma,-n}(\mathbf{r}_2)|u_{\sigma,n}|^2 f_{\sigma,n}]. \tag{3.51}$$

Considering (3.8), (3.9), and (3.11), one can conclude that there exists a one-to-one correspondence between functions ρ, κ and functions ρ_1, κ. Therefore, the functional F can be treated as a functional of ρ, κ or as a functional of ρ_1, κ. If one needs to calculate only the densities ρ, κ, then it is reasonable to employ the functional $F[\rho,\kappa]$. On the other hand, the functional $F[\rho_1,\kappa]$ may be used if it is necessary to gain a deeper insight into the problem, e.g., to calculate the single-particle excitation spectra, the functions ϕ_n, etc.

We demonstrate now that the functions ϕ_n are not defined by a local operator [9, 10]. For the sake of simplicity, we omit the spin variables and put $T=0$. Minimization of F with respect to ϕ_n leads to the eigenfunction problem

$$|v_n|^2\frac{\delta F[\rho,\kappa]}{\delta\rho(\mathbf{r}_1)}\phi_n(\mathbf{r}_1)+\int u_n v_n^*\frac{\delta F[\rho,\kappa]}{\delta\kappa(\mathbf{r}_1,\mathbf{r}_2)}\phi_{-n}(\mathbf{r}_2)d^3\mathbf{r}_2,=\lambda_n\phi_n(\mathbf{r}_1) \tag{3.52}$$

with λ_n being the Lagrangian multiplier, preserving the normalization of the eigenfunction ϕ_n. We see that if the anomalous density κ were zero, then $|v_n|^2=1$ below the Fermi level, and $|v_n|^2=0$ above the Fermi level, while functions ϕ_n would be given by a local operator.

Now let us take F as a functional of ρ_1, κ. The functions ϕ_l subject to the orthogonality constraint

$$\int\phi^*_l(\mathbf{r}_1)\phi_m(\mathbf{r}_1)d^3\mathbf{r}_1\equiv<l|m>=\delta_{lm}. \tag{3.53}$$

The constraints (3.13) may be introduced by replacing F with

$$F_1[\rho_1,\kappa]=F[\rho_1,\kappa]+\sum_{l,m}a_{lm}<l|m>, \tag{3.54}$$

where a_{lm} are Lagrangian parameters. Minimizing (3.14) with respect to ϕ_l, one obtains

$$|v_l|^2\int f^0(\mathbf{r}_3,\mathbf{r}_2)\phi_l(\mathbf{r}_2)d^3\mathbf{r}_2+v_l^*u_l\int f^1(\mathbf{r}_3,\mathbf{r}_2)\phi_{-l}(\mathbf{r}_2)d^3\mathbf{r}_2+\sum_m a_{lm}\phi_m(\mathbf{r}_3)=0, \tag{3.55}$$

with f^0 and f^1 given by

$$f^0(\mathbf{r}_1,\mathbf{r}_2)=\frac{\delta F_1[\rho_1,\kappa]}{\delta\rho_1(\mathbf{r}_1,\mathbf{r}_2)};\quad f^1(\mathbf{r}_1,\mathbf{r}_2)=\frac{\delta F_1[\rho_1,\kappa]}{\delta\kappa(\mathbf{r}_1,\mathbf{r}_2)}. \tag{3.56}$$

From (3.55), one can find that

$$\left(\frac{|v_l|^2}{v_l^*u_l}-\frac{|v_m|^2}{v_m^*u_m}\right)<m|f^0|l>=0. \tag{3.57}$$

With this result, the functions ϕ_l turn out to be solutions of the eigenvalue equation

$$\int f^0(\mathbf{r}_1, \mathbf{r}_2)\phi_m(\mathbf{r}_2)\mathrm{d}^3\mathbf{r}_2 = \varepsilon_m \phi_m(\mathbf{r}_1). \tag{3.58}$$

While the operator f^1 drops out of the problem in accordance with the results of [5]. Equation (3.58) can be rewritten as a single-particle equation

$$\left(-\frac{\nabla^2}{2} + v_{\mathrm{ext}}(\mathbf{r}_1) + \int \frac{\rho(\mathbf{r}_2)}{|\mathbf{r}_1 - \mathbf{r}_2|}\mathrm{d}^3\mathbf{r}_2\right)\phi_m(\mathbf{r}_1) \tag{3.59}$$

$$+ \int v_{\mathrm{xc}}(\mathbf{r}_1, \mathbf{r}_2)\phi_m(\mathbf{r}_2)\mathrm{d}^3\mathbf{r}_2 = \varepsilon_m \phi_m(\mathbf{r}_1),$$

with v_{xc} being a nonlocal potential

$$v_{\mathrm{xc}}(\mathbf{r}_1, \mathbf{r}_2) = \frac{\delta F_{\mathrm{xc}}}{\delta \rho_1(\mathbf{r}_1, \mathbf{r}_2)}. \tag{3.60}$$

Taking into account (3.51) and (3.58), one can also infer that

$$\frac{\delta F}{\delta |v_l|^2} =< l|f^0|l >= \varepsilon_l. \tag{3.61}$$

If V were zero, ε_l would represent the real single-particle excitation spectra of the system. The energy ε_l is perturbed by the BCS correlations, but, in fact, this perturbation is small. It is worth noting, as it will be seen below, one has to put $V = 0$ to derive equations for FC. The same consideration in the case of finite temperatures preserves (3.58) and (3.59).

In order to satisfy the constraint (3.50), it is convenient to take $v_l = \cos\theta_l$; $u_l = \sin\theta_l$. Then one finds

$$\frac{\delta F[\rho_1, \kappa]}{\delta \theta_l} = (\varepsilon_l - \mu)\tan 2\theta_l + \Delta_l = 0. \tag{3.62}$$

The gap Δ_l is given by

$$\Delta_l = -\int \frac{\delta F[\rho_1, \kappa]}{\delta \kappa(\mathbf{r}_1, \mathbf{r}_2)}\phi_l^*(\mathbf{r}_1)\phi_{-l}(\mathbf{r}_2)\mathrm{d}^3\mathbf{r}_1\mathrm{d}^3\mathbf{r}_2. \tag{3.63}$$

We have now to minimize F with respect to f_l. The minimization yields

$$f_l = \frac{1}{1 + \exp(E_l/T)}. \tag{3.64}$$

Here, the real excitation energy E_l is given as

$$\frac{\delta E}{\delta f_l} = -(\varepsilon_l - \mu)\cos 2\theta_l + \Delta_l \sin 2\theta_l. \tag{3.65}$$

It is helpful to verify (3.62) and (3.63) for homogeneous matter in the weak coupling limit $V \to 0$ at $T = 0$ using the BCS model. Recall, in this case, V enters the functional F only through the Hartree term, while F_{xc} is completely defined by the Coulomb interaction. One obtains

$$\Delta \sim \exp(-1/N(0)V), \tag{3.66}$$

where $N(0)$ being the correct density of the single-particle states at the Fermi level, given by (3.61), while the functions ϕ_k are plane waves. If the potential v_{xc} entering (3.59) were a local one, then $v_{\mathrm{xc}} = \mathrm{const}$, and the energies ε_l would be defined by the single-particle energies of a free electron gas. As a result, instead of the correct density of states in (3.65) one would have the density of states of a free electron gas. So, we are led to the conclusion that it is of crucial importance to take into account from the very beginning the nonlocality or, equivalently, velocity-dependence of the single-particle potential (3.60).

We shall now give further justification of (3.45) deduced at $T = 0$. Consider once again (3.62) in the limit $V = 0$. In this case, $\Delta_l = 0$, and (3.62) can be written as

$$(\varepsilon_l - \mu)\tan 2\theta_l = 0. \tag{3.67}$$

Equation (3.67) requires that

$$\varepsilon_l - \mu = 0, \quad \text{if} \quad |v_l|^2 \neq 0, 1, \tag{3.68}$$

with ε_l defined by (3.21). Therefore, the fermion-condensation solution is a new solution to the old equations. On the other hand, it is seen from (3.68) that the standard Kohn-Sham scheme for the single-particle equations [5] is no longer valid beyond the point of the FC phase transition, since one has to introduce the nonlocal single-particle potential (3.60), while the quasiparticle occupation numbers $|v_l|^2$ become variational parameters, minimizing the total energy. In the homogeneous limit, (3.68) takes the form of (3.45), and the single-particle energy $\varepsilon(p)$ is given by (3.45), while $n(p) = |v_p|^2$ with E being the exact functional of $n(p)$. Thus, we have proved in the framework of DFT microscopic theory [2, 5] that the Landau theory of Fermi liquid is a microscopic theory rather than a phenomenological one. We stress that there is nothing wrong with a phenomenological theory, for by now the main equation of physics cannot be derived from so-called first principles [19]. One can conclude from (3.43), (3.45), and (3.68) that FC is related to the unbounded growth of the density of states. As a result, FC serves as a source for new phase transitions which lift the degeneracy of the spectrum. We are going to analyze the situation when the superconductivity wins the competition with the other phase transitions.

Now let us switch on the interaction V. Then, as it follows from (3.47) and (3.63), $\Delta \sim V$, when V is sufficiently small, while in the BCS-case Δ, given by (3.66), is exponentially small. Inserting the result $\Delta \sim V$ into (3.47), one finds that the pairing correction $\delta E_s(T = 0)$ to the ground state energy at $T = 0$

$$\delta E_s(T = 0) \sim \frac{\rho_c}{\rho}\Delta. \tag{3.69}$$

This result is analogous to that in the strong coupling limit of the BCS theory and differs drastically from the ordinary BCS result $\delta E_0 \sim \Delta^2$. Due to this, an essential increase of the critical magnetic field, destroying superconductivity, can be expected. The further investigation of this increase and close connection of FC with the superconductivity are presented in Chap. 6. Above the critical temperature, the system under consideration is in its non-Fermi liquid state, (3.43) is valid, and one can observe the smooth segments of the spectra at the Fermi level.

3.6 Summary

We have presented the density functional formulation of the fermion condensation. As a result, it is demonstrated that the standard Kohn-Sham scheme is not valid beyond the FC phase transition, since one has to introduce the occupation numbers, which serve as variational parameters, while the single-particle potential transforms into the nonlocal one. We have demonstrated that both the Landau Fermi liquid theory and the fermion-condensation theory are microscopic theories rather than phenomenological ones. We also present both the functional equation that defines the density functional and the procedure to solve the equation. Our consideration yields a new set of the microscopic single-particle equations for superconductors which allow one to calculate the normal and anomalous densities and the real single-particle excitation spectra as well. We have also shown that the superconducting state, superimposed on the FC state, and the ordinary BCS state are considerably different, as it is shown in detail in Chap. 23.

References

1. P. Hohenberg, W. Kohn, Phys. Rev. **136**, B864 (1964)
2. W. Kohn, L.J. Sham, Phys. Rev. **140**, A1133 (1965)
3. H. Eschrig, *Fundamentals of Density Functional Theory* (Teubner, Stuttgart, 1996)
4. W. Kohn, Rev. Mod. Phys. **71**, 1253 (1999)
5. L.N. Oliveira, E.K.U. Gross, W. Kohn, Phys. Rev. Lett. **60**, 2430 (1988)
6. V.A. Khodel, V.R. Shaginyan, JETP Lett. **51**, 553 (1990)
7. V.R. Shaginyan, M.Y. Amusia, A.Z. Msezane, K.G. Popov, Phys. Rep. **492**, 31 (2010), arXiv:1006.2658
8. M.Y. Amusia, K.G. Popov, V.R. Shaginyan, W.A. Stephanowich, *Theory of Heavy-Fermion Compounds*. Springer Series in Solid-State Sciences, vol. 182 (Springer, Berlin, 2015)
9. V.A. Khodel, V.R. Shaginyan, in *Condensed Matter Theories*, vol. 12, ed. by J. Clark, V. Plant (Nova Science Publishers Inc., New York, 1997), p. 221

10. V.R. Shaginyan, JETP Lett. **68**, 527 (1998)
11. V.A. Khodel, V.R. Shaginyan, V.V. Khodel, Phys. Rep. **249**, 1 (1994)
12. D. Ceperley, B. Alder, Phys. Rev. Lett. **45**, 566 (1980)
13. M. Amusia, V. Shaginyan, Phys. Rev. A **63**, 056501 (2001)
14. D. Pines, P. Noziéres, *Theory of Quantum Liquids* (Benjamin, New York, 1966)
15. L.D. Landau, Sov. Phys. JETP **3**, 920 (1956)
16. V.A. Khodel, J.W. Clark, V.R. Shaginyan, Solid State Commun. **96**, 353 (1995)
17. M. de Llano, J.P. Vary, Phys. Rev. C **19**, 1083 (1979)
18. V.R. Shaginyan, Phys. Lett. A **249**, 237 (1998)
19. R.B. Laughlin, D. Pines, Proc. Natl. Acad. Sci. USA **97**, 28 (2000)

Chapter 4
Topological Fermion-Condensation Quantum Phase Transition

Abstract Here, we discuss the general properties of the topological fermion-condensation quantum phase transition (FCQPT) leading to the emergence of the fermion condensation (FC). Basing on the main results of Chap. 3, we continue to present a microscopic derivation of the main equations of FC and show that Fermi systems with FC form an entirely new class of Fermi liquids with its own topological structure, protecting the FC state. We construct the phase diagram, and explore the order parameter of these systems. We show that the fermion condensate has a strong impact on the observable physical properties of systems, where it is realized, up to relatively high temperatures of a few tens kelvin. Two different scenarios of the quantum-critical point (QCP), a zero-temperature instability of the Landau state, related to the divergence of the effective mass, are also briefly investigated. Flaws of the standard scenario of the QCP, where this divergence is attributed to the occurrence of some second-order phase transition, are demonstrated. We also consider other topological phase transitions, taking place in normal Fermi liquid. These are associated with the emergence of a multi-connected Fermi surface. Depending on the parameters and analytical properties of the Landau interaction, such instabilities lead to several possible types of restructuring of initial Fermi liquid ground state. This restructuring generates topologically distinct phases. One of them is the FC discussed above, and another one belongs to a class of topological transitions and will be called "iceberg" phase, where the sequence of rectangles ("icebergs") $n(p) = 0$ and $n(p) = 1$ is realized at $T = 0$. At elevated temperatures the "icebergs meltdown" and the behavior of the system becomes similar to that with the fermion-condensate state.

4.1 The Fermion-Condensation Quantum Phase Transition

As it is shown in Sect. 2.2, the Pomeranchuk stability conditions do not encompass all possible types of instabilities and that at least one related to the divergence of the effective mass given by (2.21) was overlooked [1]. This type of instability corresponds to a situation where the effective mass, the most important characteristic of a quasiparticle, can become infinitely large. As a result, the quasiparticle kinetic

M. Amusia and V. Shaginyan, *Strongly Correlated Fermi Systems*,
Springer Tracts in Modern Physics 283,
https://doi.org/10.1007/978-3-030-50359-8_4

energy is infinitely small near the Fermi surface and the quasiparticle distribution function $n(\mathbf{p})$ minimizing $E(n(\mathbf{p}))$ is determined by the potential energy. This leads to the formation of a new class of strongly correlated Fermi liquids with fermion condensate (FC) [1–4], separated from the normal Fermi liquid by the topological FCQPT [5–9].

It follows from (2.21) that at $T = 0$ and $r \to 0$ the effective mass diverges, $M^*(r) \to \infty$. Beyond the critical point $x_{\rm FC}$, the distance r becomes negative and, correspondingly, so does the effective mass. To avoid an unstable and physically meaningless state with a negative effective mass, the system must undergo a quantum phase transition at the critical point $x = x_{\rm FC}$, which, as we will see shortly, is FCQPT [5–7]. As the kinetic energy of quasiparticles near the Fermi surface is proportional to the inverse effective mass, their potential energy determines the ground state energy as $x \to x_{\rm FC}$. Hence, a phase transition reduces the energy of the system and transforms the quasiparticle distribution function. Beyond QCP $x = x_{\rm FC}$, the quasiparticle distribution is determined by the ordinary equation for a minimum of the energy functional [1]:

$$\frac{\delta E(n(\mathbf{p}))}{\delta n(\mathbf{p}, T = 0)} = \varepsilon(\mathbf{p}) = \mu; \; p_i \le p \le p_f. \tag{4.1}$$

Equation (4.1) yields the quasiparticle distribution function $n_0(\mathbf{p})$ that minimizes the ground state energy E. This function found from (4.1) differs from the step function in the interval from p_i to p_f, where $0 < n_0(\mathbf{p}) < 1$, and coincides with the step function outside this interval. In fact, (4.1) coincides with (2.3) provided that the Fermi surface at $p = p_F$ transforms into the Fermi volume at $p_i \le p \le p_f$ suggesting that the single-particle spectrum is absolutely "flat" within this interval. A possible solution $n(\mathbf{p})$ of (4.1) and the corresponding single-particle spectrum $\varepsilon(\mathbf{p})$ are depicted in Fig. 4.1. Quasiparticles with momenta within the interval $(p_f - p_i)$ have the same single-particle energies equal to the chemical potential μ and form FC, while the distribution function $n_0(\mathbf{p})$ describes the new state of the Fermi liquid with FC [1, 2, 4]. In contrast to the Landau, marginal, or Luttinger Fermi liquids [10–12], which exhibit the same topological structure of Green's function, in systems with FC, where the Fermi surface spreads into a strip, Green's function belongs to a different topological class. Green's function of systems with FC is considered in Sects. 4.1.4 and 6.1.1. The topological class of the Fermi liquid is characterized by the invariant [4, 13, 14]

$$N_t = \mathrm{tr} \oint_C \frac{\mathrm{d}l}{2\pi i} G(i\omega, \mathbf{p}) \partial_l G^{-1}(i\omega, \mathbf{p}), \tag{4.2}$$

where "tr" denotes the trace over the spin indices of Green's function and the integral is taken along an arbitrary contour C encircling the singularity of Green's function. The invariant N_t in (4.2) takes integer values even when the singularity is not of the pole type, cannot vary continuously, and is conserved in a transition from the Landau Fermi liquid to marginal liquids and under small perturbations of Green's

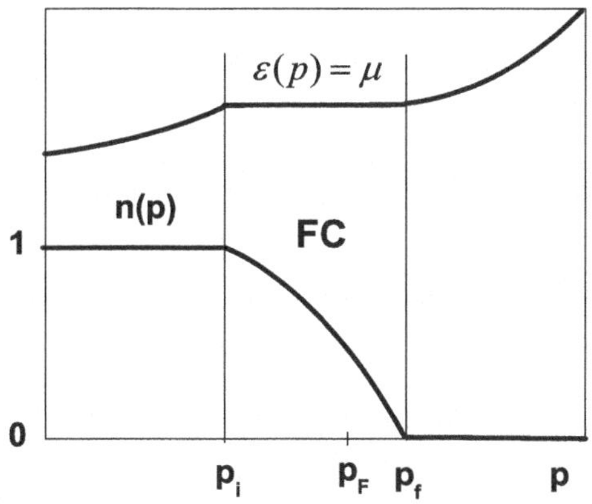

Fig. 4.1 The single-particle spectrum $\varepsilon(p)$ and the quasiparticle distribution function $n(p)$. Because $n(p)$ is a solution of (4.1), we have $n(p < p_i) = 1$, $0 < n(p_i < p < p_f) < 1$, and $n(p > p_f) = 0$, while $\varepsilon(p_i < p < p_f) = \mu$. The Fermi momentum p_F satisfies the condition $p_i < p_F < p_f$

function. As it was shown by Volovik [4, 13, 14], the situation is quite different for systems with FC, where the invariant N_t becomes a half-integer and the system with FC transforms into an entirely new class of Fermi liquids with its own topological structure, thus forming the new state of matter protected by the topological invariant given by (4.2) [15].

A few remarks are in order here. As it was shown above, the solutions of (4.1) describe the topologically protected new state of matter. Equation (4.1) represents the common equation to search the minimum of functional E. In the case of Bose's system, the equation $\delta E/\delta n(p) = \mu$ describes a common situation. In the case of Fermi systems such an equation, generally speaking, was not correct. It is the FC state, taking place behind FCQPT that makes (4.1) meaningful for Fermi systems. Thus, Fermi quasiparticles in the region $p_i < p < p_f$ can behave as Bose one, occupying the same energy level $\varepsilon = \mu$, and (4.1) yields the quasiparticle distribution function $n_0(\mathbf{p})$ that delivers a flat band and minimizes the ground state energy E. We note that it is the flat band that minimizes the ground state energy; if it were not the ground state, then one could shift the quasiparticles to minimize the energy, for $n_0(p) \leq 1$ over the region $p_i \leq p \leq p_f$, while it is seen in (4.1) that this manipulation is meaningless because the band is flat. In the absence of flat band, one would have either the LFL Fermi sphere or multi-connected Fermi sphere, see Sect. 4.2.

The FC state can be viewed as the state possessing the supersymmetry (SUSY) that interchanges bosons and fermions eliminating in some sense the difference between them. In the strong coupling limit when the Pauli principle $n(p) < 1$ is automatically met and all the quasiparticles form the FC state, p_f is determined by the condition

$$x = \int_0^{p_f} n_0(p) \frac{p^2 \mathrm{d}p}{\pi^2}. \tag{4.3}$$

In that case, SUSY is restored over all the configuration space. We shall see that SUSY emerges naturally in condensed matter systems known as HF compounds. In Chap. 18, we shall see that emerging the FC state accompanied by SUSY violates both the time invariance symmetry and the baryon symmetry of the Universe. Thus restoring one important symmetry, the FC state violates another essential symmetry.

4.1.1 The FCQPT Order Parameter

We start with visualizing the main properties of FCQPT. To this end, again consider the density functional theory for superconductors (SCDFT) [16]. SCDFT states that the thermodynamic potential Φ is a universal functional of the number density $n(\mathbf{r})$ and the anomalous density (or the order parameter) $\kappa(\mathbf{r}, \mathbf{r}_1)$, providing a variational principle to determine the densities. At the superconducting transition temperature T_c, a system undergoes the second-order phase transition into superconducting state. Our goal now is to construct a quantum phase transition which evolves from the superconducting one.

Let us recall that the vanishing of coupling constant λ_0 of the BCS-like pairing interaction [17] implies the disappearance of the superconducting gap at any finite temperature, see Chap. 6. In that case, $T_c \to 0$ and the superconducting state occurs at $T = 0$ so that at finite temperatures there is a normal state only. This means that at $T = 0$ the anomalous density

$$\kappa(\mathbf{r}, \mathbf{r_1}) = \langle \Psi_\uparrow(\mathbf{r})\Psi_\downarrow(\mathbf{r_1})\rangle \tag{4.4}$$

is finite, while the superconducting gap

$$\Delta(\mathbf{r}) = \lambda_0 \int \kappa(\mathbf{r}, \mathbf{r_1}) d\mathbf{r_1} \tag{4.5}$$

is infinitely small for $\lambda_0 \to 0$ [8, 9]. In (4.4), the field operator $\Psi_\sigma(\mathbf{r})$ annihilates an electron of spin $\sigma, \sigma = \uparrow, \downarrow$ at the position $\mathbf{r}$. For the sake of simplicity, we consider the model of homogeneous HF liquid [8, 18]. Then at $T = 0$, the thermodynamic potential Φ reduces to the ground state energy E which turns out to be a functional of the occupation number $n(\mathbf{p})$ since in that case the order parameter $\kappa(\mathbf{p}) = v(\mathbf{p})u(\mathbf{p}) = \sqrt{n(\mathbf{p})(1 - n(\mathbf{p}))}$. The latter expression relates the order parameter κ to the coefficients $u(\mathbf{p})$ and $v(\mathbf{p})$ of Bogoluybov transformation, diagonalizing the corresponding Hamiltonian in electron (fermion) creation and annihilation operators, see, e.g., [19]. More precisely,

$$n(\mathbf{p}) = v^2(\mathbf{p}), \quad \kappa(\mathbf{p}) = v(\mathbf{p})u(\mathbf{p}), \tag{4.6}$$

where for fermions the parameters $u(\mathbf{p})$ and $v(\mathbf{p})$ are normalized so as $v^2(\mathbf{p}) + u^2(\mathbf{p}) = 1$, see also [19].

Minimization of E over $n(\mathbf{p})$ leads to (4.1). If (4.1) has nontrivial solution $n_0(\mathbf{p})$, then instead of the Fermi step we have $0 < n_0(\mathbf{p}) < 1$ in certain range of momenta $p_i \leq p \leq p_f$. In this case, as $n_0(\mathbf{p})$ is neither 0 nor 1, the order parameter $\kappa(\mathbf{p}) =$

$\sqrt{n_0(\mathbf{p})(1-n_0(\mathbf{p}))}$ becomes finite in this range, while the single-particle spectrum $\varepsilon(\mathbf{p})$ is flat. Thus, the step-like Fermi occupation restructures inevitably and forms FC when (4.1) possesses the nontrivial solution at some $x = x_c$. This solution is indeed a QCP of FCQPT. In QCP point, one has $p_i \to p_f \to p_F$ so that the effective mass M^* diverges [1, 8, 18, 20]

$$\left[M^*(x \to x_c)\right]^{-1} = \frac{1}{p_F} \frac{\partial \varepsilon(\mathbf{p})}{\partial \mathbf{p}}\bigg|_{\substack{p \to p_F \\ x \to x_c}} \to 0. \tag{4.7}$$

At any small but finite temperature the order parameter (anomalous density) κ decays and this state undergoes the first-order phase transition converting into a normal state with thermodynamic potential Φ_0. Indeed, at $T \to 0$, the entropy $S = -\partial\Phi_0/\partial T$ of the normal state is given by (2.4). It is seen from (2.4) that the normal state is characterized by the temperature-independent entropy S_0 [8, 18, 20, 21]. Since the entropy of the superconducting ground state is zero, we conclude that the entropy is discontinuous at the phase transition point with the gap $\delta S = S_0$. Thus, the system undergoes the first-order phase transition. The heat q of this transition is $q = T_c S_0 = 0$ since $T_c = 0$. Because of the stability condition at the point of the first-order phase transition, we have $\Phi_0(n(\mathbf{p})) = \Phi(\kappa(\mathbf{p}))$. Obviously, the condition is satisfied since $q = 0$. We prolong more detailed discussion of the superconductivity (and related Green's functions) of the systems with FC to Chap. 6.

4.1.2 Quantum Protectorate Related to FCQPT

Like any other phase transition, the FCQPT comprises strong interparticle interaction so that there is no way to describe reliably all its details within approaches based on a perturbation theory or some schemes using Feynman-type diagrams. On the other hand, as we have seen above, SCDFT represents reliable theory [16]. Then, the way to confirm the FC existence on pure theoretical grounds is to study model systems, which admit the exact solutions. Such theoretical studies should be augmented by careful examination of experimental data that could be interpreted in favor (or to the detriment) of FC existence. Exactly solvable models unambiguously suggest that Fermi systems with FC exist (see, e.g., [22–25]). Taking the results of topological investigations into account, we can affirm that the new class of Fermi liquids with FC is nonempty, actually exists, and represents an extended family of new states of Fermi systems [4, 13, 14]. We note that theoretical and experimental studies confirm the presence of the FC state in different HF compounds [9, 25–29].

We note that the solutions $n_0(\mathbf{p})$ of (4.1) are new solutions of the well-known equations of the Landau Fermi liquid theory. Indeed, at $T = 0$, the standard solution given by a step function, $n(\mathbf{p}, T \to 0) \to \theta(p_F - p)$, proved to be not the only possible one. Anomalous solutions $\varepsilon(\mathbf{p}) = \mu$ of (2.1) can exist if the logarithmic expression on its right-hand side is finite. This is possible if $0 < n_0(\mathbf{p}) < 1$ for

$p_i \le p \le p_f$. Then, this logarithmic expression remains finite within this interval as $T \to 0$, the product $T \ln[(1 - n_0(\mathbf{p}))/n_0(\mathbf{p})]_{|T\to 0} \to 0$, and we again arrive at (4.1).

Thus, as $T \to 0$, the quasiparticle distribution function $n_0(\mathbf{p})$, which is a solution of (4.1), does not tend to the step function $\theta(p_F - p)$ and, correspondingly, in accordance with (2.4), the entropy $S(T)$ of this state tends to a finite value S_0 as $T \to 0$:

$$S(T \to 0) \to S_0. \tag{4.8}$$

As the density $x \to x_{\rm FC}$ (or as the interaction force increases), the system reaches QCP where FC is formed. This means that $p_i \to p_f \to p_F$ and that the deviation $\delta n(\mathbf{p})$ from the step function is small. Expanding the function $E(n(\mathbf{p}))$ in Taylor series in $\delta n(\mathbf{p})$ and keeping only the leading terms, we can use (4.1) to obtain the following relation that is valid for $p_i \le p \le p_f$:

$$\mu = \varepsilon(\mathbf{p}) = \varepsilon_0(\mathbf{p}) + \int F(\mathbf{p}, \mathbf{p}_1)\delta n(\mathbf{p}_1)\frac{d\mathbf{p}_1}{(2\pi)^2}. \tag{4.9}$$

Both quantities, the Landau interaction $F(\mathbf{p}, \mathbf{p}_1)$ and the single-particle energy $\varepsilon_0(\mathbf{p})$, are calculated at $n(\mathbf{p}) = \theta(p_F - p)$.

Equation (4.9) has nontrivial solutions for densities $x \le x_{\rm FC}$ if the corresponding Landau interaction, which is density-dependent, is positive and sufficiently large for the potential energy to be higher than the kinetic energy. For instance, such a state is realized in a low-density electron liquid. The transformation of the Fermi step function $n(\mathbf{p}) = \theta(p_F - p)$ into a smooth function determined by (4.9) then becomes possible [1, 2, 30].

It follows from (4.9) that the quasiparticles of FC form a collective state, because their state is determined by the macroscopic number of quasiparticles with momenta $p_i < p < p_f$. The shape of the FC single-particle spectrum is independent of the Landau interaction details, which in general is determined by the microscopic properties of the system like chemical composition, interparticle interaction, structure irregularities, and the presence of impurities. The only characteristic determined by the Landau interaction is the length of interval (from p_i to p_f) of FC existence. Of course, the interaction must be strong enough for FCQPT to occur. Therefore, we conclude that spectra related to FC have a universal shape. In Sects. 4.1.3 and 6.1, we show that these spectra depend on temperature and the superconducting gap and that this dependence is also universal. The existence of such spectra can be considered a characteristic feature of a "quantum protectorate", in which the properties of the material, including the thermodynamic properties, are determined by a certain fundamental principle [31, 32]. In our case, the state of matter with FC is also a quantum protectorate, since the new type of quasiparticles of this state determines the special universal thermodynamic and transport properties of Fermi liquids with FC.

4.1.3 The Influence of FCQPT at Finite Temperatures

According to (2.1), the single-particle energy $\varepsilon(\mathbf{p}, T)$ is linear in T for $T \ll T_f$ for $p_f < p < p_i$ [33]. Expanding $\ln((1 - n(\mathbf{p}))/n(\mathbf{p}))$ in series in $n(\mathbf{p})$ at $p \simeq p_F$, we arrive at the expression

$$\frac{\varepsilon(\mathbf{p}, T) - \mu(T)}{T} = \ln \frac{1 - n(\mathbf{p})}{n(\mathbf{p})} \simeq \left. \frac{1 - 2n(\mathbf{p})}{n(\mathbf{p})} \right|_{p \simeq p_F}, \tag{4.10}$$

where T_f is the temperature above which the effect of FC is insignificant [34]:

$$\frac{T_f}{\varepsilon_F} \sim \frac{p_f^2 - p_i^2}{2m\varepsilon_F} \sim \frac{\Omega_{\rm FC}}{\Omega_F}. \tag{4.11}$$

Here $\Omega_{\rm FC}$ is the volume occupied by FC, ε_F is the Fermi energy and $\Omega_{\rm FC}$ is the volume of the Fermi sphere. Taking as an example $\Omega_F/\Omega \sim 0.2$ and $\varepsilon_F \sim$ 20–50 meV, we obtain

$$T_f \sim (100\text{–}200)\ \text{K}, \tag{4.12}$$

and conclude that the NFL behavior can take place up to temperatures of several 100 K. Thus, as it follows from (4.12), the topological FCQPT, forming flat bands, can define the universal behavior of HF compounds up to temperatures of few hundreds of K. As a result, HF compounds form the new state of matter, exhibiting their universal behavior, see Chap. 7.

We note that for $T \ll T_f$, the occupation numbers $n(\mathbf{p})$ obtained from (4.1) are almost perfectly independent of T [33–35]. At finite temperatures, according to (4.10), the dispersionless plateau $\varepsilon(\mathbf{p}) = \mu$ shown in Fig. 4.1 is slightly rotated counterclockwise relatively to μ, so that $\varepsilon(\mathbf{p}) - \mu = T \ln[(1 - n_p)/n_p]$. As a result, the plateau is slightly tilted and rounded off at its end points. According to (2.6) and (4.10), the effective mass $M^*_{\rm FC}$ of the FC quasiparticles is given by

$$M^*_{\rm FC} \simeq p_F \frac{p_f - p_i}{4T}. \tag{4.13}$$

To derive (4.13), we approximate $dn(p)/dp \simeq -1/(p_f - p_i)$. Equation (4.13) shows clearly that for $0 < T \ll T_f$, the electron liquid with FC behaves as if it were placed at a QCP since the electron effective mass diverges as $T \to 0$. Actually, as we shall see in Sect. 6.3, the system is at a quantum-critical line as critical behavior is observed behind QCP with $x = x_{\rm FC}$ of FCQPT as $T \to 0$. Equations (4.11) and (4.13) permit to estimate the effective mass $M^*_{\rm FC}$ in the form

$$\frac{M^*_{\rm FC}}{M} \sim \frac{N(0)}{N_0(0)} \sim \frac{T_f}{T}, \tag{4.14}$$

where $N_0(0)$ is the density of states of a noninteracting electron gas and $N(0)$ is the density of states at the Fermi surface. Equations (4.13) and (4.14) yield the temperature dependence of M^*_{FC}. Multiplying both sides of (4.13) by $p_f - p_i$, we obtain an expression for the characteristic energy,

$$E_0 \simeq 4T, \tag{4.15}$$

which determines the momentum interval $p_f - p_i$ having the low-energy quasiparticles with the energy $|\varepsilon(\mathbf{p}) - \mu| \leq E_0/2$ and the effective mass M^*_{FC}. The quasiparticles that do not belong to this momentum interval have an energy $|\varepsilon(\mathbf{p}) - \mu| > E_0/2$ and an effective mass M^*_L that is weakly temperature-dependent [6, 7, 36]. Equation (4.15) shows that E_0 is independent of the condensate volume. We conclude from (4.13) and (4.15) that for $T \ll T_f$, the single-electron spectrum of FC quasiparticles has a universal shape and has the features of a quantum protectorate.

The above discussion shows that a system with FC is characterized in fact by two effective masses, M^*_{FC} and M^*_L. This manifests itself in the abrupt variation of the quasiparticle dispersion law, which for quasiparticles with energies $\varepsilon(\mathbf{p}) \leq \mu$ can be approximated by two straight lines intersecting at $E_0/2 \simeq 2T$. Figure 4.1 shows that at $T = 0$, the straight lines intersect at $p = p_i$. This also occurs when the system is in its superconducting state at temperatures $T_c \leq T \ll T_f$, where T_c is the critical temperature of the superconducting phase transition, which agrees with the experimental data of [37]. We will see in Sect. 6.1 that this behavior also agrees with the experimental data at $T \leq T_c$. At $T > T_c$, the quasiparticles are well defined as their width γ is small compared to their energy and is proportional to the temperature, $\gamma \sim T$ [34, 38]. The quasiparticle excitation curve (see Sect. 6.2) can be approximately described by a simple Lorentzian [36], which also agrees with the experimental data [37, 39–41].

We now estimate the density x_{FC} at which FCQPT occurs. We show in Sect. 6.4 that an unlimited increase of the effective mass precedes the appearance of a density wave or a charge density wave formed in electron systems at $r_s = r_0/a_B = r_{cdw}$, where r_0 is the average distance between electrons, and a_B is the Bohr radius. Hence, FCQPT certainly occurs at $T = 0$ when r_s reaches its critical value r_{FC} corresponding to x_{FC}, with $r_{FC} < r_{cdw}$ [30]. We note that the increase of the effective mass at the electron number density decrease has been observed experimentally, see Figs. 2.1 and 2.2.

Thus, the formation of FC can be considered as a general property of different strongly correlated systems rather than an exotic phenomenon corresponding to the anomalous solution of (4.1). Beyond FCQPT, the condensate volume is proportional to $r_s - r_{FC}$, with $T_f/\varepsilon_F \sim (r_s - r_{FC})/r_{FC}$, at least when $(r_s - r_{FC})/r_{FC} \ll 1$. This implies that [8, 18]

$$\frac{r_s - r_{FC}}{r_{FC}} \sim \frac{p_f - p_i}{p_F} \sim \frac{x_{FC} - x}{x_{FC}}. \tag{4.16}$$

Since a state of a system with FC is highly degenerate, FCQPT serves as a stimulator of phase transitions that could lift the degeneracy of the spectrum at the interval $p_f - p_i$. For instance, FC can stimulate the formation of spin density waves, antiferromagnetic and/or ferromagnetic state, etc., thus strongly stimulating the competition between phase transitions eliminating the degeneracy. The presence of FC facilitates a transition to the superconducting state as both phases have the same order parameter.

4.1.4 Two Scenarios of the Quantum-Critical Point

The statement that the Landau quasiparticle picture breaks down at points of second-order phase transitions has become a truism. The violation of this picture is attributed to vanishing of the quasiparticle weight z in the single-particle state. In nonsuperfluid Fermi systems, the z-factor is determined by the formula $z = [1 - (\partial\Sigma(p,\varepsilon)/\partial\varepsilon)_0]^{-1}$ where the subscript 0 indicates that the respective derivative of the mass operator Σ is evaluated at the Fermi surface. This factor enters a textbook formula

$$\frac{M}{M^*} = z\left[1 + \left(\frac{\partial\Sigma(p,\varepsilon)}{\partial\varepsilon_p^0}\right)_0\right] \tag{4.17}$$

for the ratio M^*/M of the effective mass M^* to the mass M of a free particle. As seen from this formula, where $\varepsilon_p^0 = p^2/2M$, the effective mass diverges at a critical density ρ_c, where z vanishes provided the sum $1 + \left(\partial\Sigma(p,\varepsilon=0)/\partial\varepsilon_p^0\right)_0$ has a *positive and finite value* at this point. Nowadays, when studying critical fluctuations of arbitrary wavelengths $k < 2p_F$ has become popular, this restriction is often assumed to be valid without any stipulations. For example, a standard scenario of the quantum-critical point where M^* diverges is formulated as follows: in the vicinity of an impending second-order phase transition, "quasiparticles get heavy and die" [42].

However, as seen from (4.17), M^* may diverge not only at the points of the second-order phase transitions, but also at a critical density ρ_∞, where the sum $1 + \left(\frac{\partial\Sigma(p,\varepsilon)}{\partial\varepsilon_p^0}\right)_0$ changes its sign. Furthermore, it has been demonstrated that except for the case of the ferromagnetic instability, M^* cannot diverge at ρ_c without violation of stability conditions [43]. For a detailed consideration of this problem, the reader can find in [9, 43].

4.1.5 Phase Diagram of Fermi System with FCQPT

At $T = 0$, a quantum phase transition is driven by a nonthermal control parameter like number density x. As we have seen, at QCP, $x = x_{FC}$, the effective mass diverges. It follows from (2.21) that beyond QCP, the effective mass becomes negative. As such

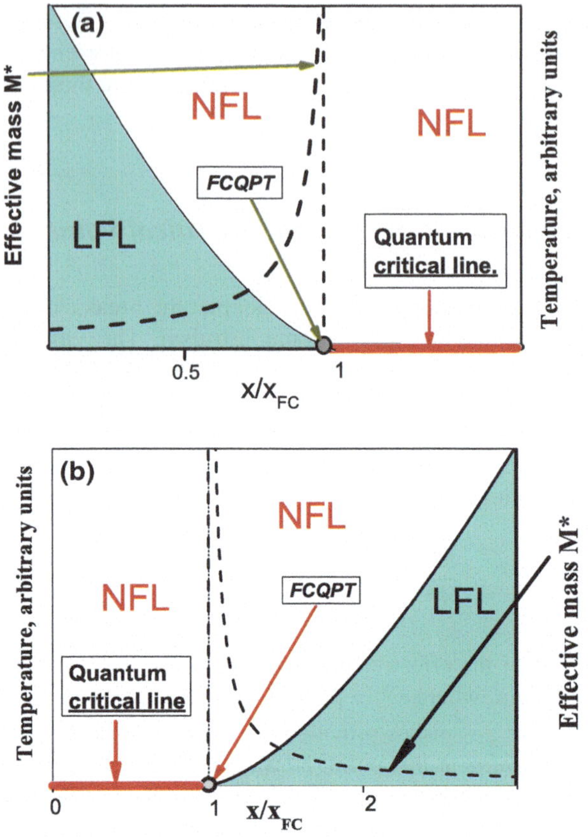

Fig. 4.2 Schematic temperature—number density phase diagrams of the systems with FC. The panel A displays the phase diagram for the case when the FCQPT takes place at growing densities, while the panel B displays the opposite case. The number density x is taken as the control parameter and depicted as x/x_{FC}. The dashed line shows $M^*(x/x_{FC})$ as the system approaches FCQPT, marked by the arrow. The shadowed area in the panel A corresponds to the case $x/x_{FC} < 1$ and sufficiently low temperatures, where the system is in the LFL phase. This case in the panel B corresponds to $x/x_{FC} > 1$. At $T = 0$ and beyond the FCQPT critical point, the system is at the quantum-critical line as shown in the legend. This critical line is characterized by the FC state with finite superconducting order parameter κ. At any finite temperature $T > T_c = 0$, κ is destroyed so that the system undergoes the first-order phase transition, possesses finite entropy S_0, and exhibits the NFL behavior at any finite temperatures $T < T_f$

a physically meaningless state cannot be realized, the system undergoes FCQPT leading to the FC formation.

The schematic phase diagrams of the systems which are driven to the FC state by variation of x are reported in Fig. 4.2. The panel A displays the case when FCQPT takes place at growing densities. As we have seen in Sect. 4.1, FCQPT occurs as soon as the potential energy of the quasiparticles near the Fermi surface determines the ground state energy. Therefore, the panel A represents the phase diagram of a system composed of particles interacting with each other by van der Waals forces with strong hardcore repulsion. At elevated densities, the potential energy overcomes the kinetic one leading to FC emergence, which is the case for 2D ^{3}He films, see Sect. 21.4. The panel B displays the opposite case which occurs in electronic systems, when the potential energy dominates at lowering densities, see Sect. 6.4 for details. The similarity of both the diagrams reflects the universal behavior of systems located near FCQPT, as it is discussed in Chap. 21.

Figure 4.2 demonstrates that upon approaching the critical density $x_{\rm FC}$ the system remains in the LFL region at sufficiently low temperatures as it is shown by the shadowed area. The temperature range of the shadowed area shrinks as the system approaches FCQPT, and $M^*(x/x_{\rm FC})$ diverges as shown by the dashed line and (2.21). At FCQPT, $x_{\rm FC}$ shown by the arrow in Fig. 4.2, the system demonstrates the NFL behavior down to the lowest temperatures. Beyond the critical point at finite temperatures, the behavior remains the NFL and is determined by the temperature-independent entropy S_0 [8, 18, 21]. In that case at $T \to 0$, the system is approaching a quantum-critical line (shown in the legend) rather than a quantum-critical point. Upon reaching the quantum-critical line from the above at $T \to 0$, the system undergoes the first-order quantum phase transition, which is FCQPT taking place at $T_c = 0$. At the same time, at temperature lowering for x before QCP (which is $x < x_{\rm FC}$ in panel A and $x > x_{\rm FC}$ in panel B), the system does not undergo a phase transition and transits smoothly from NFL to LFL phase.

It is shown in Fig. 4.2 that at finite temperatures there is no boundary (or phase transition) between the states of systems located before or after FCQPT. Therefore, at elevated temperatures the properties of systems with $x/x_{\rm FC} < 1$ or with $x/x_{\rm FC} > 1$ become indistinguishable. On the other hand, at $T > 0$ the NFL state above the critical line and in the QCP vicinity is strongly degenerate so that the degeneracy stimulates different phase transitions, which finally eliminate it. The elimination of the degeneracy means that the NFL state can be obscured by the other states like superconducting (for example, in $CeCoIn_5$ [21, 44]) or antiferromagnetic (for example, in $YbRh_2Si_2$ [45]), etc. The diversity of low-temperature phase transitions is one of the most spectacular features of the physics of many HF metals. The scenario of ordinary quantum phase transitions makes it hard to understand why they are so different and why their critical temperatures are so small. However, such diversity is endemic to systems with an FC [20].

Upon varying nonthermal tuning parameters like the number density, pressure, or magnetic field, the NFL behavior could be destroyed and the LFL behavior is restored as we shall see in Chap. 7. For example, the application of magnetic field $B > B_{c0}$ drives a system to QCP and destroys the AF state restoring the LFL behavior. Here, B_{c0} is a critical magnetic field, such that at $B > B_{c0}$ the system is driven toward its LFL state. In some cases as in the HF metal $CeRu_2Si_2$, $B_{c0} = 0$, see, e.g., [46], while in $YbRh_2Si_2$, $B_{c0} \simeq 0.06$ T [47].

4.2 Topological Phase Transitions Related to FCQPT

In Sects. 4.1 and 4.1.5, we have investigated the structure of the Fermi surface beyond FCQPT within the extended quasiparticle paradigm. We have shown that at $T = 0$ there is a scenario that entails the formation of FC, manifested by the emergence of a completely flat portion of the single-particle spectrum. On the other hand, there are different kinds of instabilities related to the emergence of a multi-connected Fermi surface, see, e.g., [20, 48–53]. In such considerations, we analyze the stability of a model fermion system with repulsive Landau interaction allowing to carry out analytical studies of the emergence of a multi-connected Fermi surface [48, 49]. We show, in particular, that the Landau interaction given by the screened Coulomb law does not generate the FC phase, but rather an "iceberg" TT phase. For this model, we plot a phase diagram in the variables "screening parameter—coupling constant" displaying two kinds of TT: a 5/2-kind similar to the known Lifshitz transitions in metals, and a 2-kind characteristic for a uniform strongly interacting system. The Lifshitz transitions are topological transitions of the Fermi surface with no symmetry breaking [54]. We note that at elevated temperatures the "iceberg" TT phase is melted and the difference between two systems one of which starts from FQCPT and the other from the multi-connected "iceberg" state vanishes, see, e.g., [8, 9].

The common ground state of isotropic LFL with density ρ_x is described at zero temperature by the stepwise Fermi function $n_F(p) = \theta(p_F - p)$, dropping discontinuously from 1 to 0 at the Fermi momentum p_F. The LFL theory states that the quasiparticle distribution function $n(p)$ and its single-particle spectrum $\varepsilon(p)$ are in all but name similar to those of an ideal Fermi gas with the substitution of real fermion mass m by the effective one M^*. These $n_F(p)$ and $\varepsilon(p)$ can become unstable under several circumstances. The best known example is Cooper pairing at arbitrarily weak attractive interaction with subsequent formation of the pair condensate and gapped quasiparticle spectrum [17]. However, a sufficiently strong repulsive Landau interaction can also generate nontrivial ground states. The first example of such restructuring for a Fermi system with model repulsive interaction is FC [1]. It reveals the existence of a critical value α_{cr} of the interaction constant α such that at $\alpha = \alpha_{cr}$ the stability criterion $s(p) = (\varepsilon(p) - E_F)/(p^2 - p_F^2) > 0$ fails at the Fermi surface $s(p_F) = 0$ (p_F-instability). We recall that in the case of this instability the single-particle spectrum $\varepsilon(p)$ possesses the inflection point at the Fermi surface. Then at $\alpha > \alpha_{cr}$ an exact solution of a variational equation for $n(p)$, following from the Landau functional $E(n(p))$ (2.15) (see also (4.18) below), can be found. This solution exhibits some finite interval $p_f - p_i$ (we recall that indices "f" and "i" stand for "final" and "initial", respectively) around p_F where the distribution function $n(p)$ varies continuously taking intermediate values between 1 and 0, while the single-particle excitation spectrum $\varepsilon(p)$ has a flat plateau. Equation (4.1) means actually that the roots of the equation $\varepsilon(p) = \mu$ form an uncountable set in the range $p_i \leq p \leq p_f$, see Fig. 4.3. It is seen from (4.1) that the occupation numbers $n(p)$ become variational parameters, deviating from the Fermi step function to minimize the energy E.

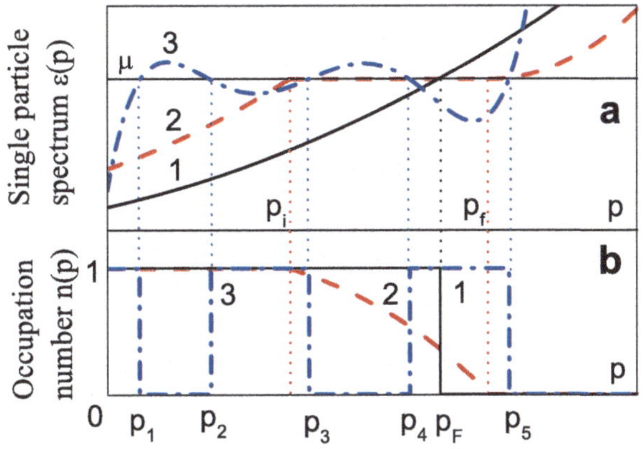

Fig. 4.3 Schematic plot of the single-particle spectrum $\varepsilon(p)$ (**a**) and occupation numbers $n(p)$ (**b**), corresponding to LFL (curves 1), FC (curves 2, dashed line) and iceberg (curves 3, dot-dashed line) phases at $T = 0$. For LFL the equation, $\varepsilon(p) = \mu$, has a single root equal to Fermi momentum p_F. For iceberg phase, the above equation has countable set of the roots $p_1 \dots p_N \dots$, for FC phase the roots occupy the whole segment $(p_f - p_i)$. We note that $p_i < p_F < p_f$ and the states, where $\varepsilon(p) < \mu$ are occupied (n = 1), while those with $\varepsilon(p) > \mu$ are empty (n = 0)

The other type of phase transition, corresponding to the so-called iceberg phase, occurs when the equation $\varepsilon(p) = \mu$ has discrete countable number of roots, either finite or infinite. This is depicted in Fig. 4.3, and related to the situation when the Fermi surface becomes multi-connected. Note that the idea of multi-connected Fermi surface, with the production of new, interior segments, had already been considered [50–52].

Let us take the Landau functional $E(n(p))$ in the form

$$E(n(p)) = \int \frac{p^2}{2M} n(p) \frac{d\mathbf{p}}{(2\pi)^3} + \frac{1}{2} \int \int n(p) U(|\mathbf{p} - \mathbf{p}'|) n(p') \frac{d\mathbf{p} d\mathbf{p}'}{(2\pi)^6}, \tag{4.18}$$

which, by virtue of (2.3), leads to the following quasiparticles dispersion law:

$$\varepsilon(p) = \frac{p^2}{2M} + \int U(|\mathbf{p} - \mathbf{p}'|) n(p') \frac{d\mathbf{p}'}{(2\pi)^3}. \tag{4.19}$$

The angular integration with subsequent passing to the dimensionless variables $x = p/p_F$, $y = y(x) = 2M\varepsilon(p)/p_F^2$, $z = 2\pi^2 ME/p_F^5$, leads to simplification of the (4.18) and (4.19)

$$z[\nu(x)] = \int \left[x^4 + \frac{1}{2} x^2 V(x) \right] \nu(x) \mathrm{d}x, \tag{4.20}$$

$$y(x) = x^2 + V(x), \tag{4.21}$$

where

$$V(x) = \frac{1}{x} \int x' \nu(x') u(x, x') \mathrm{d}x',$$

$$u(x, x') = \frac{M}{\pi^2 p_F} \int\limits_{|x-x'|}^{x+x'} u(t) t \mathrm{d}t. \tag{4.22}$$

In this chapter, we do not mention the renormalization constant z, and use z as the dimensional variable. Here $u(x) \equiv U(p_F x)$ and the distribution function $\nu(x) \equiv n(p_F x)$ is positive, obeys the normalization condition

$$\int x^2 \nu(x) \mathrm{d}x = 1/3, \tag{4.23}$$

and the Pauli principle limitation $\nu(x) \leq 1$. The latter can be lifted using, e.g., the ansatz: $\nu(x) = [1 + \tanh \eta(x)]/2$. In the latter case, the system ground state gives a minimum to the functional

$$f[\eta(x)] = \int [1 + \tanh \eta(x)] \Big\{ x^4 - \mu x^2 + x'[1 + \tanh \eta(x')] u(x, x') \mathrm{d}x' \Big\} \mathrm{d}x, \tag{4.24}$$

containing a Lagrange multiplier μ, with respect to an arbitrary variation of the auxiliary function $\eta(x)$. This allows to represent the necessary condition of extremum $\delta f = 0$ in the form

$$x^2 \nu(x)[1 - \nu(x)][y(x) - \mu] = 0. \tag{4.25}$$

This means that either $\nu(x)$ takes only the values 0 and 1 or the dispersion law is flat: $y(x) = \mu$ [1], in accordance with (4.1). The former possibility corresponds to iceberg phase, while the latter to FC. As it is seen from (4.1), the spectrum $\varepsilon(p)$ in this case cannot be an analytic function of complex p in any open domain, containing the FC interval $(p_f - p_i)$. In fact, all the derivatives of $\varepsilon(p)$ with respect to p along the strip $(p_f - p_i)$ should be zero, while this is not the case outside $(p_f - p_i)$. For instance, in the FC model with $U(p) = U_0/p$ [1], the kernel, (4.22), is nonanalytic

$$u(x, x') = \frac{M U_0}{\pi^2 p_F} (x + x' + |x - x'|), \tag{4.26}$$

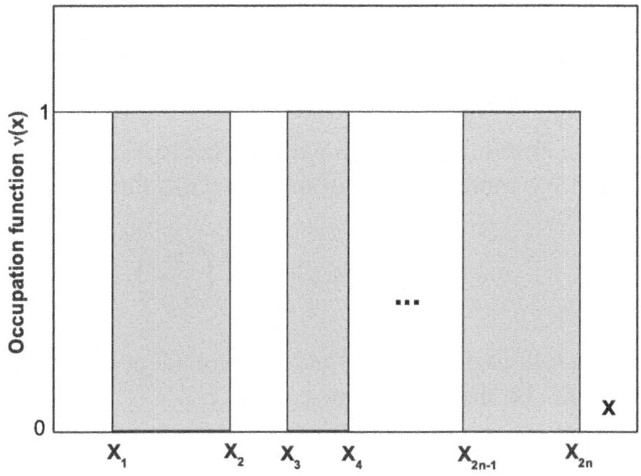

Fig. 4.4 Occupation function for a multi-connected distribution

which eventually causes nonanalyticity of the potential $V(x)$. It follows from (4.21) that the single-particle spectrum is an analytic function on the whole real axis if $V(x)$ is such a function. In this case, FC is forbidden and the only alternative to the Fermi ground state (if the stability criterion gets broken) is iceberg phase corresponding to TT between the topologically unequal states with $\nu(x) = 0, 1$ [4].

On the other hand, applying the technique of Poincaré mapping, it is possible to analyze the sequence of iterative maps generated by (2.14) for the single-particle spectrum at zero temperature [20]. If the sequence of maps converges, the multi-connected Fermi surface is formed. If it fails to converge, the Fermi surface swells into a volume that provides a measure of entropy associated with the formation of an exceptional state of the system characterized by partial occupation of single-particle states and dispersion of their spectrum proportional to temperature as seen from (4.13).

Generally, all such states related to the formation of iceberg phases are classified by the indices of connectedness (known as Betti numbers in algebraic topology [55, 56]) for the support of $\nu(x)$. In fact, for an isotropic system, these numbers simply count the separate (concentric) segments of the Fermi surface. Then, the system ground state corresponds to the following multi-connected distribution shown in Fig. 4.4:

$$\nu(x) = \sum_{i=1}^{n} \theta(x - x_{2i-1})\theta(x_{2i} - x), \tag{4.27}$$

where the parameters $0 \leq x_1 < x_2 < \cdots < x_{2n}$ obey the normalization condition

$$\sum_{i=1}^{n} (x_{2i}^3 - x_{2i-1}^3) = 1. \tag{4.28}$$

The function z, (4.20),

$$z = \frac{1}{2}\sum_{i=1}^{n}\int\limits_{x_{2i-1}}^{x_{2i}} x^2[x^2 + y(x)]\mathrm{d}x, \tag{4.29}$$

has the absolute minimum with respect to $x_1, \ldots, x_{2n-1}$ and to $n \geq 1$. To obtain the necessary condition of extremum, we use the relations

$$\frac{\partial x_{2n}}{\partial x_k} = (-1)^{k-1}\left(\frac{x_k}{x_{2n}}\right)^2, \quad 1 \leq k \leq 2n-1, \tag{4.30}$$

which follows from (4.28) and the dependence of the potential $V(x)$ in the dispersion law $y(x)$ on the parameters $x_1, \ldots, x_{2n-1}$

$$V(x) = \frac{1}{x}\sum_{i=1}^{n}\int\limits_{x_{2i-1}}^{x_{2i}} x'u(x, x')\mathrm{d}x'. \tag{4.31}$$

The differentiation of (4.29) with respect to the parameters $x_1, \ldots, x_{2n-1}$ with subsequent use of (4.30) and (4.31) yield the necessary conditions of extremum in the following form:

$$\frac{\partial z}{\partial x_k} = (-1)^k x_k^2[y(x_k) - y(x_{2n})] = 0, \quad 1 \leq k \leq 2n-1. \tag{4.32}$$

This means that a multi-connected ground state is controlled by the evident rule of unique Fermi level $y(x_k) = y(x_{2n})$ for all $1 \leq k \leq 2n-1$ (except for $x_1 = 0$). In principle, given the dispersion law $y(x)$, all the $2n-1$ unknown parameters x_k can be found from (4.32). Then, the sufficient stability conditions $\partial^2 z/\partial x_i \partial x_j = \gamma_i \delta_{ij}$, $\gamma_i > 0$ yield the generalized stability criterion. Namely, the dimensionless function

$$\sigma(x) = 2Ms(p) = \frac{y(x) - y(x_{2n})}{x^2 - x_{2n}^2} \tag{4.33}$$

should be positive within filled and negative within empty intervals, turning to zero at their boundaries in accordance with (4.32). It can be proved rigorously that, for given analytic kernel $u(x, x')$, (4.33) defines the system ground state uniquely.

Subsequently, we shall label each multi-connected state, (4.27), by an integer number related to the binary sequence of empty and filled intervals read from x_{2n} to 0. Thus, the Fermi state with a single filled interval ($x_2 = 1, x_1 = 0$) reads as unity, the state with a void at the origin (filled $[x_2, x_1]$ and empty $[x_1, 0]$) reads as $(10) = 2$, the state with a single gap: $(101) = 3$, etc. Note that all even phases have a void at the origin and odd phases have not.

For free fermions $V(x) = 0$, $y(x) = x^2$, (4.32) only yields the trivial solution corresponding to the Fermi state 1. To obtain nontrivial realizations of TT, we choose $U(p)$ to correspond to the common screened Coulomb potential:

$$U(p) = \frac{4\pi e^2}{p^2 + p_0^2}. \tag{4.34}$$

The related explicit form of the kernel,

$$u(x, x') = \alpha \ln \frac{(x + x')^2 + x_0^2}{(x - x')^2 + x_0^2}, \tag{4.35}$$

with the dimensionless screening parameter $x_0 = p_0/p_F$ and the coupling constant $\alpha = 2Me^2/\pi p_F$, evidently displays the necessary analytical properties for existence of iceberg phase. Equations (4.31) and (4.35) permit to express the potential $V(x)$ in elementary functions [48]. Then, the straightforward analysis of (4.32) shows that its nontrivial solutions appear only when the coupling parameter α exceeds a certain critical value α^*. This corresponds to the situation when the stability criterion [1] $\sigma(x) = (y_F(1) - y_F(x))/(1 - x^2) > 0$ calculated with the Fermi distribution, $y_F(x) = x^2 + V(x; 0, 1)$, fails at a certain point $0 \leq x_i < 1$ within the Fermi sphere: $\sigma(x_i) \to 0$. There are two different types of such instabilities depending on the screening parameter x_0 (Fig. 4.5). For x_0 below a certain threshold value $x_{\text{th}} \approx 0.32365$ (weak screening regime, WSR) the instability point x_i sets rather close to the Fermi surface: $1 - x_i \ll 1$, while it drops abruptly to zero at $x_0 \to x_{\text{th}}$ and equals zero for all $x_0 > x_{\text{th}}$ (strong screening regime, SSR). The critical coupling $\alpha^*(x_0)$ results in a monotonously growing function of x_0 with the asymptotics $\alpha^* \approx (\ln 2/x_0 - 1)^{-1}$ at $x_0 \to 0$ and staying analytic at $\alpha_{\text{th}} = \alpha^*(x_{\text{th}}) \approx 0.91535$, where it only exhibits an inflection point.

These two types of instabilities give rise to different types of TT from the state 1 at $\alpha > \alpha^*$: at SSR a void appears around $x = 0$ ($1 \to 2$ transition), and at WSR a gap opens around x_i ($1 \to 3$ transition). Further analysis of (4.32) shows that the point $x_{\text{th}}, \alpha_{\text{th}}$ represents a triple point in the phase diagram in the variables x_0, α (Fig. 4.5) where the phases 1, 2, and 3 meet one another. Similar to the onset of instability in the Fermi state 1, each evolution of TT to higher order phases with growing α is manifested by a zero of $\sigma(x)$, (4.33), at some point $0 \leq x_i < x_{2n}$ different from the existing interfaces. If this occurs at the very origin, $x_i = 0$, the phase number rises at TT by 1, corresponding to the opening of a void (passing from odd to even phase) or to emerging "island" (even $\to$ odd). For $x_i > 0$, either a thin spherical gap opens within a filled region or a thin filled spherical sheet emerges within a gap, so that the phase number rises by 2, living the parity unaltered. A part of the whole diagram shown in Fig. 4.6 demonstrates that with decreasing of x_0 (screening weakening) all even phases terminate at certain triple points. This, in particular, agrees with numerical studies of the considered model along the line $x_0 = 0.07$ at growing α [52], where only the sequence of odd phases $1 \to 3 \to 5 \to \cdots$ has been revealed (shown by

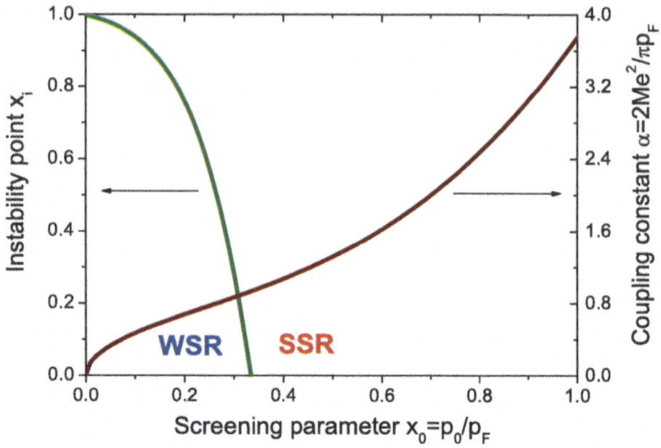

Fig. 4.5 Instability point x_i and critical coupling α^* as functions of screening. The regions of weak screening (WSR) and strong screening (SSR) are separated by the threshold value x_{th}. Note that a x_{th}, α_{th} is the triple point between the phases 1, 2, and 3 in Fig. 4.6

the arrow in Fig. 4.5). The energy gain $\Delta\,(\tau_a)$ at TT as a function of small parameter $\tau_a = \alpha/\alpha^* - 1$ is evidently proportional to τ_a times the volume of a new emerging phase region (empty or filled). Introducing a void radius δ and expanding the energy gain $\Delta(\delta) = z[n(x,\delta)] - z[n_F(x)]$ in δ, one gets $\Delta = -\beta_1\tau_a\delta^3 + \beta_2\delta^5 + O(\delta^6)$, $\beta_1, \beta_2 > 0$. As a result, the optimum void radius is $\delta \sim \sqrt{\tau_a}$. Consequently, we have $\Delta(\tau_a) \sim \tau_a^{5/2}$ indicating a resemblance to the known "$^5/_2$-kind" phase transitions in the theory of metals [55]. The peculiar feature of our situation is that the new segment of the Fermi surface opens at very small momentum values, which can dramatically change the system response to, e.g., electron-phonon interaction. On the other hand, this segment may have a pronounced effect on the thermodynamical properties of ^{3}He at low temperatures, especially in the case of P-pairing, producing excitations with extremely small momenta.

For a TT with unchanged parity, the width of a gap (or a sheet) is found to be $\sim\tau_a$ so that the energy gain is $\Delta\,(\tau_a) \sim \tau_a^2$ and such TT can be related to the second kind. It follows from the above consideration that each triple point in the $x_0 - \alpha$ phase diagram is a point of confluence of two $^5/_2$-kind TT lines into one 2-kind line. The latter type of TT has already been discussed in the literature [50, 52]. Here, we only mention that its occurrence on a whole continuous surface in the momentum space is rather specific for the systems with strong fermion-fermion interaction, while the known TTs in metals, under the effects of crystalline field, occur typically at separate points in the quasimomentum space. It is interesting to note that in the limit $x_0 \to 0$, $\alpha \to 0$, reached along the line $\alpha = kx_0$, we attain the exactly solvable model: $U(p) \to (2\pi)^3 U_0\delta(p)$ with $U_0 = k/(2Mp_F)$, which is known to display FC for all $U_0 > 0$ [1]. The analytic mechanism of this behavior is the disappearance of the poles of $U(p)$, (4.34), as $p_0 \to 0$, restoring the analytical properties necessary for FC. Otherwise, the FC regime corresponds to the phase order $\to \infty$, when the density of infinitely thin filled regions separated by the empty ones approaches some continuous function $0 < \nu(x) < 1$ [52], and the dispersion law turns flat according to (4.32). Several remarks are in place at this point.

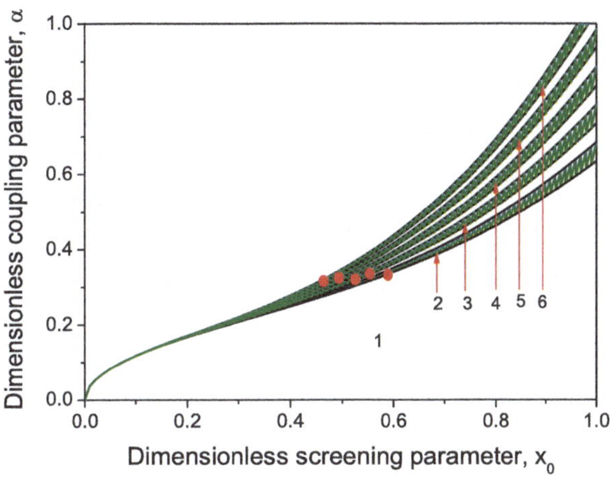

Fig. 4.6 Phase diagram in "screening-coupling" variables. Each phase with certain topology is labeled by the total number of filled and empty regions (see Fig. 4.4). Even phases (shaded) are separated from odd ones by "$^5/_2$-kind" topological transition (TT) lines, while odd phases are separated from each other by TT lines of 2 (second) kind odd phases. Triple points, where two $^5/_2$-TT and one 2-TT meet, are shown by circles

First, when the dispersion law starts to turn flat, the system of icebergs can transform into the FC state due to some admixture of a nonanalytical interaction. In fact, even in the absence of such an admixture the difference in the energy between the ground state with FC and one with the system of icebergs is negligibly small. Therefore, at lowering temperature the system is "wandering" in the energy landscape trying to find the absolute minimum of its energy. We speculate that in such a case some of HF compounds can exhibit the dynamics of glasses when the dynamics is strongly suppressed becoming very slow.

Second, the considered model formally treats x_0 and α as independent parameters, though in fact a certain relation between them can be imposed. Under such restriction, the system ground state should depend on a single parameter, say the particle density ρ_x, along a certain trajectory $\alpha(x_0)$ in the above phase diagram. For instance, with the simplest Thomas-Fermi relation for a free electron gas $\alpha(x_0) = x_0^2/2$, this trajectory stays fully within the Fermi state 1 over all the physically reasonable range of densities. Hence, a faster growth of $\alpha(x_0)$ is necessary for the realization of TT in any fermionic system with the interaction (4.34).

Third, at increasing temperatures, the stepwise form of the quasiparticle distribution is smearing. We call this the melting of icebergs. Therefore, as temperature increases from zero, the concentric Fermi spheres are taken up by FC. In fact, these arguments do not work in the case of a few icebergs. Thus, it is quite possible to observe the two separate Fermi sphere regimes related to the FC and iceberg states.

There is a good reason to mention that neither in the FC phase nor in the other TT phases, the standard Kohn-Sham scheme [57, 58] is no longer valid. This is because in the systems with FC or TT phase transitions the occupation numbers of quasiparticles are indeed variational parameters, see, for details, Chap. 3. Thus, to get a reasonable description of the system, one has to consider the ground state energy as a functional of the occupation numbers $E[(n(p))]$ rather than a functional of the density $E[\rho_x]$ [59–61].

References

1. V.A. Khodel, V.R. Shaginyan, JETP Lett. **51**, 553 (1990)
2. V.A. Khodel, V.R. Shaginyan, V.V. Khodel, Phys. Rep. **249**, 1 (1994)
3. V.A. Khodel, J.W. Clark, H. Li, M.V. Zverev, Phys. Rev. Lett. **98**, 216404 (2007)
4. G.E. Volovik, JETP Lett. **53**, 222 (1991)
5. V.R. Shaginyan, J.G. Han, J. Lee, Phys. Lett. A **329**, 108 (2004)
6. M.Y. Amusia, V.R. Shaginyan, JETP Lett. **73**, 232 (2001)
7. M.Y. Amusia, V.R. Shaginyan, Phys. Rev. B **63**, 224507 (2001)
8. V.R. Shaginyan, M.Y. Amusia, A.Z. Msezane, K.G. Popov, Phys. Rep. **492**, 31 (2010), arXiv:1006.2658
9. M.Y. Amusia, K.G. Popov, V.R. Shaginyan, W.A. Stephanowich, *Theory of Heavy-Fermion Compounds*. Springer Series in Solid-State Sciences, vol. 182 (Springer, Berlin, 2015)
10. C.M. Varma, Z. Nussionov, W. van Saarloos, Phys. Rep. **361**, 267 (2002)
11. C.M. Varma, P.B. Littlewood, S. Schmittrink, E. Abrahams, A.E. Ruckenstein, Phys. Rev. Lett. **63**, 1996 (1989)
12. C.M. Varma, P.B. Littlewood, S. Schmittrink, E. Abrahams, A.E. Ruckenstein, Phys. Rev. Lett. **64**, 497 (1990)
13. G.E. Volovik, Acta Phys. Slov. **56**, 49 (2006)
14. G.E. Volovik, in *Quantum Analogues: From Phase Transitions to Black Holes and Cosmology*. Springer Lecture Notes in Physics, vol. 718, ed. by W.G. Unruh, R. Schutzhold (Springer, Orlando, 2007), p. 31
15. V.R. Shaginyan, V.A. Stephanovich, A.Z. Msezane, P. Schuck, J.W. Clark, M.Y. Amusia, G.S. Japaridze, K.G. Popov, E.V. Kirichenko, J. Low Temp. Phys. **189**, 410 (2017)
16. L.N. Oliveira, E.K.U. Gross, W. Kohn, Phys. Rev. Lett. **60**, 2430 (1988)
17. J. Bardeen, L.N. Cooper, J.R. Schrieffer, Phys. Rev. **108**, 1175 (1957)
18. V.R. Shaginyan, M.Y. Amusia, K.G. Popov, Phys. Usp. **50**, 563 (2007)
19. E.M. Lifshitz, L.P. Pitaevskii, *Statisticheskaya Fizika (Statistical Physics), Part 2*. Course of Theoretical Physics (Pergamon Press, Oxford, 1980)
20. V.A. Khodel, J.W. Clark, M.V. Zverev, Phys. Rev. B **78**, 075120 (2008)
21. V.A. Khodel, M.V. Zverev, V.M. Yakovenko, Phys. Rev. Lett. **95**, 236402 (2005)
22. D.V. Khveshchenko, R. Hlubina, T.M. Rice, Phys. Rev. B **48**, 10766 (1993)
23. I.E. Dzyaloshinskii, J. Phys. I (France) **6**, 119 (1996)
24. D. Lidsky, J. Shiraishi, Y. Hatsugai, M. Kohmoto, Phys. Rev. B **57**, 1340 (1998)
25. V.Y. Irkhin, A.A. Katanin, M.I. Katsnelson, Phys. Rev. Lett. **89**, 076401 (2002)
26. D. Yudin, D. Hirschmeier, H. Hafermann, O. Eriksson, A.I. Lichtenstein, M.I. Katsnelson, Phys. Rev. Lett. **112**, 070403 (2014)
27. S. Link, S. Forti, A. Stöhr, K. Küster, M. Rösner, D. Hirschmeier, C. Chen, J. Avila, M.C. Asensio, A.A. Zakharov, T.O. Wehling, A.I. Lichtenstein, M.I. Katsnelson, U. Starke, Phys. Rev. B **100**, 121407(R) (2019)
28. Y. Cao, V. Fatemi, A. Demir, S. Fang, S.L. Tomarken, J.Y. Luo, J.D. Sanchez-Yamagishi, K. Watanabe, T. Taniguchi, E. Kaxiras, R.C. Ashoori, P. Jarillo-Herrero, Nature **556**, 80 (2018)
29. Y. Cao, V. Fatemi, S. Fang, K. Watanabe, T. Taniguchi, E. Kaxiras, P. Jarillo-Herrero, Nature **556**, 43 (2018)
30. V.A. Khodel, V.R. Shaginyan, M.V. Zverev, JETP Lett. **65**, 242 (1997)
31. R.B. Laughlin, D. Pines, Proc. Natl. Acad. Sci. USA **97**, 28 (2000)
32. P.W. Anderson, Science **288**, 480 (2000)
33. V.A. Khodel, J.W. Clark, V.R. Shaginyan, Solid State Commun. **96**, 353 (1995)
34. J. Dukelsky, V. Khodel, P. Schuck, V. Shaginyan, Z. Phys. **102**, 245 (1997)
35. V.A. Khodel, V.R. Shaginyan, in *Condensed Matter Theories*, vol. 12, ed. by J. Clark, V. Plant (Nova Science Publishers Inc., New York, 1997), p. 221
36. S.A. Artamonov, V.R. Shaginyan, JETP **92**, 287 (2001)
37. P.V. Bogdanov, A. Lanzara, S.A. Kellar, X.J. Zhou, E.D. Lu, W.J. Zheng, G. Gu, J.I. Shimoyama, K. Kishio, H. Ikeda, R. Yoshizaki, Z. Hussain, Z.X. Shen, Phys. Rev. Lett. **85**, 2581 (2000)

38. J.D. Koralek, J.F. Douglas, N.C. Plumb, Z. Sun, A.V. Fedorov, M.M. Murnane, H.C. Kapteyn, S.T. Cundiff, Y. Aiura, K. Oka, H. Eisaki, D.S. Dessau, Phys. Rev. Lett. **96**, 017005 (2006)
39. A. Kaminski, M. Randeria, J.C. Campuzano, M.R. Norman, H. Fretwell, J. Mesot, T. Sato, T. Takahashi, K. Kadowaki, Phys. Rev. Lett. **86**, 1070 (2001)
40. T. Valla, A.V. Fedorov, P.D. Johnson, B.O. Wells, S.L. Hulbert, Q. Li, G.D. Gu, N. Koshizuka, Science **285**, 2110 (1999)
41. T. Valla, A.V. Fedorov, P.D. Johnson, Q. Li, G.D. Gu, N. Koshizuka, Phys. Rev. Lett. **85**, 828 (2000)
42. P. Coleman, C. Pépin, Q. Si, R. Ramazashvili, J. Phys. Condens. Matter **13**, R723 (2001)
43. V.A. Khodel, JETP Lett. **86**, 721 (2007)
44. V.R. Shaginyan, JETP Lett. **77**, 99 (2003)
45. V.R. Shaginyan, M.Y. Amusia, K.G. Popov, Phys. Lett. A **373**, 2281 (2009)
46. D. Takahashi, S. Abe, H. Mizuno, D. Tayurskii, K. Matsumoto, H. Suzuki, Y. Onuki, Phys. Rev. B **67**, 180407(R) (2003)
47. P. Gegenwart, J. Custers, C. Geibel, K. Neumaier, K.T.T. Tayama, O. Trovarelli, F. Steglich, Phys. Rev. Lett. **89**, 056402 (2002)
48. S.A. Artamonov, Y.G. Pogorelov, V.R. Shaginyan, JETP Lett. **68**, 942 (1998)
49. Y.G. Pogorelov, V.R. Shaginyan, in *Condensed Matter Theories*, vol. 18 (Nova Science Publishers Inc., New York, 2003), p. 191
50. M. de Llano, J.P. Vary, Phys. Rev. C **19**, 1083 (1979)
51. M. de Llano, A. Plastino, J.P. Zabolitsky, Phys. Rev. C **20**, 2418 (1979)
52. M.V. Zverev, M. Baldo, J. Phys. Condens. Matter **11**, 2059 (1999)
53. M.V. Zverev, M. Baldo, JETP **87**, 1129 (1998)
54. I.M. Lifshitz, Sov. Phys. JETP **11**, 130 (1960)
55. I.M. Lifshitz, Sov. Phys. JETP 1130 (1960)
56. M. Nakahara, *Geometry, Topology and Physics* (IOP Publishing, Bristol, 1990)
57. W. Kohn, L.J. Sham, Phys. Rev. **140**, A1133 (1965)
58. W. Kohn, P. Vashishta, in *Theory of the Inhomogeneous Electron Gas*, ed. by S. Lundqvist, N. March (Plenum, New York, 1983)
59. V.R. Shaginyan, Phys. Lett. A **249**, 237 (1998)
60. M.Y. Amusia, V.R. Shaginyan, Phys. Lett. A **269**, 337 (2000)
61. V.R. Shaginyan, JETP Lett. **68**, 527 (1998)

Chapter 5
Rearrangement of the Single-Particle Degrees of Freedom

Abstract In this chapter to additionally familiarize the reader to the fermion-condensation state, the properties of this new phase, related to the rearrangement of the Fermi surface in Fermi systems with strongly repulsive interaction, are analyzed. Beyond the FCQPT, the system is found to spontaneously separate into two subsystems, containing normal quasiparticles and the fermion condensate localized at the Fermi surface and associated with a completely flat portion of the spectrum of single-particle excitations. Superfluid correlations in systems with the fermion condensate are also considered. Applicability of the quasiparticle pattern to the analysis of the fermion condensation is demonstrated. Mechanisms that could promote this phase transition in real physical systems are discussed.

5.1 Introduction

The Landau Fermi liquid theory (LFL) deals with energy functionals $E_0[n(\mathbf{p})]$ in the functional space $[n]$ of quasiparticle distributions $n(\mathbf{p})$ located in $[n]$ between 0 and 1 [1]. This theory is based on assumption that the single-particle spectrum of a normal Fermi liquid is similar to that of an ideal Fermi gas, differing from the latter in the value of the effective mass M^*. In the homogeneous isotropic matter, we primarily analyze here that the LFL ground state quasiparticle distribution is the Fermi step function $n_F(p) = \theta(p - p_F)$. Quasiparticles fill the Fermi sphere up to the same radius $p_F = (3\pi^2\rho)^{1/3}$ (ρ is the density and p_F is the Fermi momentum) as noninteracting particles do (the Landau-Luttinger theorem [2]). From the mathematical point of view, in the LFL, the minimum of $E_0[n]$ is supposed to always lie at a boundary point n_F of the space $[n]$. This assumption remains valid as long as the necessary stability condition

$$\delta E_0 = \int (\varepsilon[\mathbf{p}, n(\mathbf{p}, T = 0)] - \mu)\delta n(\mathbf{p}, T = 0)\frac{d^3 p}{(2\pi)^3} \geq 0,$$

with $\varepsilon[\mathbf{p}, n(\mathbf{p})] = \delta E_0/\delta n(\mathbf{p})$ being the quasiparticle energy and μ, the chemical potential, which requires that the change of E_0 for any admissible variations of

M. Amusia and V. Shaginyan, *Strongly Correlated Fermi Systems*,
Springer Tracts in Modern Physics 283,
https://doi.org/10.1007/978-3-030-50359-8_5

n_F be non-negative, holds. It is the violation of this condition that results in the rearrangement of the distribution $n_F(\mathbf{p})$. The idea of the alteration of n_F leading to a multi-connected Fermi sphere $\tilde{n}_F(p)$ that corresponds to "floating" the minimum of E_0 on the surface of $[n]$ from n_F to $\tilde{n}_F$ was quite popular in the late 70s [3, 4]. But experimental data conflicting with the LFL theory predictions were scarce, and the interest in this problem has been disengaged. The discovery around 10 years ago of high-T_c superconductors sparked it again since the LFL theory fails to explain correctly their properties. As a result, new models of strongly correlated systems were developed (the so-called theories of Luttinger liquid [5–7]) in which the quasiparticle pole in the single-particle Green function G disappears.

On one hand, however, the same experimental data that contradict the LFL theory furnish decisive proof of the presence of well-defined single-particle pole in G [8–10]. On the other hand, within the quasiparticle approach there exist unexplored possibilities. Indeed, we get used to dealing only with the quasiparticle distributions with integer occupation numbers, equal to 1 or 0. Such an assumption is adequate if the Fermi surface is non-degenerate. Otherwise, the number of available free places there can exceed the number of quasiparticles (remember, e.g., the quantum Hall problem), and the quasiparticle filling ceases to be integer. Dealing with the extended class of distributions, we make the next step [11], replacing "swimming" on the surface of $[n]$ by "scuba diving" and suggest the minimum of $E_0[n]$ to be sought deep into $[n]$, say, at a point n_0. As a function of p, such a distribution appears to be continuous with $dn_0/dp \neq 0$ in a region Ω, called in [11] the fermion condensate (FC) region which fully covers the Fermi surface. As we shall see, it is this innocent, at first sight, difference, that results in a complete alteration of the LFL theory predictions [12].

Since the article [11] much work was done to clarify the situation. In particular, the rearrangement found in [11] was proven to be the topological phase transition [13]. At the same time, several authors [14, 15] raise objections against the FC. They claimed that i) the state with the FC can be obtained only within the Hartree-Fock approximation and only in the case of long-range forces, ii) the quasiparticle concept is inapplicable to this problem. Note, however, that the latter conclusion [14] has been obtained just in perturbation theory. As a matter of fact, beyond the phase transition point perturbation theory fails while the non-perturbative phenomenological Landau approach holds allowing to avoid difficulties related to the diagram summation. That is why the analysis of this article is conducted within this approach not resorting to the Hartree-Fock approximation. We shall also see that a strong enough velocity-dependence of the effective interaction f between quasiparticles is more necessary for the fermion condensation than the presence of any long-range components in f. The existence of FC not only contradicts laws of nature but is supported by theoretical and experimental observations [16, 17].

5.2 Basic Properties of Systems with the FC

5.2.1 The Case $T_c < T < T_f^0$

Suppose $E_0[n(\mathbf{p})]$ is explicitly known, so that the quasiparticle energy $\varepsilon[\mathbf{p}, n(\mathbf{p})] = \delta E_0/\delta n(\mathbf{p})$ may be calculated. Appealing to the expression [1]

$$S = -\mathrm{Tr} \int \left[n(\mathbf{p}, T) \ln n(\mathbf{p}, T) + \left(1 - n(\mathbf{p}, T)\right) \ln\left(1 - n(\mathbf{p}, T)\right)\right] \frac{\mathrm{d}^3 p}{(2\pi)^3}, \quad (5.1)$$

for the entropy S one can construct the free energy $F = E_0 - \mu N - TS$. The distribution $n(\mathbf{p}, T)$ is then to be determined from the variational condition $\delta F/\delta n(\mathbf{p}, T) = 0$ written explicitly as

$$\varepsilon[\mathbf{p}, n(\mathbf{p}, T)] - \mu(T) - T \ln\Big((1 - n(\mathbf{p}, T))/n(\mathbf{p}, T)\Big) = 0. \quad (5.2)$$

This condition is customarily rewritten in the standard Fermi-Dirac form

$$n(\mathbf{p}, T) = \left\{\exp\left[\frac{\varepsilon[\mathbf{p}, n(\mathbf{p}, T)] - \mu}{T}\right] + 1\right\}^{-1}. \quad (5.3)$$

However, in contrast to a gas-like system, the energy ε in (5.3) is a nontrivial functional of $n(\mathbf{p}, T)$, so this expression does not constitute a solution of (5.2). In homogeneous isotropic matter, the orthodox solution $n_F(p) = \theta(p - p_F)$ is obtained assuming $d\varepsilon(p, T = 0)/\mathrm{d}p$ to be positive near the Fermi surface, henceforth the T argument will be omitted at $T = 0$, and T-dependent corrections to M^*, $\varepsilon(p)$ and other quantities start with T^2-terms [18].

However, the point $T = 0$ is a singular point of the nonlinear equation (5.3). Then at low T various types of solutions can exist and they do exist. A new class of them found in [11] possesses a shape of the spectrum linear in T

$$\varepsilon(p, T) - \mu(T) = T\nu_0(p), \quad (5.4)$$

in an interval $p_i < p < p_f$. With this result, using (5.3) we are led to a new shape of the ground state quasiparticle distribution at $T = 0$

$$n_0(p) = (1 + \exp(\nu_0(p)))^{-1}, \qquad p_i < p < p_f. \quad (5.5)$$

Looking at the formula for the above necessary stability condition, we infer that deviating n_0 from n_F can survive at $T = 0$ only if the spectrum $\varepsilon(p)$ has a plateau

$$\varepsilon[p, n_0(p)] = \mu, \qquad p_i < p < p_f, \quad (5.6)$$

lying exactly at the Fermi surface, otherwise, increasing (or decreasing) $n_0(p)$ we gain energy. Note that this conclusion is consistent with (5.4). As a result, the Fermi surface of the three-dimensional system turns to a volume while in the two-dimensional one it becomes a surface instead of the Fermi line. Remembering the definition of $\varepsilon(p)$, one can write (5.6) as a variational condition [11]

$$\frac{\delta E_0}{\delta n(p)} = \mu \qquad p_i < p < p_f, \tag{5.7}$$

which follows from $\delta F/\delta n(p) = 0$ omitting the contribution of the last term to (5.2) negligible at $T = 0$ due to the continuity of $n_0(p)$ in the FC region. In finite systems, this equation is rewritten in the form

$$\frac{\delta E_0}{\delta n_\lambda} = \mu, \qquad \lambda \in \Omega \tag{5.8}$$

where λ is a necessary set of the single-particle numbers. Thus, the occupation numbers serve as variational parameters, and E_0 is lowered by varying them. Outside the FC region, $n_0 = n_F$ independent of variations of n_0. We postpone for a while the discussion about the validity of the quasiparticle formalism in this case and consider first properties of the system with the FC. We see that the FC effective mass M^*_{FC} found from the definition $(d\varepsilon(p)/dp)_F = p_F/M^*$ turns out to be infinite at $T = 0$. At low T, (5.4) yields

$$M^*_{FC}(T) \simeq p_f(p_f - p_i)/T. \tag{5.9}$$

Here, we used the estimate $dn_0(p)/dp \simeq (p_f - p_i)^{-1}$. Thus, we conclude that the effective mass $M^*_{FC}(T)$, as well as the density of states at the Fermi surface $N_F = \rho(\varepsilon = \mu, T) = p_F M^*(T)/\pi^2$ are enormously enhanced at low T. Further, inserting (5.9) into the LFL result $S_F = p_F M^* T/3$ for the entropy S one finds that $S(T \to 0) \neq 0$. The same result

$$S_0 = -\int_{p_i}^{p_f} [n_0(p)\ln n_0(p) + (1 - n_0(p))\ln(1 - n_0(p))]\frac{p^2 dp}{\pi^2} \tag{5.10}$$

can be obtained by inserting (5.5) into (5.1). The term S_0, proportional to the FC density ρ_c, looks like a residue entropy of a system with impurities.

Thus, we infer that the rearrangement results in the separation of the Landau quasiparticle system into two subsystems. The first contains normal quasiparticles with non-degenerate energies and effective masses $\sim M^*_F$ independent of T while the second, the FC, is formed by quasiparticles localized at the Fermi surface with $M^*_{FC} \sim T^{-1}$. This also reflects on the Landau-Luttinger theorem [2]. Now it reads [19]

$$\rho = \frac{p_i^3}{3\pi^2} + \int_{p_i}^{p_f} n_0(p)\frac{p^2 dp}{\pi^2}. \tag{5.11}$$

As seen from (5.11) the situation resembles that in Bose liquid. That is why the quasiparticles localized at the Fermi surface and contributing to the last term of (5.11) have been called the fermion condensate. Further, we suggest that the FC density ρ_c is small: $\rho_c/\rho << 1$.

Unfortunately, the FC wave function is much more intricate than that of Bose liquid. So far we accounted for it in terms of the occupation numbers $n_0(p)$. The impact of all the FC structure on properties of the systems is a question which remains to be answered. The first step in this direction is to involve the FC damping $\gamma_{FC}(T)$ at $T \neq 0$. Consider, as an example, the entropy $S(T)$ given by [2]

$$S = \frac{2}{T} \int \varepsilon \frac{\partial n_F(\varepsilon, T)}{\partial \varepsilon} \ln\Big(G_R(\mathbf{p}, \varepsilon)/G_R^*(\mathbf{p}, \varepsilon)\Big) \frac{d\varepsilon d^3 p}{(2\pi)^4 i}, \tag{5.12}$$

where $n_F(\varepsilon, T) = (1 + \exp(\frac{\varepsilon}{T}))^{-1}$ and $G_R(\varepsilon, \mathbf{p}) = (\varepsilon - \varepsilon(p, T) + i\gamma(T))^{-1}$ is the retarding single-particle Green function. The damping $\gamma_{FC}(T)$ is found to be linear in T (see Sect. 5.3) similar to the distortion of $\varepsilon(p)$. With these results, replacing ε/T in (5.12) by z one can verify that the integral over the FC region is T-independent and again one obtains $S \sim \rho_c$.

The existence of the constant term S_0 in a certain interval of T results in unusual low T-behavior of the thermal expansivity α given by [18]

$$\alpha = \frac{1}{3}\left(\frac{\partial \log V}{\partial T}\right)_P = -\frac{1}{3K}\rho \frac{\partial (S/\rho)}{\partial \rho}, \tag{5.13}$$

with $K = \partial P/\partial \rho$ being the compressibility of the system. In a normal Fermi system, $|\alpha_F(T)|$ is of the order of $p_F M_F^* T/\rho$ [18]. As seen from (5.10), $|\rho \partial (S/\rho)/\partial \rho| \sim \rho_c/\rho^2$. The compressibility K is insensitive to variations of ρ_c, omitting it in the ratio $|\alpha(T)/\alpha_F(T)|$ one finds

$$|\frac{\alpha_{FC}(T)}{\alpha_F(T)}| \simeq \frac{M_{FC}^*(T)}{M_F^*} \simeq \frac{\rho_c}{\rho} \frac{M}{M_F^*} \frac{\varepsilon_F^0}{T}. \tag{5.14}$$

The ratio (5.14) is greatly enhanced due to the factor ε_F^0/T. Therefore, the measurement of the thermal expansivity is able to probe very small concentrations of the FC. Note that the above enhancement also shows up in the difference $(C_P(T) - C_V(T))$, proportional to $T\alpha^2(T)$ [18], while the specific heat $C_P = TdS/dT$ is free of this enhancement since S_0 is T-independent.

Now let us turn to the solution of (5.7). In several models, it is solved analytically [19], in others, its solution is found numerically. There exists an extended parallelism between (5.7) and the well-known equation of statistical physics $\delta E_0/\delta \rho(\mathbf{r}) = \mu$: the density $\rho(\mathbf{r})$ corresponds to the distribution $n(\mathbf{p})$, the force $\mathbf{F} = \partial \varepsilon(\mathbf{r})/\partial \mathbf{r}$, to the group velocity $\partial \varepsilon(\mathbf{p})/\partial \mathbf{p}$, etc. Both of these equations have solutions for a wide class of two-body interactions. However, solutions of (5.7) should meet an additional restriction $n_0(p) \leq 1$ absent from statistical physics.

To give a better insight into this problem, consider the Landau identity [1, 20]

$$\frac{d\varepsilon(p)}{dp} = \frac{p}{M} + \frac{1}{3}\int f_1(p, p_1)\frac{\partial n(p_1)}{\partial p_1}\frac{p_1^2 dp_1}{\pi^2}, \tag{5.15}$$

where $f_1(p, p_1)$ is the first harmonic f_1 of the Legendre expansion of the Landau interaction $f(\mathbf{p}, \mathbf{p}_1)$ characterizing its velocity-dependent part and made up of the Hartree-Fock and correlation terms. Equation (5.15) is applicable to any distribution, including unusual $n_0(p)$. For such solutions, the left term of (5.15) necessarily vanishes when $p \in \Omega$. For the r.h.s. of (5.15) to vanish, the integral term must be negative. This is possible only if $f_1 > 0$, since $dn_0(p)/dp < 0$. Thus, no unusual solutions exist in the weak coupling limit, because f_1 is then dominated by the Fock term, implying $f_1 < 0$. As the interaction strength rises, the correlation contribution, which opposes the Fock term, increases more rapidly, changing the sign of f_1 and yielding $M_F^*/M > 1$. It can be traced to the calculation of the LFL value of $M^*[n_F]$ from (5.15)

$$M^*[n_F] = M/\left(1 - \frac{1}{3}F_1^0\right), \tag{5.16}$$

where $F_1^0 = p_F M f_1(p_F, p_F)/\pi^2$. For small $F_1^0 > 0$, vanishing of the r.h.s. of (5.15) would require an implausibly large average value of $|dn_0(p)/dp|$, and n_F is expected to survive. However, in real systems, the velocity-dependent component F_1^0 can be sufficiently large: for liquid He-3, one has $F_1^0 = 2$, in strongly correlated metals its value can be still greater. As seen from (5.16), it becomes infinite at $F_{1c}^0 = 3$ and then changes sign. The similar Pomeranchuk condition is violated when $M_F^* \to 0$ [19, 21].

If F_1^0 exceeds F_{1c}^0, the necessary stability condition is violated and the Fermi step is inevitably rearranged. A quantitative evaluation of the effect can be carried out in the case of small $\delta n(p) = n_0(p) - n_F(p)$ with the help of the formula [1]

$$\delta\varepsilon[\mathbf{p}, n(\mathbf{p})] = \mathrm{Tr}\int f[\mathbf{p}, \mathbf{p}_1, n_F]\delta n(\mathbf{p}_1)\frac{d^3 p_1}{(2\pi)^3}. \tag{5.17}$$

Upon inserting (5.17) into (5.6), we arrive at the following equation:

$$\varepsilon[p, n_F] + \int f_0[p, p_1, n_F]\delta n(p_1)\frac{p_1^2 dp_1}{\pi^2} = \mu, \tag{5.18}$$

where f_0 stands for the zero harmonic of the Landau interaction f. Equation (5.18) can be rewritten in the form $e(p) + (f_0 n) = \mu$ with $e(p) = \varepsilon(p) - (f_0 n_F)$. The function $e(p)$ has no specific properties, and often in solvable models, $e(p)$ is replaced by $\varepsilon_p^0 = p^2/2M$. Then (5.18) takes the Hartree-Fock-like form

$$\frac{p^2}{2M} + \int f_0[p, p_1, n_F]n(p_1)\frac{p_1^2 dp_1}{\pi^2} = \mu. \tag{5.19}$$

Being a phenomenological quantity, the interaction $f_0[n_F]$ has nothing to do with the similar Hartree-Fock term. Because it can be arbitrarily large, we are able to apply (5.18) to the FC problems. For the illustration, we present here some results for a solvable model [14] with infinite range forces when $f(\mathbf{p}_1, \mathbf{p}_2) = U\delta(\mathbf{p}_1 - \mathbf{p}_2)$ and (5.6) is reduced to

$$\mu = \varepsilon_p^0 + U n_0(p), \qquad p_i < p < p_f. \tag{5.20}$$

It has been studied in great detail [14, 22]. From (5.20), one obtains $n_0(p) = (\mu - \varepsilon_p^0)/U$. The boundaries p_i and p_f are determined from conditions $n_0(p_i) = 1$ and $n_0(p_f) = 0$. Straightforward calculations yield $\mu_{FC} - \mu_F = -U^2/48\varepsilon_F^0$ [22]. Similar results are obtained in other solvable models [19]. Thus, the Hartree-Fock filling always loses the contest.

Now, substituting (5.4) into (5.17), let us estimate the temperature T_f^0 at which the T-dependent corrections to (5.5) become of the order of 1. Having solved this equation, we have at hand the linear in T correction to $n_0(p)$. In some models, this correction attains considerable values already at $T_f^0 \simeq (p_f - p_i)^2/M << \varepsilon_F^0$. Another restriction comes from low T: a nonzero S_0 term stemming from degeneracy of $\varepsilon(p)$ begins to conflict with the Nernst theorem. In this case, at $T \to 0$ a lot of opportunities to violate the Pomeranchuk stability conditions emerge due to increasing density of states [19]. Let one of them occur at $T = T_c$. Then it will lift degeneracy of the spectrum $\varepsilon(p)$ that, in its turn, results in vanishing of $S(T = 0)$. We shall see how it happens, treating the superfluid phase transition as an example.

5.2.2 The Case $T < T_c$. Superfluid Systems with the FC

Let the effective particle-particle (p-p) interaction in the system with the FC be attractive. In principle, a strong repulsion in the particle-hole channel, inherented in the systems with FC, can coexist with an attraction in the p-p channel. It is enough to recall the electron-electron interaction arising in solids due to the phonon exchange. In superfluid liquid, F is a functional of two order parameters $n(\mathbf{p}) =< a^+_{\mathbf{p},\beta} a_{\mathbf{p},\beta} >$ and $\kappa(\mathbf{p}) =< a^+_{\mathbf{p},\beta} a^+_{-\mathbf{p},-\beta} >= \sqrt{n(\mathbf{p})(1 - n(\mathbf{p}))}$. In this subsection, we outline some features of superconductivity facilitated by the FC state, while the FC theory of superconductivity is presented in Chap. 6. At $T = 0$, the condition $\delta F/\delta n(p) = 0$ is reduced to

$$\xi(p) - \frac{\Delta(p)(1 - 2n(p))}{\sqrt{n(p)(1 - n(p))}} = 0. \tag{5.21}$$

Here $\xi(p) = \varepsilon(p) - \mu$ with $\varepsilon(p) = \delta E_0/\delta n(p)$, while $\Delta(p) = -\delta F/\delta\kappa(p)$,

$$n(p) = \frac{1}{2}\left[1 - \frac{\xi(p)}{E(p)}\right], \quad E(p) = \sqrt{(\xi(p)^2 + \Delta^2(p)}, \quad \kappa(p) = \frac{\Delta(p)}{2E(p)}. \tag{5.22}$$

In the BCS theory, $\kappa(p) \to 0$ while $\Delta \to 0$. However, there exist entirely different solutions with $\kappa \neq 0$ even if $\Delta = 0$. Indeed, for $\Delta = 0$, $\kappa \neq 0$, the second term in (5.21) vanishes and again we arrive at (5.6). Thus, such solutions describe superfluid states of the system with the FC [11, 14, 22]. Since $\Delta/\varepsilon_F^0 << 1$, one can write $F[n,\kappa] = E_0[n] - \mu N + \delta E_s[\kappa]$ where

$$\delta E_s = \frac{1}{2}\int\int \lambda(\mathbf{p},\mathbf{p}_1)\kappa(\mathbf{p})\kappa(\mathbf{p}_1)\frac{d^3 p d^3 p_1}{(2\pi)^6}. \tag{5.23}$$

Then in the case of s-pairing, the gap $\Delta(p)$ is given by

$$\Delta(p) = -\int \lambda_0(p,p_1)\sqrt{n(p_1)(1-n(p_1))}\frac{p_1^2 dp_1}{4\pi^2}, \tag{5.24}$$

where λ_0 is the zero harmonics of $\lambda(\mathbf{p},\mathbf{p}')$ over the angle between $\mathbf{p}$ and $\mathbf{p}'$. Further, we suppose that $T_c << T_f^0$. Then pairing forces influence little FC, and replacing in (5.24) $n(p,T)$ by $n_0(p)$, we obtain

$$\Delta_0(p) = -\int \lambda_0(p,p_1)\sqrt{n_0(p_1)(1-n_0(p_1))}\frac{p_1^2 dp_1}{4\pi^2}. \tag{5.25}$$

Substituting this result into (5.22) one finds that in the FC region

$$\xi_0(p) = \frac{\Delta_0(p)(1-2n_0(p))}{\sqrt{n_0(p)(1-n_0(p))}},\ \kappa_0^2(p) = n_0(p)(1-n_0(p)),\ E_0(p) = \frac{|\Delta_0(p)|}{2\kappa_0(p)}. \tag{5.26}$$

The next step is in adding integration outside of the FC region, e.g.,

$$\Delta_1(p) = \Delta_0(p) - \int \lambda_0(p,p_1)\theta(p_1)\frac{\Delta_0(p_1)}{E_0(p_1)}\frac{p_1^2 dp_1}{4\pi^2}, \tag{5.27}$$

where $\theta(p) = 0$ within the FC region. The integral term, logarithmical divergent in the BCS theory, now is a correction of the order of $\lambda_0^2 \ln \lambda_0$ to $\Delta_0 \sim \lambda_0$ that results in the suppression of the isotopic effect in T_c. We see that the order parameter $\kappa_0(p)$ differs from zero even if $\Delta = 0$. In other words, even in the absence of pairing forces, if the situation for the fermion condensation prevails, the system undergoes a transition to the state with spontaneous breaking of gauge invariance [22] that is traditionally associated with superconductivity. We also see that the gap Δ is linear in the coupling constant λ_0 in contrast to the ordinary BCS-case where the gap is exponentially small. Since $T_c \sim \lambda_0$ [11, 14, 20, 22] even a weak, e.g., phonon attraction in the p-p channel with $L = 0$ is sufficient to provide rather high T_c. On the other hand, when λ_0 turns out to be repulsive, the components λ_L with $L = 1$ or 2 can be attractive generating pairing with nonzero L. Neutron stars present an example of such an attraction: s-pairing dies out there at $p_F \simeq 1.7\ fm^{-1}$, while p-pairing survives till $p_F \simeq 4\ fm^{-1}$ [23]. Usually, the gap's magnitude with $L \neq 0$

is suppressed compared to the s-case. There is no such a suppression in the systems with the FC. As for the entropy S of superfluid Fermi liquid, (5.1) is to be modified, as usual, by replacing $(\varepsilon(p) - \mu) \to E(p)$. Then S vanishes at $T = 0$ in accordance with the BCS theory. At the same time, serious deviations from the BCS results emerge when calculating the pairing correction ΔE_{FC} to E_0. Applying the Landau formula for the change of E_0 due to the variation $\delta n(\mathbf{p}, T) = n(\mathbf{p}, T) - n_0(\mathbf{p})$ of the occupation numbers and adding the superfluid term (5.23), we obtain the following result:

$$\Delta E_{FC} = \int (\varepsilon(\mathbf{p}, n_0(\mathbf{p})) - \mu)\delta n(\mathbf{p})\frac{d^3p}{(2\pi)^3} + \delta E_s. \tag{5.28}$$

In the usual BCS case, the first term and the second one on the right-hand side of (5.28) almost entirely compensate each other and, hence, $|\Delta E_{BCS}| \sim \Delta^2/\varepsilon_F^0$. In the system with the FC, the first term in (5.28) turns out to be zero. As a result, we are left with δE_s that yields

$$\Delta E_{FC} = -\frac{1}{2}\int_{p_i}^{p_f} \Delta_0(p)\sqrt{n_0(p_1)(1 - n_0(p_1))}\frac{p^2 dp}{2\pi^2}. \tag{5.29}$$

Accounting for the Landau interaction term $(f\delta n\delta n)$ in (5.28) does not change this result. The estimate of (5.29) is: $\Delta E_{FC} \sim \rho_c \Delta_0$ in contrast to ΔE_{BCS}, proportional to Δ^2. This difference leads to an essential increase of the critical magnetic fields H_c, destroying superfluidity.

5.3 Validity of the Quasiparticle Pattern

5.3.1 Finite Systems

The quasiparticle pattern, we proceeded from, is applicable provided the width of γ of the relevant single-particle states (in our case, the FC) and does not exceed the energy $\varepsilon(p)$. First of all, this pattern is valid for finite systems where low-lying states do not decay at all. To give an example of the fermion condensation in finite systems, let us consider a spherical atomic nucleus with closed shells and add to it $K >> 1$ particles [24]. The change in the energy E_0 of the system is given by [1]

$$\delta E_0 = \Sigma_{\lambda,\lambda_1}\left(\varepsilon_\lambda^o \delta n_\lambda \delta_{\lambda,\lambda_1} + \frac{1}{2}\Gamma_{\lambda,\lambda}^{\lambda_1,\lambda_1}\delta n_\lambda \delta n_{\lambda_1}\right), \tag{5.30}$$

the levels being degenerate with respect to the magnetic quantum number m. The energies ε_λ^o (where $\lambda = n, l, j, m$) are calculated in the nucleus with closed shells. The interaction amplitude Γ contains all rescattering processes occurring in the final nucleus. It does not differ considerably from the static scattering amplitude calculated in infinite nuclear matter. The case with only two terms with $j_i >> 1$ in the (5.30) for which $n_1 + n_2 = K < (2j_{\min} + 1)$ is especially clear. Then the single-particle energies ε_i are given by

$$\varepsilon_1 = \frac{\delta E_0}{\delta n_1} = \varepsilon_1^o + D_1 n_1 + U n_2, \quad \varepsilon_2 = \frac{\delta E}{\delta n_2} = \varepsilon_2^o + U n_1 + D_2 n_2, \tag{5.31}$$

where $i = 1, 2$ and $D_i = \Gamma_{ii}^{ii}$ and $U = \Gamma_{22}^{11} = \Gamma_{11}^{22}$. Note that the numbers n_1 and n_2 are integers in the usual case of the Hartree-Fock filling. Let us choose $\varepsilon_1^o < \varepsilon_2^o$. As seen from (5.31), the normal Hartree-Fock occupation $n_2 = 0, n_1 = K$ persists if $\varepsilon_2^o + KU > \varepsilon_1^o + KD_1$. If this inequality fails, one can try another reasonable occupation $n_1 = 0, n_2 = K$. But it also fails if $\varepsilon_2^o + KD_2 > \varepsilon_1^o + UK$. Thus, any Hartree-Fock filling fails provided

$$K(U - D_2) < (\varepsilon_2^o - \varepsilon_1^o) < K(D_1 - U). \tag{5.32}$$

In this case, we are forced to resort to the variational condition (5.8) which implies $\varepsilon_1 = \varepsilon_2$. Then after simple algebra one finds [24]

$$n_1 = [\varepsilon_2^o - \varepsilon_1^o + K(D_2 - U)]/Z \quad n_2 = [\varepsilon_1^o - \varepsilon_2^o + K(D_1 - U)]/Z, \tag{5.33}$$

with $Z = D_1 + D_2 - 2U$ being positive. We see that both numbers n_1 and n_2 differ from 0 simultaneously if (5.32) holds. The energy gained due to the rearrangement is $\delta E_{FC} = -[\varepsilon_1^o - \varepsilon_2^o + K(D_1 - U)]^2/2Z$. The analysis can be easily extended to a larger number of single-particle levels. The energy δE_{FC} is relatively small but this effect, in principle, could promote to stabilize heavy and superheavy nuclei. The main feature of the solution obtained is the forced "collapse" of distances between single-particle energies ε_i of the levels involved. It has not been assumed anywhere in the analysis that the input parameters are small, and if this is so, then the Fermi liquid approach rather than the Hartree-Fock method is applicable to the investigation of the fermion condensation. The results obtained can be applied to a completely different problem if one associates n_i in (5.31) with occupation numbers for the state i. Then omitting the diagonal contribution D_i and taking $U > 0$, we arrive at the Coulomb gap problem and get ordinary results: the density of states at the Fermi surface falls [25]. The collapse of the distances between different levels happens if we add the repulsion on the same site, i.e., suggesting $D_i \neq 0$, so to provide the fulfillment of the criterion (5.32). This case, having much in common with the Hubbard model, remains unexplored.

5.3.2 Macroscopic Systems

Turning to macroscopic systems, where T exceeds the splitting between levels, it is worth noting that the quasiparticle picture of the fermion condensation holds for superfluid systems with the FC because of the existence of the gap in the spectrum $E(p)$. Thus, only the case $T_c < T < T_f^0$ remains to be analyzed. In this case, calculations have been done in [14] and the width $\gamma_{FC}(T)$ turned out to diverge at low T as $1/T$. This result implies the conclusion $\gamma_{FC} >> \varepsilon_{FC}(p)$ that, seemingly, destroys the quasiparticle pattern of the phenomenon. Unfortunately, accounting for the huge enhancement of the effective mass $M^*_{FC} \sim T^{-1}$; the author of [14] neglected a reciprocal suppressing effect in the scattering amplitude Γ, responsible for the quasiparticle decay. Here, we briefly outline the results of the improved evaluation of the width $\gamma_{FC}(T)$ [22]. To find $\gamma(T)$ in the three-dimensional system, we employ the well-known LFL formula [26]

$$\gamma \sim M^{*3}T^2 < \frac{W(\theta,\phi)}{\cos(\theta/2)} > . \tag{5.34}$$

Here $W(\theta,\phi) \sim |\Gamma^2|$ is the transition probability depending on the angle θ between the vectors $\mathbf{p}_1$ and $\mathbf{p}_2$ of the incoming particles and the angle ϕ between the planes defined by the vectors $\mathbf{p}_1$, $\mathbf{p}_2$ and $\mathbf{p}'_1$, $\mathbf{p}'_2$. The brackets $< ... >$ denote angular averaging. In perturbation theory, Γ coincides with the interaction potential V. The result $\gamma_{FC} \sim T^{-1}$[14] follows from (5.34) substituting there (5.9) and $\Gamma = V$. However, in the systems with the FC, the difference between Γ and V is as enormous as the difference between M^*_{FC} and M^*_F. To clarify, let us consider the velocity-independent scalar part of Γ related to the Landau amplitude f as follows: [22]

$$\Gamma(q,\omega,T) = f(q)/(1 - f(q)\chi_L(q,\omega,T)), \tag{5.35}$$

the Lindhard function χ_L being evaluated for particles with the effective mass M^*. At $\omega \sim T$, both of the real and imaginary parts of $\chi_L(q,\omega \sim T)$ are of the same order of magnitude: $|\mathrm{Re}\chi_L(q,\omega \sim T)| \sim N_F = p_F M^*_{FC}/\pi^2 \sim T^{-1}$, and $|\mathrm{Im}\chi_L(q,\omega \sim T)| \sim M^{*2}\omega/\pi q \sim T^{-1}$. This fact makes it possible to neglect 1 in the denominator in (5.35). As a result, one finds that $|N_F(T)\Gamma(T)| \sim 1$. This relation holds for any strongly correlated system. Inserting it into (32) together with (5.9), we obtain

$$\gamma_{FC}(T) \sim \frac{M^{*3}_{FC}T^2}{M^{*2}_{FC}} \sim T\frac{\rho_c}{\rho}. \tag{5.36}$$

We see that $\gamma_{FC}(T)/|\varepsilon_{FC}(p,T)| \sim \rho_c/\rho$, and therefore the quasiparticle pattern survives as long as $\rho_c << \rho$. Accounting for other components of Γ does not change this conclusion [22]. The estimate (5.36) can also be obtained otherwise [24] from the formula for the width γ, referring to the decay of a quasiparticle with the energy $\varepsilon(p)$ and momentum $p > p_F$ [26],

$$\gamma = 2\pi \int |\Gamma(q,\omega)|^2 n_{k,\sigma}(1 - n_{\mathbf{k}+\mathbf{q},\sigma})\delta(\omega_0 - \omega)\frac{d^3k\, d^3q}{(2\pi)^6}, \tag{5.37}$$

where $\omega = \varepsilon(\mathbf{k}+\mathbf{q}) - \varepsilon(k)$ is the transferred energy, and $\omega_0 = \varepsilon(p) - \varepsilon(\mathbf{p}-\mathbf{q}) \sim T$ is the decrease in the quasiparticle energy as a result of rescattering processes. Therefore, a quasiparticle with the energy $\varepsilon(p)$ decays predominantly into a quasi-hole $\varepsilon(k)$ and two quasiparticles $\varepsilon(\mathbf{p}-\mathbf{q})$ and $\varepsilon(\mathbf{k}+\mathbf{q})$. It should also be kept in mind that $\mathbf{q}$ must satisfy the condition: $p > |\mathbf{p}-\mathbf{q}| > p_F$ since a quasiparticle loses momentum and energy. Integration over d^3k in (5.37) gives $\mathrm{Im}\chi_L(q,\omega)$ [24]. As a result, one finds

$$\gamma_{FC} = -\int\int |\Gamma(q,\omega_0)|^2 M^{*2}\omega \frac{q^2 \mathrm{d}q\mathrm{d}x}{2\pi^3}. \tag{5.38}$$

To evaluate this integral, we denote $t = \varepsilon(p) - \varepsilon(\mathbf{p}-\mathbf{q}) \leq T$ and express the angular variable x in terms of t that yields $dx = M^*\mathrm{d}t/pq$. Integration over t contributes an additional factor of T in (5.38), and once again we obtain

$$\gamma_{FC}(T) \sim |\Gamma|^2 M^{*3} T^2 \sim T\frac{\rho_c}{\rho}. \tag{5.39}$$

We see again that in the case $\rho_c/\rho << 1$, the FC lifetime is long in spite of the enormous density of states, and the quasiparticle pattern of the fermion condensation survives, in contrast to the conclusion of [14]. Thus, as often happens, a disaster occurs in perturbation theory only, disappearing in more accurate treatment.

5.4 Interplay Between Fermion Condensation and Density Wave Instability

Here, we discuss mechanism that could promote the FC in real systems: strongly correlated metals, dense neutron matter, and liquid He-3. We shall see that an unlimited rise of the effective mass precedes the appearance of a density wave instability in homogeneous Fermi liquid [27]. The basic idea of this effect is simple. Let the density wave instability emerge with a wave vector $q_c \geq 2p_F$, at a critical density ρ_i [19, 28, 29]. When ρ is approaching ρ_i, the interaction amplitude $\Gamma(\mathbf{p}_1,\mathbf{p}_2,\mathbf{p}_1,\mathbf{p}_2;\omega=0)$ can be written in the form

$$\Gamma(\mathbf{p}_1,\mathbf{p}_2,\mathbf{p}_1,\mathbf{p}_2;\omega=0) \sim \frac{-1}{\omega^2_{\mathrm{min}}(\rho) + b^2[|\mathbf{p}_1-\mathbf{p}_2| - q_c)]^2}, \tag{5.40}$$

where $\omega^2_{\mathrm{min}}(\rho_i) = 0$ is implied. Multiplying Γ by $x = \mathbf{p}_1\mathbf{p}_2/p_1p_2$ and integrating over x from -1 to 1, one finds the contribution of (5.40) to F_1^0

$$F_1^0 \sim \frac{1}{|1 - q_c/2p_F|}. \tag{5.41}$$

The density wave instability sets in with $q_c > 2p_F$ and as the input parameters change it approaches the point $2p_F$. At the same time, M^*, as follows from (5.16), tends to infinity. Experts in theory of superfluid He-3 can easily reveal a similarity between these arguments and those by Anderson and Brinkman, [30] who calculated the paramagnon contribution to the p-p channel attraction in this system. We can arrive at the same result (5.41) more rigorously by evaluating the effective mass $M^* = p_F(d\varepsilon(p)/dp)^{-1}$ microscopically [27]. Let us briefly outline key points of this article where we used a version of the microscopic approach developed in [19]. The formula $\varepsilon(p) = \delta E_0/\delta n(p)$ allows us to evaluate M^* in terms of the ground state energy E_0 given by [26]

$$E_0 = T_0 - \int V(q)[\mathrm{Im}\chi(q,\omega) + 2\pi\rho\delta(\omega)]\frac{d^3q\, d\omega\, dg}{(2\pi)^4 g}, \tag{5.42}$$

where T_0 is the kinetic energy of noninteracting particles, $\chi(q,\omega)$ is the linear response function of the system

$$\chi(q,\omega) = \phi(q,\omega)\chi_0(q,\omega) = \chi_0(q,\omega)/(1 - R(q,\omega,g)\chi_0(q,\omega)), \tag{5.43}$$

where χ_0 is the linear response function of noninteracting free particles. In perturbation theory, the effective interaction $R = V$. For its evaluation in the strongly correlated systems, special methods are elaborated [19]. The integration over frequency ω goes along the real axis from 0 to $+\infty$, while the integration over g goes from 0 to a real value of g_0. Upon varying E_0 with respect to $n(\mathbf{p})$ and after some tedious algebra, we are led to the formula

$$\frac{1}{M^*} = \frac{1}{M}$$

$$-\left(\frac{\partial}{2p\,\partial p}\right)_{p_F} \mathrm{Im}\int V(q)\left[\phi^2(q,\omega)\frac{\delta\chi_0(q,\omega)}{\delta n(p)} - \chi^2(q,\omega)\frac{\delta R(q,\omega,g)}{\delta n(p)}\right]\frac{dg d^3q\, d\omega}{g(2\pi)^4}. \tag{5.44}$$

Using the explicit form of χ_0 [26], one obtains

$$I_0(q,\omega) = \left(\frac{\partial}{\partial p}\frac{\delta}{\delta n_p} Im\chi_0(q,\omega)\right) = -\frac{4\pi}{p_F^2}\delta(p_F - \mid \mathbf{p}+\mathbf{q}\mid)\delta(\omega)\Big(\mathbf{p}(\mathbf{p}+\mathbf{q})\Big)|_{p=p_F}. \tag{5.45}$$

It is seen that I_0 is a singular function. This singular function will make a divergent contribution to (5.44) if it meets another singular function. Otherwise, the four-dimensional integral takes away the two-dimensional singularity, and the first term in the brackets (5.44) will be finite and comparable to the rest of the terms. Such a singular function, $\phi(q,\omega)$, emerges in the vicinity of the density wave instability since the pole structure of ϕ coincides with that of Γ. It can be shown that the second term on the r.h.s. of (5.44) has no such singularity, and therefore its contribution is negligible. Omitting it and introducing as usual the momentum transfer $q = \sqrt{2p_F^2(1-x)}$, we find

$$\frac{1}{M^*} = \frac{1}{M} + \frac{1}{4\pi^2 p_F} \int_0^{2p_F} \int_0^{g_0} V(q) \frac{q(1 - q^2/2p_F^2)dqdg}{[1 - R(q, \omega = 0, g)\chi_0(q, \omega = 0)]^2}. \quad (5.46)$$

In the weak coupling limit, the well-known Gell-Mann [31] result for M^* can be obtained by substituting here $V(q) = 4\pi e^2/q^2$. On the other hand, in the vicinity of the critical point ρ_{dwi} the integral (5.46), being negative, diverges generating the divergence of M^*. Since the fermion condensation begins as soon as the integral on the r.h.s. of (5.46) cancels the term $1/M$, it is clear that it must take place long before the density wave instability manifests itself. It is worth noting that the charge-density-wave instability emerges in 3D electron gas at $r_s \simeq 33$, and $q_c/2p_F \simeq 1.1$ [28, 29] while Wigner crystallization, at $r_s \sim 100$ [32].

Now let us consider quasi-two-dimensional (2D) liquid. This means that electrons move within 2D plane, while these cannot move along perpendicular direction. In that case, we put $V(q) = 2\pi e^2/q$. The path from 3D liquid to 2D one is clear since the form of the singular function I_0 is preserved. To get the final result, we have to omit $\sin\theta$ in the phase volume of (5.44), thus obtaining

$$\frac{1}{M^*} = \frac{1}{M} + \frac{1}{4\pi^2 p_F^2} \int_0^{2p_F} \int_0^{g_0} \frac{V(q)(1 - q^2/2p_F^2)}{[1 - R(q, \omega = 0, g)\chi_0(q, , \omega = 0)]^2} \frac{qdqdg}{\sqrt{1 - q^2/4p_F^2}}. \quad (5.47)$$

We see that the fermion condensation can easily occur in 2D liquid than in the 3D case due to the presence of the factor $1/\sqrt{1 - q^2/4p_F^2} \geq 1$. On the other hand, the density wave instability in the 2D case is also expected to take place "more easily" as compared to the 3D case. Indeed, the Monte Carlo calculations predict Wigner crystallization of 2D electron gas at $r_s \simeq 37$ [33]. But before crystallization the density waves should take place, and that is the case: the charge-density-wave instability is shown to occur at $r_s \sim 10$ and $q_c \simeq 2p_F$, in parallel electron layers separated by potential barriers [34]. Thus, as we have shown above, the fermion condensation will inevitably arise, being produced by the density wave instability. Our estimation shows that the fermion condensation in a 3D electron gas takes place at $r_s \simeq 21$, and estimation in the 2D case gives the value $r_s \simeq 8$.

Thus, we see that the interplay between different phase transitions in this problem is quite intricate. On one hand, the infinite rise of the density of states at $T \to 0$, associated with the formation of the FC, results in the emergence of any phase transition capable of lifting degeneracy of the FC spectrum at $T = 0$. On the other hand, the phase transition, related to the density wave instability, promotes the formation of the FC itself.

5.5 Discussion

First of all, let us return to objections against the FC [14]. We have already shown that the FC solutions of the Landau equations are obtained not resorting to the Hartree-Fock approximation. The next point pertains to screening effects that should have killed the fermion condensation. However, we have seen in Sect. 5.3.2 that screening existing at the low momentum transfer q is transformed to antiscreening if q attains values around $2p_F$ and the input parameters get into the density wave instability region that, in its turn, creates conditions for the FC formation.

We discussed the validity of the quasiparticle approach in great detail. We have seen that in the 3D case, the width of the FC quasiparticles is small compared to their energy if $\rho_c/\rho << 1$, so that this approach is applicable. As for one-dimensional systems, the microscopic calculations [19] demonstrate that in this case, the effective mass diverges even if one takes the coupling constant to be arbitrarily small. This divergence gives rise to a complete rearrangement of the spectrum $\varepsilon(p)$ and, hence, the quasiparticle formalism fails in the 1D case. At the same time, such a divergence in systems with $D \geq 2$ may exist only in an extraordinary case of a singular interaction when the 2D problem can be approximately reduced to the 1D one [15]. Moreover, the experimental data give evidence for the presence of the quasiparticle poles in all strongly correlated Fermi liquids including high T_c superconductors [8–10]. Therefore, for example, Bosonization theories, where the quasiparticles do not exist, are irrelevant when dealing with these data.

When conducting the analysis of these data, anisotropy effects created by the crystal lattice have to be taken into account. There are several state-of-the-art codes for computation of the electronic structure of solids that proceed on the basis of the LFL theory. The best of these codes reproduce experimental data on many compounds rather well. However, these computational approaches fail to explain the emergence of the flat portion of the spectrum $\varepsilon(\mathbf{p})$ near the Fermi surface. The approach outlined here copes with this problem so that "the patient" is rather alive than dead. Certainly, the FC solution must be anisotropic in real crystals. As it was shown first in [35], in the Brillouin zone, there are points which are most liable to the FC formation. These are so-called van Hove points at which the quasiparticle group velocity vanishes due to symmetry requirements. Their presence makes the formation of the FC easier that could result in lowering the critical parameters needed for the FC emergence. Experimental data obtained in angular resolved photoemission studies (ARPES) of strongly correlated metals confirm the conclusion: flattening in $\varepsilon(\mathbf{p})$ exhibits itself only close to the van Hove points.

As for the existence of the FC in other systems, it could even exist in such unusual systems as fermions confined in vortex cores in superfluid Fermi liquids [36]. On the other hand, we have seen that well before forming crystal structures a Fermi liquid becomes unstable against small-amplitude density fluctuations, which could promote the formation of the FC. Liquid 2D ^{3}He exhibits the NFL behavior if its density is increased and tends to the critical, see Sect. 12.2. Much greater compressions are needed to solidify neutron matter [37, 38]. Therefore, the fermion condensation may display itself not only in dilute electron gas but also in liquid 2D ^{3}He, dense neutron, and nuclear matter. Little is known about solidification of quark matter, but if it undergoes this phase transition, then the fermion condensation could precede it.

We have carefully considered rather unusual properties of strongly correlated Fermi systems which could undergo the phase transition related to the rearrangement of the single-particle degrees of freedom and called the fermion condensation. We have argued that this phenomenon could take place in any Fermi liquid (electron gas, nuclear matter, neutron matter, HF compounds, etc.). Thus, the fermion condensation can be treated as a widespread phenomenon rather than an uncommon and "anomalous" effect. Then, the predicted charge density waves [27] that make emerge FC are observed in many HF compounds like HF metals and superconductors, see, e.g., [39].

To sum up, we conclude that the fermion-condensate state presents a good foundation for considering real strongly correlated Fermi systems beyond the phase transition point related to the collapse of the single-particle degrees of freedom in normal Fermi liquid, thus transforming the liquid into the new state of matter.

References

1. L.D. Landau, JETP **30**, 1058 (1956). JETP **32**, 59 (1957)
2. A.A. Abrikosov, L.P. Gor'kov, I.E. Dzialoshinskii, *Methods of Quantum Field Theory in Statistical Physics* (Engelwood Cleffs, 1963)
3. M. de Llano, J.P. Vary, Phys. Rev. C **19**, 1083 (1979)
4. M. de Llano, A. Plastino, J.G. Zabolitsky, Phys. Rev. C **20**, 2418 (1979)
5. F.D. Haldane, Phys. Lett. A **81**, 153 (1981)
6. H. Chen, D. Mattis, Modern. Phys. Lett. B **7**, 723 (1993)
7. Y. Ren, P.W. Anderson, Phys. Rev. B **48**, 16662 (1993)
8. Z.-X. Shen, D.S. Dessau, Phys. Rep. **253**, 1 (1995)
9. D.H. Lu, M. Schmidt, T.R. Cummins, S. Schuppler, F. Lichtenberg, J.G. Bednorz, Phys. Rev. Lett. **76**, 4845 (1996)
10. T. Yokoya, A. Chainani, T. Takahashi, H. Katayama-Yoshida, M. Kasai, Y. Tokura, Phys. Rev. B **54**, 13311 (1996)
11. V.A. Khodel, V.R. Shaginyan, JETP Lett. **51**, 553 (1990)
12. V.A. Khodel, J.W. Clark, V.R. Shaginyan, Solid Stat. Commun. **96**, 353 (1995)
13. G.E. Volovik, JETP Lett. **53**, 208 (1991)
14. P. Nozières, J. Phys. France I **2**, 443 (1992)
15. T.M. Rice, D.V. Khveshchenko, R. Hlubina, Phys. Rev. B **48**, 10766 (1993)
16. V.R. Shaginyan, M.Y. Amusia, A.Z. Msezane, K.G. Popov, Phys. Rep. **492**, 31 (2010)
17. M.Y. Amusia, K.G. Popov, V.R. Shaginyan, W.A. Stephanowich, *Theory of Heavy-Fermion Compounds*. Springer Series in Solid-State Sciences, vol. 182 (Springer, Berlin, 2015)
18. L.D. Landau, E.M. Lifshitz, *Statistical Physics* (Pergamon Press, Oxford, 1980)
19. V.A. Khodel, V.R. Shaginyan, V.V. Khodel, Phys. Rep. **249**, 1 (1994)
20. V.A. Khodel, J.W. Clark, V.R. Shaginyan, Solid St. Commun. **96**, 353 (1995)
21. I.Y. Pomeranchuk, JETP **35**, 524 (1958)
22. J. Dukelsky, V.A. Khodel, P. Schuck, V.R. Shaginyan, Z. Phys. B **102**, 245 (1997)
23. T. Takatsuka, Prog. Theor. Phys. **71**, 1432 (1984)
24. V.A. Khodel, V.R. Shaginyan, P. Schuck, JETP Lett. **63**, 752 (1996)
25. A.L. Efros, B.I. Shklovskii, J. Phys. C **8**, L49 (1975)

26. D. Pines, P. Nozières, *Theory of Quantum Liquids* (W.A. Benjamin, New York, 1966)
27. V.A. Khodel, V.R. Shaginyan, M.V. Zverev, JETP Lett. **65**, 253 (1997)
28. H.B. Shore, E. Zaremba, J.H. Rose, L. Sander, Phys. Rev. B **18**, 6506 (1978)
29. M. Levy, J.P. Perdew, Phys. Rev. B **48**, 11638 (1993)
30. P.W. Anderson, W.F. Brinkman, Phys. Rev. Lett. **30**, 1108 (1970)
31. M. Gell-Mann, Phys. Rev. **106**, 369 (1957)
32. D.M. Ceperley, B.J. Alder, Phys. Rev. Lett. **45**, 566 (1980)
33. B. Tanatar, D.M. Ceperley, Phys. Rev. B **39**, 5005 (1989)
34. L. Swierkowski, D. Neilson, J. Szymański, Phys. Rev. Lett. **67**, 240 (1991)
35. G.E. Volovik, JETP Lett. **59**, 830 (1994)
36. G.E. Volovik, JETP Lett **63**, 729 (1996)
37. D.M. Ceperley, G.V. Chester, M.H. Kalos, Phys. Rev. B **16**, 3081 (1971)
38. G. Baym, C.J. Pethick, Ann. Rev. Nucl. Sci. (1975)
39. R. Arpaia, S. Caprara, R. Fumagalli, G. De Vecchi, Y.Y. Peng, E. Andersson, D. Betto, G.M. De Luca, N.B. Brookes, F. Lombardi, M. Salluzzo, L. Braicovich, C. Di Castro, M. Grilli, G. Ghiringhelli, Science **365**, 906 (2019)

Chapter 6
Topological FCQPT in Strongly Correlated Fermi Systems

Abstract As high-T_c superconductors are represented primarily by 2D layered structures, in Sect. 6.1 we discuss the superconducting state of a 2D liquid of heavy electrons, and within the framework of Gor'kov microscopic equations construct Green's functions of FC state. On the other hand, our study can easily be generalized to 3D case. To show that there is no fundamental difference between the 2D and 3D cases, we derive Green's functions for 3D case in Sect. 6.1.1. In Sect. 6.2, we consider the dispersion law and lineshape of single-particle excitations. Section 6.3 is devoted to the behavior of heavy-electron liquid with FC in magnetic field. In Sect. 6.4, we analyze conditions which lead to the emergence of FC in Fermi systems composed of different fermions such as ^{3}He atoms and electrons.

6.1 The Superconducting State with FC at $T = 0$

As we have seen in Sect. 4.1.1, the ground state energy $E_{gs}(\kappa(\mathbf{p}), n(\mathbf{p}))$ of a 2D electron liquid is a functional of the superconducting order parameter $\kappa(\mathbf{p})$ and of the quasiparticle occupation numbers $n(\mathbf{p})$. This energy is determined by the well-known Bardeen-Cooper-Schrieffer (BCS) equations and in the weak coupling superconductivity theory is given by [1–3]

$$E_{gs}(\kappa(\mathbf{p}), n(\mathbf{p})) = E(n(\mathbf{p})) + \lambda_0 \int V(\mathbf{p}_1, \mathbf{p}_2) \times \kappa(\mathbf{p}_1)\kappa^*(\mathbf{p}_2)\frac{d\mathbf{p}_1 d\mathbf{p}_2}{(2\pi)^4}. \quad (6.1)$$

Here

$$\kappa(\mathbf{p}) = \sqrt{n(\mathbf{p})(1 - n(\mathbf{p}))} \quad (6.2)$$

with $n(\mathbf{p}) = v^2(\mathbf{p})$, $u(\mathbf{p})$ and $v(\mathbf{p})$ are parameters of the Bogolubov transformation, which obey the normalization rule $v^2(\mathbf{p}) + u^2(\mathbf{p}) = 1$ for fermions. It is assumed that the constant λ_0, which determines the magnitude of the pairing interaction $\lambda_0 V(\mathbf{p}_1, \mathbf{p}_2)$, is small. The superconducting gap is then defined as

M. Amusia and V. Shaginyan, *Strongly Correlated Fermi Systems*, Springer Tracts in Modern Physics 283,
https://doi.org/10.1007/978-3-030-50359-8_6

$$\Delta(\mathbf{p}) = -\lambda_0 \int V(\mathbf{p}, \mathbf{p}_1)\kappa(\mathbf{p}_1)\frac{d\mathbf{p}_1}{4\pi^2}. \tag{6.3}$$

The equation (6.2) shows that the order parameter $\kappa(\mathbf{p})$ can be expressed equally well through coefficient $v(\mathbf{p})$ of the corresponding Bogolyubov transformation. The latter fact shows that the energy E_{gs} in (6.1) is a functional of $v(\mathbf{p})$. Minimizing this energy in $v(\mathbf{p})$ and using (6.3), we arrive at equations that relate the single-particle energy $\varepsilon(\mathbf{p})$ to $\Delta(\mathbf{p})$ and $E(\mathbf{p})$

$$\varepsilon(\mathbf{p}) - \mu = \Delta(\mathbf{p})\frac{1-2v^2(\mathbf{p})}{2\kappa(\mathbf{p})}, \quad \frac{\Delta(\mathbf{p})}{E(\mathbf{p})} = 2\kappa(\mathbf{p}). \tag{6.4}$$

Here the single-particle energy $\varepsilon(\mathbf{p})$ is determined by (2.3), and

$$E(\mathbf{p}) = \sqrt{\xi^2(\mathbf{p}) + \Delta^2(\mathbf{p})} \tag{6.5}$$

with $\xi(\mathbf{p}) = \varepsilon(\mathbf{p}) - \mu$. Substituting the expression for $\kappa(\mathbf{p})$ from (6.4) in (6.3), we obtain the well-known equation for $\Delta(\mathbf{p})$ of the BCS theory

$$\Delta(\mathbf{p}) = -\frac{\lambda_0}{2} \int V(\mathbf{p}, \mathbf{p}_1)\frac{\Delta(\mathbf{p}_1)}{E(\mathbf{p}_1)}\frac{d\mathbf{p}_1}{4\pi^2}. \tag{6.6}$$

As $\lambda_0 \to 0$, the maximum value Δ_1 of the superconducting gap $\Delta(\mathbf{p})$ tends to zero and each equation in (6.4) reduces to (4.1)

$$\frac{\delta E(n(\mathbf{p}))}{\delta n(\mathbf{p})} = \varepsilon(\mathbf{p}) - \mu = 0, \tag{6.7}$$

if $0 < n(\mathbf{p}) < 1$, or $\kappa(\mathbf{p}) \neq 0$, in the interval $p_i \leq p \leq p_f$. Equation (6.7) shows that the function $n_0(\mathbf{p})$ is determined from the solution of the standard problem of functional $E(n(\mathbf{p}))$ minimization [4–6]. Equation (6.7) specifies the quasiparticle distribution function $n_0(\mathbf{p})$ that ensures the minimum of the ground state energy $E(\kappa(\mathbf{p}), n(\mathbf{p}))$. We can now study the relation between the state specified by (6.7) or (4.1) and the superconducting state.

At $T = 0$, (6.7) determines the specific state of a Fermi liquid with FC, the state for which the absolute value of the order parameter $|\kappa(\mathbf{p})|$ is finite in the momentum interval $p_i \leq p \leq p_f$ as $\Delta_1 \to 0$. Such a state can be considered superconducting with an infinitely small value of Δ_1. Hence, the entropy of this state at $T = 0$ is zero. Solutions $n_0(\mathbf{p})$ of (6.7) constitute a new class of solutions of both the BCS equations and the Landau Fermi liquid equations. In contrast to the ordinary solutions of the BCS equations [2], the new solutions are characterized by an infinitely small superconducting gap $\Delta_1 \to 0$, with the order parameter $\kappa(\mathbf{p})$ remaining finite. On the other hand, in contrast to the standard solution of the Landau Fermi liquid theory, the new solutions $n_0(\mathbf{p})$ determine the state of a heavy-electron liquid with a finite entropy S_0 as $T \to 0$ (see (4.8)). We arrive at an important conclusion that the solutions of

(6.7) can be interpreted as the simultaneous general solutions of the BCS equations and the Landau Fermi liquid theory equations, while (6.7) by itself can be derived either from the BCS theory or from the Landau Fermi liquid theory. Thus, as it is shown in Sect. 4.1.1, both states of the system coexist at $T \to 0$. As the system passes into a state with the order parameter $\kappa(\mathbf{p})$, the entropy vanishes abruptly and the system undergoes the first-order transition. In the vicinity of this point, the critical quantum and thermal fluctuations are suppressed so that the quasiparticles remain well-defined excitations. It follows from (4.2) that FCQPT is related to an alteration in the topological structure of Green's function and belongs to Lifshitz's topological phase transitions, which occur at absolute zero temperature [7]. This fact sets a relation between FCQPT and quantum phase transitions, under which the Fermi sphere splits into a sequence of Fermi layers [8, 9] (see Sects. 6.3 and 4.2). We note that in the state with the order parameter $\kappa(\mathbf{p})$, the system entropy $S = 0$ and the Nernst theorem holds in systems with FC.

If $\lambda_0 \neq 0$, the gap Δ_1 becomes finite, leading to a finite value of the effective mass M^*_{FC}, which may be obtained from (6.4) by differentiation its both sides over the momentum p with subsequent usage of (2.6) [10–13]:

$$M^*_{FC} \simeq p_F \frac{p_f - p_i}{2\Delta_1}. \tag{6.8}$$

It follows from (6.8) that in the superconducting state the effective mass is always finite. Its energy scale is determined by the parameter E_0:

$$E_0 = \varepsilon(\mathbf{p}_f) - \varepsilon(\mathbf{p}_i) \simeq p_F \frac{(p_f - p_i)}{M^*_{FC}} \simeq 2\Delta_1. \tag{6.9}$$

6.1.1 *Green's Function of the Superconducting State with FC at $T = 0$*

We begin with the Gor'kov equations [14], determining Green's functions $F^+(\mathbf{p}, \omega)$ and $G(\mathbf{p}, \omega)$ of a superconductor (see, e.g., [15]). For 3D case they read

$$F^+ = \frac{-\lambda_0 \Xi^*}{(\omega - E(\mathbf{p}) + i\,0)(\omega + E(\mathbf{p}) - i\,0)};$$
$$G = \frac{u^2(\mathbf{p})}{\omega - E(\mathbf{p}) + i\,0} + \frac{v^2(\mathbf{p})}{\omega + E(\mathbf{p}) - i\,0}. \tag{6.10}$$

The gap Δ and the function Ξ are given by

$$\Delta = \lambda_0 |\Xi|, \quad i\Xi = \int \int_{-\infty}^{\infty} F^+(\mathbf{p}, \omega) \frac{d\omega d\mathbf{p}}{(2\pi)^4}. \tag{6.11}$$

We recall that the function $F^+(\mathbf{p},\omega)$ has the meaning of the wave function of Cooper pairs and Ξ is the wave function of the motion of these pairs as a whole and is just a constant in a homogeneous system [15]. It follows from (6.4) and (6.11) that

$$i\Xi = \int_{-\infty}^{\infty} F_0^+(\mathbf{p},\omega)\frac{d\omega d\mathbf{p}}{(2\pi)^4} = i\int \kappa(\mathbf{p})\frac{d\mathbf{p}}{(2\pi)^3}. \tag{6.12}$$

With respect to (6.11) and (6.4), the (6.10) assume the form

$$\begin{aligned} F^+ &= -\frac{\kappa(\mathbf{p})}{\omega - E(\mathbf{p}) + i\,0} + \frac{\kappa(\mathbf{p})}{\omega + E(\mathbf{p}) - i\,0}; \\ G &= \frac{u^2(\mathbf{p})}{\omega - E(\mathbf{p}) + i\,0} + \frac{v^2(\mathbf{p})}{\omega + E(\mathbf{p}) - i\,0}. \end{aligned} \tag{6.13}$$

As $\lambda_0 \to 0$, the gap $\Delta \to 0$, but Ξ and $\kappa(\mathbf{p})$ remain finite if the spectrum becomes flat, $E(\mathbf{p}) = 0$, and (6.13) lead

$$\begin{aligned} F^+(\mathbf{p},\omega) &= -\kappa(\mathbf{p})\left[\frac{1}{\omega + i\,0} - \frac{1}{\omega - i\,0}\right]; \\ G(\mathbf{p},\omega) &= \frac{u^2(\mathbf{p})}{\omega + i\,0} + \frac{v^2(\mathbf{p})}{\omega - i\,0}, \end{aligned} \tag{6.14}$$

in the interval $p_i \le p \le p_f$. The parameters $v(\mathbf{p})$ and $u(\mathbf{p})$ are determined by the condition that the spectrum should be flat: $\varepsilon(\mathbf{p}) = \mu$. With respect to Landau equation (2.3), this condition again reduces to (4.1) and (6.7) for the minimum of the functional $E(n(\mathbf{p}))$.

We construct the functions $F^+(\mathbf{p},\omega)$ and $G(\mathbf{p},\omega)$ in the case where the constant λ_0 is finite, but small, namely, that $v(\mathbf{p})$ and $\kappa(\mathbf{p})$ can be found on the basis of the FC solutions of (4.1). Then $\kappa(\mathbf{p})$, Ξ, $E(\mathbf{p})$ and Δ are given by (6.2), (6.12), (6.11), and (6.4) respectively. Substituting the functions constructed in this manner into (6.13), we obtain $F^+(\mathbf{p},\omega)$ and $G(\mathbf{p},\omega)$ [16]. We note that (6.11) imply that the gap Δ is a linear function of λ_0 under the adopted conditions. As we shall see in Sect. 6.1.2, this gives rise to high-T_c at common values of the superconducting coupling constant.

6.1.2 The Superconducting State at Finite Temperatures

We assume that the region occupied by FC is small: $(p_f - p_i)/p_F \ll 1$ and $\Delta_1 \ll T_f$. Then, the order parameter $\kappa(\mathbf{p})$ is determined primarily by FC, i.e., by the distribution function $n_0(\mathbf{p})$ [10–12]. To be able to solve (6.6) analytically, we adopt the BCS approximation for the interaction [2]: $\lambda_0 V(\mathbf{p},\mathbf{p}_1) = -\lambda_0$ if $|\varepsilon(\mathbf{p}) - \mu| \le \omega_D$ and the interaction is zero outside this region, with ω_D being a certain characteristic energy, proportional to the Debye temperature. As a result, the superconducting gap depends only on the temperature, $\Delta(\mathbf{p}) = \Delta_1(T)$, and (6.6) becomes

$$1 = N_{FC}\lambda_0 \int_0^{E_0/2} \frac{d\xi}{\sqrt{\xi^2 + \Delta_1^2(0)}} + N_L\lambda_0 \int_{E_0/2}^{\omega_D} \frac{d\xi}{\sqrt{\xi^2 + \Delta_1^2(0)}}. \tag{6.15}$$

Here $\xi = \varepsilon(\mathbf{p}) - \mu$ and N_{FC} is the density of states in the momentum interval $p_f - p_i$ or in the energy interval E_0. It follows from (6.8) that $N_{FC} = (p_f - p_F)p_F/(2\pi\Delta_1)$. Within the energy interval $\omega_D - E_0/2$, the density of states N_L has the standard form $N_L = M_L^*/2\pi$. As $E_0 \to 0$, (6.15) becomes the BCS equation. On the other hand, assuming that $E_0 \leq 2\omega_D$ and discarding the second integral on the right-hand side of (6.15), we obtain

$$\begin{aligned}\Delta_1(0) &= \frac{\lambda_0 p_F(p_f - p_F)}{2\pi} \ln\left(1 + \sqrt{2}\right) \\ &= 2\beta\varepsilon_F \frac{p_f - p_F}{p_F} \ln\left(1 + \sqrt{2}\right),\end{aligned} \tag{6.16}$$

where $\varepsilon_F = p_F^2/2M_L^*$ is the Fermi energy and $\beta = \lambda_0 M_L^*/2\pi$ is the dimensionless coupling constant. Using the standard value of $\beta \simeq 0.3$ for ordinary superconductors and assuming that $(p_f - p_F)/p_F \simeq 0.2$, we obtain from (6.16) a large value $\Delta_1(0) \sim 0.1\varepsilon_F$. For ordinary superconductors, this gap is much smaller: $\Delta_1(0) \sim 10^{-3}\varepsilon_F$. We recollect that in BCS theory the temperature of superconducting phase transition T_c equals to $T_c \approx \Delta_1(0)/3.52$. This, along with above expressions for $\Delta_1(0)$, permits to estimate T_c in the substances with FC. Note that in FC theory this is underestimation of T_c, see (6.21). This means that with accurate account of all factors in FC approach T_c will be even larger.

As the Fermi energy order of magnitude is around 1 eV $\approx 10^4$ K, we obtain that in the substances with FC one has $T_c \simeq 300$ K, which is around room temperature. At the same time, for ordinary superconductors the expression $\Delta_1(0) \sim 10^{-3}\varepsilon_F = 10$ K gives correct estimate for T_c. This means that the FC concept explains successfully the physics of high-T_c superconductors. The above calculations show that the explanation is very simple and does not require any exotic departure from weak coupling BSC-like theory. Moreover, we recollect that BCS theory of superconductivity uses the Fermi liquid as the normal state of the system. Since the presented FC approach is a generalization (in the sense of new quasiparticle concept) of Landau Fermi liquid theory, we conclude that it is capable to describe naturally the BSC pairing of new "Fermi-condensed" quasiparticles, thus lying a foundation for successful explanation of high-T_c superconductivity phenomenon.

If the integral discarded earlier is taken into account, one finds that

$$\Delta_1(0) \simeq 2\beta\varepsilon_F \frac{p_f - p_F}{p_F} \ln\left(1 + \sqrt{2}\right) + \Delta_1(0)\beta \ln\left(\frac{2\omega_D}{\Delta_1(0)}\right), \tag{6.17}$$

where $\Delta_1(0)$ is given by (6.16). As $E_0 \to 0$ and $p_f \to p_F$, the first term on the right-hand side of (6.15) is zero. Therefore, we obtain the ordinary BCS result with $\Delta_1 \propto \exp(-1/\lambda_0)$. The correction related to the second integral in (6.15) is small because the second term on the right-hand side of (6.17) contains the additional factor β. In what follows, we show that $2T_c \simeq \Delta_1(0)$. The isotopic effect is small in this case since T_c depends on ω_D logarithmically, but the effect is restored as $E_0 \to 0$.

At $T \simeq T_c$, (6.8) and (6.9) are replaced by (4.13) and (4.15), which also hold for $T_c \leq T \ll T_f$:

$$M^*_{FC} \simeq p_F \frac{p_f - p_i}{4T_c}, \quad E_0 \simeq 4T_c, \text{ at } T \simeq T_c, \tag{6.18}$$

$$M^*_{FC} \simeq p_F \frac{p_f - p_i}{4T}, \quad E_0 \simeq 4T, \text{ at } T < T_c. \tag{6.19}$$

For finite temperatures, (6.15) is replaced by its standard generalization

$$1 = N_{FC}\lambda_0 \int\limits_0^{E_0/2} \frac{d\xi}{\sqrt{\xi^2 + \Delta_1^2}} \tanh \frac{\sqrt{\xi^2 + \Delta_1^2}}{2T}$$
$$+ N_L\lambda_0 \int\limits_{E_0/2}^{\omega_D} \frac{d\xi}{\sqrt{\xi^2 + \Delta_1^2}} \tanh \frac{\sqrt{\xi^2 + \Delta_1^2}}{2T}. \tag{6.20}$$

Since $\Delta_1(T \to T_c) \to 0$, (6.20) implies a relation that closely resembles the BCS result [17],

$$2T_c \simeq \Delta_1(0), \tag{6.21}$$

where $\Delta_1(T = 0)$ is found from (6.17). Comparing (6.8) and (6.9) with (6.18) and (6.19), we see that both M^*_{FC} and E_0 are temperature-independent for $T \leq T_c$.

6.1.3 Bogolyubov Quasiparticles

Equation (6.6) shows that the superconducting gap depends on the single-particle spectrum $\varepsilon(\mathbf{p})$. On the other hand, it follows from (6.4) that $\varepsilon(\mathbf{p})$ depends on $\Delta(\mathbf{p})$ if (6.7) has a solution that determines the existence of FC at $\lambda_0 \to 0$. We assume that λ_0 is so small that the pairing interaction $\lambda_0 V(\mathbf{p}, \mathbf{p}_1)$ leads only to a small perturbation of the order parameter $\kappa(\mathbf{p})$. Equation (6.8) implies that the effective mass and the density of states $N(0) \propto M^*_{FC} \propto 1/\Delta_1$ are finite. Thus, in contrast to the spectrum in the standard superconductivity theory, the single-particle spectrum $\varepsilon(\mathbf{p})$ depends strongly on the superconducting gap, and (2.3) and (6.6) must be solved self-consistently.

We assume that (2.3) and (6.6) have been solved and the effective mass M^*_{FC} has been found. This means that we can find the quasiparticle dispersion law $\varepsilon(\mathbf{p})$ by choosing the effective mass M^* equal to the obtained value of M^*_{FC} and then solve (6.6) without (2.3), as it is done in the standard BCS superconductivity theory [2]. Hence, the superconducting state with FC is characterized by Bogolyubov quasiparticles [18] with dispersion (6.5) and the normalization condition $v^2(\mathbf{p}) + u^2(\mathbf{p}) = 1$ for the coefficients $v(\mathbf{p})$ and $u(\mathbf{p})$. Moreover, quasiparticle excitations of the superconducting state in the presence of FC coincide with the Bogolyubov quasiparticles in the BCS theory. This means that the superconductivity in FC state resembles BCS one, which points to the applicability of the modified BCS formalism to the description of the high-T_c superconducting state. At the same time, the maximum value of the superconducting gap set by (6.17) and other exotic properties are determined by the presence of FC. These results are in good agreement with the experimental observations for the high-T_c superconductor $Bi_2Sr_2Ca_2Cu_3O_{10+\delta}$ [19].

In constructing the superconducting state with FC, we returned to the foundations of the LFL theory, where the high-energy degrees of freedom had been eliminated by the introduction of new quasiparticles. The main difference between the LFL, which forms the basis for constructing the superconducting state, and the Fermi liquid with FC is that in the latter case we should increase the number of low-energy degrees of freedom by introducing the new type of quasiparticles with the effective mass M^*_{FC} and the characteristic energy E_0 given by (6.9). Hence, the dispersion law $\varepsilon(\mathbf{p})$ in FC state is characterized by two types of quasiparticles with the effective masses M^*_L and M^*_{FC} and the scale E_0. The extended paradigm and new quasiparticles determine the properties of the superconductor, including the lineshape of quasiparticle excitations [10–12, 20], while the formal dispersion of the Bogolyubov quasiparticles has the standard form (6.5).

We note that for $T < T_c$, the effective mass M^*_{FC} and the scale E_0 are temperature-independent [20]. For $T > T_c$, the effective mass M^*_{FC} and the scale E_0 are given by (4.13) and (4.15). Obviously, we cannot directly relate these new quasiparticles (excitations) of the Fermi liquid with FC to excitations (quasiparticles) of an ideal Fermi gas, as is done in the standard LFL theory, because the system is beyond FCQPT. The properties and dynamics of quasiparticles are given by the extended paradigm and closely related to the properties of the superconducting state and are of a collective nature, formed by FCQPT and determined by the macroscopic number of FC quasiparticles with momenta in the interval $(p_f - p_i)$. Such a system cannot be perturbed by scattering on impurities and on lattice defects and, therefore, has the features of a quantum protectorate and demonstrates universal scaling behavior, forming a new state of matter [10–12, 21–24], see Chap. 7.

Several remarks concerning the quantum protectorate and the universal behavior of superconductors with FC are in order. Similar to the Landau Fermi liquid theory, the theory of high-T_c superconductivity based on FCQPT deals with quasiparticles that are elementary low-energy excitations. The theory provides general qualitative description of the superconducting and the normal states of high-T_c superconductors and HF metals. Of course, the proper choice of the phenomenological parameters like pairing coupling constant can yield a quantitative description of superconductivity similar to the case of Landau theory for an ordinary Fermi liquid like ^{3}He. Hence, any formalism capable of FC description and compatible with the BCS theory yields the same qualitative picture of normal and superconducting states in a substance with FCQPT. Obviously, both approaches may be coordinated on the level of numerical results by choosing the appropriate parameters. For instance, because the formation of FC is possible in the Hubbard model [25], it allows reproducing the results of the theory based on FCQPT.

6.1.4 The Dependence of Superconducting Phase Transition Temperature T_c on Doping

We examine the maximum value of the superconducting gap Δ_1 as a function of the number density x of mobile charge carriers, which is proportional to the degree of doping. Using (4.16), we can rewrite (6.16) as

$$\frac{\Delta_1}{\varepsilon_F} \sim \beta \frac{(x_{FC} - x)x}{x_{FC}}. \tag{6.22}$$

Here, we take into account that the Fermi level $\varepsilon_F \propto p_F^2$ and the number density $x \sim p_F^2/(2M^*)$ with the result $\varepsilon_F \propto x$. It is realistic to assume that $T_c \propto \Delta_1$, because the curve $T_c(x)$ obtained in experiments with high-T_c superconductors [26] must be a smooth function of x. Hence, we can approximate $T_c(x)$ by a smooth bell-shaped function [27]:

$$T_c(x) \propto \beta(x_{FC} - x)x. \tag{6.23}$$

To illustrate the application of the above analysis, we examine the main features of a superconductor that can hypothetically exist at room temperature. Such a superconductor should consist of two-dimensional layers similar to high-T_c superconducting cuprates. Equation (6.16) implies that $\Delta_1 \sim \beta\varepsilon_F \propto \beta/r_s^2$. Bearing in mind that FCQPT occurs at $r_s \sim 20$ in 3D systems and at $r_s \sim 8$ in 2D systems [28], we can expect that in 3D systems Δ_1 amounts to 10% of the maximum size of the superconducting gap in 2D systems, which in our case amounts to 60 mV for weakly doped cuprates with $T_c = 70$ K [29]. On the other hand, (6.16) implies that Δ_1 may be even larger, $\Delta_1 \sim 75$ mV. We can expect that $T_c \sim 300$ K in the case of s-wave pairing, as the simple relation $2T_c \simeq \Delta_1$ implies. Indeed, we can take $\varepsilon_F \sim 500$ mV, $\beta \sim 0.3$, and $(p_f - p_i)/p_F \sim 0.5$. Thus, the hypothetical room-temperature superconductor could have simple s-wave pairing. We note that the number density x of mobile charge carriers have to satisfy the condition $x \leq x_{FC}$ and should be varied to reach the optimum degree of doping $x_{opt} \simeq x_{FC}/2$.

6.1.5 The Gap and Heat Capacity Near T_c

We now calculate the gap and the heat capacity at temperatures $T \to T_c$. Our analysis is valid at $T_p^* \simeq T_c$ as otherwise the discontinuities in the heat capacity considered below are smeared over the temperature interval between T_p^* and T_c where T_p^* is the temperature at which the pseudogap is closed. Since the origin of the pseudogap is controversial and still subject to debate in the condensed matter community, we do not consider here this phenomenon. To simplify matters, we calculate the leading contribution to the gap and heat capacity related to FC. We use (6.20) to find the function $\Delta_1(T \to T_c)$ simply by expanding the first integral on its right-hand side in powers of Δ_1 and dropping the contribution from the second integral. This procedure leads to the equation [20]

$$\Delta_1(T) \simeq 3.4T_c\sqrt{1 - \frac{T}{T_c}}. \tag{6.24}$$

Therefore, the gap in the single-particle excitations spectrum behaves in the ordinary BCS manner.

To calculate the heat capacity, we can use the standard expression for the entropy S [2]:

$$S(T) = -2\int\Big[f(\mathbf{p})\ln f(\mathbf{p}) + (1 - f(\mathbf{p}))\ln(1 - f(\mathbf{p}))\Big]\frac{d\mathbf{p}}{(2\pi)^2}, \tag{6.25}$$

where

$$f(\mathbf{p}) = \left(1 + \exp\frac{E(\mathbf{p})}{T}\right)^{-1}, \quad E(\mathbf{p}) = \sqrt{(\varepsilon(\mathbf{p}) - \mu)^2 + \Delta_1^2(T)}. \tag{6.26}$$

The heat capacity C is given by

$$C(T) = T\frac{dS}{dT} \simeq 4\frac{N_{FC}}{T^2}\int\limits_0^{E_0} f(E)(1 - f(E))\left[E^2 + T\Delta_1(T)\frac{d\Delta_1(T)}{dT}\right]d\xi$$
$$+ 4\frac{N_L}{T^2}\int\limits_{E_0}^{\omega_D} f(E)(1 - f(E))\left[E^2 + T\Delta_1(T)\frac{d\Delta_1(T)}{dT}\right]d\xi. \tag{6.27}$$

In deriving (6.27), we again use the variables ξ and $E = \sqrt{\xi^2 + \Delta_1^2(T)}$ and the above notation for the densities of states N_{FC} and N_L. Equation (6.27) describes a jump in heat capacity, $\delta C(T) = C_s(T) - C_n(T)$, where $C_s(T)$ and $C_n(T)$ are, respectively, the heat capacities of the superconducting and normal states at T_c; the jump is determined by the last two terms in the square brackets on the right-hand side of this equation. Using (6.24) to calculate the first term on the right-hand side of (6.27), we find [20]

$$\delta C(T_c) \simeq \frac{3}{2\pi^2}(p_f - p_i)p_F^n, \tag{6.28}$$

where $n = 1$ in the 2D case and $n = 2$ in the 3D case. This result differs from the ordinary BCS result, according to which the discontinuity in the heat capacity is a linear function of T_c. The jump $\delta C(T_c)$ is independent of T_c because, as (6.19) shows, the effective mass M^*_{FC} varies in inverse proportion to T_c. We note that deriving (6.28) we took into account the leading contribution coming from FC. This contribution disappears as $E_0 \to 0$, and the second integral on the right-hand side of (6.27) yields the standard result.

As we will show in Chap. 7, the heat capacity of a system with FC behaves as $C_n(T) \propto M * T \propto \sqrt{T/T_f}$. The jump in the heat capacity given by (6.28) is temperature-independent. As a result, we find that

$$\frac{\delta C(T_c)}{C_n(T_c)} \sim \sqrt{\frac{T_f}{T_c}} \frac{(p_f - p_i)}{p_F}. \tag{6.29}$$

In contrast to the case of normal superconductors, where $\delta C(T_c)/C_n(T_c) = 1.43$ [15], in our case (6.29) implies that the ratio $\delta C(T_c)/C_n(T_c)$ is not constant and may be very large when $T_f/T_c \gg 1$ [20, 30]. It is instructive to apply this analysis to $CeCoIn_5$, where T_c =2.3 K [30]. In this material [31], $\delta C/C_n \simeq 4.5$ is substantially higher than the BCS value, in agreement with (6.29).

6.2 The Dispersion Law and Lineshape of Single-Particle Excitations

Recently discovered gap in the quasiparticles dispersion at energies between 40 and 70 meV, resulting in quasiparticles velocity altering in this energy range [32–35], can hardly be explained by the marginal Fermi liquid theory as it contains no additional energy scales or parameters that would allow taking the gap into account [36, 37]. One could assume that the gap, defining new energy scale, occurs due to the interaction of electrons and collective excitations. In this case, however, we would have to discard the idea of a quantum protectorate, which in its turn would contradict the experimental data [21, 22].

As shown in Sects. 4.1.3 and 6.1, a system with FC has two effective masses: M^*_{FC}, which determines the single-particle spectrum at low energies, and M^*_L, which determines the spectrum at high energies. The fact that there are two effective masses manifests itself in the form of a kink in the quasiparticle dispersion law. The dispersion law can be approximated by two straight lines intersecting at the binding energy $E_0/2$ [see (4.15) and (6.9)]. The kink in the dispersion law occurs at temperatures much lower than $T \ll T_f$, when the system is in the superconducting or normal state. Such behavior is in good agreement with the experimental data [35]. It is pertinent to note that at temperatures below $T < T_c$, the effective mass M^*_{FC} is independent of the momenta p_F, p_f, and p_i, as shown by (6.8) and (6.16):

$$M^*_{FC} \sim \frac{2\pi}{\lambda_0}. \tag{6.30}$$

This formula implies that M^*_{FC} is only weakly dependent on x, if a dependence of λ_0 on x is allowed. This result is in good agreement with the experimental data [38–40]. The same is true for the dependence of the Fermi velocity $v_F = p_F/M^*_{FC}$ on x because the Fermi momentum $p_F \sim \sqrt{n}$ is weakly dependent on the electron number density $n = n_0(1-x)$ [38, 39]. Here, n_0 is the single-particle electron number density at half-filling.

Since λ_0 is the coupling constant that determines the magnitude of the pairing interaction, e.g., the electron-phonon one, we can expect a kink in the quasiparticle dispersion law to be caused by the electron-phonon interaction. The electron-phonon scenario could explain the constancy of the kink at $T > T_c$ as phonon contribution is temperature-independent. On the other hand, it was found that the quasiparticle dispersion law distorted by the interaction with phonons has a tendency to restore itself to the ordinary single-particle dispersion law when the quasiparticle energy becomes higher than the phonon energy [41]. However, there is no experimental evidence that such restoration of the dispersion law actually takes place [35].

The quasiparticle excitation curve $L(q,\omega)$ is a function of two variables. Measurements at a constant energy $\omega = \omega_0$, where ω_0 is the single-particle excitation energy, determine the curve $L(q, \omega = \omega_0)$ as a function of the momentum q. We have shown above that M^*_{FC} is finite and constant at temperatures not exceeding T_c. Hence, at excitation energies $\omega < E_0$, the system behaves as an ordinary superconducting Fermi liquid with the effective mass determined by (6.8) [10–13]. At $T_c \leq T$, the effective mass M^*_{FC} is also finite and is given by (4.13). In other words, at $\omega < E_0$, the system behaves as a Fermi liquid whose single-particle spectrum is well defined and the width of the single-particle excitations is of the order of T [5, 10–12]. Such behavior has been observed in measurements of the quasiparticle excitation curve at fixed energy [33, 42, 43].

The quasiparticle excitation curve can also be described as a function of ω, at a constant momentum $q = q_0$. For small values of ω, the behavior of this function is similar to that described above, with $L(q = q_0, \omega)$ having a characteristic maximum and width. For $\omega \geq E_0$, the contribution provided by quasiparticles of mass M^*_L becomes significant and leads to an increase in the function $L(q = q_0, \omega)$. Thus, $L(q = q_0, \omega)$ has a certain structure of maxima and minima directly determined by the existence of two effective masses, M^*_{FC} and M^*_L [10–13]. We conclude that, in contrast to Landau quasiparticles, these quasiparticles have a more complicated spectral lineshape.

To calculate the imaginary part $\mathrm{Im}\,\Sigma(\mathbf{p},\varepsilon)$ of the self-energy $\Sigma(\mathbf{p},\varepsilon)$, we use the Kramers-Kronig relations. For that we first calculate its real part $\mathrm{Re}\,\Sigma(\mathbf{p},\varepsilon)$, which determines the effective mass M^* [44],

$$\frac{1}{M^*} = \left(\frac{1}{m} + \frac{1}{p_F}\frac{\partial \mathrm{Re}\Sigma}{\partial p}\right) \Big/ \left(1 - \frac{\partial \mathrm{Re}\Sigma}{\partial \varepsilon}\right). \tag{6.31}$$

The corresponding momenta p and energies ε satisfy the inequalities $|p - p_F|/p_F \ll 1$, and $\varepsilon/\varepsilon_F \ll 1$. We take Re $\Sigma(\mathbf{p}, \varepsilon)$ in the simplest possible form that ensures the correct variation of the effective mass at the energy $E_0/2$,

$$\begin{aligned}\mathrm{Re}\Sigma(\mathbf{p}, \varepsilon) &= -\varepsilon\frac{M^*_{FC}}{m} + \left(\varepsilon - \frac{E_0}{2}\right)\frac{M^*_{FC} - M^*_L}{m} \\ &\times \left[\theta\left(\varepsilon - \frac{E_0}{2}\right) + \theta\left(-\varepsilon - \frac{E_0}{2}\right)\right],\end{aligned} \tag{6.32}$$

where $\theta(\varepsilon)$ is the step function. To ensure a smooth transition from the single-particle spectrum characterized by M^*_{FC} to the spectrum characterized by M^*_L, we must replace the step function by a smoothed one. Substituting (6.32) in (6.31), we see that $M^* \simeq M^*_{FC}$ within the interval $(-E_0/2,\ E_0/2)$, while $M^* \simeq M^*_L$ outside this interval. Applying the Kramers-Kronig relation to Re$\Sigma(\mathbf{p}, \varepsilon)$, we express the imaginary part of the self-energy as [20]

$$\mathrm{Im}\Sigma(\mathbf{p}, \varepsilon) \sim \varepsilon^2\frac{M^*_{FC}}{\varepsilon_F m} + \frac{M^*_{FC} - M^*_L}{m}\left[\varepsilon\ln\left|\frac{2\varepsilon + E_0}{2\varepsilon - E_0}\right| + \frac{E_0}{2}\ln\left|\frac{4\varepsilon^2 - E_0^2}{E_0^2}\right|\right]. \tag{6.33}$$

Clearly, with $\varepsilon/E_0 \ll 1$, the imaginary part is proportional to ε^2; at $2\varepsilon/E_0 \simeq 1$, we have Im$\Sigma \sim \varepsilon$, and for $E_0/\varepsilon \ll 1$, the main contribution to the imaginary part is approximately constant.

It follows from (6.33) that as $E_0 \to 0$, the second term on its right-hand side vanishes and the single-particle excitations become well defined, which resembles the situation with a normal Fermi liquid, while the pattern of minima and maxima eventually disappears. Now the quasiparticle renormalization factor $z(\mathbf{p})$ is given by the equation [44]

$$\frac{1}{z(\mathbf{p})} = 1 - \frac{\partial \mathrm{Re}\Sigma(\mathbf{p}, \varepsilon)}{\partial \varepsilon}. \tag{6.34}$$

Consequently, the (6.33) and (6.34) show that for $T \le T_c$, the interaction of a quasiparticle on the Fermi surface increases as the characteristic energy E_0 decreases. Equations (6.9) and (6.23) imply that $E_0 \sim (x_{FC} - x)/x_{FC}$. When $T > T_c$, it follows from (6.32) and (6.34) that the quasiparticle interaction increases as the effective mass M^*_{FC} decreases. So, from (4.13) and (4.16) $M^*_{FC} \sim (p_f - p_i)/p_F \sim (x_{FC} - x)/x_{FC}$. As a result, we conclude that the interaction increases with the doping level x and the single-particle excitations are better defined in heavily doped samples. As $x \to x_{FC}$, the characteristic energy $E_0 \to 0$ and the quasiparticles become normal excitations of LFL. We note that such behavior has been observed in experiments with heavily doped Bi2212, which demonstrates high-T_c superconductivity with the gap of about 10 mV [45]. The size of the gap suggests that the region occupied by FC is small because $E_0/2 \simeq \Delta_1$. For $x > x_{FC}$ and low temperatures, the HF liquid behaves as LFL (see Fig. 4.2 and Sect. 4.1.5). Experimental data show that, as expected, the LFL state exists in super-heavily doped nonsuperconducting $La_{1.7}Sr_{0.3}CuO_4$ [46, 47].

6.3 Electron Liquid with FC in Magnetic Fields

The behavior of heavy-electron liquid with FC in magnetic field is considered. At low temperatures and under the application of weak magnetic field, the distribution function is to be reconstructed so that the order parameter vanishes, and a new distribution function is to deliver the same ground state energy. We show that the new distribution emerges as a result of topological phase transition, and is represented by multiply connected Fermi spheres. This topological phase transition and multiply connected Fermi spheres (icebergs) are considered in Chap. 4.

6.3.1 Phase Diagram of Electron Liquid in Magnetic Field

Let us assume that the coupling constant is nonzero, $\lambda_0 \neq 0$, but is infinitely small. We found in Sect. 6.1 that at $T = 0$ the superconducting order parameter $\kappa(\mathbf{p})$ is finite in the region occupied by FC and that the maximum value of the superconducting gap $\Delta_1 \propto \lambda_0$ is infinitely small. Hence, any weak magnetic field $B \neq 0$ is critical and destroys the order parameter and FC. Simple energy arguments suffice to determine the type of rearrangement of the FC state. On one hand, since FC state is destroyed, the energy gain $\Delta E_B \propto B^2$ vanishes as $B \to 0$. On the other hand, the function $n_0(\mathbf{p})$, which occupies the finite interval $(p_f - p_i)$ in the momentum space and is specified by (4.1) or (6.9), leads to a finite gain in the ground state energy compared to that of a normal Fermi liquid [4]. Thus, the distribution function is to be reconstructed so that the order parameter is to vanish, while a new distribution function is to deliver the same ground state energy. Thus, in weak magnetic fields, the new ground state without FC must have almost the same energy as the state with FC. As shown in Chap. 4, such a state is formed by multiply connected Fermi spheres resembling an onion, in which a smooth distribution function of quasiparticles, $n_0(\mathbf{p})$, is replaced in the interval $(p_f - p_i)$ with the distribution function [8, 48]

$$\nu(\mathbf{p}) = \sum_{k=1}^{n} \theta(p - p_{2k-1})\theta(p_{2k} - p). \tag{6.35}$$

where the parameters $p_i \leq p_1 < p_2 < \ldots < p_{2n} \leq p_f$ are chosen so as to satisfy the conditions of normalization and conservation of the particles number:

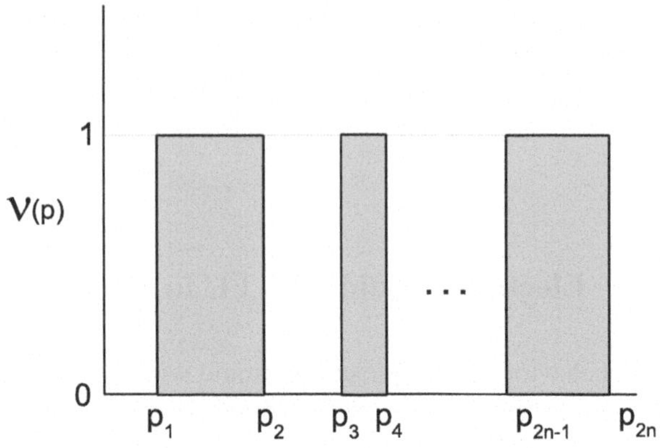

Fig. 6.1 The function $\nu(\mathbf{p})$ for the multiply connected distribution that replaces the function $n_0(\mathbf{p})$ in the region $(p_f - p_i)$ occupied by FC. The momenta satisfy the inequalities $p_i < p_F < p_f$. The outer Fermi surface at $p \simeq p_{2n} \simeq p_f$ has the shape of a Fermi step, and therefore the system behaves like LFL at sufficiently low temperatures

$$\int_{p_{2k-1}}^{p_{2k+3}} \nu(\mathbf{p})\frac{d\mathbf{p}}{(2\pi)^3} = \int_{p_{2k-1}}^{p_{2k+3}} n_0(\mathbf{p})\frac{d\mathbf{p}}{(2\pi)^3}.$$

Figure 6.1 shows the corresponding multiply connected distribution. For definiteness, we present the most interesting case of a three-dimensional system. The two-dimensional case can be examined similarly. We note that the possibility of the existence of multiply connected Fermi spheres have been studied in [49–52].

We assume that the thickness of each inner slice of the Fermi sphere, $\delta p \simeq p_{2k+1} - p_{2k}$, is determined by the magnetic field B. Using the well-known rule for estimating errors of numerical calculation of definite integrals, we find that the minimum loss of the ground state energy due to slice formation is approximately $(\delta p)^4$. This becomes especially clear if we take into account the fact that the continuous FC functions $n_0(\mathbf{p})$ ensure the minimum value of the energy functional $E[n(\mathbf{p})]$, while the approximation of $\nu(\mathbf{p})$ by steps of width δp leads to a minimal error of the order of $(\delta p)^4$. Recalling that the gain due to the magnetic field is proportional to B^2 and equating the two contributions, we obtain

$$\delta p \propto \sqrt{B}. \tag{6.36}$$

Therefore, as $T \to 0$, with $B \to 0$, the slice thickness δp also tends to zero and the behavior of a Fermi liquid with FC is replaced with that of LFL with the Fermi momentum p_f. Equation (6.7) implies that $p_f > p_F$ and the electron number density x remains constant, with the Fermi momentum of the multiply connected Fermi sphere $p_{2n} \simeq p_f > p_F$ (see Fig. 6.1). These observations play an important role in studying the behavior of both the Hall coefficients $R_H(B)$ and the second-order phase transitions in HF metals as a function of B at low temperatures [53].

To calculate the effective mass $M^*(B)$ as a function of the applied magnetic field B, we first note that at $T = 0$ the field B splits the FC state into Landau levels, suppresses the superconducting order parameter $\kappa(\mathbf{p})$, and destroys FC, which leads to restoration of LFL [54, 55]. The Landau levels near the Fermi surface can be approximated by separate slices whose thickness in momentum space is δp.

Approximating the quasiparticle dispersion law within a single slice, $\varepsilon(p) - \mu \sim (p - p_f + \delta p)(p - p_f)/M^*$, we find the effective mass $M^*(B) \sim M^*/(\delta p/p_f)$. The energy increment ΔE_{FC} caused by the transformation of the FC state can be estimated based on using the Landau formula [15]

$$\Delta E_{FC} = \int (\varepsilon(\mathbf{p}) - \mu)\delta n(\mathbf{p}) \frac{d\mathbf{p}^3}{(2\pi)^3}. \tag{6.37}$$

The region occupied by the variation $\delta n(\mathbf{p})$ has the thickness δp, with $(\varepsilon(\mathbf{p}) - \mu) \sim (p - p_f)p_f/M^*(B) \sim \delta p p_f/M^*(B)$. As a result, we find that $\Delta E_{FC} \sim p_f^3 \delta p^2/M^*(B)$. On the other hand, there is one more term in the energy E expression, $\Delta E_B \sim (B\mu_B)^2 M^*(B) p_f$, caused by the applied magnetic field, which decreases the energy and is related to the Zeeman splitting. Equating ΔE_B and ΔE_{FC} and recalling that $M^*(B) \propto 1/\delta p$ in this case, we obtain the chain of relations

$$\frac{\delta p^2}{M^*(B)} \propto \frac{1}{(M^*(B))^3} \propto B^2 M^*(B), \tag{6.38}$$

which implies that the effective mass $M^*(B)$ diverges as

$$M^*(B) \propto \frac{1}{\sqrt{B - B_{c0}}}, \tag{6.39}$$

where B_{c0} is the critical magnetic field, which places HF metal at the magnetic-field-tuned quantum-critical point and nullifies the respective Nèel temperature, $T_{NL}(B_{c0}) = 0$ [55]. In our simple model of HF liquid, the quantity B_{c0} is a parameter determined by the properties of the specific metal with heavy fermions. We note that in some cases $B_{c0} = 0$, e.g., the HF metal $CeRu_2Si_2$ has no magnetic order, exhibits no superconductivity, and does not behave like a Landau Fermi liquid even at the lowest reached temperatures [56].

Formula (6.39) and Fig. 6.1 show that the application of a magnetic field $B > B_{c0}$ brings the FC system back to the LFL state with the effective mass $M^*(B)$ that depends on the magnetic field. This means that the following characteristics of LFL are restored: $C/T = \gamma_0(B) \propto M^*(B)$ for the heat capacity and $\chi_0(B) \propto M^*(B)$ for the magnetic susceptibility. The coefficient $A(B)$ determines the temperature-dependent part of the resistivity, $\rho(T) = \rho_0 + \Delta\rho$, where ρ_0 is the residual resistivity and $\Delta\rho = A(B)T^2$. Since this coefficient is directly determined by the effective mass, $A(B) \propto (M^*(B))^2$ [57], (6.39) yields

$$A(B) \propto \frac{1}{B - B_{c0}}. \tag{6.40}$$

Thus, the empirical Kadowaki-Woods relation [58] $K = A/\gamma_0^2 \simeq const$ is valid in our case [57]. Furthermore, K may depend on the quasiparticles degree of degeneracy. With this degeneracy, the Kadowaki-Woods relation provides a good description of the experimental data for a broad class of HF metals [59, 60]. In the simplest case, where HF liquid is formed by spin-$1/2$ quasiparticles with the degeneracy degree 2, the value of K turns out to be close to the empirical value [57] known as the Kadowaki-Woods ratio [58]. Hence, the system, subjected to a magnetic field, returns to the LFL state with constant Kadowaki-Woods relation. We note that when the system is located near FCQPT, the application of magnetic field brings it to the LFL state due to the Zeeman splitting, as it is discussed in Chap. 7.

At finite temperatures, the system remains in the LFL state, but when $T > T^*(B)$, the NFL behavior is restored. As regards to finding the function $T^*(B)$, we reminder that the transition region characterized by the function $T^*(B)$ can be determined in different measurements on HF compounds such as measurements of the maximums of the heat capacity C/T or maximums of the magnetic susceptibility χ, etc. In the considered case, $T^*(B)$ is determined in measurements of the resistivity $\rho(T)$ in magnetic fields. We note that the effective mass M^*, characterizing the single-particle spectrum, cannot change at $T^*(B)$ because no phase transition occurs at this temperature. To calculate $M^*(T)$, we equate the effective mass $M^*(T)$ in (4.13) to $M^*(B)$ in (6.39), $M^*(T) \sim M^*(B)$,

$$\frac{1}{M^*(T)} \propto T^*(B) \propto \frac{1}{M^*(B)} \propto \sqrt{B - B_{c0}}. \tag{6.41}$$

Let us note that

$$T^*(B) \propto \sqrt{B - B_{c0}}. \tag{6.42}$$

At temperatures $T \geq T^*(B)$, the system returns to the NFL behavior with the effective mass M^* specified by (4.13). Thus, expression (6.42) determines the line in the $T - B$ phase diagram that separates the region where the effective mass depends on B and the heavy-Fermi liquid behaves like a Landau Fermi liquid from the region where the effective mass is temperature-dependent. At $T^*(B)$, the temperature dependence of the resistivity ceases to be quadratic and becomes linear.

A schematic $T - B$ phase diagram of HF liquid with FC in magnetic field is shown in Fig. 6.2. At magnetic field $B < B_{c0}$, the FC state can be captured by FM, AFM, and/or SC states lifting the degeneracy of the FC state. It follows from (6.42) that at a certain temperature $T^*(B) \ll T_f$, the heavy-electron liquid transits from its NFL state to LFL one acquiring the properties of LFL at $(B - B_{c0}) \propto (T^*(B))^2$. At temperatures below $T^*(B)$, as shown by the horizontal arrow in Fig. 6.2, the heavy-electron liquid demonstrates an increasingly metallic behavior as the magnetic field B increases. This is because the effective mass decreases, see (6.39). Such behavior of the effective mass can be observed, for instance, in measurements of the heat capacity, magnetic susceptibility, resistivity, and Shubnikov-de Haas oscillations. The $T - B$ phase diagram in Fig. 6.2 shows that a unique possibility emerges where a magnetic field can be employed to control the variations in the physical nature and type of behavior of the electron liquid with FC.

We briefly discuss the case where the system is extremely close to FCQPT on the ordered size of this transition, and hence $\delta p_{FC} = (p_f - p_i)/p_F \ll 1$. Because

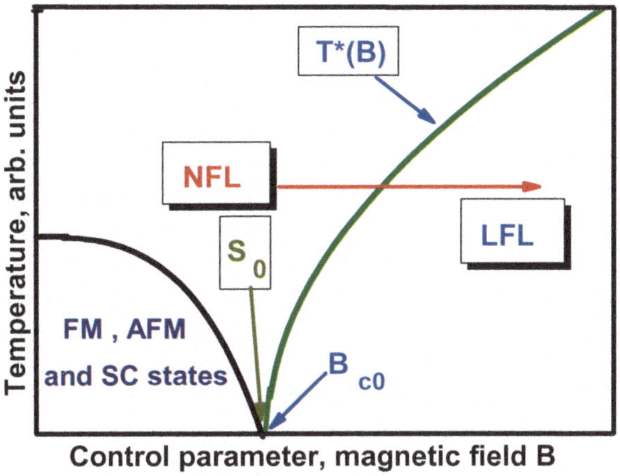

Fig. 6.2 Schematic $T - B$ phase diagram of heavy-electron liquid. B_{c0} denotes the magnetic field at which the effective mass diverges according to (6.39). The horizontal arrow illustrates the system moving in the NFL-LFL direction along B at fixed temperature. As shown by the dashed curve, at $B < B_{c0}$ the system can be in its ferromagnetic (FM), antiferromagnetic (AFM), or superconducting (SC) states. The NFL state is characterized by the entropy S_0 given by (4.8). The solid curve $T^*(B)$ separates the NFL state and the weakly polarized LFL one and represents the transition regime

$\delta p \propto M^*(B)$, it follows from (6.36) and (6.39) that

$$\frac{\delta p}{p_F} \sim a_c \sqrt{\frac{B - B_{c0}}{B_{c0}}}, \tag{6.43}$$

where $a_c \sim 1$. As the magnetic field B increases, $\delta p/p_F$ becomes comparable to δp_{FC}, and the distribution function $\nu(\mathbf{p})$ disappears, being absorbed by the ordinary Zeeman splitting. As a result, we are dealing with HF liquid located on the disordered side of FCQPT. Equation (6.43) implies that the relatively weak magnetic field B_{cr},

$$B_{red} \equiv \frac{B - B_{c0}}{B_{c0}} = (\delta p_{FC})^2 \sim B_{cr}, \tag{6.44}$$

where B_{red} is the reduced field, takes the system from the ordered side of the phase transition to the disordered if $\delta p_{FC} \ll 1$.

6.3.2 *Magnetic Field Dependence of the Effective Mass in HF Metals and High-T_c Superconductors*

Observations have shown that in the normal state both heavily $Tl_2Ba_2CuO_{6+\delta}$ [61] and optimally doped cuprates $Bi_2Sr_2CuO_{6+\delta}$ [62] exhibit no significant violations of the Wiedemann-Franz law. The normal state has been obtained by applying a magnetic field whose strength is higher than the maximum critical field B_{c2} that destroys superconductivity. Studies of the electron-doped superconductor $Pr_{0.91}LaCe_{0.09}Cu0_{4-y}$ (T_c=24 K) revealed that when a magnetic field destroyed superconductivity in this material, the spin-lattice relaxation constant $1/T_1$ obeyed the relation $T_1T = const$, known as the Korringa law, down to temperatures about $T \simeq 0.2$ K [63, 64]. At higher temperatures and in magnetic fields up to 15.3 T perpendicular to the CuO_2 plane, the ratio $1/T_1T$ remains constant as a function of T for $T \leq 55$ K. In the temperature range from 50 to 300 K, the ratio $1/T_1T$ decreases as the temperature increases [64]. Measurements involving the heavily doped nonsuperconducting material $La_{1.7}Sr_{0.3}CuO_4$ have shown that the resistivity ρ varies with T as T^2 and that the Wiedemann-Franz law holds [46, 47].

As Korringa and Wiedemann-Franz laws strongly indicate the presence of the LFL state, experiments show that the observed elementary excitations are indistinguishable from Landau quasiparticles in high-T_c superconductors. This places severe restrictions on models describing hole- or electron-doped high-T_c superconductors. For instance, for a Luttinger liquid [65, 66], for spin-charge separation [67], and in some $t - J$ models [68], a violation of the Wiedemann-Franz law was predicted, which is in contradiction with experimental evidence and points to the limited applicability of these models.

If the constant λ_0 is finite, then an HF liquid with FC is in the superconducting state. We examine the behavior of the system in magnetic fields $B > B_{c2}$. In this case, the system becomes LFL induced by the magnetic field, and the elementary excitations become quasiparticles that cannot be distinguished from Landau quasiparticles, with the effective mass $M^*(B)$ given by (6.39). As a result, the Wiedemann-Franz law holds as $T \to 0$, which agrees with the experimental data [61, 62]. Note that violations of the Wiedemann-Franz law in HF metals are considered in Chap. 20.

The low-temperature properties of the system depend on the effective mass; in particular, the resistivity $\rho(T)$ obeys (2.19) with $A(B) \propto (M^*(B))^2$. Assuming that for high-T_c superconductors the critical field $B = B_{c0}$, we deduce from (6.39) that

$$\gamma_0\sqrt{B - B_{c0}} = const. \tag{6.45}$$

Taking (6.40) and (6.45) into account, we find that

$$\gamma_0 \sim A(B)\sqrt{B - B_{c0}}. \tag{6.46}$$

At finite temperatures, the system remains LFL, but for $T > T^*(B)$ the effective mass becomes temperature-dependent, $M^* \propto 1/T$, and the resistivity becomes a linear function of the temperature, $\rho(T) \propto T$ [69]. Such behavior of the resistivity has been observed in the high-T_c superconductor $Tl_2Ba_2CuO_{6+\delta}$ ($T_c < 15$ K) [70]. At $B < 10$ T, the resistivity is a linear function of the temperature in the range from 120 mK to 1.2 K, and at $B = 10$ T the temperature dependence of the resistivity can be presented in the form $\rho(T) \propto T^2$ in the same temperature range [70, 71], clearly demonstrating that the LFL state is restored under the application of magnetic fields.

In LFL, the spin-lattice relaxation parameter $1/T_1$ is determined by the quasiparticles near the Fermi level, whose population is proportional to M^*T, whence $1/T_1T \propto M^*$, and is a constant quantity [63, 64]. When the superconducting state disappears as a magnetic field is applied, the ground state can be regarded as a field-induced LFL with field-dependent effective mass. As a result, $T_1T = const$, which implies that the Korringa law holds. According to (6.39), the ratio $1/T_1T \propto M^*(B)$ decreases as the magnetic field increases at $T < T^*(B)$, whereas in the case of a Landau Fermi liquid it remains constant, as noted above. On the other hand, at $T > T^*(B)$, the ratio $1/T_1T$ is a decreasing function of the temperature, $1/T_1T \propto M^*(T)$. These results are in good agreement with the experimental data [64]. Since $T^*(B)$ is an increasing function of the magnetic field [see (6.42)], the Korringa law remains valid even at higher temperatures and in stronger magnetic fields. Hence, at $T_0 \leq T^*(B_0)$ and high magnetic fields $B > B_0$, the system demonstrates distinct metallic behavior, because the effective mass decreases as B increases, see (6.39).

The existence of FCQPT can also be verified experimentally, for at number densities $x > x_{FC}$ or beyond the FCQPT point, the system must become LFL at sufficiently low temperatures [54]. Experiments have shown that such a liquid indeed exists in the heavily doped nonsuperconducting compound $La_{1.7}Sr_{0.3}CuO_4$ [46, 47]. It is remarkable that for $T < 55$ K, the resistivity exhibits a T^2-behavior without an additional linear term and the Wiedemann-Franz law holds [46, 47]. At temperatures above 55 K, experimentalists have detected significant deviations from the LFL behavior. Observations [48, 72, 73] are in accord with these experimental findings showing that the system can again be returned to the LFL state by applying sufficiently strong magnetic fields (also see Chap. 7).

6.3.2.1 Common QCP in the High-T_c $Tl_2Ba_2CuO_{6+x}$ and the HF Metal $YbRh_2Si_2$

Under the application of magnetic fields $B > B_{c2} > B_{c0}$ and at $T < T^*(B)$, a high-T_c superconductor or HF metal can be driven to the LFL state with its resistivity given by (2.19). In that case, measurements of the coefficient A give information on its field dependence.

Precise measurements of $A(B)$ on the high-T_c compound $Tl_2Ba_2CuO_{6+x}$ [75] allow us to establish relationships between the physics of both high-T_c superconductors and HF metals and clarify the role of the extended quasiparticle paradigm. The $A(B)$ coefficient, being proportional to the quasiparticle-quasiparticle scattering cross section, is found to be $A \propto (M^*(B))^2$ [57, 74]. With respect to (6.39), this implies that

$$A(B) \simeq A_0 + \frac{D}{B - B_{c0}}, \tag{6.47}$$

where A_0 and D are fitting parameters.

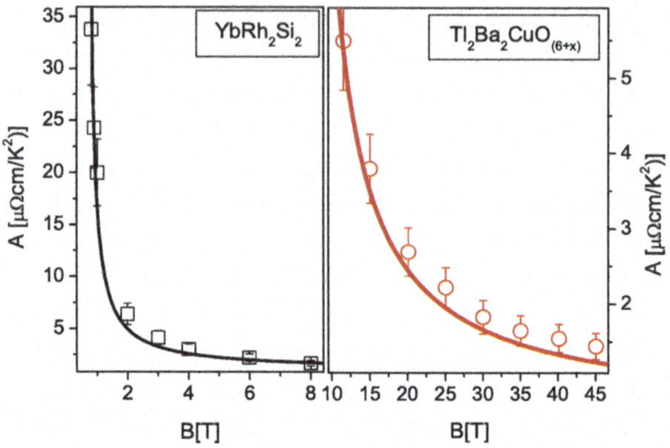

Fig. 6.3 The charge transport coefficient $A(B)$ as a function of magnetic field B obtained in measurements on $YbRh_2Si_2$ [74] and $Tl_2Ba_2CuO_{6+x}$ [75]. The different field scales are clearly seen. The solid curves represent our fit by (6.47)

Figure 6.3 reports the fit of our theoretical dependence (6.47) to the experimental data for the measurements of the coefficient $A(B)$ for two different classes of substances: HF metal $YbRh_2Si_2$ (with $B_{c0} = 0.06$ T, left panel) [74] and high-T_c $Tl_2Ba_2CuO_{6+x}$ (with $B_{c0} = 5.8$ T, right panel) [75]. In Fig. 6.3, left panel, $A(B)$ is shown as a function of magnetic field B, applied both along and perpendicular to the c-axis. For the latter, the B values have been multiplied by a factor of 11 [74]. The different scales of field B_{c0} are clearly seen and demonstrate that B_{c0} has to be taken as an input parameter. Indeed, the critical field of $Tl_2Ba_2CuO_{6+x}$ with $B_{c0} = 5.8$ T is 2 orders of magnitude larger than that of $YbRh_2Si_2$ with $B_{c0} = 0.06$ T.

Figure 6.3 displays good coincidence of the theoretical dependence (6.40) with the experimental facts [75, 76]. This means that the physics underlying the field-induced reentrance into the LFL behavior is the same for both classes of substances. To further corroborate this point, we rewrite (6.47) in the reduced variables A/A_0 and B/B_{c0}. Such rewriting immediately reveals the scaling nature of the behavior of these two substances—both of them are driven to common QCP related to FCQPT and induced by the application of magnetic field. As a result, (6.47) takes the form

$$\frac{A(B)}{A_0} \simeq 1 + \frac{D_N}{B/B_{c0} - 1}, \tag{6.48}$$

where $D_N = D/(A_0 B_{c0})$ is a constant. From (6.48), it is seen that upon applying the scaling to both coefficients $A(B)$ for $Tl_2Ba_2CuO_{6+x}$ and $A(B)$ for $YbRh_2Si_2$, they are reduced to a function depending on the single variable B/B_{c0}, thus demonstrating universal behavior. To support (6.48), we replot both dependencies in reduced variables A/A_0 and B/B_{c0} in Fig. 6.4. Such replotting immediately reveals the universal scaling nature of the behavior of these two substances. It is shown in Fig. 6.4 that close to the magnetic induced QCP there are no "external" physical scales revealing the scaling. Therefore, the normalization by the scales A_0 and B_{c0} immediately uncovers the common physical nature of these substances, allowing us to get rid of the specific properties of the system that define the values of A_0 and B_{c0}. Based on the above analysis of the A coefficients, we conclude that there is at least one quantum

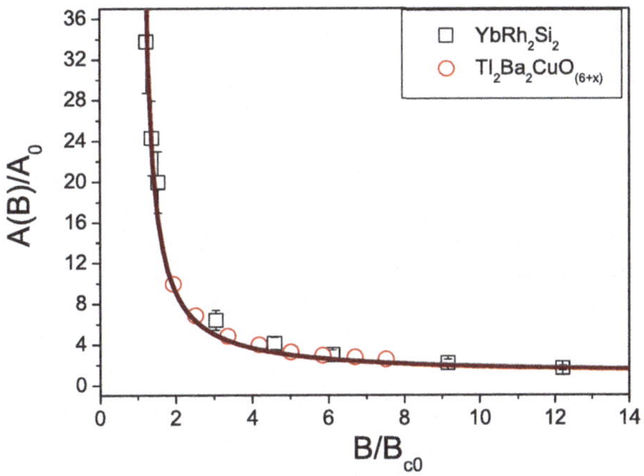

Fig. 6.4 Normalized coefficient $A(B)/A_0 \simeq 1 + D_N/(y-1)$ given by (6.48) as a function of normalized magnetic field $y = B/B_{c0}$ shown by squares for $YbRh_2Si_2$ and by circles for high-T_c $Tl_2Ba_2CuO_{6+x}$. D_N is the only fitting parameter

phase transition inside the superconducting dome of high-T_c superconductors, and this transition is FCQPT [77].

6.4 Appearance of FCQPT in HF Compounds

We. call the Fermi systems approaching QCP from a disordered side as highly correlated systems. We do that in order to distinguish them from strongly correlated systems (or liquids) that are already beyond FCQPT placed at the quantum-critical line as shown in Fig. 4.2. In the present section, we discuss the behavior of such systems as $x \to x_{FC}$. The experimental data for high-density 2D ^{3}He [78–81] show that the effective mass becomes divergent when the density, at which the 2D liquid ^{3}He begins to solidify, is reached [79]. Also observed was a sharp increase in the effective mass in the metallic 2D electron system as the density x decreases and tends to the critical density of the metal-insulator transition [82]. We note that there is no ferromagnetic instability in the Fermi systems under consideration, and the corresponding Landau interaction obeys the inequality $F_0^a > -1$ [79, 82], which agrees with the model of nearly localized fermions [83–85].

In Sect. 5.4, we have examined the divergence of the effective mass in 2D and 3D highly correlated Fermi liquids at $T = 0$ as the density $x \to x_{FC}$ while approaching FCQPT from the disordered phase. As in Sects. 2.3.1 and 5.4, we examine the divergence of the effective mass in 2D and 3D highly correlated Fermi liquids at $T = 0$ as the density $x \to x_{FC}$ while approaching FCQPT from the disordered phase. Both the divergence of M^* related to the generation of density wave and the emergence of the density wave in various HF compounds have been predicted [28]. We note that the emergence of density waves is observed in many experiments on high-T_c superconductors, see e.g., [86]. As $x \to x_{FC}$, the effective mass M^* can be approximately written as

$$\frac{1}{M^*} \simeq \frac{1}{m} + \frac{1}{4\pi^2} \int_{-1}^{1} \int_{0}^{g_0} \frac{y dy dg}{\sqrt{1-y^2}} \frac{v(q(y))}{[1 - R(q(y), g)\chi_0(q(y))]^2}. \tag{6.49}$$

Here we use the notation $p_F\sqrt{2(1-y)} = q(y)$, where $q(y)$ is the momentum, $v(y)$ is the interaction, the integral over coupling constant g is taken from zero to the actual value g_0, $\chi_0(q, \omega)$ is the linear response function for the noninteracting Fermi liquid, and $R(q, \omega)$ is the effective interaction, with both functions taken at $\omega = 0$. The quantities R and χ_0 determine the response function for the system,

$$\chi(q, \omega, g) = \frac{\chi_0(q, \omega)}{1 - R(q, \omega, g)\chi_0(q, \omega)}. \tag{6.50}$$

Near the instability related to the generation of density wave at the density x_{cdw}, the singular part of the response function χ has the well-known form, see, e.g., [26]

$$\chi^{-1}(q, \omega, g) \simeq a(x_{cdw} - x) + b(q - q_c)^2 + c(g_0 - g), \tag{6.51}$$

where a, b, and c are constants and $q_c \simeq 2p_F$ is the momentum of the density wave. Substitution of (6.51) in (6.49) and integration permits to represent the equation for the effective mass M^* in the form

$$\frac{1}{M^*(x)} = \frac{1}{m} - \frac{c}{\sqrt{x - x_{cdw}}}, \tag{6.52}$$

where c is a positive constant. It follows from (6.52) that $M^*(x)$ diverges as a function of the difference $(x - x_{FC})$ and $M^*(x) \to \infty$ as $x \to x_{FC}$ [87, 88]

$$\frac{M^*(x)}{m} \simeq a_1 + \frac{a_2}{x - x_{FC}}, \tag{6.53}$$

where a_1 and a_2 are constants. We note that (6.52) and (6.53) do not explicitly contain the interaction $v(q)$, although $v(q)$ affects a_1, a_2, and x_{FC}. This result agrees with (2.21), which determines the same universal type of divergence (i.e., a divergence that is independent explicitly of the interaction type). Hence, both (2.21) and (6.53) can be applied to 2D ^{3}He, the electron liquid, and other Fermi liquids. We also see that FCQPT. Precedes the formation of density waves (or charge density waves) in Fermi systems. As we have seen in Sect. 6.1, the high-T_c superconductivity is explained within the framework of the fermion-condensation theory. Therefore, one expect to observe charge density waves in high-T_c compounds, and this expectation is in accordance with experimental facts, see, e.g., [89]. We note that the difference $(x - x_{FC})$ must be positive in both cases, since the density x approaches x_{FC} when the system is on the disordered side of FCQPT with finite effective mass $M^*(x) > 0$. In the case of ^{3}He, FCQPT occurs as the density increases, when the potential energy begins to dominate the ground state energy due to the strong repulsive short ranged

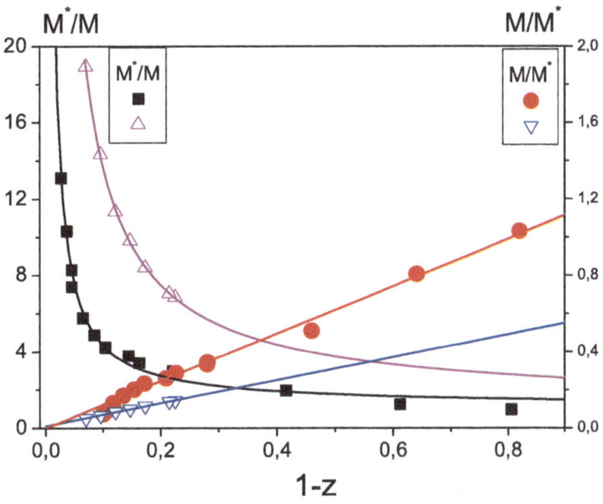

Fig. 6.5 The dependence of the effective mass $M^*(\mathrm{z})$ on dimensionless density $\mathrm{z} = x/x_{FC}$. Experimental data from [79] are shown by circles and squares and those from [81] are shown by triangles. The effective mass is fitted as $M^*(\mathrm{z})/M \propto b_1 + b_2/(1-\mathrm{z})$ (also see (2.21)), while the reciprocal one as $M/M^*(\mathrm{z}) \propto b_3\,\mathrm{z}$, where b_1, b_2, and b_3 are constants

part of the interparticle interaction. Thus, for the 2D ^{3}He liquid, the difference $(x - x_{FC})$ on the right-hand side of (6.53) must be replaced by $(x_{FC} - x)$. Experiments have shown that the effective mass indeed diverges at high densities for 2D ^{3}He and at low ones for 2D electron systems [79, 82].

Experiments show that the effective mass indeed diverges at high densities for 2D ^{3}He and at low densities for 2D electron systems [79, 82]. In Fig. 6.5, we present the experimental values of the effective mass $M^*(\mathrm{z})$ obtained by the measurements on ^{3}He monolayer [79]. These measurements, in coincidence with those from [81, 81], show the divergence of the effective mass at $x = x_{FC}$. To show that our FCQPT approach is able to describe the above data, we represent the fit of $M^*(\mathrm{z})$ by the rational expression $M^*(\mathrm{z})/M \propto b_1 + b_2/(1-\mathrm{z})$ and the reciprocal effective mass by the linear fit $M/M^*(\mathrm{z}) \propto b_3\,\mathrm{z}$. We note here that the linear fit has been used to describe the experimental data for bilayer ^{3}He [81, 90], and we use this function here for the sake of illustration. It is shown in Fig. 6.5 that the data of [81] (^{3}He bilayer) can be equally well approximated by both linear and rational functions, while the data in [79] cannot. For instance, both fitting functions give for the critical density in bilayer $x_{FC} \approx 9.8$ nm^{-2}, while for monolayer [79] these values are different—$x_{FC} = 5.56$ nm^{-2} for linear fit and $x_{FC} = 5.15$ nm^{-2}. It is shown in Fig. 6.5 that the linear fit is unable to properly describe the experiment [79] at small $1-\mathrm{z}$ (i.e., near $x = x_{FC}$), while this fit describes the experiment very well. This means that more detailed measurements are necessary in the vicinity $x = x_{FC}$ [91].

The effective mass as a function of the electron density x in a silicon MOSFET is shown in Fig. 2.1. The divergence of the effective mass $M^*(x)$ discovered in measurements involving 2D ^{3}He [78, 79, 81] is illustrated in Figs. 2.2 and 6.5. Figures 2.1, 2.2, and 6.5 show that the description provided by (6.52) and (6.53) is in good agreement with the experimental data. The results of Sect. 5.4 shows that there is no essential difference between them, although they describe different cases,

2D and 3D. In the 3D case, we can derive equations similar to (6.52) and (6.53) just as we did in the 2D case, but the numerical coefficients are different, because they depend on the dimensionality. The only difference between 2D and 3D electron systems is that FCQPT occurs in 3D systems at densities much lower than in those corresponding to 2D systems. No such transition occurs in massive 3D ^{3}He because of the FCQPT. Transition is presumably absorbed by the first-order liquid-solid phase transition [78, 79].

References

1. L.N. Oliveira, E.K.U. Gross, W. Kohn, Phys. Rev. Lett. **60**, 2430 (1988)
2. J. Bardeen, L.N. Cooper, J.R. Schrieffer, Phys. Rev. **108**, 1175 (1957)
3. D.R. Tilley, J. Tilley, *Superfluidity and Superconductivity* (Hilger, Bristol, 1985)
4. V.A. Khodel, V.R. Shaginyan, JETP Lett. **51**, 553 (1990)
5. J. Dukelsky, V. Khodel, P. Schuck, V. Shaginyan, Z. Phys. **102**, 245 (1997)
6. V.A. Khodel, V.R. Shaginyan, in *Condensed Matter Theories 12*, ed. by J. Clark, V. Plant (Nova Science Publishers Inc., New York, 1997), p. 221
7. G.E. Volovik, JETP Lett. **53**, 222 (1991)
8. S.A. Artamonov, Y.G. Pogorelov, V.R. Shaginyan, JETP Lett. **68**, 942 (1998)
9. Y.G. Pogorelov, V.R. Shaginyan, in *Condensed Matter Theories 18* (Nova Science Publishers Inc., New York, 2003), p. 191
10. M.Y. Amusia, V.R. Shaginyan, JETP Lett. **73**, 232 (2001)
11. M.Y. Amusia, V.R. Shaginyan, Phys. Rev. B **63**, 224507 (2001)
12. V.R. Shaginyan, Phys. B **312-313C**, 413 (2002)
13. S.A. Artamonov, V.R. Shaginyan, JETP **92**, 287 (2001)
14. L.P. Gor'kov, Sov. Phys. JETP **7**, 505 (1958)
15. E.M. Lifshitz, L.P. Pitaevskii, *Statisticheskaya Fizika (Statistical Physics), part 2 Course of Theoretical Physics* (Pergamon Press, Oxford, 1980)
16. V.R. Shaginyan, A.Z. Msezane, V.A. Stephanovich, E.V. Kirichenko, Europhys. Lett. **76**, 898 (2006)
17. V.I. Belyavsky, Y.V. Kopaev, Phys. Usp. **49**, 441 (2006)
18. N.N. Bogoliubov, Nuovo Cimento **7**, 794 (1958)
19. H. Matsui, T. Sato, T. Takahashi, S.C. Wang, H.B. Yang, H. Ding, T. Fujii, T. Watanabe, A. Matsuda, Phys. Rev. Lett. **90**, 217002 (2003)
20. M.Y. Amusia, S.A. Artamonov, V.R. Shaginyan, JETP Lett. **74**, 396 (2001)
21. R.B. Laughlin, D. Pines, Proc. Natl. Acad. Sci. USA **97**, 28 (2000)
22. P.W. Anderson, Science **288**, 480 (2000)
23. M.Y. Amusia, V.R. Shaginyan, Contrib. Plasma Phys. **53**, 721 (2013)
24. D. Yudin, D. Hirschmeier, H. Hafermann, O. Eriksson, A.I. Lichtenstein, M.I. Katsnelson, Phys. Rev. Lett. **112**, 070403 (2014)
25. V.Y. Irkhin, A.A. Katanin, M.I. Katsnelson, Phys. Rev. Lett. **89**, 076401 (2002)
26. C.M. Varma, Z. Nussionov, W. van Saarloos, Phys. Rep. **361**, 267 (2002)
27. M.Y. Amusia, V.R. Shaginyan, JETP Lett. **76**, 651 (2002)
28. V.A. Khodel, V.R. Shaginyan, M.V. Zverev, JETP Lett. **65**, 242 (1997)
29. N. Miyakawa, J. Zasadzinski, L. Ozyuzer, P. Guptasarma, D.G. Hinks, C. Kendziora, K.E. Gray, Phys. Rev. Lett. **83**, 1018 (1999)
30. V.A. Khodel, M.V. Zverev, V.M. Yakovenko, Phys. Rev. Lett. **95**, 236402 (2005)
31. C. Petrovic, P.G. Pagliuso, M.F. Hundley, R. Movshovich, J.L. Sarrao, J.D. Thompson, Z. Fisk, P. Monthoux, J. Phys. Condens. Matter **13**, L337 (2001)

32. A. Kaminski, M. Randeria, J.C. Campuzano, M.R. Norman, H. Fretwell, J. Mesot, T. Sato, T. Takahashi, K. Kadowaki, Phys. Rev. Lett. **86**, 1070 (2001)
33. T. Valla, A.V. Fedorov, P.D. Johnson, B.O. Wells, S.L. Hulbert, Q. Li, G.D. Gu, N. Koshizuka, Science **285**, 2110 (1999)
34. T. Valla, A.V. Fedorov, P.D. Johnson, Q. Li, G.D. Gu, N. Koshizuka, Phys. Rev. Lett. **85**, 828 (2000)
35. P.V. Bogdanov, A. Lanzara, S.A. Kellar, X.J. Zhou, E.D. Lu, W.J. Zheng, G. Gu, J.I. Shimoyama, K. Kishio, H. Ikeda, R. Yoshizaki, Z. Hussain, Z.X. Shen, Phys. Rev. Lett. **85**, 2581 (2000)
36. C.M. Varma, P.B. Littlewood, S. Schmittrink, E. Abrahams, A.E. Ruckenstein, Phys. Rev. Lett. **63**, 1996 (1989)
37. C.M. Varma, P.B. Littlewood, S. Schmittrink, E. Abrahams, A.E. Ruckenstein, Phys. Rev. Lett. **64**, 497 (1990)
38. A. Ino, C. Kim, M. Nakamura, T. Yoshida, T. Mizokawa, A. Fujimori, Z.X. Shen, T. Kakeshita, H. Eisaki, S. Uchida, Phys. Rev. B **65**, 094504 (2002)
39. X.J. Zhou, T. Yoshida, A. Lanzara, P.V. Bogdanov, S.A. Kellar, K. Shen, W.L. Yang, F. Ronning, T. Sasagawa, T. Kakeshita, T. Noda, H. Eisaki, S. Uchida, C.T. Lin, F. Zhou, J.W. Xiong, W.X. Ti, Z.X. Zhao, A. Fujimori, Z. Hussain, Z.X. Shen, Nature **423**, 398 (2003)
40. W.J. Padilla, Y.S. Lee, M. Dumm, G. Blumberg, S. Ono, K. Segawa, S. Komiya, Y. Ando, D.N. Basov, Phys. Rev. B **72**, 060511 (2005)
41. T. Valla, A.V. Fedorov, P.D. Johnson, S.L. Hulbert, Phys. Rev. Lett. **83**, 2085 (1999)
42. J.D. Koralek, J.F. Douglas, N.C. Plumb, Z. Sun, A.V. Fedorov, M.M. Murnane, H.C. Kapteyn, S.T. Cundiff, Y. Aiura, K. Oka, H. Eisaki, D.S. Dessau, Phys. Rev. Lett. **96**, 017005 (2006)
43. D.L. Feng, A. Damascelli, K.M. Shen, N. Motoyama, D.H. Lu, H. Eisaki, K. Shimizu, J.I. Shimoyama, K. Kishio, N. Kaneko, M. Greven, G.D. Gu, X.J. Zhou, C. Kim, F. Ronning, N.P. Armitage, Z.X. Shen, Phys. Rev. Lett. **88**, 107001 (2002)
44. A.B. Migdal, *Theory of Finite Fermi Systems and Applications to Atomic Nuclei* (Wiley, New York, 1967)
45. Z.M. Yusof, B.O. Wells, T. Valla, A.V. Fedorov, P.D. Johnson, Q. Li, C. Kendziora, S. Jian, D.G. Hinks, Phys. Rev. Lett. **88**, 167006 (2002)
46. S. Nakamae, K. Behnia, N. Mangkorntong, M. Nohara, H. Takagi, S.J.C. Yates, N.E. Hussey, Phys. Rev. B **68**, 100502(R) (2003)
47. N.E. Hussey, M. Abdel-Jawad, A. Carrington, A.P. Mackenzie, L. Balicas, Nature **425**, 814 (2003)
48. Y.G. Pogorelov, V.R. Shaginyan, JETP Lett. **76**, 532 (2002)
49. M. de Llano, J.P. Vary, Phys. Rev. C **19**, 1083 (1979)
50. M. de Llano, A. Plastino, J.P. Zabolitsky, Phys. Rev. C **20**, 2418 (1979)
51. M.V. Zverev, M. Baldo, J. Phys. Condens. Matter **11**, 2059 (1999)
52. V.A. Khodel, J.W. Clark, M.V. Zverev, Phys. Rev. B **78**, 075120 (2008)
53. M.Y. Amusia, K.G. Popov, V.R. Shaginyan, W.A. Stephanowich, *Theory of Heavy-Fermion Compounds*. Springer Series in Solid-State Sciences, vol. 182 (Springer, Berlin, 2015)
54. V.R. Shaginyan, A.Z. Msezane, M.Y. Amusia, Phys. Lett. A **338**, 393 (2005)
55. V.R. Shaginyan, JETP Lett. **79**, 286 (2004)
56. D. Takahashi, S. Abe, H. Mizuno, D. Tayurskii, K. Matsumoto, H. Suzuki, Y. Onuki, Phys. Rev. B **67**, 180407(R) (2003)
57. V.A. Khodel, P. Schuck, Z. Phys. B: Condens. Matter **104**, 505 (1997)
58. K. Kadowaki, S.B. Woods, Solid State Commun. **58**, 507 (1986)
59. N. Tsujii, H. Kontani, K. Yoshimura, Phys. Rev. Lett. **94**, 057201 (2005)
60. A.C. Jacko, J.O. Fjærestad, B.J. Powell, Nat. Phys. **5**, 422 (2009)
61. C. Proust, E. Boaknin, R.W. Hill, L. Taillefer, A.P. Mackenzie, Phys. Rev. Lett. **89**, 147003 (2002)
62. R. Bel, K. Behnia, Y. Nakajima, K. Izawa, Y. Matsuda, H. Shishido, R. Settai, Y. Onuki, Phys. Rev. Lett. **92**, 217002 (2004)
63. J. Korringa, Physica (Utrecht) **16**, 601 (1950)
64. G.Q. Zheng, T. Sato, Y. Kitaoka, M. Fujita, K. Yamada, Phys. Rev. Lett. **90**, 197005 (2003)

65. C.L. Kane, M.P.A. Fisher, Phys. Rev. Lett. **76**, 3192 (1996)
66. C.L. Kane, M.P.A. Fisher, Phys. Rev. B **55**, 15832 (1997)
67. T. Senthil, M.P.A. Fisher, Phys. Rev. B **62**, 7850 (2000)
68. A. Houghton, S. Lee, J.B. Marston, Phys. Rev. B **65**, 22503 (2002)
69. V.A. Khodel, M.V. Zverev, JETP Lett. **85**, 404 (2007)
70. A.P. Mackenzie, S.R. Julian, D.C. Sinclair, C.T. Lin, Phys. Rev. B **53**, 5848 (1996)
71. M.Y. Amusia, V.R. Shaginyan, Phys. Lett. A **315**, 288 (2003)
72. V.R. Shaginyan, M.Y. Amusia, K.G. Popov, Phys. Usp. **50**, 563 (2007)
73. V.R. Shaginyan, M.Y. Amusia, A.Z. Msezane, K.G. Popov, Phys. Rep. **492**, 31 (2010), arXiv:1006.2658
74. P. Gegenwart, J. Custers, C. Geibel, K. Neumaier, K.T.T. Tayama, O. Trovarelli, F. Steglich, Phys. Rev. Lett. **89**, 056402 (2002)
75. T. Shibauchi, L. Krusin-Elbaum, M. Hasegawa, Y. Kasahara, R. Okazaki, Y. Matsuda, Proc. Natl. Acad. Sci. USA **105**, 7120 (2008)
76. V.R. Shaginyan, K.G. Popov, JETP Lett. **88**, 183 (2008)
77. V.R. Shaginyan, M.Y. Amusia, K.G. Popov, V.A. Stephanovich, Phys. Lett. A **373**, 686 (2009)
78. A. Casey, H. Patel, J. Nyeki, B.P. Cowan, J. Saunders, Low Temp. Phys. **113**, 293 (1998)
79. A. Casey, H. Patel, J. Cowan, B.P. Saunders, Phys. Rev. Lett. **90**, 115301 (2003)
80. K.D. Morhard, C. Bauerle, J. Bossy, Y. Bunkov, S.N. Fisher, H. Godfrin, Phys. Rev. B **53**, 2658 (1996)
81. M. Neumann, J. Nyéki, B. Cowan, J. Saunders, Science **317**, 1356 (2007)
82. A.A. Shashkin, S.V. Kravchenko, V.T. Dolgopolov, T.M. Klapwijk, Phys. Rev. B **66**, 073303 (2002)
83. D. Vollhardt, Rev. Mod. Phys. **56**, 99 (1984)
84. M. Pfitzner, P. Wölfle, Phys. Rev. B **33**, 2003 (1986)
85. D. Vollhardt, P. Wölfle, P.W. Anderson, Phys. Rev. B **35**, 6703 (1987)
86. R. Arpaia, S. Caprara, Fumagalli, G.D. Vecchi, Y.Y. Peng, E. Andersson, D. Betto, G.M.D. Luca, N.B. Brookes, F. Lombardi, M. Salluzzo, L. Braicovich, C.D. Castro, M. Grilli, G. Ghiringhelli, Science **365**, 906 (2019)
87. V.M. Yakovenko, V.A. Khodel, JETP Lett. **78**, 850 (2003)
88. V.R. Shaginyan, JETP Lett. **77**, 99 (2003)
89. R. Comin, A. Frano, M.M. Yee, Y. Yoshida, H. Eisaki, E. Schierle, E. Weschke, R. Sutarto, F. He, A. Soumyanarayanan, Y. He, M.L. Tacon, I.S. Elfimov, J.E. Hoffman, G.A. Sawatzky, B. Keimer, A. Damascelli, Science **343**, 390 (2014)
90. A. Benlagra, C. Pépin, Phys. Rev. Lett. **100**, 176401 (2008)
91. V.R. Shaginyan, A.Z. Msezane, K.G. Popov, V.A. Stephanovich, Phys. Rev. Lett. **100**, 096406 (2008)

Chapter 7
Effective Mass and Its Scaling Behavior

Abstract In this chapter, we consider the scaling behavior of effective mass M^* that determines the general scaling behavior of strongly correlated Fermi systems. To derive the corresponding equation determining the effective mass, we employ both the model of a homogeneous HF liquid and the density functional theory of the fermion condensation (see Chap. 3) which allows us to consider the ground state energy E as a functional of the occupations numbers $n(\mathbf{p})$. As a result, the ground state energy of the normal state E becomes the functional of the occupation numbers and the function of the number density x, $E = E(n(\mathbf{p}), x)$. We obtain the well-known equation, which gives the single-particle spectrum and the effective mass M^* and allows us to reveal its scaling behavior. We also explain how to compare the observed scaling behavior with experimental facts collected on heavy-fermion compounds.

7.1 Scaling Behavior of the Effective Mass Near the Topological FCQPT

One of the main experimental manifestations of the topological FCQPT. Phenomenon is the scaling behavior of the physical properties of HF compounds located near this phase transition, see Sect. 1.4, Figs. 1.3 and 1.4, and [1, 2] for details. To theoretically understand this scaling behavior, we begin with a description of such a behavior exhibited by the effective mass M^* in the frame of the model of a homogeneous HF liquid, see Sects. 1.5 and 4.1.1. This model avoids the complications associated with the anisotropy of solids and considers both the thermodynamic properties and NFL behavior by calculating the effective mass $M^*(T, B)$ as a function of temperature T and magnetic field B. To study the behavior of the effective mass $M^*(T, H)$, we use the Landau equation for the quasiparticle effective mass. The only modification is that in our formalism the effective mass is no longer approximately constant but depends on temperature, magnetic field, and other parameters like pressure, etc. For the model of homogeneous HF liquid at finite temperatures and magnetic fields, this equation takes the form [1, 3–5]

M. Amusia and V. Shaginyan, *Strongly Correlated Fermi Systems*, Springer Tracts in Modern Physics 283,
https://doi.org/10.1007/978-3-030-50359-8_7

$$\frac{1}{M^*_\sigma(T,H)}=\frac{1}{M}+\sum_{\sigma_1}\int\frac{\mathbf{p}_F\mathbf{p}}{p_F^3}F_{\sigma,\sigma_1}(\mathbf{p_F},\mathbf{p})\frac{\partial n_{\sigma_1}(\mathbf{p},T,H)}{\partial p}\frac{d\mathbf{p}}{(2\pi)^3}. \quad (7.1)$$

The single-particle spectrum $\varepsilon(\mathbf{p},T)$ is a variational derivative of the system energy $E[n_\sigma(\mathbf{p})]$ with respect to the quasiparticle distribution function or occupation numbers n,

$$\varepsilon_\sigma(\mathbf{p})=\frac{\delta E[n(\mathbf{p})]}{\delta n_\sigma(\mathbf{p})}, \quad (7.2)$$

can be expressed as

$$n_\sigma(\mathbf{p},T)=\left\{1+\exp\left[\frac{(\varepsilon(\mathbf{p},T)-\mu_\sigma)}{T}\right]\right\}^{-1}. \quad (7.3)$$

In our case, the chemical potential μ depends on the spin due to Zeeman splitting $\mu_\sigma=\mu\pm\mu_B B$, where μ_B is Bohr magneton. Note that magnetic field B enters (7.3) as the ratio $\mu_\sigma/T=\mu\pm\mu_B B/T$. We note that (7.1) and (7.2) are exact, as it is shown in Chap. 3. In our case, the Landau interaction F is fixed by the condition that the system is situated at FCQPT. The sole role of the Landau interaction is to bring the system to the topological FCQPT point, where $M^*\to\infty$ at $T=0$ and $B=0$, and the Fermi surface alters its topology being transformed into Fermi volume (see Chap. 4) so that the effective mass acquires temperature and field dependence [1, 3, 4, 6]. Provided that the Landau interaction is an analytical function, at the Fermi surface the momentum-dependent part of the Landau interaction can be taken in the form of truncated power series $F=aq^2+bq^3+cq^4+\cdots$, where $\mathbf{q}=\mathbf{p}_1-\mathbf{p}_2$, a, b, and c are fitting parameters which are defined by the condition that the system is at FCQPT.

A direct inspection of (7.1) shows that at $T=0$ and $B=0$, the sum of the first term and the second one on the right side vanishes, since $1/M^*(T\to 0)\to 0$ provided that the system is located at FCQPT, see Chap. 2, Sect. 2.3.1. In case of analytic Landau interaction with respect to the momenta variables, at finite T the right-hand side is proportional $F'(M^*)^2T^2$, where F' is the first derivative of $F(q)$ with respect to q at $q\to 0$. Calculations of the corresponding integrals can be found in textbooks, see, e.g., in [7]. Thus, we have $1/M^*\propto(M^*)^2T^2$ and obtain [1, 2]

$$M^*(T)\simeq a_T T^{-2/3}. \quad (7.4)$$

At finite temperatures, the application of magnetic field $\mu_B B\gg k_B T$ drives system to the LFL region with

$$M^*(B)\simeq a_B B^{-2/3}. \quad (7.5)$$

Here a_T and a_B are parameters, and μ_B and k_B are the Bohr magneton and the Boltzmann constants, respectively. If the system is located before FCQPT, the effective mass is finite $M^*=M_0$, as it is shown in Fig. 4.2, and (7.5) is to be transformed, since at $B\to 0$ the effective mass M^* does not diverge,

$$M^*(B) \simeq a_B(B_0 + B)^{-2/3}. \tag{7.6}$$

It follows from (7.6) that at some $B = B_0$ the effective mass becomes $M^* = M_0$. Therefore, at growing magnetic fields $B \gg B_0$ M^* starts to depend on the magnetic field in accordance with (7.6), since the contribution coming from B defines the behavior of M^*. As a result, we can replace (7.5) by the estimate equation

$$M^*(B) \simeq M_0 + a_B B^{-2/3}. \tag{7.7}$$

In the case of HF liquid, the above observations permit to construct the approximate solution of (7.1) in the form $M^* = M^*(B, T)$ that satisfies (7.4) and (7.5). Since we are going to describe the thermodynamic, Transport, and relaxation properties of HF compounds, we have to construct the equation that can be employed to calculate the universal effective mass M^* of these compounds. The introduction of "internal" scales simplifies the problem under consideration, for in that case we get rid of microscopic structure of HF compounds under consideration [1, 2]. To construct the "internal" scales, we observe that near FCQPT, the effective mass $M^*(B, T)$ reaches maximum M^*_M at certain temperature $T_M \propto B$, see comments to (7.3) and Fig. 7.1. Now we face a question: What are the units to measure the effective mass? To measure the effective mass and temperature, it is convenient to introduce the scales M^*_M and T_M. In this case, we have new variables $M^*_N = M^*/M^*_M$ (normalized effective mass) and $T_N = T/T_M$ (normalized temperature). In the FCQPT vicinity, the normalized effective mass $M^*_N(T_N)$ can be well approximated by a certain universal function [1, 2], interpolating the system properties between the LFL and NFL states

$$M^*_N(T_N) \approx c_0 \frac{1 + c_1 T_N^2}{1 + c_2 T_N^{8/3}}. \tag{7.8}$$

Here, $T_N = T/T_M$, $c_0 = (1 + c_2)/(1 + c_1)$, where c_1 and c_2 are free parameters. We note that values M^*_M and T_M are defined by the microscopic structure of the HF compound in question, while the normalized values M^*_N and T_N demonstrate the universal scaling behavior exhibited by HF compounds located near the topological FCQPT, for this behavior is formed by this phase transition.

We remark that the Landau interaction $F_{\sigma,\sigma_1}(q)$, entering (7.1), can lead to the general topological form of the spectrum $\varepsilon(p) - \mu \propto (p - p_b)^2(p - p_F)$ with $(p_b < p_F)$ and $(p_F - p_b)/p_F \ll 1$, which leads to $M^* \propto T^{-1/2}$ and creates a quantum-critical point [8]. As we shall see below, the same critical point is generated by the interaction $F(q)$ represented by an integrable over x nonanalytic function with $q = \sqrt{p_1^2 + p_2^2 - 2xp_1p_2}$ and $F(q \to 0) \to \infty$ [1, 9]. Both cases lead to $M^* \propto T^{-1/2}$, and (7.4) becomes

$$M^*(T) \simeq a_T T^{-1/2}. \tag{7.9}$$

In the same way, we obtain

$$M^*(B) \simeq a_B B^{-1/2}, \tag{7.10}$$

with a_T and a_B are parameters. Taking into account that (7.9) leads to the spiky DOS with the vanishing of spiky structure with increasing temperature T [10], as it is observed in quasicrystals [11, 12], we assume that the general form of $\varepsilon(p)$ produces the behavior of M^*, given by (7.9) and (7.10). This is realized in quasicrystals, which can be viewed as a generalized form of common crystals [13]. We note that the behavior $1/M^* \propto \chi^{-1} \propto T^{1/2}$ is in good agreement with $\chi^{-1} \propto T^{0.51}$ observed experimentally [12]. Our explanation is consistent with the robustness of the exponent 0.51 against the hydrostatic pressure [12] since the robustness is guaranteed by the unique singular DOS of QCs that survives under the application of pressure [11, 12, 14–16]. Then, the nonanalytic Landau interaction $F(q)$ can also serve as a good approximation, generating the observed behavior of the effective mass. We speculate that the nonanalytic interaction is generated by the nonconservation of the quasimomentum in QCs, making the Landau interaction $F(q)$ a nonlocal function of momentum q. Such a function can be approximated by a nonanalytic one.

To calculate M/M^*, we use a model Landau functional [1, 9]

$$E[n(p)] = \int \frac{\mathbf{p}^2}{2M}\frac{d\mathbf{p}}{(2\pi)^3} + \frac{1}{2}\int V(\mathbf{p}_1 - \mathbf{p}_2)n(\mathbf{p}_1)n(\mathbf{p}_2)\frac{d\mathbf{p}_1 d\mathbf{p}_2}{(2\pi)^6},$$

with the Landau interaction

$$V(\mathbf{q}) = g_0 \frac{\exp(-\beta_0\sqrt{\mathbf{q}^2+\gamma^2})}{\sqrt{\mathbf{q}^2+\gamma^2}}, \tag{7.11}$$

where the parameters g_0 and β_0 are fixed by the requirement that the system is located at FCQPT. At $\gamma = 0$, the interaction becomes nonanalytic function of q. Note that the other investigated nonanalytic interactions lead to the same behavior of M/M^*, see, e.g., [9]. To demonstrate this, we apply (7.2) to construct $\varepsilon(p)$ using the function (7.11). Taking into account that $\varepsilon(p \simeq p_F) - \mu \simeq p_F(p - p_F)$ and integrating over the angular variables, we obtain

$$\frac{1}{M^*} = \frac{1}{M} + \frac{\partial}{\partial p}\int [\Phi(p+p_1) - \Phi(|p-p_1|)]\frac{n(p_1,T)p_1 dp_1}{2\pi^2 p_F^2}. \tag{7.12}$$

Here, the derivative on the right-hand side of (7.12) is taken at $p = p_F$ and

$$\int_{|p-p_1|}^{p+p_1} V(z, \gamma = 0) z dz = \Phi(p+p_1) - \Phi(|p-p_1|). \tag{7.13}$$

The derivative $\partial\Phi(|p-p_1|)/\partial p|_{p\to p_F} = (p_F - p_1)/(|p_F - p_1|)\partial\Phi(z)/\partial z$ becomes a discontinuous function at $p_1 \to p_F$, provided that $\partial\Phi(z)/\partial z$ is finite (or integrable if the function tends to infinity) at $z \to 0$. As a result, the right-hand side of (7.12) becomes proportional M^*T and (7.12) reads $1/M^* \propto M^*T$, making $M^* \propto T^{-1/2}$.

At finite B and T near FCQPT, the solutions of (7.1) $M^*(T,B)$ can be well approximated by a simple universal interpolating function taking into account (7.10) and (7.9). In the same way, as we have derived (7.8), we obtain that near FCQPT. The normalized solution of (7.1) $M^*_N(T_N)$ with a nonanalytic Landau interaction can be well approximated by a simple universal interpolating function. The interpolation occurs between the LFL ($M^* \propto a + bT^2$) and NFL ($M^* \propto T^{-1/2}$) regimes and represents the universal scaling behavior of $M^*_N(T_N)$

$$M^*_N(T_N) \approx c_0 \frac{1 + c_1 T_N^2}{1 + c_2 T_N^{5/2}}. \tag{7.14}$$

Here a and b are constants, $c_0 = (1 + c_2)/(1 + c_1)$, c_1 and c_2 are fitting parameters.

It follows from (7.3), (7.8) and (7.14) that

$$T_M \simeq a_1 \mu_B B; \quad T_N = \frac{T}{T_M} = \frac{T}{a_1 \mu_B B} \propto \frac{T}{B}, \tag{7.15}$$

where a_1 is a dimensionless factor, and μ_B is the Bohr magneton. Expression (7.15) shows that (7.8) and (7.14) determine the effective mass scaling in terms of T and B. To be more specific, the curves $M^*_N(T,B)$ amalgamate into a single one $M^*_N(T_N)$, $T_N = T/T_M$, as it is shown in Fig. 7.1. Since $T_M \propto B$, as it follows from (7.15), we conclude that the curves $M^*_N(T,B)$ are merged into a single one $M^*_N(T_N = T/B)$, $T_N = T/T_M = T/B$, demonstrating the scaling behavior widespread among HF metals, see, e.g., [1, 2]. Such a behavior is also shown in Fig. 7.1. We note that (7.8) and (7.15) allow one to describe the strongly correlated quantum spin liquid (SCQSL) of different frustrated magnets [1, 2].

One more important feature of the FC state is that apart from the fact that the Landau quasiparticle effective mass starts to depend on external parameters like temperature and magnetic field, all relations, inherent in LFL approach, formally hold in it. Namely, the famous LFL relation [1, 2, 5, 7],

$$M^*(B,T) \propto \chi(B,T) \propto \frac{C(B,T)}{T}, \tag{7.16}$$

holds; while the resistivity $\rho(B,T)$ is given by

$$\rho(B,T) = \rho_0 + (M^*(B,T))^2 T^2. \tag{7.17}$$

Here ρ_0 is the residual resistivity. The expression (7.16) is valid in the case of HF compounds located near the topological FCQPT, where the specific heat C, magnetic susceptibility χ, and effective mass M^* depend on T and B. Taking (7.16) into account, we obtain that normalized values of C/T and χ are of the form [1, 2]

$$M^*_N(B,T) = \chi_N(B,T) = \left(\frac{C(B,T)}{T}\right)_N. \tag{7.18}$$

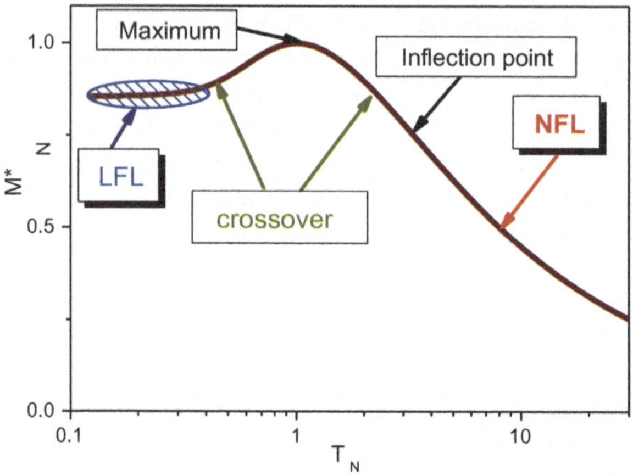

Fig. 7.1 Scaling of the thermodynamic properties governed by the normalized effective mass M^*_N, see (7.16). In the case of the application of magnetic field $T_N \propto T/B$. Thus, M^*_N is a function of T/B, as it follows from (7.15). Solid curve depicts M^*_N versus normalized temperature T_N as a function of magnetic field. It is seen that at finite $T_N < 1$ the normal Fermi liquid properties take place. At $T_N \sim 1$, it enters crossover state and at growing temperatures exhibits the NFL behavior

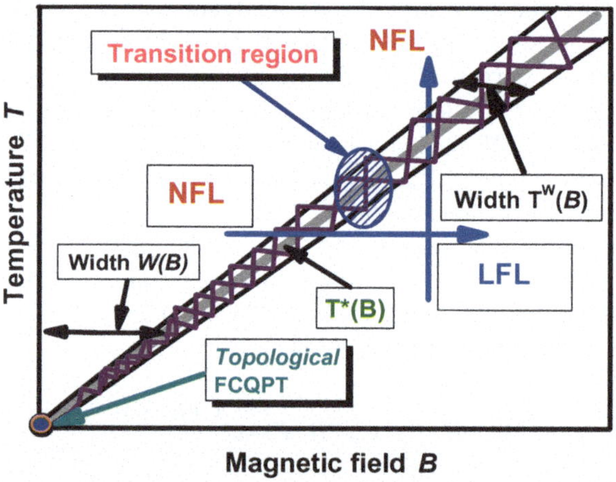

Fig. 7.2 Schematic $T - B$ SCQSL phase diagram. Magnetic field B is the control parameter. The hatched area corresponds to the crossover domain at $T_M(B)$, given by (7.15). At fixed magnetic field and elevating temperature (vertical arrow), there is an LFL-NFL crossover. The horizontal arrow indicates an NFL-LFL transition at fixed temperature and elevating magnetic field. The topological FCQPT (shown in the panel) occurs at $T = 0$ and $B = 0$, where M^* diverges

It is seen from (7.18) that the abovementioned thermodynamic properties have the same scaling behavior depicted in Fig. 7.1. Moreover, below we shall see that the thermodynamic properties of HF metals, SCQSL of frustrated magnets, and the other HF compounds have the same scaling behavior. Based on (7.8) and Fig. 7.1,

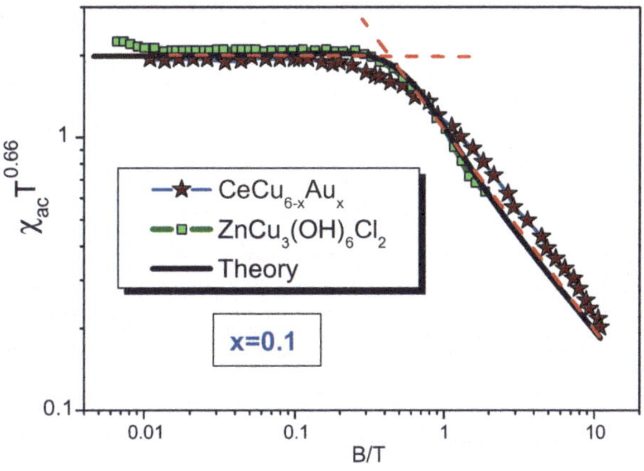

Fig. 7.3 The universal B/T scaling behavior of strongly correlated Fermi systems. The scaling behavior of the HF metal $CeCu_{6-x}Au_x$ is extracted from data [17], and that of $ZnCu_3(OH)_6Cl_2$ is extracted from data [18]. At $B/T \ll 1$, the systems demonstrate the NFL behavior and $\chi \propto M^*$ is given by (7.4), that is, $T^{2/3}\chi \propto const$. At $B/T \gg 1$, the systems demonstrate the LFL behavior and χ is given by (7.5), being a diminishing function of B/T

we can construct the general schematic $T - B$ phase diagram of SCQSL, reported in Fig. 7.2. We assume here that at $T = 0$ and $B = 0$ the system is approximately located at FCQPT. At fixed temperatures, the system is driven along the horizontal arrow (from the NFL to LFL parts of the phase diagram) by the magnetic field B. In turn, at fixed B and elevating T, the system moves from the LFL to the NFL regimes along the vertical arrow. The hatched area indicating the crossover between LFL and NFL phases separates the NFL state from the paramagnetic slightly polarized LFL state. The crossover temperature $T_M(B)$ is given by (7.15).

7.2 T/B Scaling in Heavy-Fermion Compounds

Scaling behavior of HF compounds (or strongly correlated Fermi systems) is given by (7.8) and (7.14), for their thermodynamic and transport properties is given by (7.16) and (7.18). This scaling behavior is displayed in Fig. 7.1. The challenge for theories of HF compounds is to explain the scaling behavior of the normalized effective mass $M^*_N(T_N) \propto C(B,T)/T$ displayed in Fig. 1.3, for most of the theories analyze only the critical exponents that characterize $M^*_N(T_N)$ at $T_N \gg 1$, and consider a part of the problem, missing the LFL behavior and the transition regime.

The heavy-fermion compounds or strongly correlated Fermi systems are represented by HF metals, high-T_c superconductors, quasicrystals, SCQSL of frustrated magnets, and two-dimensional liquids like ^{3}He. HF compounds with their very different compositions and microscopic structures demonstrate very different thermodynamic, transport, and relaxation properties. Nonetheless, these properties exhibit the uniform qualitative behavior, as it is shown in Fig. 1.2, where the application

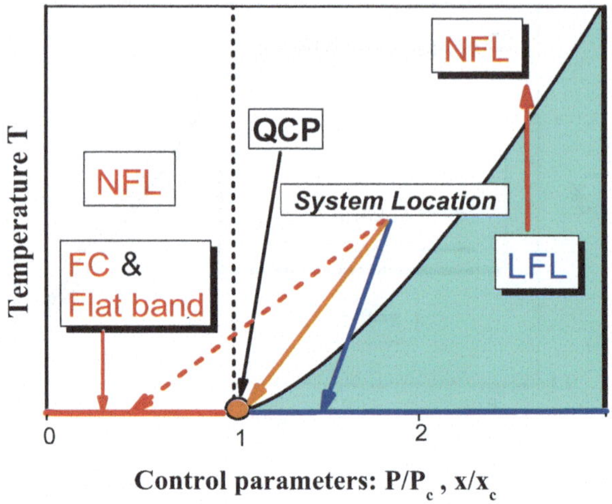

Fig. 7.4 Location of HF compound with respect to FCQPT. Schematic temperature—control parameter (number density, pressure, etc.) phase diagrams of the systems. At $T = 0$ and beyond the FCQPT critical point, the system is at the quantum-critical line with its flat band. At any finite temperature $T < T_f$, the system exhibits the NFL behavior, see (4.11). The blue arrow points to the system located before FCQPT that exhibits at $T \to 0$ the LFL behavior with the effective mass $M^* = M_0$. At elevated magnetic field, the behavior of the effective mass is given by (7.7), and the scaling behavior is restored at $B > B_0$

of magnetic field changes the value of C/T, while the qualitative behavior remains unaltered. To demonstrate the universal scaling behavior of HF compounds, we have introduced internal scales to normalize the observable variables, as it is done when we consider the scaling behavior of the effective mass, see Sect. 7.1 and Figs. 1.3 and 1.4, as examples of normalization. This uniform behavior comes from the fact that HF compounds are located near the topological FCQPT, generating their uniform scaling behavior of the effective mass M^*, see Sect. 7.1 and Fig. 7.1. As an example, Fig. 7.3 displays the universal T/B scaling behavior of the HF metal $CeCu_{6-x}Au_x$ and SCQSL of the frustrated insulator herbertsmithite $ZnCu_3(OH)_6Cl_2$. Such a universal scaling behavior, exhibited by very different strongly correlated Fermi systems, allows one to conclude that HF compounds represent a new state of matter [19]. In contrast to ordinary quantum phase transition, see Sect. 1.4, this scaling behavior induced by FCQPT takes place up to high temperatures, as it is shown in Sect. 4.1.3, thus resolving many puzzles of physics of strongly correlated fermion systems. Among these puzzles, resolved in the book, is the universal behavior of HF compounds, which form the new state of matter, see Chaps. 7, 4, and Sect. 4.1.3.

A few remarks are in order here. A strongly correlated Fermi system can be placed before quantum-critical point (QCP) represented by FCQPT, behind it on the ordered side, or very near it, as it is shown in Fig. 7.4. In case of ordinary quantum phase transition, one may expect that the scaling behavior is defined by some reasons that are not related to the presence of QCP, leading to both the divergence of M^* and the

T/B scaling, see, e.g., [20]. On the other hand, if the system in question is located *before* FCQPT, as shown by the blue arrow in Fig. 7.4, it exhibits the LFL behavior even in the absence of magnetic field B, or at low $B \to 0$. At elevated magnetic fields, as magnetic field becomes $B \gg B_0$, (7.5) and (7.10) are valid and the scaling behavior restores in accordance with (7.7). Thus, to see the presence of both the scaling behavior and the divergency of the effective mass in measurements on HF compounds, one has to carry out the measurements at sufficiently low temperatures and magnetic fields. For example, the HF metal $CeRu_2Si_2$ exhibits the NFL behavior at low temperatures (down to 170 mK) and small magnetic fields ($B \simeq 0.02$ mT) [21]. The measurements carried out under the applications of magnetic fields lead to incorrect statement that $CeRu_2Si_2$ demonstrates the LFL behavior at low temperatures, see [21] and references therein. Thus, we conclude that a theory is a good tool that allows one to clearly understand what one measures. For example, the scaling behavior at $T \to 0$ without QCP could be induced by a simple misunderstanding of experimental data.

References

1. V.R. Shaginyan, M.Y. Amusia, A.Z. Msezane, K.G. Popov, Phys. Rep. **492**, 31 (2010), arXiv:1006.2658
2. M.Y. Amusia, K.G. Popov, V.R. Shaginyan, W.A. Stephanowich, *Theory of Heavy-Fermion Compounds*. Springer Series in Solid-State Sciences, vol. 182 (Springer, Berlin, 2015)
3. V.R. Shaginyan, Phys. Atom. Nuclei **74**, 1107 (2011)
4. V.A. Khodel, J.W. Clark, M.V. Zverev, Phys. Atom. Nuclei **74**, 1237 (2011)
5. L.D. Landau, Zh Eksp, Teor. Fiz. **30**, 1058 (1956)
6. J.W. Clark, V.A. Khodel, M.V. Zverev, Phys. Rev. B **71**, 012401 (2005)
7. E.M. Lifshitz, L. Pitaevskii, *Statistical Physics. Part 2*. (Butterworth-Heinemann, Oxford, 2002)
8. V.A. Khodel, M.V. Zverev, J.W. Clark, JETP Lett. **81**, 315 (2005)
9. V.A. Khodel, V.R. Shaginyan, V.V. Khodel, Phys. Rep. **249**, 1 (1994)
10. J. Dukelsky, V. Khodel, P. Schuck, V. Shaginyan, Z. Phys. **102**, 245 (1997)
11. R. Widmer, P. Gröning, M. Feuerbacher, O. Gröning, Phys. Rev. B **79**, 104202 (2009)
12. K. Deguchi, S. Matsukawa, N.K. Sato, T. Hattori, K. Ishida, H. Takakura, T. Ishimasa, Nat. Mater. **11**, 1013 (2012)
13. R. Lifshitz, Isr. J. Chem. **51**, 1156 (2011)
14. T. Fujiwara, T. Yokokawa, Phys. Rev. Lett. **66**, 333 (1991)
15. T. Fujiwara, in *Physical Properties of Quasicrystals*. Springer Series in Solid-State Sciencesed. by Z.M. Stadnik (Springer, Berlin, 1999)
16. T. Fujiwara, S. Yamamoto, G.T. de Laissardière, Phys. Rev. Lett. **71**, 4166 (1993)
17. A. Schröder, G. Aeppli, R. Coldea, M. Adams, O. Stockert, H.v. Löhneysen, E. Bucher, R. Ramazashvili, P. Coleman, Nature **407**, 357 (2000)
18. J.S. Helton, K. Matan, M.P. Shores, E.A. Nytko, B.M. Bartlett, Y. Qiu, D.G. Nocera, Y.S. Lee, Phys. Rev. Lett. **104**, 147201 (2010)
19. V.R. Shaginyan, V.A. Stephanovich, A.Z. Msezane, P. Schuck, J.W. Clark, M.Y. Amusia, G.S. Japaridze, K.G. Popov, E.V. Kirichenko, J. Low Temp. Phys. **189**, 410 (2017)
20. A. Sakai, K. Kitagawa, K. Matsubayashi, M. Iwatani, P. Gegenwart, Rev. Rev. B. **94**, 041106(R) (2016)
21. D. Takahashi, S. Abe, H. Mizuno, D. Tayurskii, K. Matsumoto, H. Suzuki, Y. Onuki, Phys. Rev. B **67**, 180407(R) (2003)

Chapter 8
Quantum Spin Liquid in Geometrically Frustrated Magnets and the New State of Matter

Abstract In this Chapter, we consider QSL formed by spinons that are chargeless fermionic quasiparticles with spin 1/2, filling up the Fermi sphere up to the Fermi momentum p_F. The excitations of QSL are spinons, which are chargeless fermionic quasiparticles with spin 1/2. We expose a state of the art in the investigations of physical properties of geometrically frustrated magnets with QSL. As the QSL excitations are fermions, the most appropriate description of observed phenomena should be based on some fermionic formalism rather than on different forms of standard spin-wave approaches. So, one more purpose of this chapter centers on a theory employed, demonstrating its range of applicability to a novel expression of magnetic behavior. We elucidate the nature of the used QSL in terms of both experimental facts and the theory of fermion condensation (FC). Our analysis, based on the FC theory, permits to describe the multitude of experimental results regarding the thermodynamic and transport properties of QSL in geometrically frustrated magnets like herbertsmithite $ZnCu_3(OH)_6Cl_2$, the organic insulators $EtMe_3Sb[Pd(dmit)_2]_2$ and $\kappa - (BEDT - TTF)_2Cu_2(CN)_3$, quasi-one-dimensional spin liquid in the $Cu(C_4H_4N_2)(NO_3)_2$ insulator, and QSL formed in two-dimensional ^3He. Transport properties of the compounds shed light on the nature of quantum spin liquid. Analysis of the heat conductivity detects its scaling behavior resembling those of both the spin-lattice relaxation rate and the magnetoresistivity. It reveals a strong magnetic field dependence of the spinons effective mass. As a result, the strongly correlated electrical insulator gains also a new magnetic feature of the matter, for the spins represented by the deconfined QSL get mobility. Based on the experimental facts and the theory, we show that the considered magnets exhibit the universal scaling behavior resembling that of heavy-fermion metals, including T/B scaling with T being temperature and B magnetic field. We also show that QSL as a member of strongly correlated Fermi systems represents the new state of matter.

M. Amusia and V. Shaginyan, *Strongly Correlated Fermi Systems*,
Springer Tracts in Modern Physics 283,
https://doi.org/10.1007/978-3-030-50359-8_8

8.1 Introduction

The exotic substances have exotic properties. One class of such substances is geometrically frustrated magnets, where correlated spins reside in the cites of triangular or kagomé lattice. In some cases such magnet would not have long-range magnetic order. Rather, its spins tend to form kind of pairs, called valence bonds. At $T \to 0$ these highly entangled quantum objects condense in the form of a liquid, called quantum spin liquid (QSL). The observation of a gapless QSL in actual materials is of fundamental significance both theoretically and technologically, as it could open a path to creation of topologically protected states for quantum information processing and computation [1, 2].

It is well known that frustration appears when a person cannot achieve some predefined specific goal. One of possible reactions in this case is to attain not exactly the above impossible goal but something close to it, which can still be considered to be more or less satisfactory. The same is true for nonliving, physical systems, where frustration either in the interparticle interactions or in their positions (so-called geometrical frustration) generates a number of unusual properties. These properties may have to do with nonexistence of well-defined ground state, as in spin glasses [3, 4], where interacting spins are located randomly in the cites of some host crystal lattice. In this case, the unique ground state is substituted by plethora of metastable, extremely slowly relaxing ones. Such configuration is definitely free of any long-range magnetic order [3, 4] and demonstrates many exotic properties like Nernst theorem violation, i.e., finite entropy at $T = 0$. One more object, where Nernst theorem is violated, is Fermi system with the fermion condensation (FC), which has strongly degenerate ground state [5–10]. The other example is so-called geometrically frustrated magnets, where the frustration occurs on purely geometric grounds, see e.g., [11]. The example here is Ising spins, having only two possible orientations and trying to align antiferromagnetically on the sites of triangular lattice.

This situation is depicted in Fig. 8.1. It is seen that while (for negative exchange constant $J < 0$) two of the spins can be antiparallel, the third one cannot. This typical geometric frustration immediately implies the degenerate ground state consisting of six possible "up-down-up" configurations of a single triangular plaquette, see

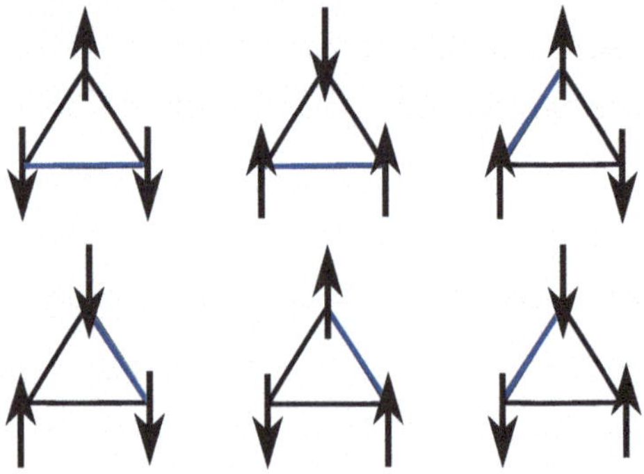

Fig. 8.1 Ising spins on triangular lattice. Six possible configurations for ground state are shown. Arrows indicate the spin direction. Blue lines denote the frustrated bonds, along which spins are parallel

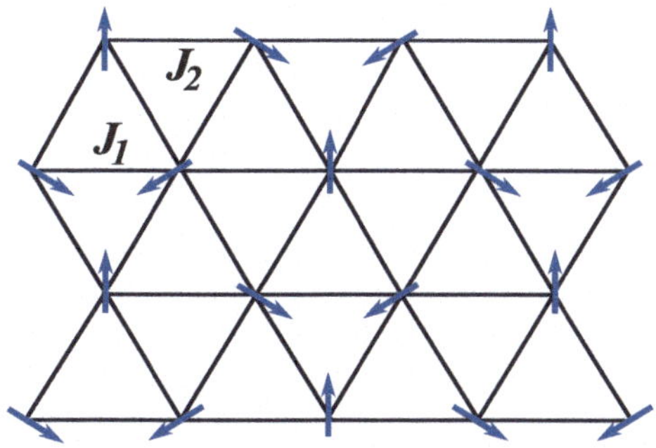

Fig. 8.2 Heisenberg spins (blue arrows) on triangular lattice forming typical 120° frustrated pattern. Different possible exchange constants J_1 and J_2, are shown

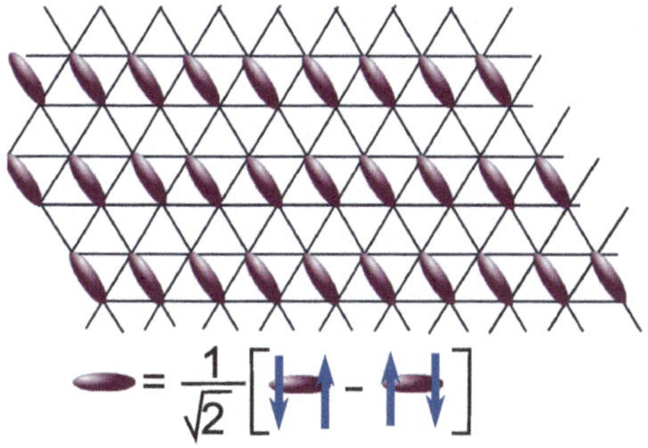

Fig. 8.3 Regular pattern of entangled spin pairs (valence bonds) in a frustrated antiferromagnet, realizing valence-bond solid on the triangular lattice. The structure of a valence bond (spin singlet state), denoted by oval, is shown below the panel

Fig. 8.1. Such degeneracy enhances fluctuations and suppresses magnetic ordering even in 3D case. Although in Heisenberg model the situation differs in the sense that there are all three components of the spin, the ground state degeneracy in the above geometrically frustrated case persists, giving rise either to exotic 120° frustrated pattern (magnetic order, Fig. 8.2) or precluding such order at all. We note here that for 2D geometrically frustrated lattices the evidence for ground state depends heavily on theoretical interpretation. For example, to form 120° pattern for Heisenberg spins on a triangular lattice (Fig. 8.2), there should be two different exchange interactions J_1 and J_2, as shown in the Fig. 8.2. This generates many possible ground states, which should be distinguished for each specific substance (having its own magnetic ions, lattice imperfections, etc.) experimentally. Similarly, in this chapter, we will avoid any attempt to classify the magnetic ground states, as we feel this obscures the nature of the problem. Rather, we will describe the universal properties of the ground states of two-dimensional (2D) frustrated magnets, related primarily to the realization of so-called quantum spin liquid in them, see below. In other words, we can safely assume, that the typical ground state of a geometrically frustrated magnet is nonmagnetic with resulting zero spin. The reasonable component, from which such ground state can be formed, is a bond of two antiparallel spins, aligned so because of antiferromagnetic exchange.

This bond, shown in Fig. 8.3, forms a spin singlet with resulting spin zero and called a *valence bond*. It is stable as two spins, which form it, are *maximally entan-*

gled. The quantum entanglement concept had been introduced by Scrödinger [12]. It asserts that two or more quantum states are *entangled* if corresponding objects (like spin in our case) are strongly bounded in such a way that they form new entity. That is to say, the entangled quantum state of each specific object cannot be considered independently of the state of the other(s), even when they are spatially separated. Instead, such quantum state must be ascribed to the system as a whole. It is interesting to note, that in many body quantum systems, the characteristic length of entanglement may exceed the ordinary correlation length and even be divergent [13]. The phenomenon of quantum entanglement is the primary resource in emerging technologies of quantum computation [14]. Below we shall see that this is also the reason for increasing interest in the investigations of geometrically frustrated magnets and those on kagomé lattice in particular. Quantum entanglement also plays an important role in other branches of condensed matter physics; it can be detected experimentally, see e.g., [15].

Note that triplet excitation (i.e., that with spin 1) of a singlet bond is nothing but bosonic magnon (spin-wave), which can undergo Bose-Einstein condensation [16]. The valence bonds are localized on the corresponding lattice sites, forming some regular structure. This state is known as a valence-bond solid (VBS) [17]. This state is shown schematically in Fig. 8.3. It is seen that the arrangement of valence bonds in a VBS is not unique. For instance, in the triangular lattice of Fig. 8.3 all bonds could be aligned either horizontally or along other side of triangles. This shows that VBS state breaks the initial symmetry of triangular lattice. To lift this symmetry breaking, it is reasonable to suppose that VBS can "melt" forming something like valence-bond liquid. Indeed, such state exists and is called *quantum spin liquid* (QSL). The main difference between QSL and VBS is that former state is characterized by long-range quantum entanglement. This implies that the valence bonds can be of arbitrary range, see Fig. 8.4. In other words, the quantum fluctuations of valence bonds in QSL lead to the ground state consisting of a superposition of different partitioning of spins into valence bonds, see Fig. 8.4, where this is shown. The wave function of such ground state had been suggested by Anderson in 1973 [18] and is called resonating valence bond (RVB) state. It had been shown theoretically, that RVB wave function realizes ground state of many geometrically frustrated magnets in two and three dimensions [19–21]. This fact will also be used below.

As QSL supports all possible partitioning of spins into valence bonds, its ground state is highly (if not infinitely) degenerate. This is an additional consequence of initial geometrical frustration. This means also that excitation spectrum of QSL is very rich and may contain different exotic species. To understand the meaning of exotic excitation, we recollect that typically in condensed matter physics the excitations (quasiparticles) are related to, say, electrons, i.e., fermions with spin 1/2 and charge $\pm e$ and phonons (or magnons), i.e., bosons with integer spin ($S = 1$ for magnons and $S = 0$ for phonons). In QSLs, the most well-known exotic excitations are *spinons*, which in two dimensions can be regarded as an "empty place," i.e., a spin, which is not paired in a valence bond. Such spinon can move by rearranging nearby valence bonds at low-energy cost, see Fig. 8.4. It carries spin $S = 1/2$ (i.e., it is a fermion) and has no electric charge. Note that initially spinons were identified

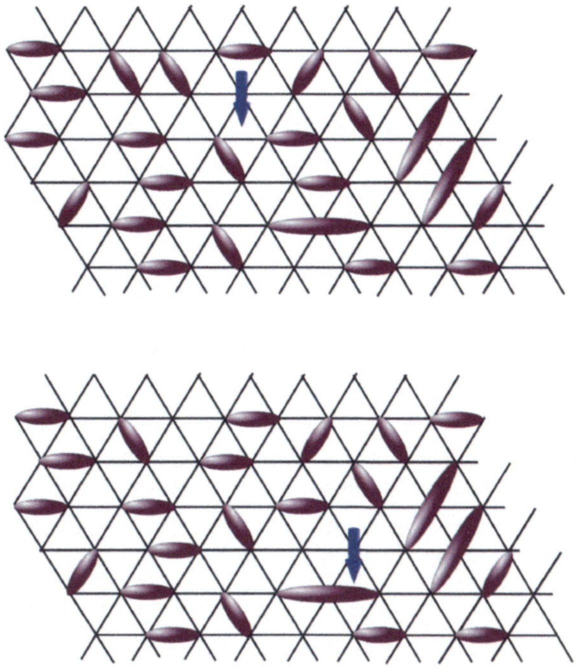

Fig. 8.4 Schematics of spinon creation and motion in 2D spin liquid. Contrary to VBS state (see Fig. 8.3), the bonds in spin liquid are located chaotically in a lattice. The spinon in this case is an unpaired spin. It moves by adjusting the nearby valence bonds, compare upper and lower panels. As valence bonds can be of arbitrary range, couple of longer bonds are shown as an example

in one-dimensional (1D) systems, where they have the form of domain-wall like objects, separating two antiferromagnetic (AFM) ground states. In this situation, it is extremely difficult for a spinon to move (either inside a chain or between chains) as such motion requires the coherent flip of infinite (or very large) number of spins. This demonstrates that in two dimensions such spinons are much more mobile then in 1D. We note that, in contrast to metals, spinons cannot support charge current but they, for instance, can carry heat, as electrons of metals do.

The presence of above exotic excitations requires "exotic" theoretical description. One of the aims of the present chapter is to show that one of the most appropriate descriptions of experimentally observed features in geometrically frustrated magnets should be based on "fermionic formalism" rather than on different standard approaches ranging from Heisenberg to Hubbard Hamiltonian. More specifically, our aim is to elucidate the QSL nature in terms of the phenomenon of FC, that generates flat bands; FC was theoretically predicted in 1990 [5] and described in details in [6, 8, 10]. The necessary ingredient for FC realization in quantum frustrated magnets is so-called flat bands, i.e., dispersionless parts of QSL forming quasiparticles spectra [22, 23]. It turns out that the best candidates for such flat bands realization are heavy-fermion metals (HF), high-T_c superconductors, see e.g., [6, 8, 10, 22, 23], and quantum frustrated magnets on a lattice of corner sharing triangles, known as kagomé lattice, see e.g., [24–30]. This lattice, named after specific Japanese weaving pattern, is displayed in Fig. 8.5 along with typical antiferromagnetic frustrated

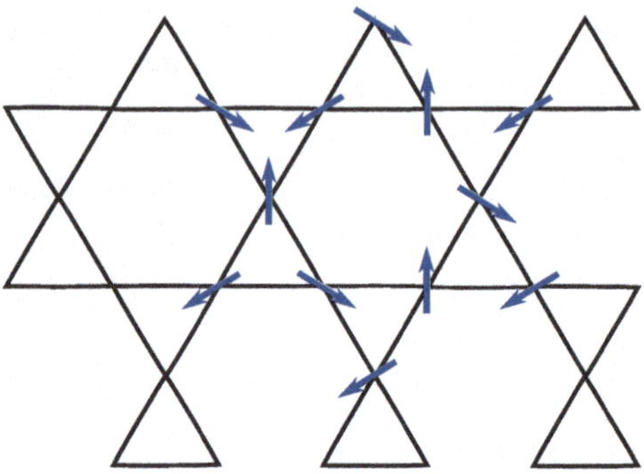

Fig. 8.5 Heisenberg spins (blue arrows) on corner sharing triangular (kagomé) lattice, forming frustrated pattern. The spins are shown on a part of the kagomé structure

configuration. Over the years, a variety of materials, belonging to the above class, have been identified. The most prominent family (i.e., best studied on one side and more or less easy to fabricate on the other) is herbertsmithite $ZnCu_3(OH)_6Cl_2$ and its prototypes like clinoatacamite $Cu_4(OH)_6Cl_2$ [31]. A simple kagome lattice may have a dispersionless topologically protected branch of the spectrum with zero excitation energy that is the flat band [24–26, 30]. In that case a topological fermion condensation quantum phase transition [6, 10] (FCQPT) forming flat bands can be considered as QCP of the $ZnCu_3(OH)_6Cl_2$ quantum spin liquid, see Sects. 8.2 and 8.3. As a result, we can safely assume that a deconfined Fermi quantum spin liquid with essentially gapless excitations formed by neutral fermions, that is spinons, is realized in quantum frustrated magnets like $ZnCu_3(OH)_6Cl_2$ and located very near QCP. Thus, $ZnCu_3(OH)_6Cl_2$ turns out to be located at its QCP without tuning this substance to QCP using a control parameter such as magnetic field, pressure, or chemical composition. Our main focus here will be herbertsmithite as in its 3D rhombohedral lattice the copper ions form kagomé structure [32]. The structure of herbertsmithite along with kagomé pattern for Cu ions is reported in Fig. 8.6.

The experiments made on $ZnCu_3(OH)_6Cl_2$ have not found any traces of magnetic order in it. Nor they have found spin freezing down to temperatures of around 50 mK. In these respects, herbertsmithite is the best candidate among quantum magnets to contain the QSL described above [33–38]. These assessments are supported by theoretical investigations of QSL and explanations of experimental facts collected on quantum frustrated magnets, indicating, particularly, that the ground state of this kagomé antiferromagnet is a gapless spin liquid [8, 10, 26–29, 39, 40]. At low temperatures T, a QSL may have or may not have a gap in the excitation spectrum of spinons, which, being spin 1/2 fermions, occupy the corresponding Fermi sphere with momentum p_F. The influence of a putative gap on the properties is huge. For instance, if the gap is present, the low temperature properties of herbertsmithite are similar to those of common insulators, while the properties resemble those of metals, provided that the gap is absent. Latter state can be thought of as something like *spinon metal* [28, 40], which differs from ordinary metal in that it cannot support the electric current. On the other hand, similarly to ordinary metals, it can conduct heat in the same way. Thus, it is an experimental challenge to establish whether the

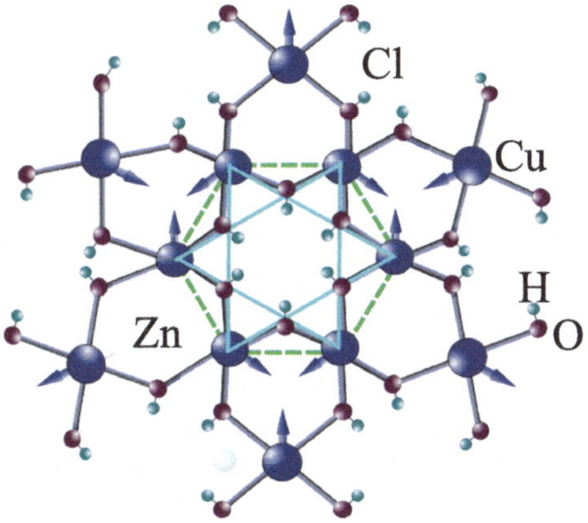

Fig. 8.6 Crystal structure of the herbertsmithite $ZnCu_3(OH)_6Cl_2$ along the hexagonal c-axis. Copper atoms are blue. Arrows show spins of Cu ions in a typical frustrated configuration. The kagome structure is shown by both two triangles and the broken line, visualizing hexagon. Groups OH^- are shown and explained in the panel. Few of Zn and Cl atoms are shown schematically not to clutter the picture. Zn atoms may be situated below or above Cu planes

gap is present and if yes, what is its value. So far, both theoretical and experimental studies of herbertsmithite are controversial, and cannot give definite answer to this puzzle, see [41–43].

The balance of electrostatic forces for the Cu^{2+} ions in the kagome structure of herbertsmithite is such that they occupy distorted octahedral sites. The magnetic planes formed by the Cu^{2+} $S = 1/2$ ions are interspersed with nonmagnetic Zn^{2+} layers, see Fig. 8.6. On the average, $x \simeq 15\%$ of the Zn sites are occupied by Cu, thus introducing randomness and inhomogeneity into the lattice [44]. As we shall see, the role of above randomness as well as impurities in the QSL formation is not clearly understood up to now. Below, we show that the influence of impurities and disorder on the properties of $ZnCu_3(OH)_6Cl_2$ can be tested by varying Zn content x.

The main goal of this chapter is to expose a state of the art of the investigations of ground state properties of perspective geometrically frustrated magnets like herbertsmithite that can hold QSLs without gap in its single-particle spectrum of excitations. Here we offer the uniform theoretical explanation of diverse experimental facts gathered on frustrated magnets in the framework of the FC theory [5, 6, 8, 10]. We also suggest a number of clarifying experiments that can allow one to unambiguously test the presence of gapless QSL. For that we attract attention to experimental studies of $ZnCu_3(OH)_6Cl_2$ that have the potential of revealing both the underlying physics of the QSL and the presence or absence of a gap in spinon excitations. Namely, we recommend measurements of heat transport, low-energy inelastic neutron scattering and optical conductivity $\overline{\sigma}$ in $ZnCu_3(OH)_6Cl_2$ single crystals subjected to an external magnetic field at low temperatures. We argue that the general physical picture inherent in the QSL can be "purified" of the non-universal microscopic contributions coming from phonons, impurities, and above discussed "zoology" of magnetic ground states.

It is worth noting that in this Book, this chapter and Chaps. 9, 11, and 12 are devoted to consideration of SCQSL. This chapter is organized as follows. In the following Sect. 8.2, we present short exposition of the FC approach. In Sect. 8.3, the scaling behavior of quantum liquids in geometrically frustrated magnets is considered. Here we show that quantum geometrically frustrated magnets can exhibit the behavior of heavy fermion metals, with one exception: they cannot support electric current. Section 8.4 is consecrated to the properties of the herbertsmithite $ZnCu_3(OH)_6Cl_2$, and includes the Subsection devoted to the thermodynamic properties Sect. 8.4.1. Chapter 9 deals with quasi-one-dimensional spin liquid in the $Cu(C_4H_4N_2)(NO_3)_2$ insulator. We demonstrate that $Cu(C_4H_4N_2)(NO_3)_2$ behaves like HF metals and quasicrystals. Chapters devoted to dynamic spin susceptibility, Chap. 10, spin-lattice relaxation rate, Chap. 11, and optical conductivity, Sect. 11.2. Here, we also suggest that number of experiments are to be performed in order to unambiguously establish the QSL properties in it. Chapter 12 deals with the organic insulator $EtMe_3Sb[Pd(dmit)_2]_2$ which is also (along with herbertsmithite) an excellent candidate for QSL realization. In Sect. 12.2, we demonstrate that the two-dimensional ^{3}He liquid can serve as the system that strongly resembles quantum frustrated magnets with QSL occupying the Fermi sphere. Sections 12.3 and 12.4 summarize the main results of the consideration of QSLs, demonstrate that these form the new state of matter, and outline the perspectives in searches for QSLs.

8.2 Fermion Condensation

Although the concept of fermion condensation has been described several times earlier (see, e.g., [6, 8, 10]) for the reader convenience, here we recapitulate briefly this methodology. The usual approach to describe the ensembles of correlated delocalized particles obeying Fermi statistics is the renowned Landau Fermi liquid theory [45, 46]. This theory represents a real structure of a solid with itinerant electrons and nuclei in terms of Fermi gas of so-called quasiparticles with weak interparticle interaction. In this case, the quasiparticles represent the excited state of above solid and are "responsible" for the low temperature physical properties of the system under consideration. These quasiparticle excitations are characterized by the effective mass M^*, which is weakly dependent on external stimuli like magnetic field, temperature, external pressure, etc. [45–47]. It is well known (see [6, 8, 10] and references therein) that the LFL theory cannot explain why sometimes the effective mass M^* starts to depend on the above stimuli. This dependence is called non-Fermi liquid (NFL) behavior and is related to the growth of the effective mass M^*, which occurs when the system approaches the transition to FC state with flat bands. This transition point is named topological fermion-condensation quantum phase transition (FCQPT). Beyond FCQPT the system obtains a flat band, formed by FC, and is characterized by the topological charge that is different from the topological charges of the Landau Fermi liquid (LFL) and marginal Fermi liquid, see [6, 8, 10, 48, 49] for detailed explanation. As a result, the stability of FC is given by its topological properties, and it can be destroyed only by first-order phase transition [8, 10]. We note that these unique properties of FC generate a new state of matter, formed by QSL, HF metals, quasicrystals and high-T_c superconductors, so that 1D, 2D, and 3D strongly correlated Fermi systems exhibit the same scaling behavior irrespective of their microscopic structure [8–10, 50].

The main ingredient of FC formalism is the existence of one more instability channel (additionally to those of Pomeranchuk, see e.g., [46]) for Landau Fermi liquid. Namely, under some conditions the effective mass of LFL quasiparticle may tend to infinity, see e.g., [8, 10]. In this case, to maintain the finite and positive effective mass at finite temperatures $T > 0$, the Fermi surface changes its topological class (hence the *topological* phase transition) generating the effective mass dependence on temperature and magnetic field. Without loss of generality [8, 10], here we suppose that the Fermi liquid is homogeneous. That is to say, in our model we account for most important and common features only, leaving aside the small effects, related to the crystalline anisotropy of solids [8]. The Landau equation, for the quasiparticle effective mass M^* reads [8, 45]

$$\frac{1}{M^*_\sigma(B,T)} = \frac{1}{M} + \sum_{\sigma_1} \int \frac{\mathbf{p}_F \mathbf{p}}{p_F^3} F_{\sigma,\sigma_1}(\mathbf{p}_\mathrm{F}, \mathbf{p}) \times \frac{\partial n_{\sigma_1}(\mathbf{p}, T, B)}{\partial p} \frac{d\mathbf{p}}{(2\pi)^3}, \tag{8.1}$$

where M is the free electron mass, $F_{\sigma,\sigma_1}(\mathbf{p}_\mathrm{F}, \mathbf{p})$ is the interaction function, introduced by Landau [46]. It has the form of infinite series over spherical harmonics (so-called Landau amplitudes) with coefficients, taken from the best fit to experiment. This function depends on Fermi momentum p_F, momentum p and spin indices σ, σ_1. The fermion occupation number n in the Fermi-Dirac statistics reads

$$n_\sigma(\mathbf{p}, T) = \left\{1 + \exp\left[\frac{(\varepsilon_\sigma(\mathbf{p}, T) - \mu_\sigma)}{T}\right]\right\}^{-1}, \tag{8.2}$$

where $\varepsilon(\mathbf{p}, T)$ is the single-particle spectrum. Here μ_σ is a spin-dependent chemical potential. Latter dependence occur at nonzero magnetic fields. It reads $\mu_\sigma = \mu \pm \mu_B B$, μ_B is Bohr magneton. The above dependence occur due to Zeeman splitting.

The ordinary procedure to obtain the single-particle spectrum $\varepsilon_\sigma(\mathbf{p}, T)$ in Landau theory is to vary the system energy $E[n_\sigma(\mathbf{p}, T)]$ over n

$$\varepsilon_\sigma(\mathbf{p}, T) = \frac{\delta E[n(\mathbf{p})]}{\delta n_\sigma(\mathbf{p})}. \tag{8.3}$$

In this case, to select the right coefficients in the above Landau interaction function we require that the system should be either at the FCQPT point or sufficiently close to it [8, 10]. The explicit form of the variational equation (8.3) reads

$$\frac{\partial \varepsilon_\sigma(\mathbf{p},T)}{\partial \mathbf{p}} = \frac{\mathbf{p}}{M} - \sum_{\sigma_1} \int \frac{\partial F_{\sigma,\sigma_1}(\mathbf{p},\mathbf{p}_1)}{\partial \mathbf{p}} n_{\sigma_1}(\mathbf{p}_1,T) \frac{d^3 p_1}{(2\pi)^3}, \tag{8.4}$$

Later on for simplicity we omit the spin indexes σ. In the FC phase (i.e., beyond FCQPT) at $T = 0$ the variational equation (8.3) converts to the following one [5]

$$\varepsilon(\mathbf{p}, T = 0) = \mu, \;\; p_i \le p \le p_f; \;\; 0 \le n(p) \le 1. \tag{8.5}$$

where $p_i < p_F < p_f$ stand for initial and final momenta, where condensed fermions reside. Actually, the condition (8.5) defines the aforementioned flat band as in this case the quasiparticles have no dispersion. Therefore, $T = 0$ the FC quasiparticles are condensed with the same energy $\varepsilon(\mathbf{p}, T = 0) = \mu$. As this is similar to the case of Bose condensation, the corresponding phenomenon is called fermion condensation [5]. It has been shown by Volovik [48, 49] that as soon as the FC state takes place, the Fermi surface of ordinary LFL is topologically transformed into the volume that sets in the range $p_i < p_F < p_f$. As a result, Fermi liquids with FC represent a new state of liquid with its special topological charge, so that the liquid acquires properties that are very different from those of ordinary Landau Fermi liquids. Namely, the part (or all) of the quasiparticles in a Fermi liquid are in FC state now. The Fermi liquid with fermion condensate forms a new, topologically protected (and thus "extremely stable") state of matter. This means that if FC is formed in a substance, it will define its properties at $T = 0$ and at low temperatures, as shown below. Figure 8.7 visualizes (at $T = 0$) the consequences of the FCQPT on the Fermi surface, spectrum, and occupation number of a Fermi liquid. The transformation from panel a) (normal Fermi liquid) to panel b) (Fermi liquid with FC) consists in altering the Fermi surface (sketched as a sphere in Fig. 8.7) topology in that the stripe of finite length $p_f - p_i$ appears instead of a single point $p = p_F$ (Fermi momentum) in the normal Fermi liquid. This immediately implies the emergence of flat part of the spectrum, defined by condition (8.5), where all condensed fermions reside. This, in turn, generates the gradual (instead of abrupt on the panel a) decrease of occupation numbers $n(p)$ from $n = 1$ at $p < p_i$ to $n = 0$ at $p > p_f$.

Equations (8.1)–(8.5) make possible to determine the energy spectrum $\varepsilon_\sigma(\mathbf{p}, T)$ and occupation number $n_\sigma(\mathbf{p}, T)$ self-consistently. These quantities, in turn, permit to calculate the effective mass, $p_F/M^* = \partial\varepsilon(p)/\partial p|_{p=p_F}$. We emphasize that both magnetic field and temperature dependences of the effective mass in FC phase comes from those of $\varepsilon_\sigma(\mathbf{p})$ and $n_\sigma(\mathbf{p})$ on T and B. The calculated (by (8.1)–(8.5)) spectrum and occupation numbers [8] in FC phase are reported in Fig. 8.8. At (almost) zero temperature the flat portion of the spectrum is clearly seen at $p_i < p < p_f$. This shape of the spectrum generates $n(p)$ (lower panel) in the form of "two steps," gradually decreasing from 1 to 0 similar to qualitative picture sketched in Fig. 8.7.

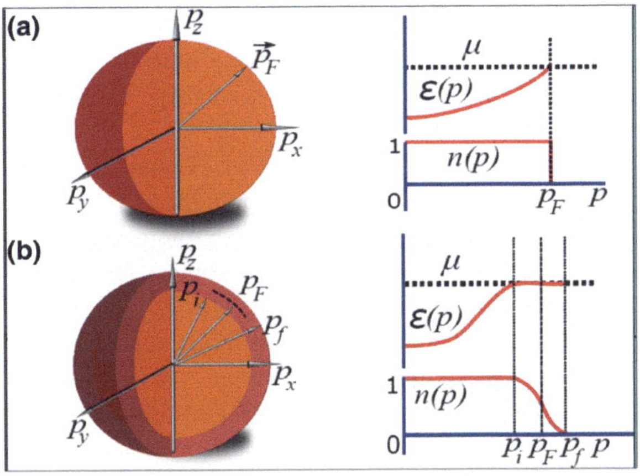

Fig. 8.7 Sketch of flat bands emergence at FCQPT Panel **a** shows normal Fermi sphere and corresponding quasiparticles spectrum and occupation number. Spectrum is of usual form $\varepsilon(p) = p^2/(2M)$ with $n(p)$ being a step function. Panel **b** shows the system in FC state after FCQPT In this case the Fermi sphere alters its topology, which is shown schematically as appearance of spherical layer of the thickness $p_f - p_i$. In this case, the Fermi momentum p_F (shown by broken line) is hidden inside "FC strip," which is defined by the condition $\varepsilon(p) = \mu$ (8.5). Latter condition defines the flat band, shown as dispersionless part of the spectrum in the right side of panel (**b**). In this case, function $n(p)$ decreases gradually from $n = 1$ to $n = 0$ without violation of Pauli exclusion principle

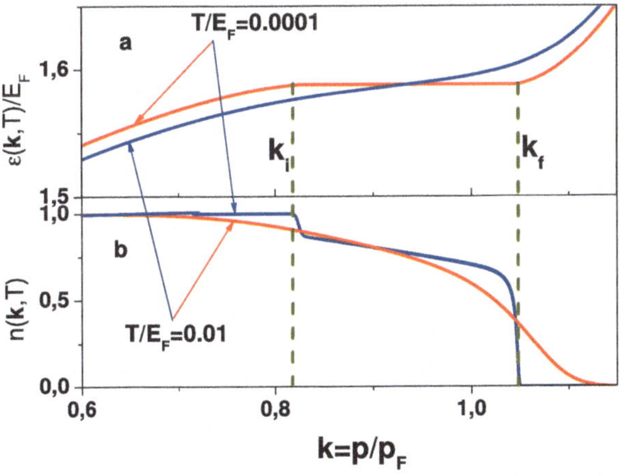

Fig. 8.8 Calculated flat bands (spectrum) and occupation numbers in FC phase [8]. The cases of almost zero ($T = 10^{-4}E_F$; red curves) and finite ($T = 10^{-2}E_F$; blue curves) temperatures are shown. It is seen that finite temperatures "smear" the dispersionless part of the spectrum, making the situation almost indistinguishable from normal Fermi liquid, while at $T \to 0$ in the region $k_i - k_f$ the band becomes flat, where $k_i = p_i/p_F$ and $k_f = p_f/p_F$

At the same time, at relatively high temperatures (equal to $T/E_F = 0.01$, which at $E_F \sim 1\,\text{eV}$ implies $T \simeq 100\,\text{K}$) this part is already blurred. This shows, that finite temperatures erode the FC state, giving rise to something similar to ordinary Fermi liquid.

To gain more insights in the physical nature of FC state, it is helpful to explore the system properties at $T \to 0$. It had been shown [5, 10] that the ground state of a system with FC is highly degenerate. In this case, the occupation numbers $n_0(\mathbf{p})$ of the FC state quasiparticles (i.e., having dispersionless spectrum or belonging to the flat band) vary gradually from $n = 1$ to $n = 0$ at $T = 0$. This variation occurs at $p_i \leq p \leq p_f$. It is clear that such property of the occupation numbers is in stark contrast with usual Fermi-Dirac function property at $T = 0$, namely the step between $n = 1$ and $n = 0$ at $p = p_F$, where p_F stands for Fermi momentum, see Fig. 8.7.

High (actually infinite) degeneracy of the ground state leads to a T-independent entropy term [6, 8], which also remains finite at $T = 0$ in violation of Nernst theorem

$$S_0 = -\sum_{\mathbf{p}} [n_0(\mathbf{p}) \ln n_0(\mathbf{p}) + (1 - n_0(\mathbf{p})) \ln(1 - n_0(\mathbf{p}))]. \tag{8.6}$$

In other words, the infinite degeneracy of the FC ground state is generated by the flat band presence, see [10] for comprehensive discussion. We note that one more system, where Nernst theorem is violated due to the ground state degeneracy, is above discussed spin glass [3, 4]. It is well known that in a normal Fermi liquid the function $n(\mathbf{p})$ at finite temperatures looses its step-like feature at $p = p_F$, becoming continuous around this point. The same is valid for a Fermi liquid with flat bands (FC state), which follows from (8.2). This means that at small but finite temperatures $T \neq 0$ the above ground state is no more degenerate so that single-particle energy $\varepsilon(\mathbf{p}, T \neq 0)$ acquires a small dispersion [51] (i.e., the corresponding band is no more flat)

$$\varepsilon(\mathbf{p}, T \to 0) = T \ln \frac{1 - n_0(\mathbf{p})}{n_0(\mathbf{p})}. \tag{8.7}$$

It is seen from (8.7) that the dispersion is proportional to T as the occupation numbers n_0 remain the same as at $T = 0$. This means that the entropy in this case still remains to be equal to S_0. This situation also jeopardizes the preservation of the Nernst theorem, see above. To avoid this unphysical situation, the (almost) flat bands constituting FC state should gain the dispersion in a way that the excessive entropy S_0 should "dissolve" as $T \to 0$. This situation occurs by virtue of some additional (to FCQPT) phase transitions (like ferromagnetic and/or superconductive) or crossover [6, 8, 10]. In fact, at $T = 0$ the FC state is represented by the superconducting state with the superconducting gap $\Delta = 0$, while the superconducting order parameter $\kappa = \sqrt{n(p)(1 - n(p))}$ is finite in the region $(p_i - p_f)$ [8, 52], because in this region $n(p) < 1$, as it is shown in Fig. 8.8.

8.3 Scaling of the Physical Properties

As the main experimental manifestation of FC phenomenon is the scaling behavior of the physical properties (see Chap. 7 for general consideration), we begin with description of the scaling behavior in case of QSLs. If QSL is formed by approximately dispersionless spinons, occupying a nearly flat band, this state of matter becomes strongly correlated quantum spin liquid (SCQSL). This is because the properties of this state strongly resemble those of HF metals [8, 10, 26, 28, 50]. As we have seen in Sect. 8.1, SCQSL emerges in quantum frustrated magnets including that with the kagomé lattice, for its geometrical frustration generates a dispersionless topologically protected spectral branch, which is frequently referred to as a flat band [24, 25, 30]. In this case, the SCQSL in herbertsmithite undergoes the topologically driven quantum phase transition (i.e., the FCQPT) into the FC state. The topological FCQPT is indeed the quantum critical point for the SCQSL in the above substance. Therefore, the main physical features of geometrically frustrated magnets, which are insulators, coincide with those of HF metals with one exception. Namely, as typical insulators, they do not conduct the electric current [26–28].

Consider now the approximate solutions of (8.1). At $B = 0$ this quantity becomes strongly temperature dependent, which is a typical NFL feature [8, 10]. This is due to the fact that in normal Fermi liquid (LFL case) the effective mass M^* is nearly constant. The former NFL regime is governed by the expression

$$M^*(T) \simeq a_T T^{-2/3}. \tag{8.8}$$

At low temperatures T and under the application of growing magnetic field, the system transits to the LFL phase with the effective mass being almost temperature independent

$$M^*(B) \simeq a_B B^{-2/3}. \tag{8.9}$$

Here, a_T and a_B are constants. Above observations permit to construct the approximate solution of (8.1) in the form $M^* = M^*(B, T)$. The introduction of "internal" scales simplifies the problem under consideration, for in that case we get rid of microscopic structure of compound in the question [8, 10]. We observe, that near FCQPT the effective mass $M^*(B, T)$ reaches a maximum M^*_M at certain temperature $T_M \propto B$ [8]. Accordingly, to measure the effective mass and temperature, it is convenient to introduce the scales M^*_M and T_M. In this case, we have new variables $M^*_N = M^*/M^*_M$ (normalized effective mass) and $T_N = T/T_M$ (normalized temperature). In the FCQPT vicinity, the normalized effective mass $M^*_N(T_N)$ can be well approximated by a certain universal function [8, 10], interpolating the system properties between the LFL and NFL states [8]

$$M^*_N(T_N) \approx c_0 \frac{1 + c_1 T_N^2}{1 + c_2 T_N^{8/3}}. \tag{8.10}$$

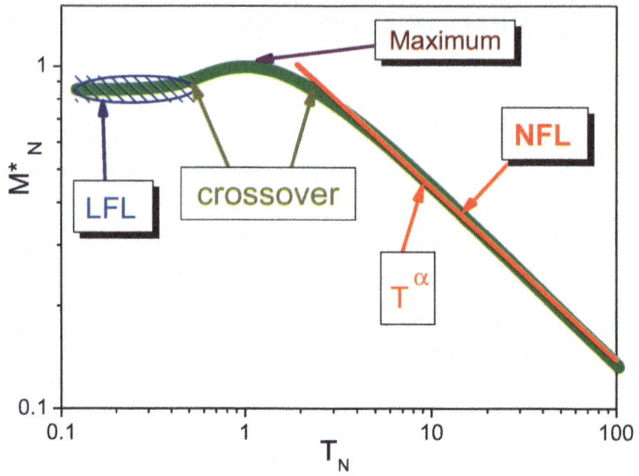

Fig. 8.9 Scaling of the thermodynamic properties governed by the normalized effective mass M_N^*. In the case of the application of magnetic field $T_N \propto T/B$, as it follows from (8.11). Solid curve depicts M_N^* versus normalized temperature T_N. It is seen that at finite $T_N < 1$ the normal Fermi liquid properties take place. At $T_N \sim 1$ it enters crossover state, and at growing temperatures exhibits the NFL behavior

Here, $T_N = T/T_M$, $c_0 = (1 + c_2)/(1 + c_1)$, where c_1 and c_2 are free parameters. As magnetic field B enters as $\mu_B B/T$, the maximum temperature $T_M \sim \mu_B B$. It follows from (8.10) that

$$T_M \simeq a_1 \mu_B B; \quad T_N = \frac{T}{T_M} = \frac{T}{a_1 \mu_B B} \propto \frac{T}{B}, \tag{8.11}$$

where a_1 is a dimensionless factor, μ_B is the Bohr magneton. Expression (8.11) shows that (8.10) determines the effective mass scaling in terms of T and B. To be more specific, the curves $M_N^*(T, B)$ amalgamate into a single one $M_N^*(T_N)$, $T_N = T/T_M$, as it is shown in Fig. 8.9. Since $T_M \propto B$, as it follows from (8.11), we conclude that the curves $M_N^*(T, B)$ are merged into a single one $M_N^*(T_N = T/B)$, $T_N = T/T_M = T/B$, demonstrating the scaling behavior widespread among HF metals. Such a behavior is also shown in Fig. 8.9. We note that (8.10) and (8.11) allow one to describe strongly correlated Fermi systems including SCQSL of different frustrated magnets, see e.g., [8, 10, 53–56].

One more important feature of the FC state is that apart from the fact that the Landau quasiparticle effective mass starts to depend on external stimuli like temperature and magnetic field, all relations, inherent in LFL approach, formally hold in it. Namely, the famous LFL relation [46],

$$M^*(B, T) \propto \chi(B, T) \propto \frac{C(B, T)}{T} \tag{8.12}$$

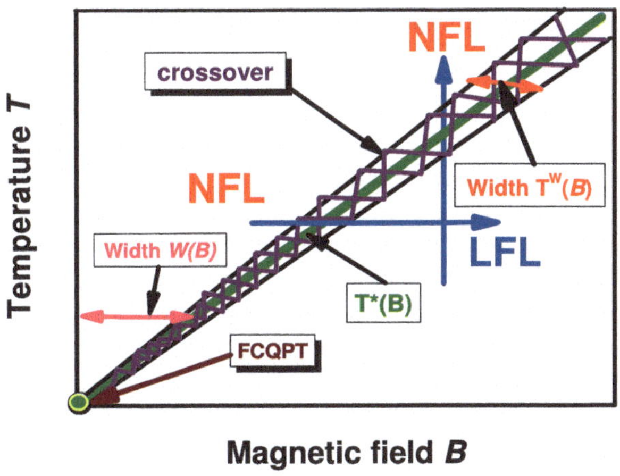

Fig. 8.10 Schematic $T - B$ SCQSL phase diagram. Magnetic field B is a control parameter. The hatched area corresponds to the crossover domain at $T_M(B)$, given by (8.11). At fixed magnetic field and elevating temperature (vertical arrow) there is a LFL-NFL crossover. The horizontal arrow indicates a NFL-LFL transition at fixed temperature and elevating magnetic field. The FCQPT (shown in the panel) occurs at $T = 0$ and $B = 0$, where M^* diverges

holds. The expression (8.12) has been rendered to the FC case, where the specific heat C, magnetic susceptibility χ and effective mass M^* depend on T and B. Taking (8.12) into account, we obtain that normalized values of C/T and χ are of the form [8, 10]

$$M^*_N(B,T) = \chi_N(B,T) = \left(\frac{C(B,T)}{T}\right)_N. \tag{8.13}$$

It is seen from (8.13) that the above thermodynamic properties have the same scaling behavior depicted in Fig. 8.9. Moreover, below we shall see that the thermodynamic properties of HF metals and SCQSL of frustrated magnets have the same scaling behavior. Based on (8.10) and Fig. 8.9, we can construct the schematic $T - B$ phase diagram of SCQSL, reported in Fig. 8.10. We assume here that at $T = 0$ and $B = 0$ the system is approximately located at FCQPT without tuning. At fixed temperatures, the system is driven along the horizontal arrow (from the NFL to LFL parts of the phase diagram) by the magnetic field B. In turn, at fixed B and elevating T, the system moves from the LFL to the NFL regimes along the vertical arrow. The hatched area indicating the crossover between LFL and NFL phases, separates the NFL state from the paramagnetic slightly polarized LFL one. The crossover temperature $T_M(B)$ is given by (8.11).

8.4 The Frustrated Insulator Herbertsmithite $ZnCu_3(OH)_6Cl_2$

If the quasiparticles forming the QSL are approximately dispersionless spinons (uncharged fermions), this state of matter becomes a SCQSL, [10, 26, 28, 50, 50]. In the case of an ideal two-dimensional lattice of an insulating compound, a SCQSL can emerge if the lattice geometrical frustration results in a dispersionless topologically protected spectral branch with zero excitation energy, i.e., a flat band as it takes place in herbertsmithite with its kagome planes [5, 25, 48, 49]. Then, the FCQPT can be considered as the quantum critical point of the SCQSL, composed of chargeless heavy spinons with $S = 1/2$ and effective mass M^*, occupying the corresponding Fermi sphere with the Fermi momentum p_F, for the one frustrated valence spin is taken away to populate the approximately flat band, forming the flat spinon band. Such a behavior strongly resembles that of HF metals whose valence electrons form conduction flat bands. At the same time, the charges $(-e)$ cannot form any band because of the large charge gap of herbertsmithite or another geometrically frustrated insulating magnet. This observation is in accordance with the experimental facts [57]. Consequently, the properties of insulating compounds coincide with those of HF metals with one exception, namely the typical insulator prevents the electric current [26–28]. Thus, both frustrated insulators and HF metals, exposing the universal behavior, forms a new state of matter [50].

According to (8.3), in the case of QSL the behavior of the specific heat $(C_{\mathrm{mag}}/T)_N$ must coincide with that of χ_N. To separate C_{mag} contribution, we approximate the general specific heat $C(T)$ at $T > 2$ K by the function

$$C(T) = a_1 T^3 + a_2 T^{1/3}. \tag{8.14}$$

Here the first term proportional to a_1 presents the lattice (phonon) contribution, while the second one is determined by the QSL when it exhibits the NFL behavior, as it follows from (8.8). It is shown in Fig. 8.11, panel b, that the approximation (8.14) is valid in a wide temperature range. We note that the value a_1 is almost independent of a_2, the presence of which allows us to achieve a better approximation for C.

The obtained heat capacity C_{mag}/T,

$$\frac{C_{\mathrm{mag}}}{T} = \frac{C - a_1 T^3}{T}, \tag{8.15}$$

is displayed in Fig. 8.12 and in the panel (a) of Fig. 8.15, while the left panel of Fig. 8.17 demonstrates the maximum temperature as a function of the magnetic field B. It is shown in Fig. 8.15 that $C_{\mathrm{mag}}/T \propto M^*$ behaves like $\chi \propto M^*$. The normalized $(C_{\mathrm{mag}}/T)_N$ and χ_N are depicted in Figs. 8.12, 8.13 and 8.16. It is seen from these figures that the results obtained on the different samples (powder samples and single-crystal samples) and the measurements [34, 35, 37, 38] exhibit

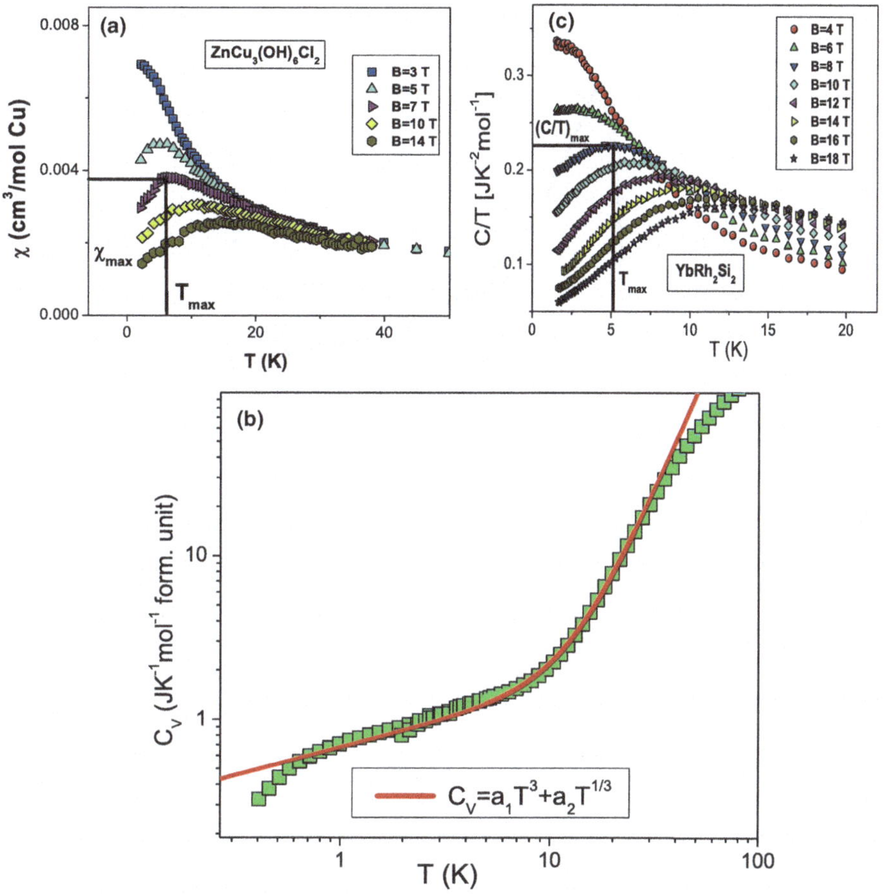

Fig. 8.11 Panel **a**: T-dependence of the magnetic susceptibility χ at different magnetic fields B [33] shown in the legend. The values of $\chi_{\max}$ and $T_{\max}$ at $B = 7$ T are also shown. Our calculations at $B = 0$ are depicted by the solid curve $\chi(T) \propto T^{-\alpha}$ with $\alpha = 2/3$. Panel **b**: The heat capacity measured on $ZnCu_3(OH)_6Cl_2$ at zero magnetic field [35] is shown by squares. Solid curve corresponds to our theoretical approximation based on the function $C = a_1 T^3 + a_2 T^{1/3}$ with fitting parameters a_1 and a_2, see (8.14). Panel c reports the T-dependence of the electronic specific heat C/T of $YbRh_2Si_2$ at different magnetic fields [58] as shown in the legend. The values of $(C/T)_{\max}$ and $T_{\max}$ at $B = 8$ T are also shown

similar properties. As it is shown in Figs. 8.12 and 8.13 that in accordance with (8.3), $(C_{\rm mag}/T)_N \simeq \chi_N$ displays the same scaling behavior as $(C/T)_N$ measured on the HF metal $YbRh_2Si_2$. This observation rules out a scenario suggesting that extra Cu spins outside the kagome planes considered as weakly interacting impurities could be responsible for the divergent behavior of the low-temperature susceptibility. In that case the supposition is to lead to explanations of the observed scaling behavior of χ and C/T in strong magnetic fields shown in Figs. 8.12, 8.13, and 8.16. Obviously,

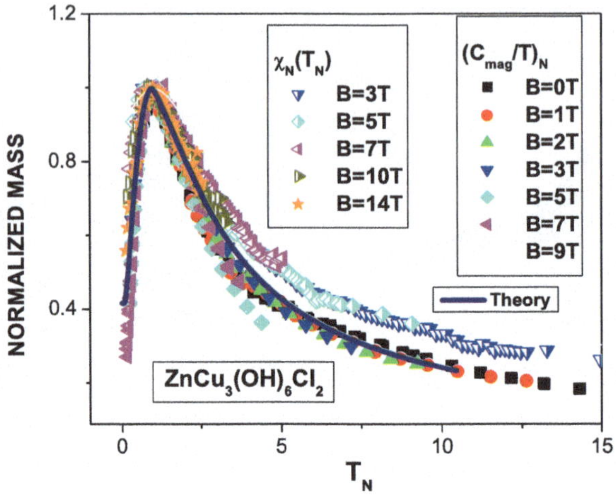

Fig. 8.12 The normalized susceptibility $\chi_N = \chi/\chi_{max} = M^*_N$ and the normalized specific heat $(C_{mag}/T)_N = M^*_N$ of QSL effective mass versus normalized temperature T_N as a function of the magnetic fields shown in the legends. χ_N is extracted from the measurements of the magnetic susceptibility χ in magnetic fields B [33] shown in the panel A of Fig. 14.1. The normalized specific heat is extracted from the data displayed in Fig. 8.11, the panel (b). Our calculations are depicted by the solid curve tracing the scaling behavior of M^*_N

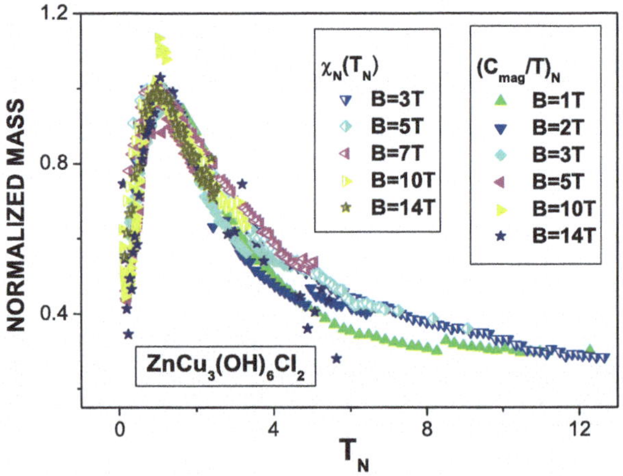

Fig. 8.13 The normalized susceptibility χ_N and the normalized specific heat $(C_{mag}/T)_N$ of QSL effective mass versus normalized temperature T_N as a function of the magnetic fields shown in the legends. χ_N and $(C_{mag}/T)_N$ are extracted from the data of [33] and [34], respectively

it is impossible, since weakly interacting impurities would be polarized by relatively weak magnetic field, and would not contribute at higher magnetic fields. As a result, the scaling behavior were destroyed at higher fields. Therefore, the scaling behavior of the thermodynamic functions of herbertsmithite is the intrinsic feature of the compound and has nothing to do with possible contribution coming from the magnetic impurities. Moreover, as it is shown in Fig. 8.16, the measurements on powder samples and on single-crystal samples, demonstrating similar results, rule out noticeable contributions to the thermodynamic properties coming from defects of the lattice and impurities.

Thus, the kagome lattice of $ZnCu_3(OH)_6Cl_2$ can be viewed as a strongly correlated Fermi system whose thermodynamics is defined by the quantum spin liquid located at FCQPT. We conclude that in herbertsmithite the entire bulk susceptibility and the heat capacity obey the scaling behavior. The scaling behavior of the thermodynamic properties coincides with that observed in HF metals and 2D ^{3}He. Herbertsmithite $ZnCu_3(OH)_6Cl_2$ exhibits the LFL, NFL and the transition behavior as HF metals and 2D ^{3}He do.

8.4.1 Thermodynamic Properties

To examine the thermodynamic properties, we first refer to the experimental behavior of the magnetic susceptibility χ of herbertsmithite in order to elucidate the possibility of gap in the excitation spectra of spinons, because the gap affects tremendously on the properties of the herbertsmithite. The solid line in Fig. 8.14 shows that the magnetic susceptibility diverges as $\chi(T) \propto T^{-2/3}$ at $B \leq 1$ T in accordance with (8.8). It had been suggested [44, 59, 60] that at low temperatures the magnetic susceptibility behaves according to a Curie–Weiss law $\chi_{CW}(T) \propto 1/(T+\theta)$, where θ is a small Weiss temperature. The latter law reflects the case of weak interaction between impurities. At the same time the dependence $\chi(T) \propto T^{-2/3}$, see (8.8), shows that not only the Curie–Weiss approximation describes the experiment [33], it also contradicts the established theoretical results [10, 26, 28]. It is also shown in Fig. 8.15 that the specific heat C/T (panel (a)) of $ZnCu_3(OH)_6Cl_2$ measured on powder samples coincide with that of single-crystal sample. Then, Fig. 8.15a, b demonstrate that C/T of $ZnCu_3(OH)_6Cl_2$ behaves as that of the archetypical HF compound $YbRh_2Si_2$ [10, 28]. These observations allow us to conclude that the contribution coming from impurities cannot be separated from that of the kagomé host crystal lattice. Moreover, Fig. 8.16 shows that normalized spin susceptibility behaves like the normalized heat capacity extracted from the high magnetic field measurements on $YbRh_2Si_2$ [58]. This observation confirms the absence of the spin gap in herbertsmithite and invalidity of procedure that artificially separates the contributions coming from the impurities and host lattice, and is in accordance with experimental facts [61]. Thus, to adequately explain the observed behavior of $\chi(T, B=0)$, shown in Fig. 8.14, one should deal with the impurities and host kagome planes as an integral entity, forming the properties of SCQSL [8, 10, 26–29, 40, 62]. These observations coincide with the theoretical results suggesting that the inhomogeneity in frustrated magnets generates robust liquid-like behavior [63, 64].

Figure 8.17, the right and left panels correspondingly, demonstrates that (8.11) and (8.9) are in a good agreement with the experimental facts.

The impurity model has been used to derive the following intrinsic scattering measure: $S_{kag}(\omega) = S_{tot}(\omega) - a S_{imp}(\omega)$. In the above expression, a is a free parameter, $S_{imp}(\omega)$ is the impurity scattering contribution and $S_{tot}(\omega)$ is the total scattering [44].

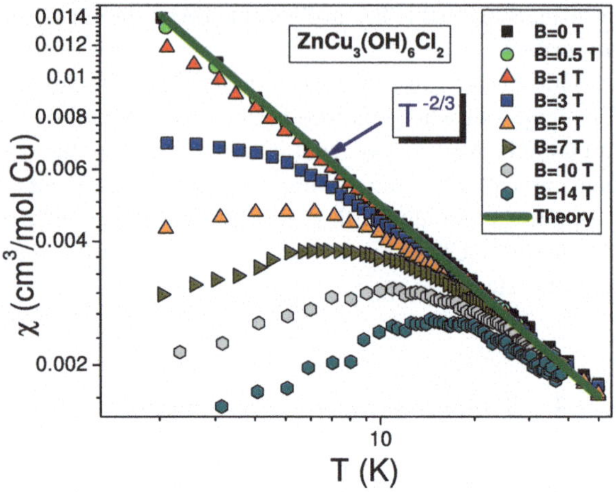

Fig. 8.14 Measured temperature dependence $\chi(T)$ of $ZnCu_3(OH)_6Cl_2$ taken from [33]. The magnetic fields are shown in the legend. Solid line $\chi(T) \propto T^{-\alpha}$ $(\alpha = 2/3)$ indicates our calculation at $B = 0$ [10, 26, 29]

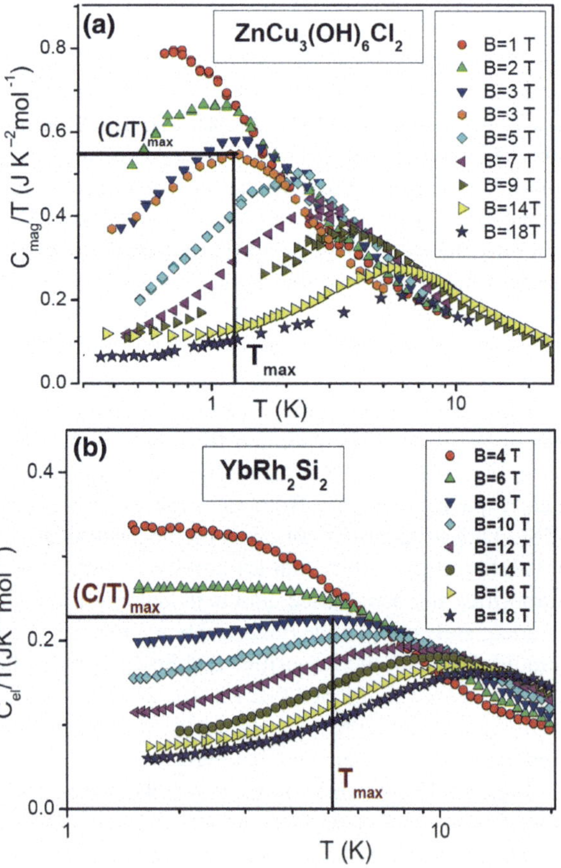

Fig. 8.15 Panel **a**. Temperature dependence of spinon specific heat $C_{\mathrm{mag}}(T)/T$ measured on powder [34, 35] and single-crystal [36–38] samples of herbertsmithite. Magnetic fields B are shown in the legend. Panel **b**. The electron contribution to the specific heat C_{el}, taken from the measurements on $YbRh_2Si_2$ in magnetic fields B [58], is shown in the legend. Both panels report the examples (for representative single curve) of $(C/T)_{\mathrm{max}}$ and T_{max} which are used for the normalization

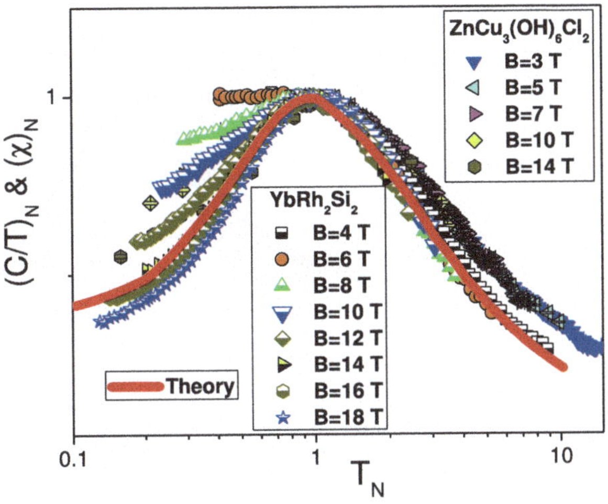

Fig. 8.16 Normalized susceptibility $\chi_N = \chi/\chi_{max} = M_N^*$ versus normalized temperature T_N (see (8.10)). The data are distilled from magnetic field measurements of χ on $ZnCu_3(OH)_6Cl_2$ [33] shown in Fig. 8.14. Normalized data of $(C/T)_N = M_N^*$ are obtained from the specific heat C/T (Fig. 8.15b) of $YbRh_2Si_2$ in magnetic fields B (legend) [58]. Solid curve corresponds to the theoretical calculations at $B \simeq B^*$ when the quasiparticles are fully polarized. Both experimental markers and theoretical curve are due to scaling of M_N^* [10, 28]. It follows from (8.11) that T_N can be replaced by the variable T/B

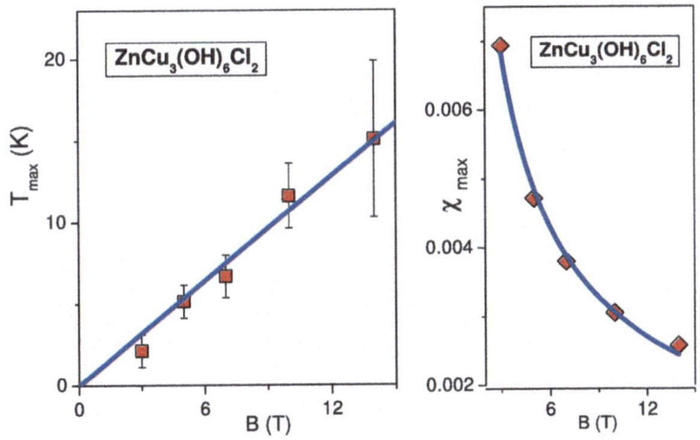

Fig. 8.17 Left panel: The temperatures $T_{max}(B)$ at which the maxima of χ are located, see Fig. 8.14. The solid line represents the function $T_{max} \propto aB$, a is a fitting parameter, see (8.11). Right panel: the maxima χ_{max} of the functions $\chi(T)$ versus magnetic field B, see Fig. 8.14. The solid curve is approximated by $\chi_{max}(B) = a_B B^{-2/3}$, see (8.9), a_B is a fitting parameter

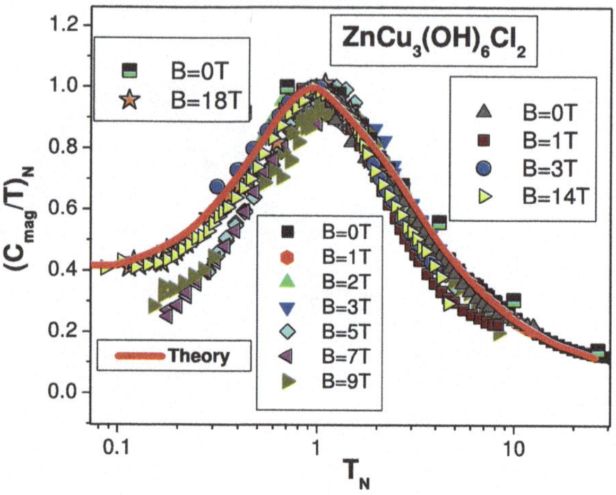

Fig. 8.18 The normalized temperature dependence of specific heat $(C_{\rm mag}(T)/T)_N$ at different B's, shown in the legend [10, 40]. Solid curve depicts the theoretical result [10, 26]

The outcome was the claim of the gap presence in the spinon excitation spectrum. We have demonstrated, however, that the above analysis shows the presence of a spurious gap based on the χ behavior analysis. Both conclusions rely completely on the hypothesis that the impurities and the host kagomé lattice should be considered separately rather then as a single entity.

Let us proceed in the discussion of the inadequacy of the spin gap existence assertion in the light of recent experimental data. Figure 8.14 shows that normal Fermi liquid behavior of the magnetic susceptibility χ occurs at low temperatures and at least for $B \geq 3$ T. At such T and B values, the impurities magnetic moments are completely aligned with magnetic field direction. In the framework of the impurity model, fully polarized (i.e., aligned with the external field direction) localized spin does not exhibit Curie–Weiss behavior. This means that such aligned impurities do not contribute to the magnetic susceptibility χ. In this case, the entire susceptibility comes from host kagomé lattice: $\chi_{\rm kag}(T) = \chi(T)$. Figure 8.15 shows that the heat capacity $C_{\rm mag}/T$ behaves similarly to the magnetic susceptibility. Namely, high magnetic fields $B \geq 3$ T restore normal Fermi liquid behavior. This permits us to claim that at low temperatures and magnetic fields $B \geq 3$ T one can safely neglect the impurity contribution to the above thermodynamic characteristics. This means that in the above temperature and magnetic field ranges, the entire thermodynamics of herbertsmithite (and similar magnetically frustrated substances) comes from the host kagomé lattice, which has a gap in the spinon excitation spectrum [44, 59, 60]. In this case both $\chi(T) \sim C_{\rm mag}(T)/T \to 0$ at $T \to 0$ and $B \geq 3$ T. It is clear from Figs. 8.14, 8.15 and 8.18, that this is not true. Namely, at $B \lesssim 14$ T and $T \to 0$ both magnetic susceptibility and heat capacity are nonzero. Furthermore, Fig. 8.18 shows the uniform behavior of $C_{\rm mag}/T$, which confirms the gap absence. We can also observe from Fig. 8.15 that for herbertsmithite the single-crystal and powder measurements look very much alike [34–38]. The above analysis of experimental results relevant to herbertsmithite thermodynamics permits us to conclude (i) it is

SCQSL that is responsible for overwhelming majority of observable properties of herbertsmithite samples (both crystalline and powdered) under study; (ii) the spectra of spinon excitations do not have noticeable gap. These theoretical conclusions [1, 2] are in agreement with low-temperature NMR contrast experiments on high-quality single crystals of the herbertsmithite that firmly conclude that $ZnCu_3(OH)_6Cl_2$ does not has any spin gap [61]. Moreover, this gap would not emerge even in high magnetic fields of 18 T. The latter conclusion is in conformity with experimental observations about the low-temperature plateau in local susceptibility, which shows that the spin-liquid ground state is indeed gapless [65]. Calculations have also confirmed the gap absence [39]. Moreover, we suggest that all factors (like growing x, elevating randomness and inhomogeneity of the lattice) facilitating the geometric frustration, can stabilize SCQSL. This observation can be tested in magnetic field experiments on herbertsmithite samples with different x.

References

1. V.R. Shaginyan, A.Z. Msezane, M.Ya. Amusia, J.W. Clark, G.S. Japaridze, V.A. Stephanovich, Y.S. Leevik, Condens. Matter **4**, 75 (2019)
2. V.R. Shaginyan, V.A. Stephanovich, A.Z. Msezane, G.S. Japaridze, J.W. Clark, M.Ya. Amusia, E.V. Kirichenko, J. Mater. Sci. **55**, 2257 (2020)
3. K. Binder, A.P. Young, Rev. Mod. Phys. **58**, 801 (1986)
4. M. Mézard, G. Parisi, M.A. Virasoro, *World Scientific Lecture Notes in Physics*, vol. 9 (2004)
5. V.A. Khodel, V.R. Shaginyan, JETP Lett. **51**, 553 (1990)
6. V.A. Khodel, V.R. Shaginyan, V.V. Khodel, Phys. Rep. **249**, 1 (1994)
7. V.R. Shaginyan, M.Ya. Amusia, K.G. Popov, Physics-Uspekhi **50**, 563 (2007)
8. V.R. Shaginyan, M.Ya. Amusia, A.Z. Msezane, K.G. Popov, Phys. Rep. **492**, 31 (2010)
9. V.R. Shaginyan, K.G. Popov, V.A. Khodel, Phys. Atomic Nuclei **77**, 1063 (2014)
10. M.Ya. Amusia, K.G. Popov, V.R. Shaginyan, V.A. Stephanovich, *Theory of Heavy-Fermion Compounds*. Springer Series in Solid-State Sciences, vol. **182** (2014)
11. L. Balents, Nature **464**, 199 (2010)
12. E. Schrödinger, Naturwissenschaften **23**, 807 (1935)
13. F. Verstraete, M.A. Martín-Delgado, J.I. Cirac, Phys. Rev. Lett. **92**, 087201 (2004)
14. C.H. Bennett, D.P. DiVincenzo, Nature **404**, 247 (2000)
15. O. Gühne, G. Tóth, Phys. Rep. **474**, 1 (2009)
16. T. Nikuni, M. Oshikawa, A. Oosawa, H. Tanaka, Phys. Rev. Lett. **84**, 5868 (2000)
17. I. Affleck, T. Kennedy, E.H. Lieb, H. Tasaki, Phys. Rev. Lett. **59**, 799 (1987)
18. P.W. Anderson, Mater. Res. Bull. **8**, 153 (1973)
19. L. Balents, M.P.A. Fisher, S.M. Girvin, Phys. Rev. B **65**, 224412 (2002)
20. A. Kitaev, Ann. Phys. (Leipzig) **321**, 2 (2006)
21. R. Moessner, S.L. Sondhi, Phys. Rev. Lett. **86**, 1881 (2001)
22. VYu. Irkhin, A.A. Katanin, M.I. Katsnelson, Phys. Rev. Lett. **89**, 076401 (2002)
23. D. Yudin, D. Hirschmeier, H. Hafermann, O. Eriksson, A.I. Lichtenstein, M.I. Katsnelson, Phys. Rev. Lett. **112**, 070403 (2014)
24. D.L. Bergman, C. Wu, L. Balents, Phys. Rev. B **78**, 125104 (2008)
25. D. Green, L. Santoz, C. Chamon, Phys. Rev. B **82**, 075104 (2010)
26. V.R. Shaginyan, A.Z. Msezane, K.G. Popov, Phys. Rev. B **84**, 060401(R) (2011)
27. V.R. Shaginyan, A.Z. Msezane, K.G. Popov, V.A. Khodel, Phys. Lett. A **376**, 2622 (2012)
28. V.R. Shaginyan, A.Z. Msezane, K.G. Popov, G.S. Japaridze, V.A. Stephanovich, Europhys. Lett. **97**, 56001 (2012)

29. V.R. Shaginyan, A.Z. Msezane, K.G. Popov, G.S. Japaridze, V.A. Khodel, Europhys. Lett. **103**, 67006 (2013)
30. H.J. Changlani, D. Kochkov, K. Kumar, B.K. Clark, E. Fradkin, Phys. Rev. Lett. **120**, 117202 (2018)
31. J.L. Lambor, J.E. Dutrizac, A.C. Roberts, J.D. Grice, J.T. Szymanski, Can. Mineral. **34**, 61 (1996)
32. M.P. Shores, E.A. Nytko, B.M. Bartlett, D.G. Nocera, J. Amer. Chem. Soc. **127**, 13462 (2005)
33. J.S. Helton, K. Matan, M.P. Shores, E.A. Nytko, B.M. Bartlett, Y. Qiu, D.G. Nocera, Y.S. Lee, Phys. Rev. Lett. **104**, 147201 (2010)
34. J.S. Helton, K. Matan, M.P. Shores, E.A. Nytko, B.M. Bartlett, Y. Yoshida, Y. Takano, A. Suslov, Y. Qiu, J.-H. Chung, D.G. Nocera, Y.S. Lee, Phys. Rev. Lett. **98**, 107204 (2007)
35. M.A. deVries, K.V. Kamenev, W.A. Kockelmann, J. Sanchez-Benitez, A. Harrison, Phys. Rev. Lett. **100**, 157205 (2008)
36. T.H. Han, J.S. Helton, S. Chu, A. Prodi, D.K. Singh, C. Mazzoli, P. Müller, D.G. Nocera, Y.S. Lee, Phys. Rev. B **83**, 100402(R) (2011)
37. T.H. Han, S. Chu, Y.S. Lee, Phys. Rev. Lett. **108**, 157202 (2012)
38. T.H. Han, R. Chisnell, C.J. Bonnoit, D.E. Freedman, V.S. Zapf, N. Harrison, D.G. Nocera, Y. Takano, Y.S. Lee(2014), arXiv:1402.2693
39. H.J. Liao, Z.Y. Xie, J. Chen, Z.Y. Liu, H.D. Xie, R.Z. Huang, B. Normand, T. Xiang, Phys. Rev. Lett. **118**, 137202 (2017)
40. V.R. Shaginyan, M.Ya. Amusia, A.Z. Msezane, K.G. Popov, V.A. Stephanovich, Phys. Lett. A **379**, 2092 (2015)
41. M.R. Norman, Rev. Mod. Phys. **88**, 041002 (2016)
42. P. Mendels, F. Bert, C.R. Physique **17**, 455 (2016)
43. Yi Zhou, K. Kanoda, and T.-K. Ng, Rev. Mod. Phys. **89**, 025003 (2017)
44. T.H. Han, M.R. Norman, J.-J. Wen, J.A. Rodriguez-Rivera, J.S. Helton, C. Broholm, Y.S. Lee, Phys. Rev. B **94**, 060409(R) (2016)
45. L.D. Landau, Sov. Phys. JETP **3**, 920 (1956)
46. E.M. Lifshits, L.P. Pitaevsky, *Statistical Physics, Part 2* (Butterworth-Heinemann, Oxford, 2014)
47. D. Pines, P. Noziéres, *Theory of Quantum Liquids* (Benjamin, New York, 1966)
48. G.E. Volovik, JETP Lett. **53**, 222 (1991)
49. G.E. Volovik, Lect. Notes in Physics **718**, 31 (2007)
50. V.R. Shaginyan, V.A. Stephanovich, A.Z. Msezane, P. Schuck, J.W. Clark, M.Ya. Amusia, G.S. Japaridze, K.G. Popov, E.V. Kirichenko, J. Low Temp. Phys. **189**, 410 (2017)
51. P. Nozières, Properties of Fermi liquids with a finite range interaction. J. Phys. I France **2**, 443 (1992)
52. V.R. Shaginyan, V.A. Stephanovich, A.Z. Msezane, G.S. Japaridze, K.G. Popov, Phys. Chem. Chem. Phys. **19**, 21964 (2017)
53. V.R. Shaginyan, V.A. Stephanovich, K.G. Popov, E.V. Kirichenko, S.A. Artamonov, Ann. Phys. (Berlin) **528**, 483 (2016)
54. T. Isono, T. Terashima, K. Miyagawa, K. Kanoda, S. Uji, Nat. Commun. **7**, 13494 (2016)
55. T. Isono, S. Sugiura, T. Terashima, K. Miyagawa, K. Kanoda, and S. Uji, Nat. Comm. **9**, Article Number 1509 (2018). https://doi.org/10.1038/s41467-018-04005-1
56. W.M. Farmer, S.F. Skinner, L.W. ter Haarb, AIP Adv. **8**, 101404 (2018)
57. Z.A. Kelly, M.J. Gallagher, T.M. McQueen, Phys. Rev. X **6**, 041007 (2016)
58. P. Gegenwart, Y. Tokiwa, T. Westerkamp, F. Weickert, J. Custers, J. Ferstl, C. Krellner, C. Geibel, P. Kerschl, K.-H. Müller, F. Steglich, New J. Phys. **8**, 171 (2006)
59. T. Imai, M. Fu, T.H. Han, Y.S. Lee, Phys. Rev. B **84**, 020411(R) (2011)
60. M. Fu, T. Imai, T.H. Han, Y.S. Lee, Science **350**, 655 (2015)
61. P. Khuntia, M. Velazquez, Q. Barthélemy, F. Bert, E. Kermarrec, A. Legros, B. Bernu, L. Messio, A. Zorko, P. Mendels, Gapless ground state in the archetypal quantum kagome antiferromagnet $ZnCu_3(OH)_6Cl_2$. Nat. Phys. **16**, 469 (2020)
62. V.R. Shaginyan, V.A. Stephanovich, K.G. Popov, E.V. Kirichenko, JETP Lett. **103**, 32 (2016)

63. H. Yamaguchi, M. Okada, Y. Kono, S. Kittaka, T. Sakakibara, T. Okabe, Y. Iwasaki, Y. Hosokoshi, Scient. Report **7**, Article Number 16144 (2017)
64. K. Uematsu, H. Kawamura, J. Phys. Soc. Jpn. **86**, 044704 (2017)
65. M. Gomilsek, M. Klanjsek, M. Pregelj, H. Luetkens, Y. Li, Q.M. Zhang, A. Zorko, Phys. Rev. B **94**, 024438 (2016)

Chapter 9
One-Dimensional Quantum Spin Liquid

Abstract We analyze measurements of the magnetization, differential susceptibility, and specific heat on one-dimensional Heisenberg antiferromagnet $Cu(C_4H_4N_2)(NO_3)_2$ (CuPzN) subjected to strong magnetic fields. Using the mapping between magnons (bosons) in CuPzN and fermions, we demonstrate that magnetic field tunes the insulator toward quantum critical point related to the topological fermion-condensation quantum phase transition (FCQPT) at which the resulting fermion effective mass diverges kinematically. We show that the FCQPT concept permits to reveal the scaling behavior of thermodynamic characteristics, describe the experimental results quantitatively, and derive the $T - H$ (temperature—magnetic field) phase diagram, that contains Landau Fermi liquid, crossover and non-Fermi liquid parts, thus resembling that of heavy-fermion compounds. Thus, the analyzed behavior of CuPzN allows us to conclude that 1D quantum spin liquids form the new state of matter.

9.1 Introduction

Measurements of the thermodynamic properties at low-temperatures T under the application of magnetic field H on the quasi-one-dimensional (Q1D) insulator $Cu(C_4H_4N_2)(NO_3)_2$ (CuPzN) have been performed [1]. The observed thermodynamic properties of CuPzN is very unusual and nobody expects that it might belong to the class of HF compounds, including quasicrystals (QC) [2], insulators with quantum spin liquid (QSL), and heavy-fermion (HF) metals [3–6]. Similar Q1D clean HF metal $YbNi_4P_2$ was recently experimentally studied, that reveals it has a Q1D electronic structure and strong correlation effects dominating the low-temperature properties, while its thermodynamic properties resembles those of HF metals, including the formation of Landau Fermi liquid (LFL) ground state [7–9]. These observations show that both CuPzN and $YbNi_4P_2$ can demonstrate a new type of Q1D Fermi liquid whose thermodynamic properties resemble that of HF compounds rather than the Tomonaga-Luttinger system [7, 8]. One of the hallmark features of geometrically frustrated insulators is spin-charge separation. The behavior of Q1D Fermi liquid (Q1DFL) is the subject of ongoing intensive experimental research in condensed

M. Amusia and V. Shaginyan, *Strongly Correlated Fermi Systems*,
Springer Tracts in Modern Physics 283,
https://doi.org/10.1007/978-3-030-50359-8_9

matter physics, see, e.g., [1] and references therein. Q1DFL survives up to the saturation magnetic field H_s, where the quantum critical point (QCP) occurs giving way to a gapped, field-induced paramagnetic phase [1]. In other words, at $H = H_s$ both antiferromagnetic (AFM) sublattices align in the field direction, i.e., the magnetic field fully polarizes Q1DFL spins. We will see below that in Q1DFL the fermion condensation quantum phase transition (FCQPT) plays a role of QCP, at which the energy band for spinons becomes almost flat at $H = H_s$ and the effective mass M^* of spinons diverges due to kinematic mechanism. Note, in case of the relatively strong interaction of spinons the fermion condensation takes place due to the interaction, and exhibits the non-Fermi liquid behavior without tuning with magnetic fields to its QCP at $H = H_s$ [10].

In the considered case, the bare interaction of spinons is weak [1]. In that case the original Tomonaga-Luttinger system can exactly be mapped on a system of free spinons, which low-temperature behavior in magnetic fields can be viewed as the LFL one [11]. Thus CuPzN offers a unique possibility to observe a new type of Q1D QSL whose thermodynamic properties resemble that of HF compounds like HF metals, including Q1D HF metal $YbNi_4P_2$ [9], quantum spin liquids of herbertsmithite $ZnCu_3(OH)_6Cl_2$ [6], and liquid ^3He [12]. Theory of Q1D liquids is still under construction and recent results show that the liquids can exhibit LFL, non-Fermi liquid (NFL) and crossover behavior [11, 13, 14].

In this chapter, we show that, contrary to ordinary wisdom, the thermodynamic properties of CuPzN are defined in fact by weakly interacting 1D QSL formed with spinons, and are similar to those of HF compounds. Here spinons are chargeless fermionic quasiparticles with spin 1/2. We demonstrate that its temperature T—magnetic field H phase diagram contains LFL, crossover and non-Fermi liquid parts, thus resembling that of HF compounds [8]. To unveil the similarity between CuPzN and HF compounds, we study the scaling behavior of its thermodynamic properties which are independent of the interparticle interaction. We demonstrate that CuPzN exhibits the universal scaling behavior, characteristic for quasicrystals, which, in turn have the properties very similar to those of HF compounds, see Chaps. 14 and 21.

9.2 General Considerations

Upon transition to fermionic description, CuPzN is indeed represented by weakly interacting fermions. The description of weakly interacting fermion gas gives magnetization in terms of fermion number per spin $N/L = \int_0^\infty D(\varepsilon)f(\varepsilon - \mu(H))d\varepsilon$, where L is the number of spins in Q1D chain, $D(\varepsilon)$ is the density of states, corresponding to free fermion spectrum $\varepsilon = p^2/(2m_0)$ with p is the momentum and m_0 is the bare mass. The chemical potential $\mu(H) = H_s - H$, and $f(x) = (e^x + 1)^{-1}$ is the Fermi distribution function [1, 15, 16]. The magnetization can be expressed as $M = M_s - N$ (M_s is the saturation magnetization) or explicitly

$$M(H,T) = M_s - \frac{\sqrt{2m_0T}}{\pi}\int_0^\infty \frac{dx}{e^{\left(x^2 - \frac{H_s - H}{T}\right)} + 1}. \tag{9.1}$$

Equation (9.1) will be used below to calculate the differential magnetic susceptibility $\chi(T,H) = \frac{\partial M(T,H)}{\partial H}$. To calculate the specific heat $C(T,H)$, we need the internal energy E, which in the above approach can be calculated as follows

$$E(T,H) = \int_0^\infty \varepsilon D(\varepsilon) f(\varepsilon - \mu(H)) d\varepsilon, \tag{9.2}$$

so that $C(T,H) = \frac{\partial E}{\partial T}$. Within the framework of fermionic description, as it is seen from (9.1), CuPzN is indeed a weakly interacting fermions with simplest possible spectrum $\varepsilon = p^2/(2m_0)$, where $\hbar = c = 1$. Near QCP taking place at $H = H_s$ and $T = 0$, the fermion spectrum becomes almost flat, and the fermion (spinon) effective mass diverges, $M^* \propto m_0/p_F \to \infty$, due to kinematic mechanism, for the Fermi momentum $p_{FH} \to 0$ of becoming empty subband. In case of weak repulsion between spinons the divergence is associated with the onset of a topological transition at finite value of p_{FH} signaling that $M^*(T) \propto T^{-1/2}$, see Chap. 7 and (7.9), [3, 17–20]. In accordance with [11], we suggest that the weakly interacting Q1DFL in CuPzN could be thought of as QSL, formed with fermionic spinons, constituting the Fermi sphere (line) with the Fermi momentum p_F, and carrying spin 1/2 and no charge. For QCP occurs at $M^* \to \infty$, as we have seen above, we propose that QCP is the topological FCQPT, at which the corresponding band becomes approximately flat [3–6].

In fermion representation the ground state energy $E(n)$ can be viewed as the Landau functional depending on the spinon distribution function $n_\sigma(\mathbf{p})$, where $\mathbf{p}$ is the momentum. Near FCQPT point, the effective mass M^* is governed by the Landau equation [4, 5, 21]

$$\begin{aligned}\frac{1}{M^*(T,H)} &= \frac{1}{M^*(T=0,H=0)} \\ &+ \frac{1}{p_F^2}\sum_{\sigma_1}\int \frac{\mathbf{p}_F\mathbf{p}_1}{p_F} F_{\sigma,\sigma_1}(\mathbf{p}_F,\mathbf{p}_1)\frac{\partial \delta n_{\sigma_1}(\mathbf{p}_1)}{\partial p_1} dv,\end{aligned} \tag{9.3}$$

where dv is the volume element. Here, we have rewritten the spinon distribution function as $\delta n_\sigma(\mathbf{p}) \equiv n_\sigma(\mathbf{p},T,B) - n_\sigma(\mathbf{p},T=0,B=0)$. The sole role of the Landau interaction $F(\mathbf{p}_1,\mathbf{p}_2) = \delta^2 E/\delta n(\mathbf{p}_1)\delta n(\mathbf{p}_2)$ is to bring the system to FCQPT point, where $M^* \to \infty$ at $T = 0$, and the Fermi surface alters its topology so that the effective mass acquires temperature and field dependences, while the proportionality of the specific heat C/T and the magnetic susceptibility χ to M^* holds: $C/T \sim \chi \sim M^*(T,H)$ [4, 5, 22, 23]. This feature can be used to separate the solutions of (9.3), corresponding to specific experimental situation. Namely, the experiment on CuPzN shows that near QCP at $H = H_s$, the specific heat $C(T)/T \propto \chi(T) \propto T^{-1/2}$ [1] which means that M^* is responsible for the observed behavior, while QCP is formed by kinematic mechanism. It has been shown that near FCQPT, $M^*(T) \propto T^{-1/2}$, while the application of H drives the system to the LFL region with $M^*(H) \propto (H_s - H)^{-1/2}$ [3–5]. At finite H and T near FCQPT, the solutions of (9.3) $M^*(T, H)$ can be well approximated by a simple universal interpolating function, see Chap. 7 for details, [3–5]. The interpolation occurs between the LFL ($M^* \propto a + bT^2$) and NFL ($M^* \propto T^{-1/2}$) regimes and represents the universal scaling behavior of $M^*_N(T_N)$ independent of spatial dimension of the considered system

$$M^*_N = \frac{1+c_2}{1+c_1}\frac{1+c_1T_N^2}{1+c_2T_N^{5/2}}, \tag{9.4}$$

where c_1 and c_2 are fitting parameters, $M^*_N = M^*/M^*_M$ and $T_N = T/T_M$ are the normalized effective mass and temperature, respectively. Here,

$$M^*_M \propto (H_s - H)^{-1/2}, \tag{9.5}$$

$$T_M \propto (H_s - H), \tag{9.6}$$

are the maximum value of the effective mass M^*_M and temperature T_M, corresponding to the maximum of $(\mathrm{d}M/\mathrm{d}T)_{\max}(H)$ and/or $\chi_{\max}(H)$ [3–5]. Below (9.4) is used along with (9.1) to describe the experiment in CuPzN.

9.3 Scaling of the Thermodynamic Properties

We begin tracing the scaling behavior of the function $(M_s - M)/\sqrt{H_s - H}$ on $T/(H_s - H)$ [7, 8]. The corresponding theoretical dependence on one hand can be seen from (9.1) and on the other hand is obtained utilizing the fact that magnetization $M = \int \chi \mathrm{d}H$, where $\chi \sim m^*(T, H)$ is a magnetic susceptibility. Then, having the effective mass $m^*(T, H)$ from the interpolative solution (9.4) of Landau equation, and using (9.5), we obtain corresponding theoretical dependence. This result is in good agreement with the experimental facts, as it is shown in Fig. 9.1 that reports the plot of the scaling behavior of the magnetization

$$\frac{M_c}{\sqrt{B}} = a + \frac{M_s - M}{\sqrt{H_s - H}}, \tag{9.7}$$

as a function of $T/B = T/(H_s - H)$, with a is a constant added for better presentation of the data. This scaling behavior, given by (9.7) and shown by the solid curve, is indistinguishable from the dependence $M_c/B^{1/2}$ versus T/B extracted from the experimental data [1, 24], which means pretty good coincidence. It is shown in Fig. 9.1 that LFL regime occurs at $T << H_s - H$, the crossover at $T \sim H_s - H$ and

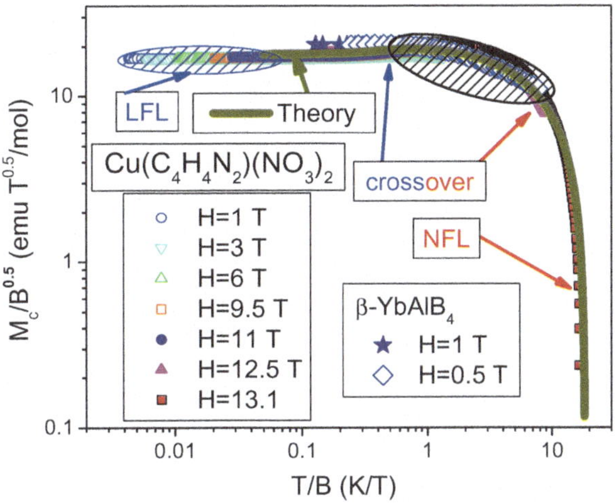

Fig. 9.1 The scaling behavior of the magnetization $M_c/B^{0.5}$ versus T/B at different magnetic fields H, shown in the legend, with $M_c = a + (M - M_s)$ and $B = H_s - H$. The theoretical curve is depicted by the solid curve. The LFL, crossover and NFL regions are shown by the arrows. The data are extracted from measurements [1, 24]

the NFL one at $T >> H_s - H$, which is the case for HF compounds, behaving like HF superconductor $\beta-$ $YbAlB_4$ [7, 24] or the quasicrystal $Au_{51}Al_{34}Yb_{15}$ [3–5].

Taking into account that $M = \int \chi dH$ and (9.4), (9.5), and (9.6), we obtain that $(M - M_s)/\sqrt{H_s - H}$ as a function of the variable $T/(H_s - H)$ exhibits scaling behavior. This result is in good agreement with the experimental facts, as it is shown in Fig. 9.1 that reports the plot of the scaling behavior of the magnetization $M_c/B^{0.5} = a + (M - M_s)/(H_s - H)^{0.5}$ as a function of $T/B = T/(H_s - H)$, with a is a constant added to a better presentation of the Figure. It is shown in Fig. 9.1, that the LFL behavior takes place at $T \ll B$, the crossover at $T \sim B$, and the NFL one at $T \gg B$, as it is in the case of HF compounds [4, 5]. It is instructive to note that the same scaling behavior exhibits M_c obtained in measurements on $YbAlB_4$ under the application of magnetic field [24], see Fig. 9.1 and Chap. 14.

Figure 9.2a and b portray the comparison between χ_N extracted from the experiments on CuPzN, panel (a) [1], $Au_{51}Al_{34}Yb_{15}$ quasicrystal panel (b), and the theory. Here χ_N is the normalized magnetic susceptibility, while the normalization is done in the same way as it is done in the case of M^*_N, see Chap. 7 [4, 5]. It is seen that for more then three decades in normalized temperature there is very good agreement between the theory and the experimental data. The double log scale, used in panels (a) and (b), reveals the universal dependence $\chi_N \sim T_N^{-0.5}$. The comparison between Fig. 9.2a and b indicates that χ_N of both CuPzN and the quasicrystal $Au_{51}Al_{34}Yb_{15}$ has three regions: low-temperature LFL part, medium-temperature crossover region where the maximum occurs, and high-temperature NFL part with the distinctive temperature dependence $T_N^{-0.5}$. Note that the dependences from Fig. 9.2a and b qualitatively resemble that of Q1D HF metal $YbNi_4P_2$ [9], and demonstrate the scaling behavior, and are similar to those of heavy-fermion compounds [3, 5]. We recall that the absolute values of the thermodynamic functions obviously depend on the interparticle interaction amplitude, therefore, to reveal the universal properties we have to employ the normalization procedure [3–5].

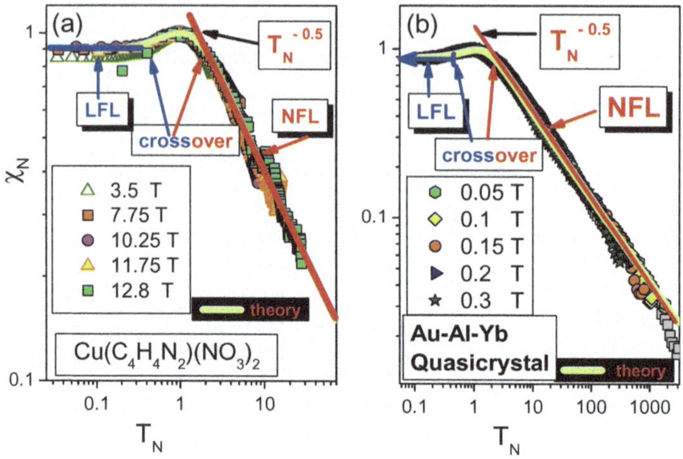

Fig. 9.2 Panels **a** and **b**. The normalized magnetic susceptibility χ_N extracted from measurements in magnetic fields H (shown in the legend) on CuPzN [1] and on Au_{51} $Al_{34}Yb_{15}$ quasicrystal [26]. Our theoretical curves, merged in the scale of the figure and plotted on the base of (9.1) and (9.4), are reported by the solid lines tracing the scaling behavior. Panels **a** and **b** show that dependence $\chi_N(T_N)$ for CuPzN and the quasicrystal Au_{51} $Al_{34}Yb_{15}$ [3] has three distinctive regions: LFL, crossover, and NFL, where $\chi_N \sim T_N^{-0.5}$ (orange curve). Panels **c** and **d** report the normalized specific heat $(C/T)_N$ vs normalized temperature T_N, calculated on the base of (9.2) for $H < H_s$ (panel (**c**)) and $H > H_s$ (panel (**d**)). The LFL, NFL, and crossover regions shown on panel (**c**) ($H < H_s$) are the same as those on panels (**a**) and (**b**). The crossover region is not shown on panel (**d**) because for complete spin polarization at $H > H_s$ it covers large T_N interval, see text for explanation. The excellent coincidence is seen, showing that actually the Fermi liquid theory relation $\chi_N = (C/T)_N = m^*$ holds for weakly interacting 1DFL

We note, that the approach of weakly interacting Q1DFL (9.1) gives for C/T and χ the same high-temperature asymptotics $T^{-1/2}$. Figure 9.3a shows the normalized temperature dependence $(\mathrm{d}M/\mathrm{d}T)_N$ of the quantity $\mathrm{d}M/\mathrm{d}T$, revealing the scaling behavior, while the normalization is done in the same way as it is done in the case of M_N^* or χ_N. The black solid theoretical curve corresponds to the temperature derivative $\mathrm{d}M/\mathrm{d}T$ of the magnetization (9.1). Good coincidence with experiment on CuPzN is seen everywhere. Such a good agreement shows that $\mathrm{d}M/\mathrm{d}T$ has the universal scaling behavior, which can also be described by taking into account that $M = \int \chi \mathrm{d}H$. In panel (b) of Fig. 9.3 the maximum values $(\mathrm{d}M/\mathrm{d}T)_{\max}$ of $(\mathrm{d}M/\mathrm{d}T)$ versus $H_s - H$ are displayed. The theoretical curve given by $(\mathrm{d}M/\mathrm{d}T)_{\max} \propto (H_s - H)^{-1/2}$ is in good agreement with experimental facts extracted from measurement of the magnetization [1], and demonstrates that the effective mass of spinons does diverge at $H \to H_s$.

Figure 9.3a shows the normalized temperature dependence $(\mathrm{d}M/\mathrm{d}T)_N$ of the quantity $\mathrm{d}M/\mathrm{d}T$, exhibiting the scaling behavior. The black solid theoretical curve corresponds to the temperature derivative $\mathrm{d}M/\mathrm{d}T$ of the magnetization (9.1). Pretty good coincidence with experiment on CuPzN is seen everywhere. Such good agree-

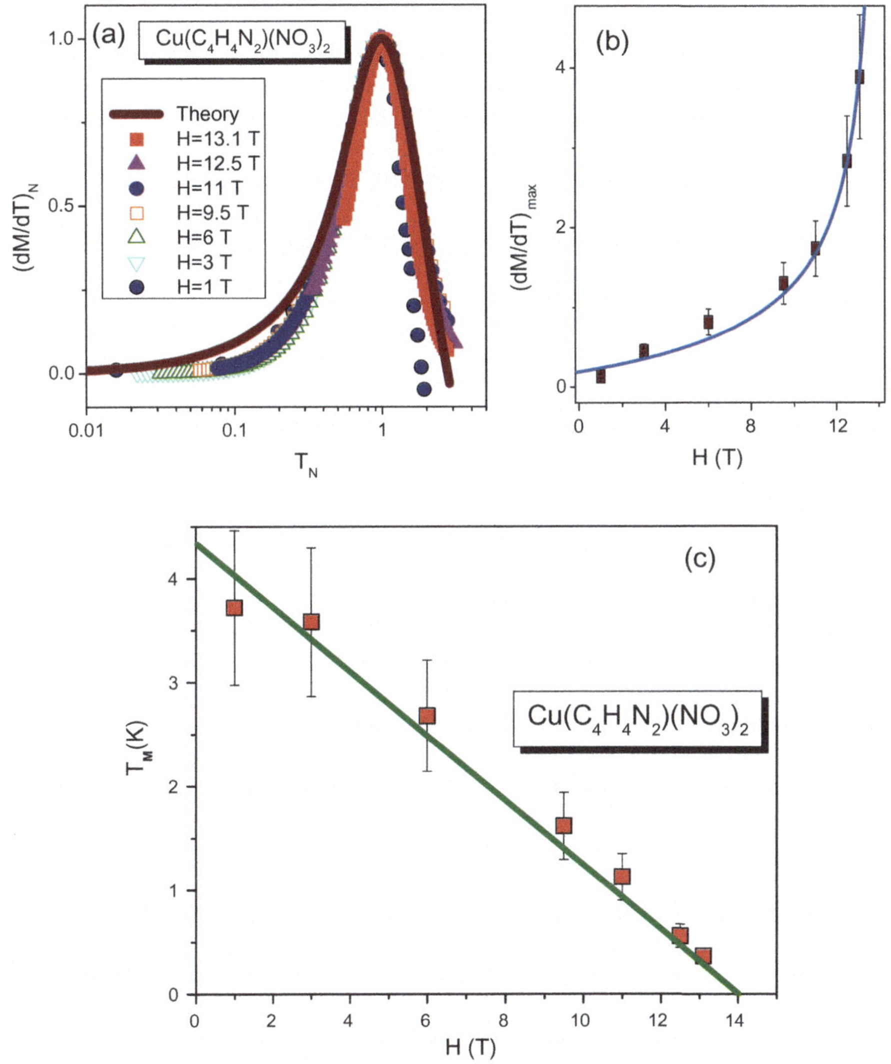

Fig. 9.3 Panel **a**: The normalized $(\mathrm{d}M/\mathrm{d}T)_N$ extracted from measurements in magnetic fields on CuPzN [1]. Theoretical curve based on (9.1) is also reported. Panel **b** reports the magnetic field dependence of the maximum values $(\mathrm{d}M/\mathrm{d}T)_{\max}$ of $(\mathrm{d}M/\mathrm{d}T)$. The theoretical curve is given by $(\mathrm{d}M/\mathrm{d}T)_{\max} \propto (H_s - H)^{-1/2}$, see (9.5). Panel **c** reports the magnetic field dependence of peak temperature T_M of $\mathrm{d}M/\mathrm{d}T$. The theoretical linear dependence, see (9.6), $T_M \propto H_s - H$, is also shown

ment shows that $\mathrm{d}M/\mathrm{d}T$ has universal scaling properties and thus can also be described by (9.4). In panel (b) of Fig. 9.3 the maximum values $(\mathrm{d}M/\mathrm{d}T)_{\max}$ of $\mathrm{d}M/\mathrm{d}T$ versus $H_s - H$ are displayed. The theoretical curve given by $(\mathrm{d}M/\mathrm{d}T)_{\max} \propto (H_s - H)^{-1/2}$ is in good agreement with experimental data extracted from measurement of the magnetization [1], and demonstrates that the effective mass of spinons does diverge at $H \to H_s$. Figure 9.3c reports the magnetic field dependence of maximum temperature T_{max}, at which the maximum of dM/dT occurs. Once more, very good coincidence of theoretical linear dependence $T_{max} \propto H_s - H$ is seen, confirming the above statement of the scaling behavior of dM/dT. Figure 9.3d shows the scaling behavior of the superconductor β-$YbAlB_4$. The fusion of the curves, corresponding to different magnetic fields into one is clearly seen. The coincidence of thus obtained "combined" experimental curve with theoretical one (green solid line) shows that this substance can also be well described by our FC theory and hence exhibits the properties of HF compound.

9.4 $T - H$ Phase Diagram of 1D Spin Liquid

The above thermodynamic properties reported in Figs. 9.2, 9.3, 9.4, and 9.5 coincide with those of HF compounds, and permit us to construct the $T - H$ phase diagram of CuPzN, is shown in Fig. 9.6. To do so, in Fig. 9.5a we report the peak temperature T_M of magnetic susceptibility as a function of H. It is seen, that the peak temperature T_M goes to zero as H approaches H_s. In panel (b) of Fig. 9.5 the maximum values $\chi_{\max}$ of χ versus $H_s - H$ are displayed. It is seen that theoretical curve given by (9.5) is in good agreement with experimental data extracted from measurement of the magnetization [1], and demonstrates that the effective mass of spinons does diverge at $H \to H_s$, as it is at FCQPT. The $T - H$ phase diagram reported in Fig. 9.6 demonstrates that peak dependence T_M takes place over wide range of variation of H, for $T_M \propto (H_s - H)$. This shows, that main property of these lines is that they are straight lines, representing energy scales typical for HF metals located at their QCP [27, 28]. Since FCQPT takes place at $H = H_s$, the phase diagram is almost symmetric with respect to the point $H = H_s$, and consists of the LFL, gapped Fermi liquid, crossover, and NFL regions. The crossover regions in Fig. 9.6 are shown by arrows, and are formed by the straight lines, which are the magnetic field dependencies of temperatures of approximate LFL and NFL boundaries as well as by that of T_M. NFL state occurs at relatively high temperatures with the distinct temperature dependence $\sim T_N^{-1/2}$. At the same time LFL regions occur at low T, where the spinon effective mass is almost constant, as is the case for LFL behavior. At $H > H_s$ the QSL becomes a gapped field-induced paramagnetic spin liquid, as shown in Fig. 9.6. At rising temperatures and fixed magnetic field H, the system transits through the crossover, and enters the NFL region, as it is shown in Fig. 9.6. It is also seen, that the crossover becomes wider, as the systems moves from FCQPT shown by the filled circle. We conclude that CuPzN exhibits the behavior typical for HF compounds [28] that leads to the formation of the corresponding $T - H$ phase diagram displayed in Fig. 9.6.

To verify our statement that the magnetic susceptibility and other thermodynamic quantities have three regions: LFL, crossover, and NFL, in Fig. 9.4, we display the quantities M/H (upper panel) and dM/dH (lower panel) for CuPzN, measured at

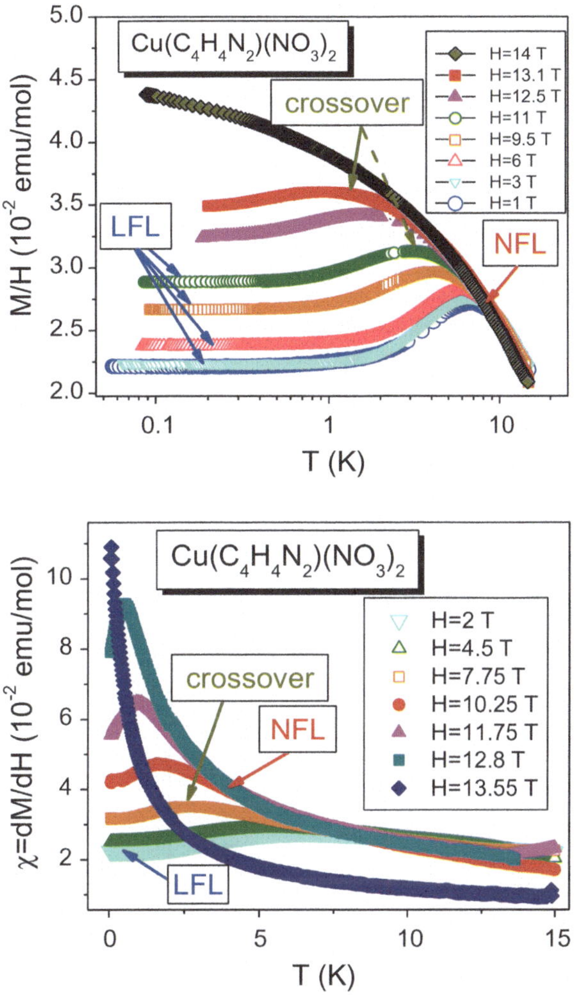

Fig. 9.4 Temperature dependence of M/H (upper panel) and $\mathrm{d}M/\mathrm{d}H$ (lower panel) for CuPzN. The experimental data are taken from [1]. The magnetic fields are shown in the legends and the three typical LFL, NFL, and crossover regions are shown, see Chap. 7 and Fig. 7.1

different magnetic fields (shown in the legends). The qualitative behavior, similar to that on Fig. 9.2, is seen. Namely, at low temperatures and fields H not very close to H_s we have LFL regions, then in the vicinity of maximum we have crossover ones and at high temperatures NFL behavior occurs. The above thermodynamic characteristics depicted in Figs. 9.1, 9.2, 9.3, 9.4 and 9.5 coincide with those of HF compounds, and permit to highlight the detailed structure of the $T - H$ phase diagram of CuPzN, shown in Fig. 9.6. To do so, in Fig. 9.5a, we report the peak temperature T_M of magnetic susceptibility as a function of H. It has been shown in [1] that this dependence is linear only near QCP at $H \to H_s$: $T_M = 0.76238(H_s - H)$. Taking into account (9.5), the linear dependencies can be revealed over a wide range of variation of H. This shows that the main characteristic of these lines is that they

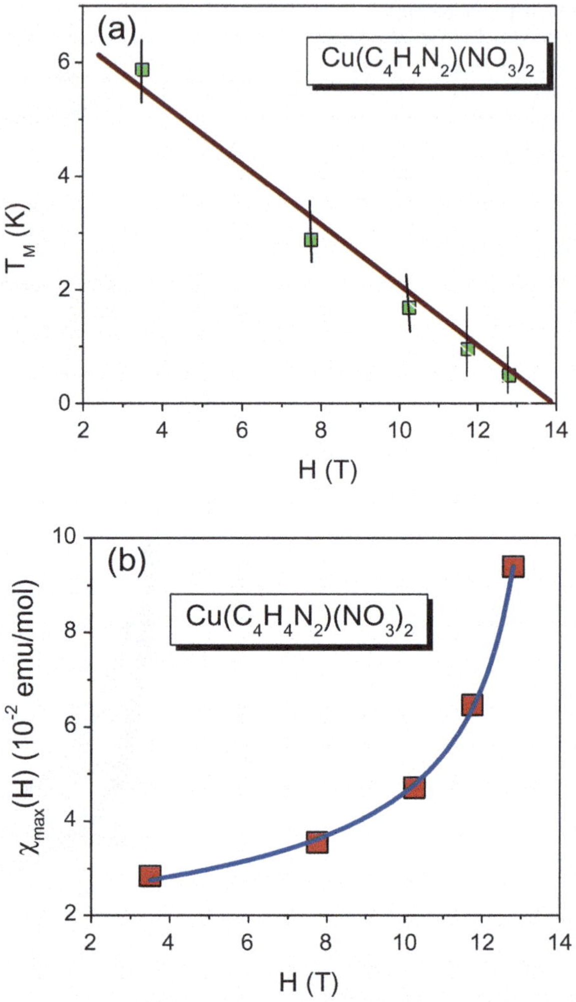

Fig. 9.5 Panel **a** reports the magnetic field dependence of peak temperature T_M of the magnetic susceptibility χ of CuPzN, gathered from experimental data [1]. The calculated straight line $T_M = a(H_s - H)$ given by (9.5) demonstrates good agreement with the experimental data. Panel **b** reports the magnetic field dependence of the maximum values $\chi_{\max}(H)$. The theoretical curve is given by $\chi_{\max} \propto (H_s - H)^{-1/2}$, see (9.5)

are straight lines. These lines represent energy scales typical for HF metals located at their QCP [27, 28]. In the panel (b) of Fig. 9.5 the maximum values $\chi_{\max}$ of χ versus $H_s - H$ are displayed. It is seen that the theoretical curve given by (9.5) is in good agreement with experimental data extracted from the measurement of the magnetization [1], and demonstrates that the effective mass of spinons does diverge at $H \to H_s$. At $H > H_s$ the QSL becomes a gapped field-induced ferromagnetic spin liquid, as shown in Fig. 9.6. At rising temperatures $T_N > 1$ the system enters the NFL region, as it is shown in Figs. 9.2d and 9.6. As a result, we conclude that CuPzN exhibits the behavior typical for HF compounds that leads to the formation of the corresponding $T - H$ phase diagram displayed in Fig. 9.6. As Fig. 9.2c, d shows the phase diagram is almost symmetric with respect to QCP point $H_s = 2J$

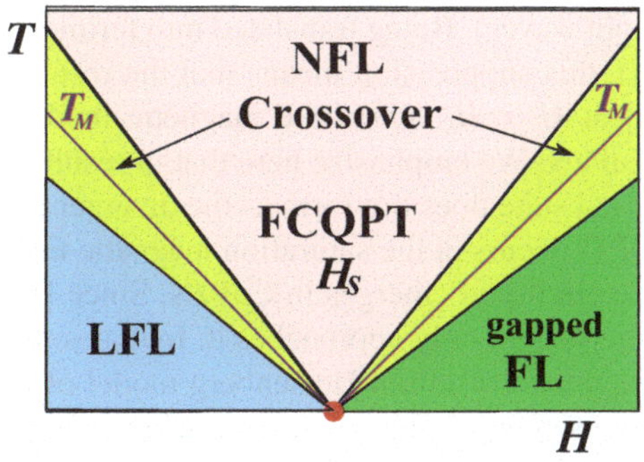

Fig. 9.6 Schematic $(T - H)$ magnetic field - temperature phase diagram of CuPzN, based on data from the panels (**a**) and (**b**) of Fig. 9.2 for $H \lessgtr H_s$. Straight lines on both sides of H_s, which is FCQPT point, indicate, respectively, the lines of LFL boundary (the lowest temperature), the temperatures of maxima (middle line, marked "T_M") and the end of crossover region (the highest temperature at which the system enters the NFL regime), see Fig. 9.2a. The right sector labeled as "gapped FL" denotes the gapped field-induced paramagnetic spin liquid

and consists of LFL, gapped Fermi liquid, crossover, and NFL regions. The crossover regions in Fig. 9.6 are shown by arrows and are formed by the straight lines, which are the magnetic field dependencies of temperatures of approximate LFL and NFL boundaries as well as by that of T_M. NFL state occurs at relatively high temperatures with the distinct temperature dependence $\sim T_N^{-1/2}$ [1]. At the same time LFL regions occur at low T, where the spinon effective mass is almost constant, as is the case for a normal Fermi liquid.

9.5 Discussion and Summary

In summary, in this chapter, we have shown that the scaling behavior and thermodynamic properties of CuPzN are defined by weakly interacting QSL, thus giving the complete theoretical explanation of corresponding experimental data [1]. Our main idea is that magnetic field drives CuPzN toward its QCP, which occurs at the saturation magnetic field H_s and this QCP is FCQPT, when the effective mass of spinons diverges as $H \to H_s$ at $T = 0$. Our analysis have shown that FCQPT in CuPzN occurs due to kinematic mechanism—the band becomes approximately flat not due to interaction between fermions but rather due to their spins polarization by the magnetic field.

Spin polarization can be easily "translated" into the language of initial Heisenberg spins. Namely, as the magnetic field tends also to align those initial spins along its direction, this means that it suppresses spin excitations, i.e., magnons. This is because as spin becomes progressively more aligned (or polarized) by the external magnetic field, it's Zeeman energy becomes more than that of thermal fluctuations, generating magnons (quantized spin waves). Being translated into fermionic language [10] (see also discussion above), this suppression means that the resulting fermion effective mass diverges [5, 10] at $H = H_s$, when the magnetic field fully polarizes initial Heisenberg spins in CuPzN. We emphasize here that in resulting 1D fermion space, the fully fermi condensed state does not form as the divergence of fermion effective mass, signifying FCQPT, occurs at the saturation magnetic field $H = H_s$, when the gapped ferromagnetic spin liquid emerges in CuPzN. Since FCQPT is precursor of FC state, at $H \leq H_s$ the system only approaches it. In that sense, the usual mapping of spin excitations (magnons) of initial Heisenberg model of CuPzN into fermions ensemble can be safely used here. This picture coincides with experimental facts as it gives their quantitative description.

Based on our approach, we have constructed the $T - H$ phase diagram of CuPzN and show that it is approximately symmetric with respect to QCP and has the LFL part, crossover regime, gapped Fermi liquid, and the NFL part. Our other important finding is that CuPzN exhibits the universal scaling behavior, characteristic for quasicrystals (see Fig. 9.2) which, in turn have the properties very similar to those of HF compounds, including the HF superconductor $\beta-$ $YbAlB_4$, 1D HF metal $YbNi_4P_2$ [9], quantum spin liquids like that in herbertsmithite $ZnCu_3(OH)_6Cl_2$ [6], and liquid ^3He [12], see Chaps. 13, 14, 15, 16 and 21. As a result, we conclude that 1D CuPzN belongs to the new state of matter, considered throughout all this book, and being formed by the topological FCQPT, see also Chap. 4 and Sect. 4.1.3.

References

1. Y. Kono, T. Sakakibara, C.P. Aoyama, C. Hotta, M.M. Turnbull, C.P. Landee, Y. Takano, Phys. Rev. Lett. **114**, 037202 (2015)
2. D. Shechtman, I. Blech, D. Gratias, J.W. Cahn, Phys. Rev. Lett. **53**, 1951 (1984)
3. V.R. Shaginyan, A.Z. Msezane, K.G. Popov, G.S. Japaridze, V.A. Khodel, Phys. Rev. B **87**, 245122 (2013)
4. V.R. Shaginyan, M.Ya. Amusia, A.Z. Msezane, K.G. Popov, Phys. Rep. **492**, 31 (2010)
5. M.Ya. Amusia, K.G. Popov, V.R. Shaginyan, V.A. Stephanovich, Theory of Heavy-Fermion Compounds. Springer Series in Solid-State Sciences, vol. 182 (2014)
6. V.R. Shaginyan, A.Z. Msezane, K.G. Popov, Phys. Rev. B **84**, 060401(R) (2011)
7. V.R. Shaginyan, V.A. Stephanovich, K.G. Popov, E.V. Kirichenko, JETP Lett. **103**, 30 (2016)
8. V.R. Shaginyan, V.A. Stephanovich, K.G. Popov, E.V. Kirichenko, S.A. Artamonov, Ann. Phys. (Berlin) **528**, 483 (2016)
9. C. Krellner, S. Lausberg, A. Steppke, M. Brando, L. Pedrero, H. Pfau, S. Tencé, H. Rosner, F. Steglich, C. Geibel, New J. Phys. **13**, 103014 (2011)
10. V.A. Khodel, V.R. Shaginyan, V.V. Khodel, Phys. Rep. **249**, 1 (1994)
11. A.V. Rozhkov, Eur. Phys. J. B **47**, 193 (2005)
12. V.R. Shaginyan, A.Z. Msezane, K.G. Popov, V.A. Stephanovich, Phys. Rev. Lett. **100**, 096406 (2008)
13. A.V. Rozhkov, Phys. Rev. Lett. **112**, 106403 (2014)
14. A.G. Lebed, Phys. Rev. Lett. **115**, 157001 (2015)
15. Y. Maeda, C. Hotta, M. Oshikawa, Phys. Rev. Lett. **99**, 057205 (2007)
16. T. Nikuni, M. Oshikawa, A. Oosawa, H. Tanaka, Phys. Rev. Lett. **84**, 5868 (2000)
17. I.M. Lifshitz, Sov. Phys. JETP **11**, 1130 (1960)

18. G.E. Volovik, Lect. Notes Phys. **718**, 31 (2007)
19. G.E. Volovik, Phys. Scr. T **164**, 014014 (2015)
20. V.A. Khodel, J.W. Clark, K.G. Popov, V.R. Shaginyan, JETP Lett. **101**, 413 (2015)
21. L.D. Landau, Sov. Phys. JETP **3**, 920 (1956)
22. J.W. Clark, V.A. Khodel, M.V. Zverev, Phys. Rev. B **71**, 012401 (2005)
23. V.A. Khodel, J.W. Clark, M.V. Zverev, Phys. Rev. B **78**, 075120 (2008)
24. Y. Matsumoto, S. Nakatsuji, K. Kuga, Y. Karaki, N. Horie, Y. Shimura, T. Sakakibara, A.H. Nevidomskyy, P. Coleman, Science **331**, 316 (2011)
25. Y. Matsumoto, S. Nakatsuji, K. Kuga, Y. Karaki, Y. Shimura, T. Sakakibara, A.H. Nevidomskyy, P. Coleman, J. Phys.: Conf. Ser. **391**, 012041 (2012)
26. K. Deguchi, S. Matsukawa, N.K. Sato, T. Hattori, K. Ishida, H. Takakura, T. Ishimasa, Nat. Mater. **11**, 1013 (2012)
27. M. Brando, L. Pedrero, T. Westerkamp, C. Krellner, P. Gegenwart, C. Geibel, F. Steglich, Phys. Status Solidi B **459**, 285 (2013)
28. V.R. Shaginyan, A.Z. Msezane, K.G. Popov, G.S. Japaridze, V.A. Khodel, Europhys. Lett. **106**, 37001 (2014)

Chapter 10
Dynamic Magnetic Susceptibility of Quantum Spin Liquid

Abstract In this chapter, we consider the dynamic magnetic susceptibility of quantum spin liquid, based on the theory of fermion condensation. The obtained results show that the dynamic magnetic susceptibility behaves as that of HF metals. Therefore, two known classical types of magnetism (ferro- and antiferromagnetism) can be augmented by one more, caused not by the order of the magnetic moments of atoms, ions, or electrons, but by the "liquid" behavior of spins. A new magnetic state of matter emerges, which is characterized by a spins flow. This flow is described by means of virtual chargeless particles—spinons, behaving as HF liquid. The theory of the dynamic magnetic susceptibility allows us to reveal that at low-temperature quasiparticles excitations, or spinons, form a continuum, and populate an approximately flat band crossing the Fermi level. The obtained results are in good agreement with experimental facts collected on herbertsmithite $ZnCu_3(OH)_6Cl_2$ and as well as on HF metals, and allow us to predict a new scaling in magnetic fields in the dynamic susceptibility. Under the application of strong magnetic fields, quantum spin liquid becomes completely polarized. We show that this polarization can be viewed as a manifestation of gapped excitations when investigating the spin-lattice relaxation rate.

10.1 Dynamic Spin Susceptibility of Quantum Spin Liquids and HF Metals

The key point of the LFL theory is the existence of fermionic quasiparticles defining the thermodynamic, relaxation, and dynamic properties of the considered material. However, strongly correlated Fermi systems encompassing a variety of systems that display behavior not easily understood within the LFL frame are called NFL. An important example of the NFL behavior is represented by HF metals. As we have demonstrated in Chap. 8, exotic QSL is formed by such hypothetic particles as fermionic spinons. The experimental studies of herbertsmithite $ZnCu_3(OH)_6Cl_2$ have discovered gapless excitations, analogous to excitations near the Fermi surface in HF metals, indicating that $ZnCu_3(OH)_6Cl_2$ is a quite promising system in investigating its QPTs and QSLs. The observed behavior of the thermodynamic properties

M. Amusia and V. Shaginyan, *Strongly Correlated Fermi Systems*, Springer Tracts in Modern Physics 283,
https://doi.org/10.1007/978-3-030-50359-8_10

of $ZnCu_3(OH)_6Cl_2$ strongly resembles that in HF metals, since the kagome lattice being strongly frustrated has a dispersionless topologically protected branch of the spectrum with zero excitation energy. This indicates that QSL formed by the ideal kagome lattice is located near the ordered side of FCQPT that is characterized by the presence of the spectrum with zero excitation energy. This observation allows us to establish a close similarity between QSL and HF metals whose HF systems are located near FCQPT and, therefore, exhibiting an universal scaling behavior. Although, as we have seen in Chap. 8, experimental facts on the thermodynamic properties give conclusive evidence that the QSL does exist, the theoretical interpretation of other bunch of data, namely those on inelastic neutron scattering spectrum and spin-lattice relaxation rates on herbertsmithite can deliver additional information on close relationship between QSL and HF metals. Here, we employ the Landau transport equation to construct the dynamical spin susceptibility. We elucidate how the calculated susceptibility is affected by magnetic field and describe the experimental data for herbertsmithite and HF metals.

10.2 Theory of Dynamic Spin Susceptibility of Quantum Spin Liquid and Heavy-Fermion Metals

To construct the dynamic spin susceptibility $\chi(\mathbf{q},\omega,T)=\chi'(\mathbf{q},\omega,T)+i\chi''(\mathbf{q},\omega,T)$ as a function of momentum $\mathbf{q}$, frequency ω, and temperature T, we use the model of homogeneous HF liquid located near FCQPT. To deal with the dynamic properties of Fermi systems, one can use the transport equation describing a slowly varying disturbance $\delta n_\sigma(\mathbf{q},\omega)$ of the quasiparticle distribution function $n_0(\mathbf{p})$, therefore $n=\delta n+n_0$. We consider the case when the disturbance is induced by the application of external magnetic field $B=B_0+\lambda B_1(\mathbf{q},\omega)$ with B_0 being a static field and λB_1 being a ω-dependent field with $\lambda\to 0$. As long as the transferred energy ω obeys the inequality, $\omega<qp_F/M^*<<\mu$, where M^* is the effective mass and μ is the chemical potential, and the quasiparticle distribution function $n(\mathbf{q},\omega)$ satisfies the transport equation [1, 2]

$$(\mathbf{q}\mathbf{v}_\mathbf{p}-\omega)\delta n_\sigma-\mathbf{q}\mathbf{v}_\mathbf{p}\frac{\partial n_0}{\partial\varepsilon_p}\sum_{\sigma_1\mathbf{p}_1}F_{\sigma,\sigma_1}(\mathbf{p}\mathbf{p}_1)\delta n_{\sigma_1}(\mathbf{p}_1)$$
$$=\mathbf{q}\mathbf{v}_\mathbf{p}\frac{\partial n_0}{\partial\varepsilon_p}\sigma\mu_B(B_0+\lambda B_1). \tag{10.1}$$

We assume that B_0 is finite but not strong enough to lead to the total polarization of the corresponding quasiparticle band. In the field B_0, the two, the spin up and spin down, Fermi surfaces are displaced by opposite values of energy, $\pm B_0\mu_B$, and the magnetization is given by $\mathscr{M}=\mu_B(\delta n_+-\delta n_-)$, where the two spin orientations with respect to the magnetic field are denoted by $\pm$, and $\delta n_\pm=\sum_p\delta n_\pm(\mathbf{p})$. The spin susceptibility χ is given by $\chi=\partial\mathscr{M}/\partial B|_{B=B_0}$. The transport equation (10.1) is reduced to two equations, which can be solved for each direction $\pm$ independently and allows to calculate $\delta n_{\pm}$ and the magnetization. The response to $\lambda B_1(\mathbf{q},\omega)$ can be found by expanding the solution of (10.1) in a power series with respect to $M^*\omega/qp_F$. As a result, we obtain the imaginary part of the spin susceptibility

$$\chi''(\mathbf{q},\omega) = \mu_B^2 \frac{\omega(M^*)^2}{2\pi q}\frac{1}{(1+F_0^a)^2}, \tag{10.2}$$

where F_0^a is the dimensionless spin antisymmetric quasiparticle interaction [1]. The interaction F_0^a is found to saturate at $F_0^a \simeq -0.8$ so that $1+F_0^a$ is positive. It is seen from (10.2) that the second term is an odd function of ω. Therefore, it does not contribute to the real part χ' and forms the imaginary part χ''. Taking into account that at relatively high frequencies $\mu \gg \omega \geq qp_F/M^*$ in the hydrodynamic approximation one has $\chi' \propto 1/\omega^2$ [1], we conclude that the equation

$$\chi(\mathbf{q},\omega) = \frac{\mu_B^2}{\pi^2(1+F_0^a)}\frac{M^* p_F}{1+i\pi\frac{M^*\omega}{qp_F(1+F_0^a)}}, \tag{10.3}$$

produces the simplest approximation for the susceptibility χ and satisfies the Kramers-Kronig relation connecting the real and imaginary parts of χ.

To understand how can χ'' and χ given by (10.2) and (10.3), respectively, depend on temperature T and magnetic field B, we recollect that near FCQPT point the effective mass M^* depends on T and B, and is given by the Landau equation, while magnetic field B enters it only in the combination $\mu_B B/k_B T$, making $k_B T_{\max} \simeq \mu_B B$, see Chap. 7. Thus, in the presence of magnetic fields the variable y becomes $y = T/T_{\max} \simeq k_B T/\mu_B B$. Since the variables T and B enter symmetrically (see (8.10), Chap. 7) is valid for $y = \mu_B B/k_B T$.

The schematic $T-B$ phase diagram of QSL and HF liquid is presented in Fig. 10.1. Magnetic field B and temperature T play the role of control parameters, the variation of which leads to the transition from the NFL to LFL regions, as it is shown by the vertical and horizontal arrows. At fixed B and increasing T, the system transits along the vertical arrow from LFL to NFL crossing the transition region. On the contrary, at fixed T the increase of B leads the system along the horizontal arrow from NFL region to LFL region.

10.3 Scaling Behavior of the Dynamic Susceptibility

The usual tool to probe experimentally the properties of the quasiparticle excitations is neutron-scattering measurements. These measurements have been performed for herbertsmithite. Namely, the complex dynamic spin susceptibility $\chi(\mathbf{q},\omega,T) = \chi'(\mathbf{q},\omega,T) + i\chi''(\mathbf{q},\omega,T)$ (where q is the transferred momentum and ω is the frequency) had been measured. The results obtained for low T show that the quasiparti-

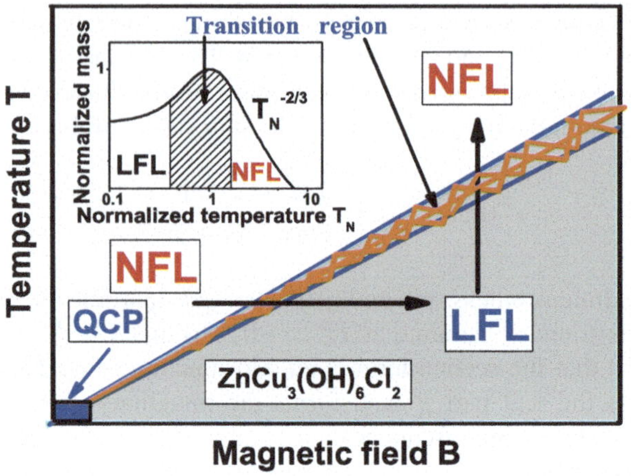

Fig. 10.1 Phase diagram of $ZnCu_3(OH)_6Cl_2$. The vertical and horizontal arrows show LFL-NFL and reverse transitions at fixed B and T, respectively. Inset shows a plot of the normalized effective mass versus the normalized temperature. Transition region, where M_N^* reaches its maximum at $T_N = T/T_{\max} = 1$, is shown by the arrows and hatched area in both the main panel and in the inset

cles are indeed gapless spinons with almost dispersionless spectrum, which crosses the Fermi level [4, 6, 7]. The imaginary part $\chi''(T, \omega_1)$ obeys the equation [4, 5]

$$T^{2/3}\chi''(T, \omega_1) \simeq \frac{a_1\omega_1}{1 + a_2\omega_1^2}, \tag{10.4}$$

where a_1 and a_2 are constants and $\omega_1 = \omega T^{-2/3}$. It is seen from (10.4) that $T^{2/3}\chi''(T, \omega_1)$ has a maximum $(T^{2/3}\chi''(T, \omega_1))_{\max}$ at some $\omega_{\max}$ and depends on the only variable ω_1. Equation (10.4) supports the scaling of $\chi'' T^{0.66}$, which has been studied experimentally in [6]. Similar to (8.10), we use the following dimensionless quantities: $(T^{2/3}\chi'')_N = T^{2/3}\chi''/(T^{2/3}\chi'')_{\max}$ and $\omega_N = \omega_1/\omega_{\max}$. In this case, (10.4) reduces to

$$(T^{2/3}\chi'')_N \simeq \frac{b_1\omega_N}{1 + b_2\omega_N^2}, \tag{10.5}$$

where b_1 and b_2 are fitting parameters. Their role is to permit the (10.5) to satisfy the requirement $(T^{2/3}\chi'')_{N\ max} = 1$ at $\omega_N = 1$. It is anticipated that the function $(T^{2/3}\chi'')_N$ (10.5) has scaling properties in terms of normalized variable ω_N [4, 5]. We speculate that for measurements of $\chi''(B, T)$ family with T being a parameter, the function $B^{2/3}\chi''(\omega_N)$ will also demonstrate the scaling with respect to ω_N

$$(B^{2/3}\chi'')_N \simeq \frac{d_1\omega_N}{1 + d_2\omega_N^2}. \tag{10.6}$$

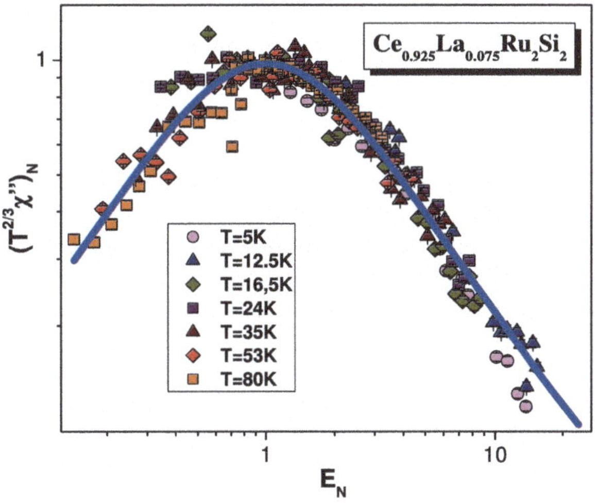

Fig. 10.2 The normalized dynamic spin susceptibility $(T^{2/3}\chi'')_N$ versus dimensionless variable E_N. The experimental data have been taken for the heavy-fermion metal $Ce_{0.925}La_{0.075}Ru_2Si_2$ [3]. Model calculations (10.5) are shown by the solid curve [4, 5]

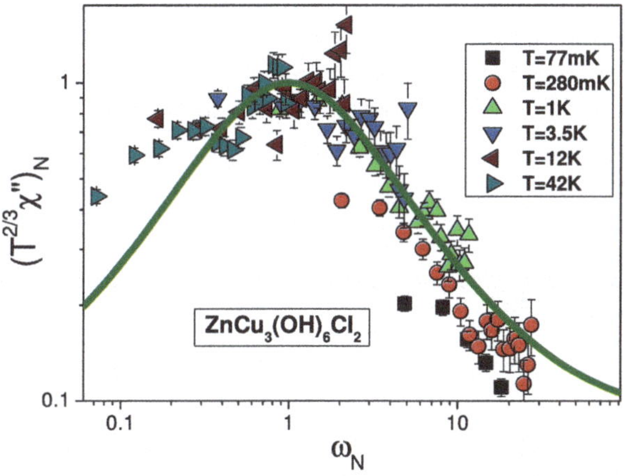

Fig. 10.3 The same as in Fig. 10.2 for the herbertsmithite. The experimental data are extracted from the measurements on $ZnCu_3(OH)_6Cl_2$ [6]

Similarly, d_1 and d_2 are fitting parameters chosen from the condition that the function $(B^{2/3}\chi'')_N = 1$ at $\omega_N = 1$.

The normalized imaginary part of dynamic susceptibility $(T^{2/3}\chi'')_N$, extracted from inelastic neutron-scattering spectrum of the HF compound $Ce_{0.925}La_{0.075}Ru_2Si_2$ [3], is reported in Fig. 10.2. The same quantity for herbertsmithite [6] is displayed in Fig. 10.3. The solid curves presenting the theoretical results from [4] give a pretty good description of the experimental data over almost three orders in the scaled variable E_N. We infer that the spin excitations in $ZnCu_3(OH)_6Cl_2$ demonstrate the same itinerant behavior as the electronic quasiparticles in the HF compound $Ce_{0.925}La_{0.075}Ru_2Si_2$ and hence do not have a gap in their excitation spectra. This clearly signals the presence of a gapless SCQSL in herbertsmithite [4, 5, 8].

The solid lines in Figs. 10.2 and 10.3 show that the calculations deliver a pretty good description of the experimental data. This means that it is the SCQSL which shapes the low-temperature physical properties of herbertsmithite, and makes them be similar to that of HF metals. It is not difficult to deduce now, that the kagomé layer excitations (spinons) do not have a spin gap as its presence would contradict the vast majority of firmly established experimental facts. This denies the existence of the above spin gap. Note that the presence or absence of the spin gap is not significant as it does not determine the thermodynamics of $ZnCu_3(OH)_6Cl_2$. This is because neither χ nor C_{mag}/T vanish in the high magnetic fields, eliminating the contribution coming from the impurities, see Figs. 8.14, 8.15, 8.16, 8.17 and 8.18. Here $C_{mag}/T \propto M^*_{mag}$ is the spinons contribution to the specific heat, with M^*_{mag} being the spinon effective mass. It is shown in Fig. 10.3 that the gap does not contribute to the imaginary part of the magnetic susceptibility that does not vanish at lowest temperatures. Rather, the SCQSL represents the underlying mechanism to explain the above observable properties.

A few remarks regarding the imaginary part $\chi''(T,\omega)$ are in order here. Equation (10.4) is valid provided that the system approaches FCQPT from the disordered side as shown in the phase diagram Fig. 10.1. If the system is located on the ordered side then at $B = 0$ the behavior of the effective mass is a function of T, as it is discussed in Chap. 7, and is given by the equation

$$M^*(T) \simeq a_\tau T^{-1}, \tag{10.7}$$

where a_τ is a constant. Upon taking into account (10.7) and acting in the same way as it was done in deriving (10.4), we obtain that the imaginary part $\chi''(T,\omega)$ is given by the equation

$$T\chi''(T,\omega) \simeq \frac{a_5 E}{1 + a_6 E^2}, \tag{10.8}$$

where a_5 and a_6 are constants, and $E = \omega/k_B T$. It is seen from (10.8) that $T\chi''(T,\omega)$ depends on the only variable $E = \omega/k_B T$. Thus, (10.4) and (10.8) establish two types of scaling behavior of $\chi''(\omega, T)$. Since the scaling behavior of $\chi''(\omega, T)$ is defined by the dependence of M^* on T, one may expect new types of scaling, especially at the transition region shown in Figs. 15.1 and 10.1.

Figure 10.4 presents the dynamic susceptibility ($T\chi''$) extracted from measurements of the inelastic neutron-scattering spectrum on the HF metal $YbRh_2Si_2$ [9]. The data ($T\chi''$) exhibit the scaling over three decades of the variation of both the function and the variable, thus confirming the validity of (10.8). The scaled data obtained in measurements on such quite different strongly correlated systems as $ZnCu_3(OH)_6Cl_2$, $Ce_{0.925}La_{0.075}Ru_2Si_2$, and $YbRh_2Si_2$ collapse fairly well onto a single curve over almost three decades of the scaled variables. It is seen that our calculations shown by the solid curves are in good agreement with the experimental data. Thus, the spin excitations in $ZnCu_3(OH)_6Cl_2$ exhibit the same behavior as electron excitations of the HF metal $Ce_{0.925}La_{0.075}Ru_2Si_2$, and, therefore form a continuum. This observation of the continuum is of great importance since it clearly

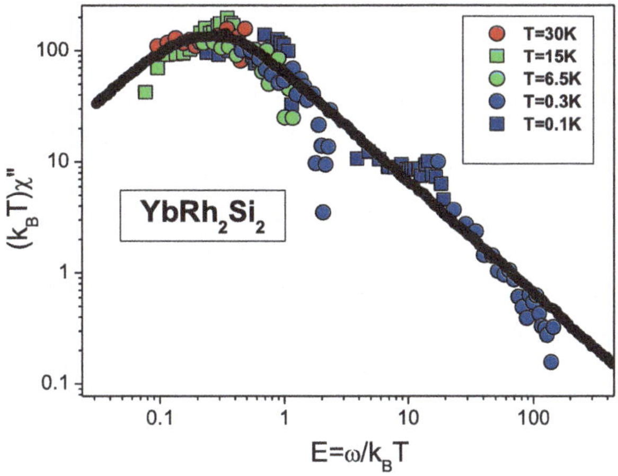

Fig. 10.4 $T\chi''$ plotted against $E = \omega/k_B T$. The data are extracted from measurements on $YbRh_2Si_2$ [9]. The solid curve is the fit by the function (10.8)

reveals the presence of SCQSL in herbertsmithite. Thus, in contrast to ferromagnet (with parallel direction of all atomic spins) and antiferromagnet (with opposite directions of atomic spins in its two sublattices), quantum spin liquid has the third type of magnetism—the direction of constituting spins is not fixed, but constantly rotated in time and space, being never ordered finally. In this chapter, we have also seen that quantum spin liquid, being represented by SCQSL, exhibits the properties of HF metals, and forms the new state of matter.

References

1. D. Pines, P. Noziéres, *Theory of Quantum Liquids* (Benjamin, New York, 1966)
2. M. Neumann, J. Nyéki, B. Cowan, J. Saunders, Science **317**, 1356 (2007)
3. W. Knafo, S. Raymond, J. Flouquet, B. Fák, M.A. Adams, P. Haen, F. Lapierre, S. Yates, P. Lejay, Phys. Rev. B **70**, 174401 (2004)
4. V.R. Shaginyan, A.Z. Msezane, K.G. Popov, V.A. Khodel, Phys. Lett. A **376**, 2622 (2012)
5. M.Y. Amusia, K.G. Popov, V.R. Shaginyan, W.A. Stephanowich, *Theory of Heavy-Fermion Compounds*. Springer Series in Solid-State Sciences, vol. 182 (Springer, Berlin, 2015)
6. J.S. Helton, K. Matan, M.P. Shores, E.A. Nytko, B.M. Bartlett, Y. Qiu, D.G. Nocera, Y.S. Lee, Phys. Rev. Lett. **104**, 147201 (2010)
7. T.H. Han, J.S. Helton, S. Chu, D.G. Nocera, J.A. Rodriguez-Rivera, C. Broholm, Y.S. Lee, Nature **492**, 406 (2012)
8. V.R. Shaginyan, K.G. Popov, V.A. Stephanovich, V.I. Fomichev, E.V. Kirichenko, Europhys. Lett. **93**, 17008 (2011)
9. C. Stock, C. Broholm, F. Demmel, J.V. Duijn, J.W. Taylor, H.J. Kang, R. Hu, C. Petrovic, Phys. Rev. Lett. **109**, 127201 (2012)

Chapter 11
Spin-Lattice Relaxation Rate and Optical Conductivity of Quantum Spin Liquid

Abstract In this chapter, we present the spin-lattice relaxation rate of quantum spin liquid. Also, we analyze optical conductivity with the goal to demonstrate experimental manifestation of the new state of matter.

11.1 Spin-Lattice Relaxation Rate of Quantum Spin Liquid

A strong deviation from LFL behavior has been revealed in the magnetic susceptibility χ and muon and ^{63}Cu nuclear spin-lattice relaxation rates $1/T_1$ of $YbCu_{5-x}Au_x$ ($x = 0.6$). The anomalies in these characteristics along with magnetic-field-induced reentrance of LFL properties are determined by the magnetic field B and temperature T dependence of the quasiparticle effective mass M^*. They demonstrate also that violations of the Korringa law also come from $M^*(B, T)$ dependence, as this dependence is discussed in Sect. 6.3. Our theoretical analysis of experimental data on the base of FCQPT approach permits not only to explain the above two experimental facts in a unified manner, but to unveil their universal properties, relating the peculiar features of both longitudinal magnetoresistance and specific heat in $YbRh_2Si_2$ to the behavior of spin-lattice relaxation rates.

Consider the effect of B on the spin-lattice relaxation rate $1/T_1T$ determined by χ'' given by (10.2)

$$\frac{1}{T_1T} = \frac{3}{4\mu_B^2}\sum_{\mathbf{q}} A_{\mathbf{q}}A_{-\mathbf{q}}\frac{\chi''(\mathbf{q},\omega)}{\omega}|_{\omega\to 0} \propto (M^*)^2, \tag{11.1}$$

where $A_{\mathbf{q}}$ is the hyperfine coupling constant of the muon (or nuclei) with the spin excitations at wave vector $\mathbf{q}$ [1–3]. Figure 11.1 displays the normalized value $(1/T_1T)_N$ and the normalized longitudinal magnetoresistance ρ_N at fixed temperatures versus the normalized magnetic field B_N. It is shown in Fig. 11.1 that the increase of magnetic field progressively reduces $1/T_1T$ and the longitudinal magnetoresistance (LMR), and these both as a function of B possess an inflection point at $B = B_{inf}$ shown by the arrow. The normalized LMR obeys the equation [3]

M. Amusia and V. Shaginyan, *Strongly Correlated Fermi Systems*,
Springer Tracts in Modern Physics 283,
https://doi.org/10.1007/978-3-030-50359-8_11

$$\rho_N(B_N) = \frac{\rho(B_N) - \rho_0}{\rho_{inf}} = \left(\frac{1}{T_1 T}\right)_N = (M_N^*)^2, \tag{11.2}$$

where ρ_0 is the residual resistance, ρ_{inf} is LMR taken at the inflection point, ρ is LMR, and $B_N = B/B_{inf}$. We normalize $(1/T_1 T)$ and LMR by their values at the inflection point, and the magnetic field is normalized by B_{inf}. In accordance with the phase diagram 10.1, at $B > B_{inf}$, as it is shown in Figs. 11.1 and 11.1 inset (B), QSL enters the LFL region with B-dependence of the effective mass defined by (7.5). It follows from (11.1) and (11.2) that $(1/T_1 T)_N = \rho_N = (M_N^*)^2$ where $(M_N^*)^2$ is defined by (8.10), which shows that different strongly correlated Fermi systems have to exhibit the same scaling as $(M_N^*)^2$. It is shown in Fig. 11.1 and from the inset, that $YbCu_{5-x}Au_x$, herbertsmithite $ZnCu_3(OH)_6Cl_2$ and $YbRh_2Si_2$ demonstrate similar behavior of $(M_N^*)^2$ resulting in the scaling of LMR and $1/T_1 T$. Thus, (10.2), (11.1) and (11.2) determine the close relationship existing between the quite different dynamic properties and different HF compounds such as herbertsmithite and organic insulators (see Chap. 12) with QSL and HF metals, revealing their scaling behavior at FCQPT. We note that one may be confused when applying (11.1) to describe $(1/T_1 T)$ in strong magnetic fields. In that case both QSL and HF metals become fully polarized due to Zeeman splitting [1, 2, 4, 5]. As a result, one subband becomes empty, while the energy ε_F of spinons at the Fermi surface of the other subband lies below the chemical potential μ determined by the magnetic field B_0. It follows from (10.1) that $\chi'' = 0$ and (11.1) is not valid. The difference $\delta = \mu - \varepsilon_F$ can be viewed as a gap that makes $1/T_1 T \propto \exp -(\delta/k_B T)$. At temperatures $k_B T \sim \delta$, the subbands are populated by spinons and the validity of (11.1) is restored. Thus, the presence of δ can be interpreted as the existence of gapped excitations. On the other hand, if there would be the gapped excitations, then the heat capacity would demonstrate the exponential decay rather than a linear T -dependence at low temperatures. Indeed, the analysis based on experimental data shows the presence of linear T -dependence even under the application of high magnetic fields [4], while measurements on $ZnCu_3(OH)_6Cl_2$ of $1/T_1 T$ suggest that the excitations have a gap [6]. To clarify whether the gapped excitations could occur in $ZnCu_3(OH)_6Cl_2$, accurate experimental measurements of the low-temperature heat capacity in magnetic fields are necessary. Measurements under the application of magnetic field up to 18 T [7] show that the gap in the heat capacity C/T does not exist, while both the effective mass of spinons, $M^*_{mag} \propto C_{mag}/T$, and the normalized effective mass of spinons, $M_N^* = (C_{mag}/T)_N$, remain finite at the lowest accessible temperatures, as it is shown in Figs. 8.15, 8.16 and 8.18.

As we have seen above, the properties of SCQSL in herbertsmithite are quite similar to those of the electron liquid in HF compounds. The only difference is that the charge of the "SCQSL electron" (i.e., actual spinon) should be zero. As a result, the SCQSL thermal resistivity w is given by the expression [5, 8, 9]

$$w - w_0 = W_r T^2 \propto \rho - \rho_0 \propto (M^*)^2 T^2, \tag{11.3}$$

where $W_r T^2$ is due to spinon-spinon scattering, which is similar to the contribution AT^2 from electron-electron scattering to charge transport. Here ρ is the longitudinal magnetoresistivity (LMR). Also, w_0 and ρ_0 are the residual thermal and electric resistivity, respectively. It follows from (8.9) and (11.3) that under the application of magnetic field the thermal resistivity of herbertsmithite diminishes like LMR of the archetypical HF metal $YbRh_2Si_2$, as it is shown in the inset (B) of Fig. 11.1 by the solid curve. We note that this behavior is observed in measurements on the organic insulators $EtMe_3Sb[Pd(dmit)_2]_2$ and $\kappa-(BEDT-TTF)_2Cu_2(CN)_3$, see Sect. 12.1.

Let us finally study the magnetic field effect on the spin-lattice relaxation rate $1/(T_1T)$. Figure 11.1 (A) reports the normalized quantity $1/(T_1T)_N$ as a function of magnetic field B keeping temperature as a parameter. It is seen that $1/(T_1T)$ decreases with B increase. Also, the curve in Fig. 11.1 (A) has an inflection point (arrow) at some $B = B_{\rm inf}$. To reveal the possible scaling, we normalize both abscissa and ordinate values to those in the inflection point. The ordinary LFL relation $1/(T_1T)_N \propto (M^*)^2$ implies that our system is located near its QCP [3, 5, 8, 9]. Significantly, Fig. 11.1(A) shows that the herbertsmithite $ZnCu_3(OH)_6Cl_2$ [10] and the HF metal $YbCu_{5-x}Au_x$ [11] do exhibit the same behavior for the normalized spin-lattice relaxation rate. As it is shown in Fig. 11.1 (A) for $B \leq B_{\rm inf}$ (or $B_N \leq 1$), the quantity $1/(T_1T)_N$ is

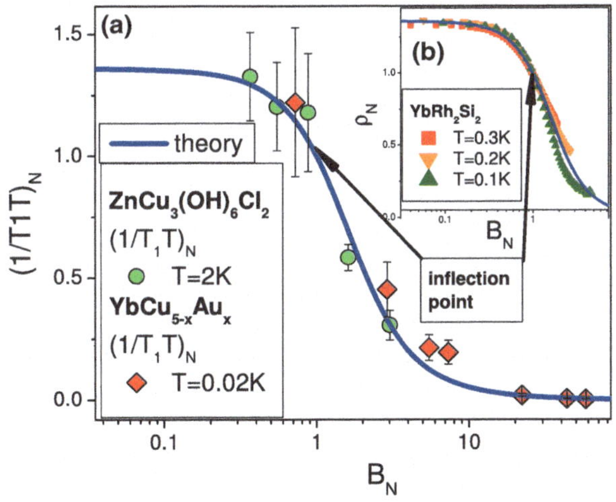

Fig. 11.1 The universal scaling behavior of the normalized muon spin-lattice relaxation rate and the thermal resistivity of $ZnCu_3(OH)_6Cl_2$. Main panel (A). Magnetic field dependence of spin-lattice relaxation rate $(1/T_1T)_N$ at fixed temperature in normalized coordinates. Circles show the data extracted from the measurements on herbertsmithite [10] at $T = 2$ K. Diamonds correspond to $YbCu_{5-x}Au_x$ at $x = 0.4$ and $T = 0.02$ K [11]. Inset (panel (B)) reports the normalized magnetoresistance $\rho_N(B_N)$, extracted from the data for $YbRh_2Si_2$ at different temperatures [12] listed in the legend. Arrows on main panel and inset show the inflection points used for the normalization. Also, in both panels, the calculated solid curves demonstrate the scaling behavior of the herbertsmithite's thermal resistivity $W_r \propto (M^*)^2$, (11.3) and (11.4))

almost independent of B at small fields and then diminishes quite rapidly at the field increase [3, 5, 8, 9] according to the relation

$$W_r \propto 1/(T_1 T)_N \propto (M^*)^2 \propto B^{-4/3}. \tag{11.4}$$

We thus predict that the increasing magnetic field $B_N > 1$ generates the crossover between the NFL and LFL phases of $ZnCu_3(OH)_6Cl_2$. This, in turn, lowers the relaxation rate, thermal resistivity $W_r(B) \propto (M^*(B))^2$ of $ZnCu_3(OH)_6Cl_2$, and LMR of $YbRh_2Si_2$ [12], see both panels of Fig. 11.1. The magnetic field measurements of low-energy inelastic neutron scattering on herbertsmithite single crystals are also desirable. This is because they permit to discern directly the existence of a possible gap in spinon excitation spectrum. The reason for that is the absence of impurity contribution, similar to the above case of the spin susceptibility χ.

11.2 Optical Conductivity

Measurements of the low-frequency optical conductivity collected on the geometrically frustrated insulator herbertsmithite provide important experimental evidence of the nature of its quantum spin liquid composed of spinons. To analyze measurements of the herbertsmithite optical conductivity at different temperatures, we employ a model of strongly correlated quantum spin liquid located near the fermion condensation phase transition. We consider an experimental manifestation of a new state of matter realized in quantum spin liquids, for, in many ways, they exhibit typical behavior of heavy-fermion metals. We also analyze dependence of the optical conductivity on magnetic fields.

Our next step is the analysis of the herbertsmithite optical conductivity $\overline{\sigma}$ at low frequencies. We consider low frequencies ω (and also temperatures T) to eliminate the phonon absorption contribution into the conductivity. It is known that latter contribution becomes substantial at high frequencies and temperatures [13–15]. The Hamiltonian of a particle with a spin reads

$$\mathcal{H} = \frac{\hbar^2}{2M}\left(\mathbf{p} - \frac{e}{c}\mathbf{A}\right)^2 + e\phi - \frac{\mu_B}{s}\mathbf{sB}, \tag{11.5}$$

where M is the spinon mass, $\mathbf{s}$ is the spin (s is its modulus). Here, ϕ and $\mathbf{A}$ are scalar and vector potentials, respectively. The spinon momentum operator $\mathbf{p} = i\nabla$, e is its (fictitious as spinon is chargeless) charge. The natural property of vector potential $\nabla\mathbf{A} = 0$ yields the relation $[\mathbf{A}, \mathbf{p}] = \mathbf{Ap}$, where [...] denotes a commutator. As spinons are chargeless ($e = 0$), only the term $\frac{\mu_B}{s}\mathbf{sB}$ in Ex. (11.5) gives nonzero contribution to $\overline{\sigma}$.

Taking latter fact into account, we express the energy transfer ε_B from the magnetic field $B(\omega)$ to the system in the form [15, 16]

$$\varepsilon_B = 2\pi\hbar\omega\left[\frac{\mu_B}{s}\mathbf{sB}\right]\chi''(\omega) \equiv 2\pi\frac{\omega^2\mu_B^3}{s}(M^*)^2\mathbf{sB}. \tag{11.6}$$

To derive the expression (11.6), we use the relation $\chi''(\omega) \propto \omega(M^*)^2$ [17] in the units $\hbar = c = 1$.

At the same time, the corresponding energy transfer ε_E from the electric field $E(\omega)$ reads

$$\varepsilon_E = E^2(\omega)\overline{\sigma}(\omega). \tag{11.7}$$

Comparison of (11.6) and (11.7) yields

$$\overline{\sigma}(\omega) \propto \omega\chi''(\omega) \propto \omega^2(M^*)^2. \tag{11.8}$$

It follows from (11.8) that $\overline{\sigma}(\omega) \propto \omega^2$. It is seen from (8.8) and (11.8), that at elevated temperatures $\overline{\sigma}(T)$ is a decreasing function, for the effective mass $M^* \propto T^{-2/3}$, see Sect. 8.3. This fact coincides with the results of corresponding measurements [14]. The same decreasing occurs for $\overline{\sigma}(B)$ as a function of magnetic field B. Latter observation is in discord with experimental data as no tangible dependence of magnetic field has been observed [14]. The clue to this puzzle is the exploration of temperature and magnetic field ranges, where the above phenomenon occurs. Namely, the measurements were done at $T = 6$ K and $B \leq 7$ T [14]. In this range of external parameters, the system is in the intermediate (between the NFL and LFL phases) regime (see Fig. 7.2) and hence cannot be represented as a normal Fermi liquid, where the effective mass M^* is given by (8.9). This means that for this specific case the effective mass is determined by (8.8), rather than by (8.9). This, in turn, implies that for the considered T and B, the magnetic field dependence of the optical conductivity cannot be observed. This permits us to predict that one should look for the essential magnetic field dependence of the optical conductivity at $B \simeq 7$ T and $T \leq 1$ K. In that case, as it shown in Fig. 8.15a, at $T \leq 1$ K the effective mass $M^* \propto C_{\text{mag}}/T$ is a decreasing function of the applied magnetic field. Thus, we predict that $\overline{\sigma}(B)$ diminishes at growing magnetic fields, as it follows from (8.9) and (11.8). Since the phonon contribution does not depend on magnetic fields, we suggest that measurements of the difference $\delta\sigma = \overline{\sigma}(\omega, B) - \overline{\sigma}(\omega, B = 0)$ can reveal both the physical properties of SCQSL in herbertsmithite and those of its ground state. This also applies to other QSL holding materials. We note that the measurements of the heat transport and optical conductivity can be carried out in the samples with different x. As a result, these experiments allow one to test the influence of impurities on the gap value. We predict that at moderate $x \sim 20\%$ SCQSL remains robust, for both the inhomogeneity and randomness facilitate frustration. We also suggest that the magnetic field can drive a gapped QSL to its QCP where the gap vanishes. These conclusions can be made upon taking into account a close familiarity between frustrated magnets and HF metals. For instance, the external magnetic field drives a number of HF compounds (like the archetypical HF metal $YbRh_2Si_2$) to their QCP's [18].

In summary, in this chapter, we have shown that SCQSL and HF metals expose the same behavior, thus, belonging to the same new state of matter.

References

1. J. Korringa, Physica (Utrecht) **16**, 601 (1950)
2. T. Moriya, *Spin Fluctuations in Itinerant Electron Magnetism* (Springer, Berlin, 1985)
3. V.R. Shaginyan, M.Y. Amusia, A.Z. Msezane, K.G. Popov, Phys. Rep. **492**, 31 (2010). (arXiv:1006.2658)
4. V.R. Shaginyan, A.Z. Msezane, K.G. Popov, Phys. Rev. B **84**, 060401(R) (2011)
5. V.R. Shaginyan, K.G. Popov, V.A. Stephanovich, V.I. Fomichev, E.V. Kirichenko, Europhys. Lett. **93**, 17008 (2011)
6. M. Jeong, F. Bert, P. Mendels, F. Duc, J.C. Trombe, M.A. de Vries, A. Harrison, Phys. Rev. Lett. **107**, 237201 (2011)
7. T.H. Han, S. Chu, Y.S. Lee, Phys. Rev. Lett. **108**, 157202 (2012)
8. V.R. Shaginyan, A.Z. Msezane, K.G. Popov, J.W. Clark, M.V. Zverev, V.A. Khodel, Phys. Lett. A **377**, 2800 (2013)
9. M.Y. Amusia, K.G. Popov, V.R. Shaginyan, W.A. Stephanowich, *Theory of Heavy-Fermion Compounds*. Springer Series in Solid-State Sciences, vol. 182 (Springer, Berlin, 2015)
10. T. Imai, E.A. Nytko, B.M. Bartlett, M.P. Shores, D.G. Nocera, Phys. Rev. Lett. **100**, 077203 (2008)
11. P. Carretta, R. Pasero, M. Giovannini, C. Baines, Phys. Rev. B **79**, 020401 (2009)
12. P. Gegenwart, T. Westerkamp, C. Krellner, Y. Tokiwa, S. Paschen, C. Geibel, F. Steglich, E. Abrahams, Q. Si, Science **315**, 969 (2007)
13. T.K. Ng, P.A. Lee, Phys. Rev. Lett. **99**, 156402 (2007)
14. D. Pilon, C.H. Lui, T.H. Han, D. Shrekenhamer, A.J. Frenzel, W.J. Padilla, Y.S. Lee, N. Gedik, Phys. Rev. Lett. **111**, 127401 (2013)
15. V.R. Shaginyan, A.Z. Msezane, V.A. Stephanovich, K.G. Popov, G.S. Japaridze, J. Low Temp. Phys. **191**, 4 (2018)
16. D. Pines, P. Noziéres, *Theory of Quantum Liquids* (Benjamin, New York, 1966)
17. V.R. Shaginyan, A.Z. Msezane, K.G. Popov, V.A. Khodel, Phys. Lett. A **376**, 2622 (2012)
18. N. Oeschler, S. Hartmann, A. Pikul, C. Krellner, C. Geibel, F. Steglich, Phys. B **403**, 1254 (2008)

Chapter 12
Quantum Spin Liquid in Organic Insulators and ^{3}He

Abstract In this chapter, we consider the low-temperature thermal conductivity of SCQSL. Measurements collected on insulators with geometrical frustration produce important experimental facts shedding light on the nature of quantum spin liquid composed of spinons. We employ the model of strongly correlated quantum spin liquid located near the topological fermion condensation phase transition to analyze the exciting measurements of the low-temperature thermal conductivity in magnetic fields collected on the organic insulators $EtMe_3Sb[Pd(dmit)_2]_2$ and $\kappa - (BEDT - TTF)_2Cu_2(CN)_3$. Our analysis of the conductivity allows us to reveal a strong dependence of the effective mass of spinons on magnetic fields, to detect a scaling behavior of the conductivity, and to relate it to both the spin-lattice relaxation rate and the magnetoresistivity. Our calculations and observations are in a good agreement with experimental data. On the example of two-dimensional (2D) ^{3}He, we demonstrate that the main universal features of its thermodynamic properties look like those of the heavy-fermion metals and SCQSL. Our theoretical analysis of 2D ^{3}He allows us to show that the fermion condensation theory describes experimental facts in 2D ^{3}He in unified manner, and to demonstrate that the universal behavior of effective mass M^* coincides with that observed in HF metals and SQCSL, and to conclude that 2D ^{3}He represents SCQSL. This observation opens a novel way to explore the properties of quantum frustrated magnets holding a spin liquid.

12.1 The Organic Insulators $EtMe_3Sb[Pd(dmit)_2]_2$ and $\kappa - (BEDT - TTF)_2Cu_2(CN)_3$

Organic insulators $EtMe_3Sb[Pd(dmit)_2]_2$ and $\kappa - (BEDT - TTF)_2Cu_2(CN)_3$ have 2D triangular lattices. It is experimentally established [1–3] that the specific geometric frustration prohibits spin ordering in these substances even at the lowest accessible temperatures T. Therefore, in studying insulators of this type, one can gain unique insights into the physics of quantum spin liquids. Indeed, the analysis of the heat capacity data of these insulators reveal a T-linear term, meaning that the low-energy excitation spectrum is gapless [2, 3]. These results can be obtained from the low-temperature data on thermal conductivity $\kappa(T)$: the residual term in $\kappa(T)/T$

M. Amusia and V. Shaginyan, *Strongly Correlated Fermi Systems*, Springer Tracts in Modern Physics 283,
https://doi.org/10.1007/978-3-030-50359-8_12

at $T \to 0$ that clearly shows that the excitation spectrum is gapless. Residual value is clearly resolved for $EtMe_3Sb[Pd(dmit)_2]_2$ and measurements of κ/T for the insulator $\kappa-(BEDT-TTF)_2Cu_2(CN)_3$ indicate that the low-energy excitation spectrum possibly has a gap [2, 3]. However, the existence of the gap becomes questionable if one takes two facts into account. Namely, while in the heat capacity measurements of $\kappa-(BEDT-TTF)_2Cu_2(CN)_3$ the T-linear term is present, the static spin susceptibility remains finite down to the lowest measured temperatures [4]. In accordance with that observation, recent measurements of the thermodynamic properties clearly demonstrate the absence of the gap [5]. This discrepancy is resolved by demonstrating that spinons are decoupled from the lattice phonons [6]. In that case the thermal conductivity decays rapidly with decreasing temperature, as it has been shown in the thermal conductivity measurements on cuprates [7]. Here we recall that SCQSL plays a role of HF liquid in cuprates. The thermal conductivity probes elementary itinerant excitations and is not sensitive to localized excitations such as those responsible for Schottky contributions, contaminating the heat capacity measurements at low temperatures. Thermal conductivity is formed primarily by both acoustic phonons and itinerant spinons, the latter form QSL. Since the phonon contribution is not sensitive to the applied magnetic field B, the elementary excitations of QSL can be further explored by the magnetic field dependence of κ. Thermal conductivity measurements under the application of magnetic field B in these insulators have demonstrated a strong dependence $\kappa(B,T)$ on B at fixed T [2, 3].

As it was mentioned in Chap. 8, SCQSL plays the role of HF liquid. Therefore, SCQSL in geometrically frustrated magnetic organic insulators behaves as the electronic liquid in HF metals and cuprates, provided that excitation has no charge. Another characteristic quantity of the SCQSL is its thermal resistivity w which is related to the thermal conductivity κ as

$$w=\frac{L_0T}{\kappa}=w_0+A_wT^2. \tag{12.1}$$

Here, L_0 is the Lorenz number, ρ_0 and w_0 are residual resistivity of electronic liquid and QSL, respectively. The coefficients $A_w\propto(M^*_{\rm mag})^2$ and $A_\rho\propto(M^*)^2$, [8]. In the presence of a magnetic field, the thermal resistivity w behaves like the electric magnetoresistivity $\rho_B=\rho_0+A_\rho T^2$ of the electronic liquid, see Sect. 11.1. This follows from understanding that A_w represents the contribution of spinon-spinon scattering to the thermal transport, being analogous to the contribution A_ρ to the charge transport, defined by electron-electron scattering. The above conclusion means that in the LFL region, the coefficient A_w in the expression (12.1) behaves like the spin-lattice relaxation rate shown in Fig. 11.1, see (11.4). In other words, the coefficient A_w of the SCQSL thermal resistivity, under the application of magnetic fields at fixed temperature has the same properties as the spin-lattice relaxation rate $1/(T_1T)_N$. It is shown in Fig. 11.1, that (in agreement with (11.4)) the magnetic field $B_N>1$ progressively reduces as $1/T_1T$. This quantity, as a function of B, possesses the inflection point at some $B/B_{\rm inf}=B_N=1$ shown by the arrow. Note, that the longitudinal magnetoresistivity also decreases with magnetic field B growth. It also has the inflection point

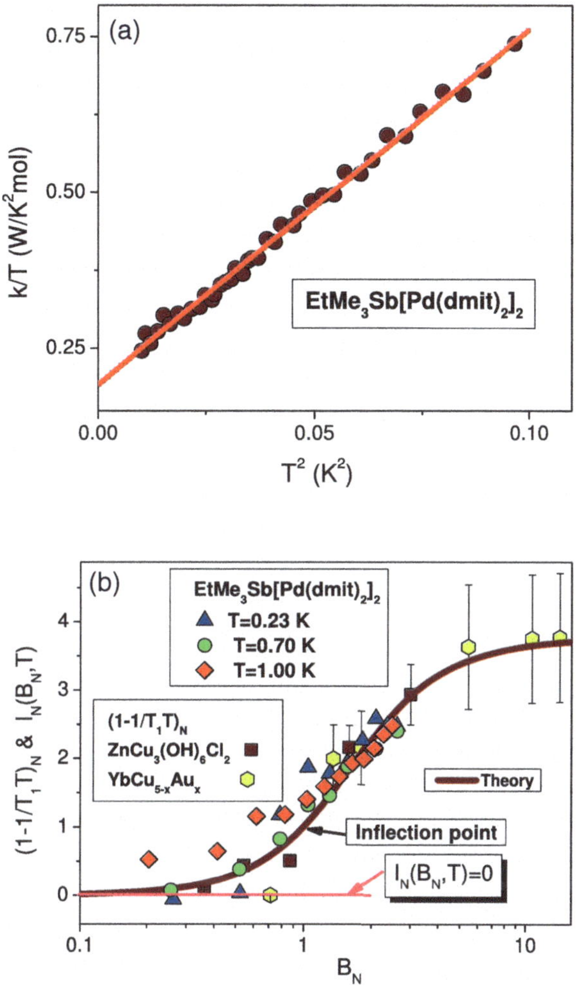

Fig. 12.1 Heat transport of $EtMe_3Sb[Pd(dmit)_2]_2$. Panel **a**. Low temperature thermal conductivity $\kappa(T)/T$ versus T^2, see (12.3). The finite value of residual conductivity is seen at $T \to 0$ [2, 3]. Panel **b**. The normalized thermal conductivity $I_N(B_N, T)$ versus B_N (markers), extracted from the data [2, 3]. The inflection point is shown by the arrow. The magnetic field dependence of the function $(1 - 1/T_1T)_N$ is extracted from measurements of $(1/T_1T)_N$ shown in Fig. 11.1a. The horizontal line shows that $I_N(B_N, T) \geq 0$. The solid line is obtained from the theoretical curve in Fig. 11.1

shown by the arrow in panel (b) (inset) of Fig. 11.1. This is consistent with the phase diagram displayed in Fig. 7.2: with growing magnetic fields the NFL region at first moves into the crossover region and then transforms into the LFL.

Study of the thermal resistivity w given by (11.3) allows one to reveal spinons as itinerant excitations. It is important that w is not contaminated by contributions coming from localized excitations. The temperature dependence of thermal resistivity w represented by the finite term w_0 directly demonstrates that the behavior of SCQSL is similar to that of metals, and there is a finite residual term κ/T in the zero-temperature limit of κ. The presence of this term immediately proves that there are gapless excitation associated with the property of normal and HF metals, in which gapless electrons govern the heat and charge transport. Finite w_0 means that in QSL both κ/T and $C_{\rm mag}/T \propto M^*_{\rm mag}$ remain nonzero at $T \to 0$. Therefore, specific heat and the transport are governed by gapless spinons, which form the Fermi surface. Further key information on the nature of spinons is provided from the B-dependence of the coefficient A_w. The particular B-dependence of $(1/T_1T)_N \propto (M^*_{\rm mag})^2$, shown in Fig. 11.1a, and given by (8.9), shows that the QSL in this case is indeed SCQSL. Note that the heat transport is usually contaminated by the phonon contributions. On the other hand, phonon contribution is hardly affected by the magnetic field B. Therefore, we expect that the B-dependence of the thermal conductivity should be determined by $A_w(B,T)$. Indeed, consider relation [8],

$$1-\frac{A_w(B,T)}{A_w(0,T)}=1-\left(\frac{M^*(B,T)_{\rm mag}}{M^*(0,T)_{\rm mag}}\right)^2$$
$$\simeq a(T)\frac{\kappa(B,T)-\kappa(0,T)}{\kappa(0,T)}\equiv a(T)I(B,T). \quad (12.2)$$

To derive (12.2), we use (12.1), and obtain

$$\frac{\kappa}{L_0T}=\frac{1}{w_0+A_wT^2}+bT^2. \quad (12.3)$$

Here, the bT^2 term describes the phonon contribution to the heat transport. Carrying out simple algebra and assuming that $[1-A_w(B,T)/A_w(0,T)]<1$, we come to (12.2). It is shown in both panels of Fig. 11.1 that the effective mass $M^*_N(B) \propto M^*_{\rm mag}(B)$ is a decreasing function of magnetic field B. Then, it follows from (8.9), (8.10), and (12.2) that the function $I(B,T)=[\kappa(B,T)-\kappa(0,T)]/\kappa(0,T)$ increases at elevated field B in the LFL region, while $I(B,T)\simeq 0$ in the NFL region as this function is approximately independent of B in that case.

Recent measurements of $\kappa(B=0)$ as a function of T^2 in the organic insulators EtMe$_3$Sb[Pd(dmit)$_2$]$_2$ [2, 3] are displayed in Fig. 12.1a. The measurements show that the heat is carried by both phonons and SCQSL, for the thermal conductivity is well fitted by $\kappa/T=b_1+b_2T^2$, where b_1 and b_2 are temperature- independent constants. The finite b_1 term implies that spinon excitations are gapless in EtMe$_3$Sb[Pd(dmit)$_2$]$_2$. A simple estimation suggests that in latter substance spinons propagate ballistically [2, 3]. As seen from (12.2), the reduced thermal conductivity $I(B,T)$ depends strongly on B. Namely, it increases with the magnetic field. Such a behavior is in agreement with (8.9) and Fig. 11.1 which show that $(M^*_{\rm mag}(B))^2$ is a decreasing function of B in the LFL region.

The scaling of the heat conductivity of the organic insulators is clearly observed in terms of dimensionless quantities, when $I(B,T)$ and B are normalized to their values at the inflection points, similar to the case of $(1/T_1T)$, see Fig. 11.1. Results of our calculations of the normalized heat conductivity $I_N(B_N,T)$ based on (8.4) and (12.2) are shown in Fig. 12.1b. The normalized heat conductivity $I_N(B_N,T)$ is independent of $a(B,T)$ and no fitting parameters are required to calculate $I_N(B_N,T)$. It is shown

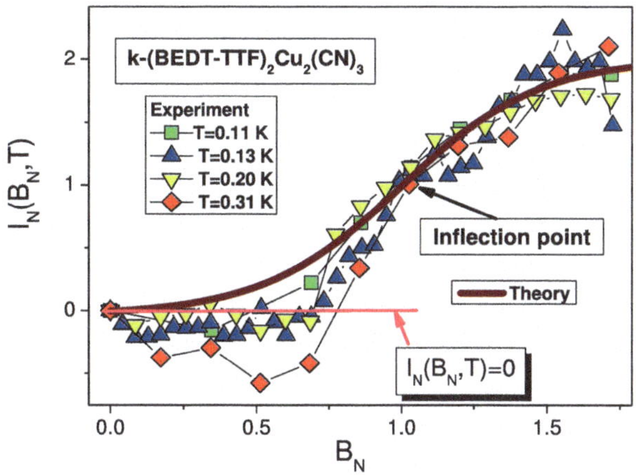

Fig. 12.2 Heat transport of $\kappa-(BEDT-TTF)_2Cu_2(CN)_3$. The normalized thermal conductivity $I_N(B_N, T)$ versus B_N shown by markers is extracted from the data [2, 3]. The horizontal line shows the level at which $I_N(B_N, T) = 0$. The solid curve coincides with that shown in Fig. 12.1b

in Fig. 12.1b, that in accordance with (8.10) $I_N(B_N, T)$ exhibits the scaling and thus becomes a function of a single variable B_N.

It is instructive to compare values of the normalized function $(1 - 1/T_1T)_N \equiv (1 - [M^*(B, T)/M^*(B = 0, T)]^2)_N$ (shown in Fig. 11.1a), extracted from the measurements of $(1/T_1T)_N$ with that of $I_N(B_N, T)$. Normalization point is chosen to be the inflection one, i.e., the magnetic field is normalized by $B_{\rm inf}$. As shown in Fig. 12.1b, $(1 - 1/T_1T)_N$ and $I_N(B_N, T)$ are in fair agreement with the solid curve depicting the theoretical function $(1 - [M^*(B, T)/M^*(B = 0, T)]^2)_N$. The latter, derived from our calculations, is represented by the solid curves in both panels of Fig. 11.1. At low B_N, since the system is in its NFL state when $B_N < 1$, the function $I_N(B_N, T) \geq 0$, and is almost B-independent constant. Thus, we expect that spinons are not decoupled from the lattice phonons at relatively high temperatures $T \geq 0.23$ K. This implies that in order to explain the growth of $I(B, T)$ at elevated B [2, 3], there is no need to introduce additional quasiparticles, which are activated when applying magnetic field. It is also shown in Figs. 11.1 and 12.1b that the organic insulators demonstrate the same behavior as $ZnCu_3(OH)_6Cl_2$, $YbCu_{5-x}Au_x$, and $YbRh_2Si_2$.

It is shown in Fig. 12.2 that at relatively high B_N the organic insulator $\kappa-(BEDT-TTF)_2Cu_2(CN)_3$ exhibits the same behavior as $EtMe_3Sb[Pd(dmit)_2]_2$ does at $B_N \gtrsim 1$. At the same time, for $B_N \lesssim 0.75$, the function $I_N(B_N, T) \leq 0$. This observation is in agreement with the experimental fact that spinons are decoupled from the phonons at low temperatures [6]. Nonetheless, at elevated magnetic fields $B_N \geq 0.75$, the function $I(B, T)$ given by (12.2) grows, so that $I_N(B_N, T)$ becomes positive. These observations are in accord with experimental data [6, 7] and demonstrate that the heat transport is supported by spinons and their effective mass diminishes under the application of magnetic field, see (8.9).

Some comments are in order here: Heat transport and thermal conductivity in $EtMe_3Sb[Pd(dmit)_2]_2$ are to be thoroughly investigated, for the results obtained in [2, 3] are under debate: on one hand, the results are confirmed [9], on the other hand, these are not supported by experimental investigations [10, 11]. We suggest that the experimental results [2, 3, 9] are correct, since the theoretical analysis of these experimental observation [8, 12] is in good agreement with them, see Fig. 12.1. Such an agreement is improbable provided that the experimental data are randomly and erroneously composed.

Measurements of the nuclear spin-lattice relaxation rate T_1^{-1} under application of magnetic field of 7.65 T indicates that $EtMe_3Sb[Pd(dmit)_2]_2$ has no classical magnetic ordering down to 19 mK. On the other hand, T_1^{-1} has a kink at 1.0 K, and demonstrates a decrease at $T < 1.0$ K proportional to T^2 [13]. One can assume a continuous phase transition taking place at 1.0 K, while the power law implies a nodal gap rather than a full gap [13]. We speculate that the observed decrease can be related to the LFL behavior of SCQSL as in that case $T_1^{-1} \sim T\chi(B,T) \sim TM^*_{mag}(B,T)$. It is now straightforward to obtain the observed behavior noting that in the LFL state the effective mass decreases, $M^*_{mag} \propto T^2$ as $T \to 0$. We suggest that measurements of the optical conductivity under the application of magnetic field could clarify the presence of the phase transition, see Sect. 11.2 for details. Thus, based on the combined experimental and theoretical grounds we can safely conclude that both $EtMe_3Sb[Pd(dmit)_2]_2$ and $\kappa - (BEDT - TTF)_2Cu_2(CN)_3$ represent frustrated quantum magnet holding SCQSL.

12.2 Quantum Spin Liquid Formed with 2D ^{3}He

We have shown above that SCQSL of both frustrated magnets and the electronic system of HF metals demonstrate the universal low-temperature behavior irrespectively of their composition [14]. Therefore it is of great importance to verify whether this behavior can be observed in 2D Fermi systems. Fortunately, such measurements on 2D ^{3}He become available [15]. These experimental results are extremely significant as they allow one to check the presence of the universal behavior in the system formed by ^{3}He atoms that are drastically different from electrons and spinons of frustrated magnets. Nonetheless, we shall see that ^{3}He atoms can represent spinons, for their electric charge is zero, their spin equals 1/2, and 2D films of ^{3}He exhibit the universal behavior. Experimental data show that 2D ^{3}He exhibits NFL behavior [15] that looks like that of HF metals [8, 16]. Thus, as we have shown in Sects. 8.2 and 8.3, it can strongly resemble that of SCQSL, that occupies approximately flat bands. Moreover, the behavior of 2D ^{3}He has its unique feature, for the bare mass M_{3He} of ^{3}He is $M_{3He} \sim 6 \times 10^3 m_e$, where m_e is the bare electron mass. Such a huge bare mass drives 2D ^{3}He to its FCQPT [8, 16]. Films of 2D ^{3}He have an important feature: it is possible to change the number density x of the film driving it toward FCQPT at which the quasiparticle effective mass M^* diverges, as it is experimentally observed [15], and explained theoretically, as it is shown in Fig. 12.3. Thus, 2D ^{3}He

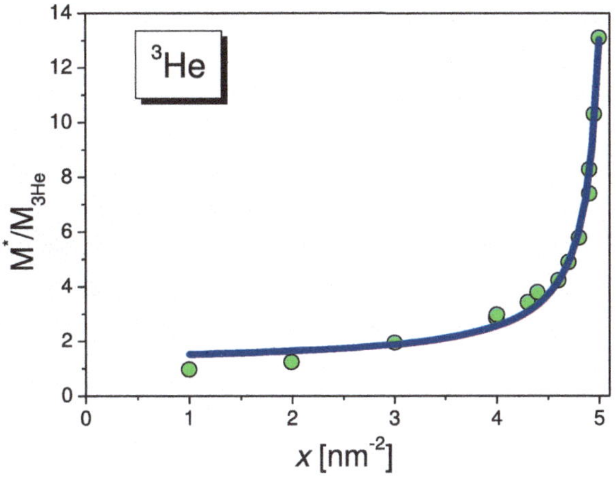

Fig. 12.3 The dependence of the dimensionless effective mass $M^*(x)/M_{3\text{He}}$ versus the number density x. Experimental data are shown by the circles [15]. The effective mass is fitted as $M^*(x)/M \propto a + b/(1 - x/x_c)$, where x_c is the critical number density at which FCQPT takes place, a and b are fitting parameters [8, 16]

could represent a unique system that allows one to check the properties of SCQSL versus its number density.

The bulk 3D liquid ^{3}He is the first object to which the LFL theory had been applied [17]. Macroscopic amount of ^{3}He represents isotropic Fermi liquid with negligible spin-orbit interaction, and is an ideal object to test the LFL theory. This 3D liquid solidifies under pressure, which prevents to study its properties as a function of the number density x. For 2D films of ^{3}He the liquid-solid phase transition occurs at relatively high densities x so that the NFL behavior can be observed. As a result, 2D films of ^{3}He have been prepared and their thermodynamic properties at different x have been extensively studied [15]. The measurements of the entropy S as a function of the number density x and temperature T reveals its NFL behavior: as shown in Fig. 12.4a, contrary to the LFL theory, at $x \geq 8.00$ nm^{-2} the entropy is no longer a linear function of T [8, 16].

It is shown in Fig. 12.4a that the behavior of the entropy $S(T)$ is different from that described by $S(T) \propto T$. One sees that at low densities $x \simeq 7$ nm^{-2} the entropy demonstrates the LFL behavior characterized by a linear function of T. The behavior changes drastically at higher densities, where $S(T)$ has an inflection point shown by the arrow in Fig. 12.4a. In order to show that the behavior of S displayed in Fig. 12.4a is of generic character, we display the normalized entropy $S_N(T_N)$ in the panel (b). Since the entropy can have no maxima, its normalization is to be performed at the inflection point $T = T_{\text{inf}}$, see above (for instance Fig. 11.1). Note that T_{inf} is x-dependent, and as it is shown in Fig. 12.4a the inflection point shifts toward lower temperatures with x increasing. The normalized entropy S_N as a function of the normalized temperature $T_N = T/T_{\text{inf}}$ is displayed in Fig. 12.4b. As it is shown in Fig. 12.4b, the normalization reveals the scaling behavior of S_N. This implies that as a function of the variable T_N, the curves at different T and x merge into a single one [8]. We have not included data with t $x \leq 8$ nm^{-2} since the corresponding curves do not contain the inflection points. Figure 12.4b shows that $S_N(T_N)$ is not a linear

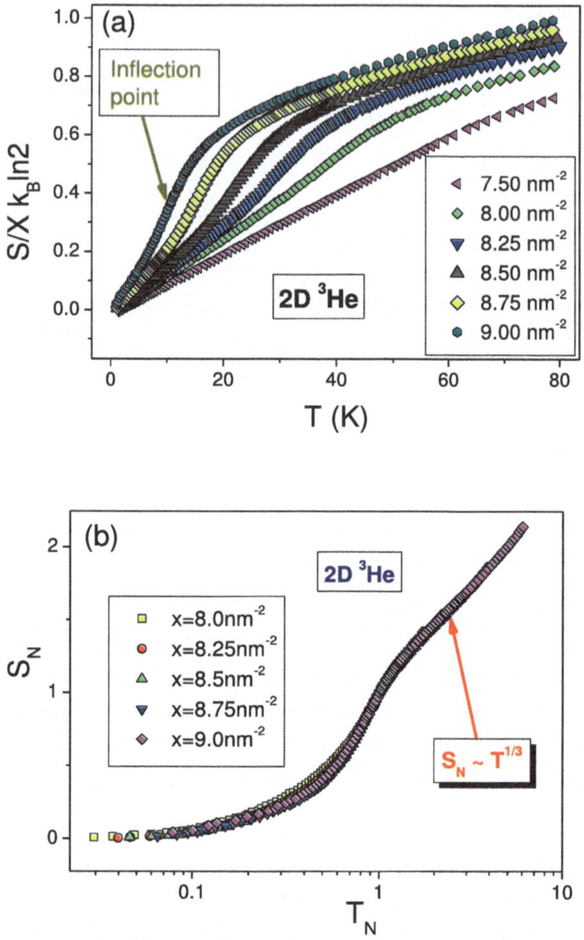

Fig. 12.4 Panel **a**. The temperature dependence of the entropy S of 2D ^{3}He at different densities x, as shown in the legend [15]. The inflection point of S at $x = 9.00$ nm^{-2} is indicated by the arrow. Panel **b**. The normalized entropy $S_N(T_N)$, extracted from the data shown in panel (**a**). The number densities x are listed in the legend

function of T_N, as it would be for a LFL, and exhibits scaling in wide range of T_N. At elevated temperatures $T_N > 1$ the normalized entropy $S_N \propto M^*T \propto T^{1/3}$, as it follows from (8.8). Here M^* is the quasiparticle effective mass of 2D ^{3}He.

It is shown in Figs. 7.1, 8.15, and 12.5 that the effective mass M^*_N (and accordingly M^*) at low T shows the LFL behavior as a function of temperature, at which M^*_N and M^* are approximately constants. Then it increases, reaching its maximum $M^*_M(T, x)$ at $T_{\max}(x)$ and then decreases with T as $T^{-2/3}$ [8]. As the peak value M^*_M increases, the $T_{\max}$ decreases. Near this temperature the last features of the LFL regime disappear, manifesting themselves in significant growth of M^*. Thus, the area around the $T_{\max}$ point can be interpreted as the crucial area of the crossover between LFL and NFL regimes.

We now employ the universal behavior, given by (8.10), to fit the data not only for 2D ^{3}He, but also for the 3D HF metals as well. The effective mass $M^*_N(T_N)$, extracted

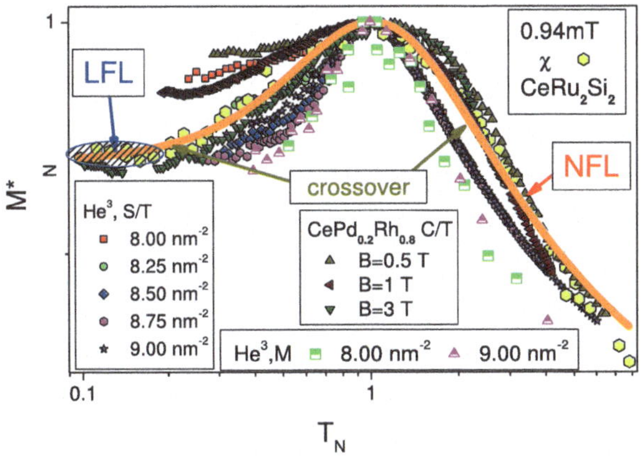

Fig. 12.5 New state of matter. The normalized effective mass M^*_N as a function of the normalized temperature T_N at densities x is shown in the lower left corner. The curve $M^*_N(T_N)$ is extracted from experimental data for $S(T)/T$ in 2D ^{3}He [15, 16] and 3D HF compounds with different magnetic ground states like $CeRu_2Si_2$ and $CePd_{1-x}Rh_x$ [18, 19]. Solid curve (similar to that in Fig. 7.1) represents our calculations that can be fitted by the universal function, given by (8.10)

from the entropy measurements on the ^{3}He film [15] at different densities x is reported in Fig. 12.5. In the same figure, the data extracted from heat capacity of ferromagnet $CePd_{0.2}Rh_{0.8}$ [18] and AC magnetic susceptibility of paramagnet $CeRu_2Si_2$ [19] are plotted for different magnetic fields. It is seen that the universal behavior of the effective mass given by (8.10) (solid curve in Fig. 12.5) is in accord with experimental data. It is seen that the behavior of the effective mass M^*_N, extracted from $S(T)/T$ in 2D ^{3}He (the entropy $S(T)$ is reported in Fig. 12.4a) looks very much like that in 3D HF compounds. We see from Fig. 7.1 that the positions of the magnetization maxima $M_0(T)$ and $S(T)/T$ in 2D ^{3}He follow nicely the interpolation formula (8.10). Thus, it follows from (8.10) that the effective mass can be described by the function of a single variable instead of four ones. Namely, the effective mass depends on magnetic field, temperature, number density, and the composition, and all these parameters can be merged in the single variable by means of an interpolating function similar to that of (8.10) [8, 16]. We conclude that despite drastic differences in microscopic features of 2D ^{3}He, 3D HF metals, and quantum frustrated magnets, their main normalized macroscopic features are universal. Therefore, we can conclude that strongly correlated Fermi systems represent a new state of matter [20]. As a result, the main properties of 2D ^{3}He can be used to investigate quantum frustrated magnets with SCQSL. This observation opens a novel way to explore the properties of quantum spin liquid.

12.3 Discussion

When analyzing QSL that defines the physics of frustrated magnets, it is of crucial importance to clarify the presence or absence of a spin gap, as it strongly influences the properties. As we have seen above, the main controversy in the assessment of QSL in geometrically frustrated magnets is "the gap or not the gap" in the spectrum of spinon excitations of such a frustrated quantum magnet as $ZnCu_3(OH)_6Cl_2$. On one hand (see the Sect. 8.4), both experiments and theory suggest that there is no gap in the spectrum. On the other hand, there has been recent suggestion [21–23], that there can exist a small spin gap in the kagome layers. However, as we have shown above, this suggestion contradicts the vast majority of experimental and theoretical results, see [24–26]. The results reported are based on both experimental findings and their theoretical interpretation in the framework of the impurity model. The experimental data involved are derived from high-resolution low-energy inelastic neutron scattering on $ZnCu_3(OH)_6Cl_2$ single crystals. The main assumption of the concomitant theoretical explanation was that the influence of the Cu impurities on the observed properties of herbertsmithite has nothing to do with the geometrical frustration inherent in kagome lattice geometry [21–23]. The model then assumes that the spin gap survives under the application of magnetic fields up to $B = 9$ T [23], while at $B = 0$ the bulk spin susceptibility χ exhibits a divergent Curie-like behavior, specifying that some of the Cu spins act like weakly coupled impurities [21–23]. Our above analysis, based on panoply of experimental data complemented by both ab initio and model calculations (and within FC formalism among them) shows that it is impossible to isolate the contributions coming from the impurities and from the kagomé planes geometry. This is because the impurities, being embedded in kagomé host lattice, form a single entity. This statement is reflected in recent experimental and theoretical studies of thermodynamic and relaxation properties of herbertsmithite [8, 12, 27–31]. Thus, when analyzing the emerging QSL physics in the geometrically frustrated magnets, it is of crucial importance to verify the existence of the abovementioned gap in spinon excitations by performing necessary experimental and theoretical studies. To analyze the QSL behavior theoretically, we employ the SCQSL model. A simple kagome lattice may host a dispersionless topologically protected branch of the quasiparticle spectrum with zero excitation energy, known as a flat band [28, 29]. In that case the topological FCQPT can be regarded as a QCP of the abovementioned geometrically frustrated magnets QSL composed of fermionic spinons. Measurements of the real part of the low-frequency optical conductivity $\overline{\sigma}$ as a function of temperature T and applied magnetic field B, collected on geometrically frustrated magnetic insulators, can yield important experimental insights to the nature of spinon-based QSL [32, 33]. In essence, we face a challenging problem of interpreting the experimental data [32] in a consistent way, including the T-dependence of the optical conductivity. As experimental data are available for a limited number of geometrically frustrated magnetic systems, here we were able to analyze two more compounds only. Namely, we discuss the properties of the organic substance $EtMe_3Sb[Pd(dmit)_2]_2$ and quasi-1D compound

$Cu(C_4H_4N_2)(NO_3)_2$. Both of them represent the insulating magnets that behave like HF metals with one exception: they do not conduct the electric current. There are other magnets that possibly hold QSL, see, e.g., [34]. We suggest that these magnets can be doped by impurities in order to stabilize QSL in them. Then, we suggest that the application of magnetic field can drive a gapped QSL to its QCP at which the gap vanishes. This conclusion can be made upon taking into account a close similarity between frustrated magnets, HF metals and 2D ^{3}He. This is because both some HF metals and 2D liquid helium can be tuned to their QCP by the application of external stimuli like magnetic field, doping, and/or pressure. Note that there is two contradictory groups of measurements regarding presence or absence of the spin gap. Namely, the thermodynamic properties demonstrate the absence of a gap, while the measurements of heat transport show its existence. This apparent contradiction can be resolved by demonstrating that spinons are decoupled from the lattice phonons and do not contribute to the heat transport [6]. As a result, the thermal conductivity vanishes rapidly with decreasing temperature, as it has been shown in measurements of the cuprates thermal conductivity [7].

12.4 Outlook

In Chaps. 8, 9, 10, 11, and 12, we consider QSLs represented by SCQSL of quantum frustrated magnets, exhibiting flat bands. The flat bands studies now attract a lot of interest both from experimentalists and theorists, see, e.g., [35, 36]. From experimental side, one of their most important features, that they can contain, being partially filled, an exponentially large number of states. As we have shown above, this macroscopic degeneracy leads to novel phenomenon of the fermion condensation, which gives new impetus to the entire physics of strongly correlated Fermi systems.

Strange as it may seem, such "non-fermionic" substances as geometrically frustrated magnets, can also be regarded as a kind of strongly correlated fermion systems. Our preceding discussion has shown convincingly that the adopted formalism can describe all experimental details of above quantum frustrated magnets. This is regardless of their microscopic details like kagomé of triangular lattice (hence the difference in the phonon spectra) and/or impurities and defects ensemble in each specific substance. Moreover, we have shown that, after scaling procedure, there is one-to-one correspondence between the experimental results obtained in 1D (see Chap. 9), 2D geometrically frustrated magnets, quasicrystals (see Sect. 21.2, and those for HF compounds. This demonstrates the similarity of the underlying physics of these very different kinds of solids, that form the new state of matter. This unifying physical mechanism is indeed the fermion condensation phenomenon and the topological FCQPT, which forms the universal properties of all these strongly correlated Fermi systems. As we have seen above, the "hidden FC" is ubiquitous and is responsible for many yet unexplained experimentally observed anomalies.

In summary, the main message of the present chapter is to unveil both the main properties of QSL in quantum frustrated magnets and attract attention to experiments that can unambiguously reveal the presence of QSL in frustrated magnets. We have shown that many seemingly unrelated and unusual experimental findings can be well explained within "hidden fermion condensation" picture. To reveal the presence of QSL, we suggest to perform the above discussed heat transport, low-energy inelastic neutron scattering, and optical conductivity $\overline{\sigma}$ measurements in 1D and 2D geometrically frustrated magnets subjected to external magnetic fields at low temperatures. These measurements are extremely important to shed light on the quantum spin-liquid physics in them.

References

1. W. Knafo, S. Raymond, J. Flouquet, B. Fák, M.A. Adams, P. Haen, F. Lapierre, S. Yates, P. Lejay, Phys. Rev. B **70**, 174401 (2004)
2. M. Yamashita, N. Nakata, Y. Senshu, M. Nagata, H.M. Yamamoto, R. Kato, T. Shibauchi, Y. Matsuda, Science **328**, 1246 (2010)
3. M. Yamashita, T. Shibauchi, Y. Matsuda, ChemPhysChem **13**, 74 (2012)
4. Y. Shimizu, K. Miyagawa, K. Kanoda, M. Maesato, G. Saito, Phys. Rev. Lett. **91**, 107001 (2003)
5. S. Yamashita, Y. Nakazawa, M. Oguni, Y. Oshima, H. Nojiri, Y. Shimizu, K. Miyagawa, K. Kanoda, Nat. Phys. **4**, 459 (2008)
6. T. Isono, S. Sugiura, T. Terashima, K. Miyagawa, K. Kanoda, S. Uji, Nat. Comm. **9**, 1509 (2018)
7. M.F. Smith, J. Paglione, M.B. Walker, L. Taillefer, Phys. Rev. B **71**, 014506 (2005)
8. M.Y. Amusia, K.G. Popov, V.R. Shaginyan, W.A. Stephanowich, *Theory of Heavy-Fermion Compounds*. Springer Series in Solid-State Sciences, vol. 182 (Springer, Berlin, 2015)
9. M. Yamashita, J. Phys. Soc. **88**, 083702 (2019)
10. P. Bourgeois-Hope, F. Laliberté, E. Lefrançois, G. Grissonnanche, S.R. de Cotret, R.K.R. Gordon, L. Taillefer, N. Doiron-Leyraud, Phys. Rev. X **9**, 041051 (2019)
11. J.M. Ni, B.L. Pan, Y. Huang, J. Zeng, Y. Yu, E. Cheng, L. Wang, R. Kato, S. Li, Phys. Rev. Lett. **123**, 247204 (2019)
12. V.R. Shaginyan, A.Z. Msezane, K.G. Popov, J.W. Clark, M.V. Zverev, V.A. Khodel, Phys. Lett. A **377**, 2800 (2013)
13. S. Maegawa, T. Itou, A. Oyamada, R. Kato, J. Phys. Confer. Ser. **320**, 012032 (2011)
14. V.R. Shaginyan, V.A. Stephanovich, A.Z. Msezane, G.S. Japaridze, J.W. Clark, M.Y. Amusia, E.V. Kirichenko, J. Mater. Sci. **55**, 2257 (2020)
15. M. Neumann, J. Nyéki, B. Cowan, J. Saunders, Science **317**, 1356 (2007)
16. V.R. Shaginyan, A.Z. Msezane, K.G. Popov, V.A. Stephanovich, Phys. Rev. Lett. **100**, 096406 (2008)
17. L.D. Landau, Sov. Phys. JETP **3**, 920 (1956)
18. N. Oeschler, S. Hartmann, A. Pikul, C. Krellner, C. Geibel, F. Steglich, Phys. B **403**, 1254 (2008)
19. D. Takahashi, S. Abe, H. Mizuno, D. Tayurskii, K. Matsumoto, H. Suzuki, Y. Onuki, Phys. Rev. B **67**, 180407(R) (2003)
20. V.R. Shaginyan, V.A. Stephanovich, A.Z. Msezane, P. Schuck, J.W. Clark, M.Y. Amusia, G.S. Japaridze, K.G. Popov, E.V. Kirichenko, J. Low Temp. Phys. **189**, 410 (2017)
21. T.H. Han, M.R. Norman, J.J. Wen, J. A., Rodriguez-Rivera, J.S. Helton, C. Broholm, Y.S. Lee, Phys. Rev. B **94**, 060409(R) (2016)
22. T. Imai, M. Fu, T.H. Han, Y.S. Lee, Phys. Rev. B **84**, 020411(R) (2011)
23. M. Fu, T. Imai, T.H. Han, Y.S. Lee, Science **350**, 655 (2015)

24. M.R. Norman, Rev. Mod. Phys. **88**, 041002 (2016)
25. Y. Zhou, K. Kanoda, T.K. Ng, Rev. Mod. Phys. **89**, 025003 (2017)
26. L. Savary, L. Balents, Rep. Prog. Phys. **80**, 016502 (2017)
27. T.H. Han, S. Chu, Y.S. Lee, Phys. Rev. Lett. **108**, 157202 (2012)
28. V.R. Shaginyan, M.Y. Amusia, A.Z. Msezane, K.G. Popov, Phys. Rep. **492**, 31 (2010). arXiv:1006.2658
29. V.R. Shaginyan, A.Z. Msezane, K.G. Popov, Phys. Rev. B **84**, 060401(R) (2011)
30. V.R. Shaginyan, A.Z. Msezane, K.G. Popov, V.A. Khodel, Phys. Lett. A **376**, 2622 (2012)
31. V.R. Shaginyan, K.G. Popov, V.A. Stephanovich, V.I. Fomichev, E.V. Kirichenko, Europhys. Lett. **93**, 17008 (2011)
32. D. Pilon, C.H. Lui, T.H. Han, D. Shrekenhamer, A.J. Frenzel, W.J. Padilla, Y.S. Lee, N. Gedik, Phys. Rev. Lett. **111**, 127401 (2013)
33. T.K. Ng, P.A. Lee, Phys. Rev. Lett. **99**, 156402 (2007)
34. Z.X. Luo, E. Lake, J.W. Mei, O.A. Starykh, Phys. Rev. Lett. **120**, 037204 (2018)
35. Y. Cao, V. Fatemi, A. Demir, S. Fang, S.L. Tomarken, J.Y. Luo, J.D. Sanchez-Yamagishi, K. Watanabe, T. Taniguchi, E. Kaxiras, R.C. Ashoori, P. Jarillo-Herrero, Nature **556**, 80 (2018)
36. Y. Cao, V. Fatemi, S. Fang, K. Watanabe, T. Taniguchi, E. Kaxiras, P. Jarillo-Herrero, Nature **556**, 43 (2018)

Chapter 13
Universal Behavior of the Thermopower of HF Compounds

Abstract In this chapter, we expose the new state of matter exhibited by strongly correlated Fermi systems represented by various HF compounds. The quantum physics of different HF compounds is universal, and emerges regardless of the microscopic structure of the compounds. This uniform behavior allows us to view it as the main characteristic of the new state of matter exhibited by HF compounds. In this chapter, we reveal and explain the universal scaling behavior of the thermopower S/T in such different heavy-fermion (HF) compounds as $YbRh_2Si_2$, β-$YbAlB_4$, and the strongly correlated layered cobalt oxide $[BiBa_{0.66}K_{0.36}O_2]CoO_2$. Using the archetypical HF metal $YbRh_2Si_2$ as an example, we demonstrate that the scaling behavior of S/T is violated by the antiferromagnetic phase transition, since both the residual resistivity ρ_0 and the density of states N experience jumps at the phase transition, making the thermopower experience two jumps and change its sign. Our elucidation is based on flattening of the single-particle spectrum that profoundly affects ρ_0 and N. To depict the main features of the S/T behavior, we construct the $T - B$ schematic phase diagram. Our calculated S/T for the HF compounds are in good agreement with experimental facts and support our observations.

13.1 Introduction

Strongly correlated Fermi systems are represented by a large variety of HF metals, insulators of new type with quantum spin liquids, quasicrystals, two-dimensional (2D) systems and liquids like ^3He, and systems like $Cu(C_4H_4N_2)(NO_3)_2$ with one-dimensional (1D) quantum spin liquid. We name these various systems HF compounds, since, as we shall see, they exhibit the behavior conforming to a type of HF metal. It is well known that three phases of matter exist: gaseous, liquid, and solid, and at high enough temperatures any of these transforms into plasma—a system of chaotically moving nuclei and electrons with a gas-type behavior. Surprisingly, at the lower end of the temperature scale, close to the absolute zero such multi-particle sys-

M. Amusia and V. Shaginyan, *Strongly Correlated Fermi Systems*,
Springer Tracts in Modern Physics 283,
https://doi.org/10.1007/978-3-030-50359-8_13

tems as HF compounds exhibit universal behavior of their thermodynamic, transport, and relaxation properties, which are governed by unique quantum phase transition called the fermion-condensation quantum phase transition (FCQPT) that creates flat bands, and leads to the specific topological state known as the fermion condensate (FC). In this chapter, we show both analytically and using arguments based entirely on the experimental grounds that all these HF compounds exhibit the universal scaling behavior formed by quasiparticles. This universal behavior, taking place at relatively low temperatures and induced by FCQPT, allows us to interpret it as the main feature of the new state of matter. Whatever mechanism drives the system to FCQPT, the system demonstrates the universal behavior, despite numerous mechanisms or tuning parameters present at zero temperature, such as pressure, number density, magnetic field, chemical doping, frustration, and geometrical frustration.

As we have shown in Chap. 7, at FCQPT the effective mass M^* of quasiparticles exhibits scaling behavior, while the very concept of quasiparticles survive FCQPT and they define the thermodynamic, transport, and relaxation properties of HF compounds. Thus, quasiparticles determine both the main properties and the universal scaling behavior, observed in HF compounds. As we already mentioned earlier in this book, contrary to the Landau paradigm, where quasiparticles with approximately constant effective mass M^* are continuously linked to quasiparticles of noninteracting gas, the effective mass of the above new quasiparticles formed at FCQPT strongly depends on temperature T, magnetic field B, pressure P, and other external parameters. As a result, we introduce and are developing in this book an extended quasiparticle paradigm, and show that new quasiparticles generate the non-Fermi liquid (NFL) behavior of the physical quantities of HF compounds that are remarkably different from those of ordinary solids or liquids described by the Landau Fermi liquid (LFL) theory. Then, we briefly consider the topological properties of FC taking place beyond FCQPT, and demonstrate how FCQPT generates the universal behavior of HF metals. In this chapter, at first, we examine the universal scaling behavior of the thermodynamic, transport and relaxation properties of HF compounds represented by 2D ^{3}He, magnets of new types with quantum spin liquid (QSL), and the recently discovered quasicrystals, respectively, and show that HF compounds present, in fact, the new state of matter. Section 13.4 summaries the main results, emphasizing the observation that the quantum physics of different strongly correlated Fermi systems is universal and emerges regardless of their underlying microscopic details. This uniform behavior, formed by flat bands, manifests the new state of matter.

13.2 Extended Quasiparticle Paradigm and the Scaling Behavior of HF Metals

To analyze the dependence of the effective mass M^* on temperature T, magnetic field B, momentum $\mathbf{p}$, number density x etc. For convenience of the reader, we use the procedure already employed in some other Chaps. 7 and 8. Namely, we use the

Landau equation for the effective mass

$$\frac{1}{M^*(T,B)}=\frac{1}{M}+\int\frac{\mathbf{p}_F\mathbf{p}_1}{p_F^3}F(\mathbf{p_F},\mathbf{p}_1,n)\frac{\partial n(\mathbf{p}_1,T,B)}{\partial \mathbf{p}_1}\frac{d\mathbf{p}_1}{(2\pi)^3}, \tag{13.1}$$

where the Fermi–Dirac distribution reads

$$n_{\pm}(\mathbf{p},T,B)=\left\{1+\exp\left[\frac{[\varepsilon(\mathbf{p},T)\pm B\mu_B-\mu]}{T}\right]\right\}^{-1}. \tag{13.2}$$

Here, M is the bare mass of constituent particles, μ is the chemical potential, B is an external magnetic field, n is quasiparticle distribution function, and μ_B is the Bohr magneton. The term $\pm B\mu_B$ entering the right-hand side of (13.2) describes Zeeman splitting. Equation (13.1) is exact and can be derived within the framework of the Density Functional Theory, see Chap. 3. Equation (13.1) allows us to calculate the behavior of M^* which now becomes a function of temperature T, external magnetic field B, number density x, pressure P, etc. It is this feature of M^* that determines both the scaling and the non-Fermi liquid (NFL) behavior observed in measurements on strongly correlated Fermi systems [1–4]. In case of finite M^* and at $T=0$ the distribution function $n(\mathbf{p},T=0)$ becomes the step function $\theta(p_F-p)$, as it follows from (13.2), and (13.1) yields the well-known result

$$\frac{M^*}{M}=\frac{1}{1-F^1/3}, \tag{13.3}$$

where $F^1=N_0f^1$, $f^1(p_F,p_F)$ is the p-wave component of the Landau interaction, and $N_0=Mp_F/(2\pi^2)$ is the density of states (DOS) of a free Fermi gas. Because the number density is $x=p_F^3/3\pi^2$, the Landau interaction can be written as $F^1(p_F,p_F)=F^1(x)$ [2]. At a certain critical point $x=x_{FC13}$, with $x\to x_{FC13}$ from below, the denominator $(1-F^1(x)/3)$ tends to zero

$$(1-F^1(x)/3)\propto(x_{FC13}-x)+a(x_{FC13}-x)^2+...\to 0, \tag{13.4}$$

and we find that

$$\frac{M^*(x)}{M}\simeq a_1+\frac{a_2}{1-x/x_{FC13}}=a_1+\frac{a_2}{1-z}, \tag{13.5}$$

where $x/x_{FC13}=z$, a_1, and a_2 are constants, and $M^*(x\to x_{FC13})\to\infty$. As a result, at $T\to 0$ and $x\to x_{FC13}$ the system undergoes a quantum phase transition [2, 4], represented by FCQPT. We note that the divergence of the effective mass given by (13.5) preserves the Pomeranchuk stability conditions, for F^1 is positive, rather than negative [5]. The divergence of the effective mass $M^*(x)$, given by (13.5) and observed in measurements on two-dimensional ^{3}He [6, 7], is illustrated in Fig. 13.1. It is seen that the calculations are in good agreement with the data [8].

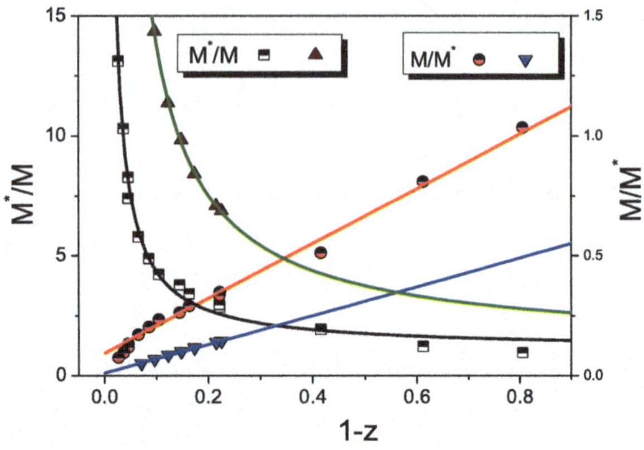

Fig. 13.1 The dependence of the effective mass $M^*(z)$ of 2D ^{3}He on dimensionless density $z = x/x_{FC13}$. Experimental data from [6] are shown by circles and squares and those from [7] are shown by triangles. The effective mass M^*/M is fitted by (13.5), while the reciprocal $M/M^*(z) \propto a_3 z$, where a_3 is a constant

13.2.1 Topological Properties of Systems with Fermion Condensate

The quasiparticle distribution function $n(\mathbf{p})$ of Fermi systems with FC is determined by the usual equation for a minimum of the Landau functional $E[n(\mathbf{p})]$. In contrast to common functionals of the number density x [9, 10], the Landau functional of the ground state energy E becomes functional of the occupation numbers n, see Chap. 3. In case of homogeneous system a common functional becomes a function of $x = \sum_p n(\mathbf{p})$, while the Landau functional $E = E[n(\mathbf{p})]$ [1, 2, 11] satisfies

$$\frac{\delta E[n(\mathbf{p})]}{\delta n(\mathbf{p})} = \varepsilon(\mathbf{p}) = \mu; \text{ when } 0 < n(\mathbf{p}) < 1. \tag{13.6}$$

The minimum of the functional E has to be found from (13.6). While in the case of Bose system the equation $\delta E/\delta n(p) = \mu$ is well established and understood, in case of Fermi system this equation, generally speaking, is not correct. It is the above constraint, $0 < n(\mathbf{p}) < 1$, valid in some region of momenta, $p_i < p < p_f$, that makes (13.6) applicable for Fermi systems. Because of the above constraint, in the region $p_i < p < p_f$, Fermi quasiparticles can behave as Bose one, occupying the same energy level $\varepsilon = \mu$, and (13.6) yields the quasiparticle distribution function $n_0(\mathbf{p})$ that minimizes the ground state energy E. Thus, at $T = 0$ the system undergoes a topological phase transition, because the Fermi surface at $p = p_F$ transforms into the Fermi volume for $p_i \le p \le p_f$ suggesting that at $T = 0$ the band is absolutely "flat" within this interval, giving rise to the spiky DOS as shown in the panel (a) of Fig. 13.2. A possible solution $n_0(\mathbf{p})$ of (13.6) and the corresponding single-particle spectrum $\varepsilon(\mathbf{p})$ are depicted in Fig. 13.2, panel (b). As shown in the panel (b), $n_0(\mathbf{p})$ differs from the step function in the interval $p_i < p < p_f$, where $0 < n_0(\mathbf{p}) < 1$, and coincides with the step function outside this interval. The existence of such flat bands, formed by interparticle interaction, was predicted for the first time in

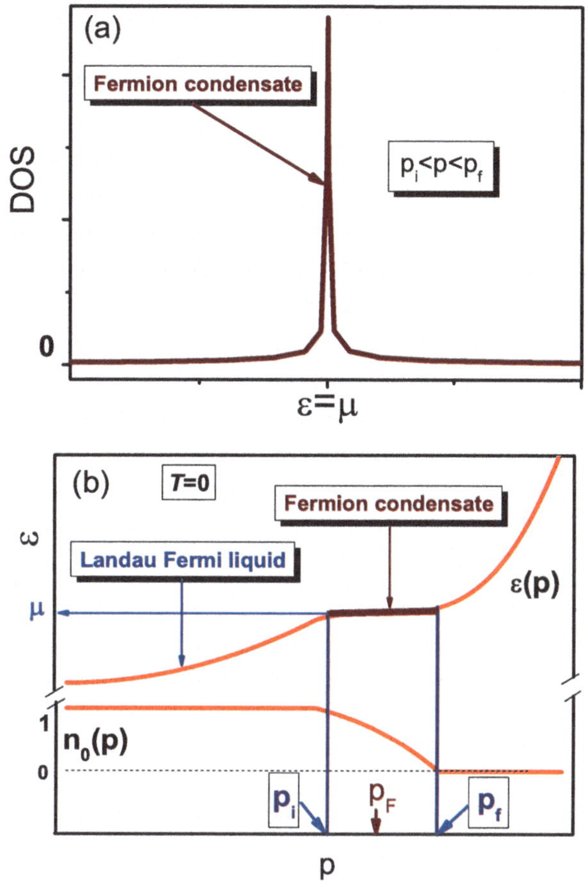

Fig. 13.2 Fermion condensate. Panel **a**: Schematic plot of the density of states (DOS) at low temperatures of quasiparticles versus ε at the momentum $p_i < p < p_f$ of a Fermi liquid with FC. Panel **b**: Schematic plot of two-component Fermi liquid at $T = 0$ with FC. The system is separated into two parts shown by the arrows: The first part is a Landau Fermi liquid with the quasiparticle distribution function $n_0(p < p_i) = 1$, and $n_0(p > p_f) = 0$; The second one is FC with $0 < n_0(p_i < p < p_f) < 1$ and the single-particle spectrum $\varepsilon(p_i < p < p_f) = \mu$. The Fermi momentum p_F satisfies the condition $p_i < p_F < p_f$

[11]. Quasiparticles with momenta within the interval $(p_i < p < p_f)$ have the same single-particle energies equal to the chemical potential μ and form FC, while the distribution $n_0(\mathbf{p})$ describes the new state of the Fermi liquid with FC, and the Fermi system is split up into two parts: LFL and the FC part, as shown in Fig. 13.2, panel (b) [4].

The theory of fermion condensation permits to construct a new class of strongly correlated Fermi liquids with the fermion condensate [2, 4]. In that case the quasiparticle system is composed of two parts, see Fig. 13.2. One of them is represented by FC located at the chemical potential μ, and giving rise to the spiky DOS, like that with $p = 0$ of Bose systems, as shown in Fig. 13.2, panel (a), that shows the DOS of a Fermi liquid with FC located at the momenta $p_i < p < p_f$ and energy $\varepsilon = \mu$. Contrary to the condensate of a Bose system occupying the $p = 0$ state, quasiparticles of FC with the energy $\varepsilon = \mu$ must be spread out over the interval $p_i < p < p_f$. Contrary to the Landau, marginal, superconducting LFL, or Tomonoga-Luttinger (marginal) Fermi liquids whose Green's functions exhibit the same topological behavior, in systems with FC, where the Fermi surface spreads into the Fermi volume, the Green's function belongs to a different topological class. The topological class of the Fermi liquid is characterized by the topological invariant [2, 4, 12]

$$N = tr \oint_C \frac{dl}{2\pi i} G(i\omega, \mathbf{p}) \partial_l G^{-1}(i\omega, \mathbf{p}), \tag{13.7}$$

where "tr" denotes the trace over the spin indices of the Green's function and the integral is calculated along an arbitrary contour C encircling the singularity of the Green's function. The invariant N in (13.7) assumes integer values even when the singularity is not a simple pole. It cannot vary continuously, and is conserved in a transition from the Landau Fermi liquid to marginal liquids and under small perturbations of the Green's function. As shown by Volovik [12], the situation is quite different for systems with FC, where the invariant N becomes a half-integer and the system with FC transforms into an entirely new class of Fermi liquids with its own topological structure, preserving the FC state, and forming the new state of matter demonstrated by many HF compounds [13].

In contrast to Bose liquid, whose entropy $S \to 0$ at temperature $T \to 0$, a Fermi liquid with FC possesses finite entropy S_0 at zero temperature [2, 14]. Indeed, as shown in Fig. 13.2, panel (b), at $T = 0$, the ground state of a system with a flat band is degenerate, and the occupation numbers $n_0(\mathbf{p})$ of single-particle states belonging to the flat band are continuous functions of momentum $\mathbf{p}$, in contrast to discrete standard LFL values 0 and 1. Such behavior of the occupation numbers leads to a T-independent entropy term $S_0 = S(T \to 0, n = n_0)$ with the entropy given by

$$S(n) = -\sum_{\mathbf{p}} [n(\mathbf{p}) \ln n(\mathbf{p}) + (1 - n(\mathbf{p})) \ln(1 - n(\mathbf{p}))]. \tag{13.8}$$

Since the state of a system with FC is highly degenerate, FC triggers phase transitions that could lift the degeneracy of the spectrum and vanishing the residual entropy S_0 of Fermi condensate at $T = 0$ in accordance with the Nernst theorem. For instance, FC can excite the formation of spin density waves, antiferromagnetic, ferromagnetic, the superconducting states, etc., thus strongly stimulating the competition between phase transitions that are eliminating the degeneracy. Contrary to LFL, where the entropy vanishes at zero temperature, the finite entropy S_0, characteristic of Fermi liquid with FC, causes the emergence of diversity of states. This observation is in accordance with the experimental phase diagrams [4, 15]. Since at $T = 0$ the entropy of ordered states is zero, we conclude that the entropy is discontinuous at the phase transition point, with its discontinuity $\delta S = S_0$. Thus, the entropy suddenly vanishes, with the system undergoing the first-order transition near which the critical quantum and thermal fluctuations are suppressed and the quasiparticles are well-defined excitations. As a result, we conclude that the main contribution to the transport and thermodynamic properties comes from quasiparticles rather than from various low-energy boson excitations. We note that the existence of FC has been convincingly demonstrated by purely theoretical arguments, see e.g., [2, 4, 16–20], and by experimental facts, see, e.g. [2, 4, 21].

There are different kinds of instabilities of normal Fermi liquids connected with several perturbations of initial quasiparticle spectrum $\varepsilon(\mathbf{p})$ and occupation numbers $n(\mathbf{p})$, associated with the emergence of a multi-connected Fermi surface, see Chap. 4. Depending on the parameters and analytical properties of the Landau interaction, such instabilities lead to several possible types of restructuring of the initial Fermi liquid ground state. This restructuring generates topologically distinct phases. One of them is the FC, and the other belongs to a class of topological phase transitions, where the sequence of rectangles $n(p) = 0$ and $n(p) = 1$ is realized at $T = 0$, see Chap. 4. In fact, at elevated temperatures the systems located at these transitions exhibit behavior typical to those located at FCQPT. Therefore, we do not consider the specific properties of these topological transitions, but focus on the behavior of systems located near FCQPT.

13.2.2 Scaling Behavior of HF Metals

It is instructive to briefly explore the behavior of M^* in order to capture its universal behavior at FCQPT; for detailed consideration see Chap. 7. Let us write the quasiparticle distribution function as $n_1(\mathbf{p}) = n(\mathbf{p}, T, B) - n(\mathbf{p})$, with $n(\mathbf{p})$ being the step function, (13.1) then becomes

$$\frac{1}{M^*(T, B)} = \frac{1}{M^*} + \int \frac{\mathbf{p}_F \mathbf{p}_1}{p_F^3} F(\mathbf{p}_F, \mathbf{p}_1) \frac{\partial n_1(p_1, T, B)}{\partial p_1} \frac{d\mathbf{p}_1}{(2\pi)^3}. \tag{13.9}$$

We now qualitatively analyze the solutions of (13.9) at $T = 0$. Application of magnetic field leads to Zeeman splitting of the Fermi surface, and the distance δp between the Fermi surfaces with spin up and spin down becomes $\delta p = p_F^{\uparrow} - p_F^{\downarrow} \sim \mu_B B M^*(B)/p_F$. We note that the second term on the right-hand side of (13.9) is proportional to $(\delta p)^2 \propto (\mu_B B M^*(B)/p_F)^2$, and therefore (13.9) reduces to [2]

$$\frac{M}{M^*(B)} = \frac{M}{M^*(x)} + c \frac{(\mu_B B M^*(B))^2}{p_F^4}, \tag{13.10}$$

where c is a constant. In the same way, one can calculate the change of the effective mass due to the variation of T [2]. For normal metals, where the electron liquid behaves like LFL with the effective mass of several bare electron masses $M^*/M \sim 1$, at temperatures even near 1000 K, the second term on the right-hand side of (13.9) is of the order of T^2/μ^2 and is much smaller than the first term. The same is true, as can be verified, when a magnetic field of reasonable strength of $B \sim 100$ T is applied. Thus, the system behaves like LFL with the effective mass that is actually independent of the temperature or magnetic field, while the resistivity $\rho(T) \propto T^2$. This means that the correction to the effective mass determined by the second term on the right-hand side of (13.10) is small. Thus, the effective mass is approximatively constant under the influence of external parameters of reasonable strength. We recall that in the case of common metals $\mu \sim 1$ eV, therefore for reasonable temperatures T and magnetic fields B one obtain $T/\mu \ll 1$ and $\mu_B B/\mu \ll 1$. As a result, the integral on the right-hand side of (13.1) represents a small correction to M^* as a function of T and B, provided that $M/M^* \sim 1$. At FCQPT, e.g., as soon as $x \to x_{FC13}$, the effective mass $M^*(x)$ diverges and $M/M^*(x) \to 0$. In that case, the first term on the right-hand side of (13.10) vanishes and the second term becomes principal, determining M^* as a function of B. In the same way, (13.9) becomes homogeneous and determines $M^*(T, B)$ as a universal function of temperature, magnetic field, and other tuning parameters that drive the system to FCQPT. As it is shown in Fig. 13.3a, b, our observations are in accordance with the experimental facts: under the application of low magnetic fields of 0.1 T the HF metal $YbRh_2Si_2$ exhibits LFL behavior at $T \simeq 0.1$ K, while the effective mass M^* is strongly dependent on both B and T. The only role of the Landau interaction F is to drive the system to FCQPT and the solutions of (13.9) are represented by some universal function of the variables T, B, x. In that case M^* strongly depends on the same variables. In contrast to the Landau quasiparticle paradigm that assumes approximate constancy of the effective mass, the extended quasiparticle paradigm is to be introduced. The main point here is that the well-defined quasiparticles determine (as for the LFL theory) the thermodynamic, relaxation, and transport properties of strongly correlated Fermi systems, while M^* becomes a function of T, B, x, etc. Moreover, the effective mass can be a divergent function of T, $M^*(T \to 0) \to \infty$ [2]. Thus, we have to introduce the modified and extended quasiparticles paradigm that permits to naturally describe the basic properties and the scaling behavior of both the M^* and HF compounds. The essence of the paradigm is that in spite of the altering of the Fermi surface topology the substance undergoes at FCQPT, and the Landau quasiparticles survive but completely change their properties [2, 4].

A deeper insight into the behavior of $M^*(T, B)$ can be achieved by using some "internal" scales. Namely, near FCQPT the solutions of (13.9) exhibit such a behavior that $M^*(T, B)$ reaches its maximum value M^*_M at some temperature $T_M \propto B$ [2]. It is convenient to introduce the internal scales M^*_M and T_M to measure the effective mass and temperature. We rescale M^* and T by M^*_M and T_M. This generates the normalized dimensionless effective mass $M^*_N = M^*/M^*_M$ and the corresponding normalized dimensionless temperature $T_N = T/T_M$. As an illustration to the above consideration, we analyze the specific heat $C/T \propto M^*$ of the HF metal $YbRh_2Si_2$ [22]. When the magnetic field B is applied, the specific heat exhibits a behavior that is described by a function of both T and B. As shown in the panel (a) of Figs. 13.3, 1.2, and 1.3, a maximum structure $(C/T)_M$ in $C/T \propto M^*_M$ at temperature T_M appears when B is applied. T_M shifts to higher T and $C/T \propto M^*_M$ diminishes as B is increased. At decreasing magnetic field B the value of C/T reaches its maximum at lower temperatures, while the maximum increases. To obtain the normalized effective mass M^*_N, we use $(C/T)_M$ and T_M as "internal" scales: the maximum structure $(C/T)_M$ was used to normalize C/T, and T was normalized by T_M. In panel (b) of Fig. 13.3, the obtained $M^*_N = (C/T)/(C/T)_M = M^*/M^*_M$, as a function of normalized temperature $T_N = T/T_M$, is shown. It is seen that the LFL state with $M^* = const$ and NFL one are separated by the crossover at which M^*_N reaches its maximum value. The

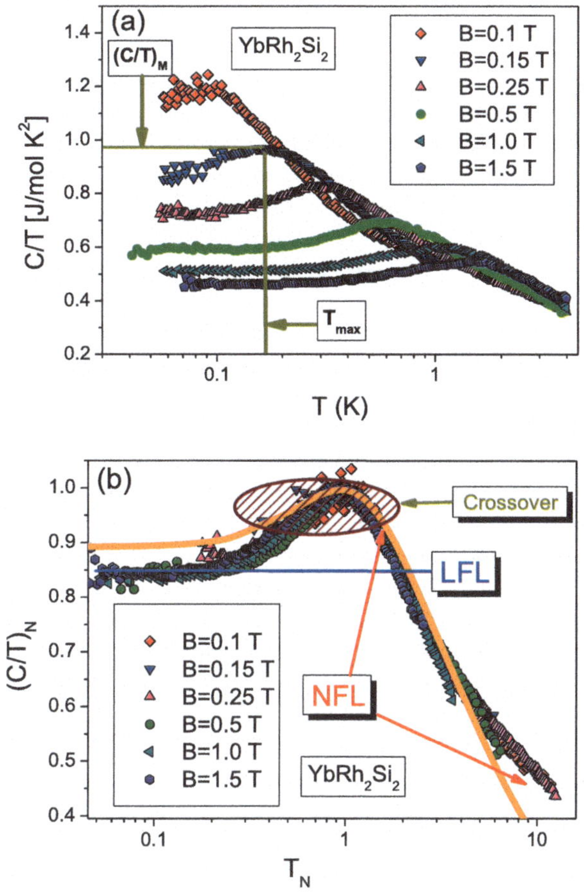

Fig. 13.3 Scaling behavior of HF metals. Panel **a**: Electronic specific heat of $YbRh_2Si_2$, C/T, versus temperature T as a function of magnetic field B [22] shown in the legend. The illustrative values of $(C/T)_M \propto M^*_M$ and T_M at $B = 0.15$ T are shown. Panel **b**: The normalized effective mass M^*_N versus normalized temperature T_N. M^*_N is extracted from the measurements of the specific heat C/T on $YbRh_2Si_2$ [22], displayed in the panel (**a**). The constant effective mass inherent in a normal Landau Fermi liquid is presented by the solid line. The schematic crossover region is indicated by the arrow and the NFL behavior is indicated by two arrows. Our calculation based on (13.1) is displayed by the solid curve

panel (b) of Fig. 13.3 reveals the scaling behavior of the normalized experimental curves: the curves at different magnetic fields B merge into a single one in terms of the normalized variable T/T_M. Our calculations of the normalized effective mass $M^*_N(T_N)$, shown by the solid line and based on (13.9), are in good agreement with the data [2, 4]. Near FCQPT the normalized solution of (13.1) $M^*_N(T_N)$ can be approximated well by a simple universal interpolating function [2]. The interpolation occurs between the LFL and NFL regimes and represents the universal scaling behavior of M^*_N, as we have demonstrated in Sect. 7.1 [2, 23]

$$M^*_N(T_N) \approx c_0 \frac{1 + c_1 T_N^2}{1 + c_2 T_N^n}. \tag{13.11}$$

Here, $c_0 = (1 + c_2)/(1 + c_1)$, c_1, c_2 are fitting parameters; and the exponent $n = 8/3$ if the Landau interaction is an analytical function, otherwise $n = 5/2$ [2]. It follows from (13.9) that

$$T_M \simeq a_1 \mu_B B, \tag{13.12}$$

where a_1 is a dimensionless number. Equation (13.12) is in good agreement with experimental facts [24]. Note that the effective mass M^* defines the thermodynamic properties of HF compounds, therefore $M^*(T) \propto C(T)/T \propto S(T)/T \propto M_0(T) \propto \chi(T)$ where $C(T)$ is the specific heat, $S(T)$ — entropy, $M_0(T)$ — magnetization, and $\chi(T)$ — AC magnetic susceptibility. For the normalized values we have

$$M_N^* = (C/T)_N = (S/T)_N = (M_0)_N = \chi_N. \tag{13.13}$$

It is seen from (13.11) and Figs. 13.3 and 13.4, that at elevated temperatures the considered HF compounds exhibit the NFL behavior, $M^*(T) \propto T^{-2/3}$. NFL behavior manifests itself in the power-law behavior of the physical quantities of strongly correlated Fermi systems located close to their QCPs [25, 26], including QCPs related to metamagnetic phase transitions [27, 28], with exponents different from those of ordinary Fermi liquids. As shown in Fig. 13.4, panels (a) and (b), M_N^* in different HF metals are the same, both in the high and low magnetic field. This observation is of utmost importance since it allows us to verify the universal behavior in HF metals, when quite different magnetic fields are applied [2, 4, 29]. Relatively small values of M_N^* observed in $URu_{1.92}Rh_{0.08}Si_2$ and $CeRu_2Si_2$ at the high fields and small temperatures can be explained by taking into account that the narrow band is completely polarized. As a result, at low temperatures the summation over up and down spin projections reduces to a single direction producing the coefficient 1/2 in front of M_N^*, thus violating the scaling at low temperatures $T_N \leq 1$. At high temperatures the polarization vanishes and the summation is restored. As shown in Fig. 13.4, panel (b), these observations are consistent with the experimental data collected in measurements under the application of high magnetic fields B on the archetypical HF metal $YbRh_2Si_2$ [30] the theoretical analysis of which is in good agreement with the experimental facts [29]. Thus, we conclude that different HF metals exhibit the same LFL, crossover, and NFL behavior in strong magnetic fields that are in agreement with the concept of the new state of matter.

It is a common belief that the main output of theory is to explain the exponents that describe the NFL behavior, e.g., $C/T \propto T^q$, which are at least dependent on the magnetic character of QCP, dimensionality of the system, and employed scenario [31]. On the contrary, the NFL behavior cannot be captured by these exponents as shown in Figs. 13.3 and 13.4. Indeed, the specific heat C/T exhibits a behavior that has to be described as a function of both temperature T and magnetic field B rather than by a single exponent. One can see that at low temperatures C/T demonstrates the LFL behavior which is changed by the crossover regime at which C/T reaches its maximum and finally, C/T decays into NFL behavior as a function of T at fixed B. It is clearly shown in Figs. 13.3 and 13.4 that, in particular, in the transition regime, these exponents may have little physical significance. Thus, it is the quasiparticles of the extended paradigm that form the universal scaling behavior, and reproduce the important experimental data [2, 4].

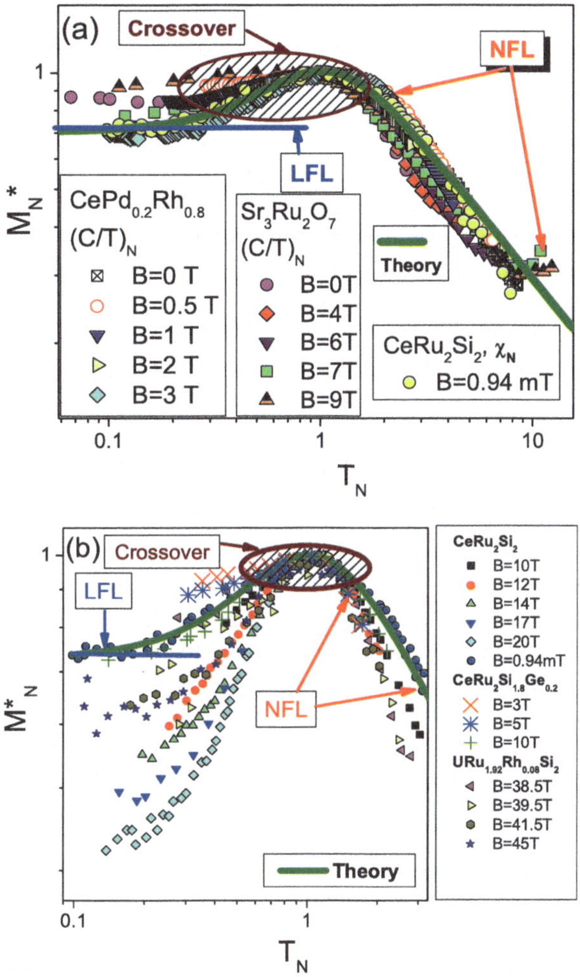

Fig. 13.4 The universal scaling behavior of the normalized effective mass M^*_N versus T_N. Panel **a**: M^*_N is extracted from the measurements of χ and C/T (in magnetic fields B shown in the legends) on $CeRu_2Si_2$ [25], $CePd_{1-x}Rh_x$ with $x = 0.80$ [22], and $Sr_3Ru_2O_7$ [26]. The LFL and NFL regimes (latter having $M^*_N \propto T_N^{-2/3}$) are shown by the arrows and straight lines. The transition regime is depicted by the shaded area. The solid curve represents our calculated universal behavior of $M^*_N(T_N)$. Panel **b**: The normalized effective mass versus the normalized temperature at different magnetic fields B, shown in the legend. $M^*_N(T_N)$ is extracted from measurements of C/T collected on $URu_{1.92}Rh_{0.08}Si_2$, $CeRu_2Si_2$ and $CeRu_2Si_{1.8}Ge_{0.2}$ at their metamagnetic phase transitions [27, 28]. The solid curve gives the universal behavior of M^*_N, (13.11)

There have been developed the Kondo breakdown theory, see, e.g., [32], the two-fluid description of heavy electron leading to quantum critical behavior, see, e.g., [33, 34], and critical quasiparticle theory, see, e.g., [35]. These theories consider hybridization between conduction electrons and local magnetic moments, e.g., f orbitals, that produces a heavy electron liquid with an associated mass enhancement. As a result, one observes the itinerant heavy electron liquid in materials that contain a Kondo lattice of localized f electrons coupled to background conduction electrons. Thus, the Kondo effect in many materials leads to quantum criticality that emerges from a competition between local momenta magnetism and the conduction electron screening of the local moments [36, 37]. On the other hand, it is observed that the heavy-fermion superconductor $\beta - YbAlB_4$ exhibits strange metallic (or NFL) behavior under the application of extensive pressure, distinctly separated from a high-pressure magnetic quantum phase transition by LFL phase [36, 37]. Moreover, the superconductor $\beta - \mathrm{YbAlB_4}$ demonstrates the robustness of the NFL behavior of the thermodynamic properties and of the anomalous $T^{3/2}$ temperature dependence of the electrical resistivity under applied pressure in zero magnetic field B. Such a behavior is at variance with the fragility of the NFL phase under application of tiny magnetic field [36, 37]. Thus, one can expect problems when applying the mentioned above theories to explain the properties of $\beta - \mathrm{YbAlB_4}$ [36, 37]. It is shown that a consistent topological basis for this combination of observations, as well as the empirical scaling laws may be found within the framework of the fermion-condensation theory in the emergence and destruction of a flat band, while the paramagnetic NFL phase takes place without magnetic criticality, thus not from quantum critical fluctuations [13, 38].

Several remarks concerning the applicability of (13.1) and (13.11) to systems with violated translational invariance are in order. We study the universal behavior of HF metals, quantum spin liquids, and quasicrystals at low temperatures using the model of a homogeneous HF liquid [2, 4]. The model is applicable because we consider the scaling behavior exhibited by the thermodynamic properties of these materials at low temperatures, a behavior related to the scaling of quantities such as the effective mass M^*, the heat capacity $C/T \propto M^*$, the magnetic susceptibility $\chi \propto M^*$, etc. The behavior of $M^*_N(T_N)$, that is used to describe quantities mentioned above, is determined by momentum transfers that are small compared to momenta of the order of the reciprocal lattice length. The high momentum contributions can therefore be ignored by substituting the lattice with the jelly model. The values of the scales, like the maximum $M^*_M(B_0)$ of the effective mass measured at some field $B = B_0$ and T_M at which M^*_M emerges, are determined by a wide range of momenta. Thus, these scales are controlled by the specific properties of the system under consideration, while the scaled thermodynamic properties of different strongly correlated Fermi systems can be described by universal function (13.11) determining $M^*_N(T_N)$. It is shown in Figs. 13.3 and 13.4, and demonstrated by numerous experimental data collected on HF compounds that this observation is in a good agreement with experimental data [2, 4], and allows one to view the universal scaling behavior as a manifestation of the new state of matter exhibited by HF compounds.

Figure 13.5a shows the resulting $(C/T)_N$ as a function of T_N, with different symbols for different magnetic field strengths B. The solid curve represents calculations of $(C/T)_N = M^*_N$ based on (8.1) and (13.11). It is seen that the LFL and NFL regions are separated by a crossover region where $(C/T)_N$ reaches its maximal value. As evident from Fig. 13.5a, $(C/T)_N$ is not a constant as would be for a LFL; it demonstrates the asserted universal scaling behavior given by (13.11) over a wide range of values of the normalized temperature T_N. This behavior coincides with that of the magnetic susceptibility $\chi \propto C/T \propto M^*$ revealed in measurements on $[\mathrm{BiBa_{0.66}K_{0.36}O_2}]\mathrm{CoO_2}$ [39] and incorporated in Fig. 13.5b. The solid curve tracks the results of the same calculations based on (8.1) that describe the universal scaling behavior $(C/T)_N(T/T_M) = M^*_N(T/T_M) \propto \chi(B - B_{c0})^{0.6}$ shown in Fig. 13.5a. We thus conclude that the solid curve drawn in Figs. 13.5a, b exhibits the universal scaling behavior that is intrinsic to HF compounds.

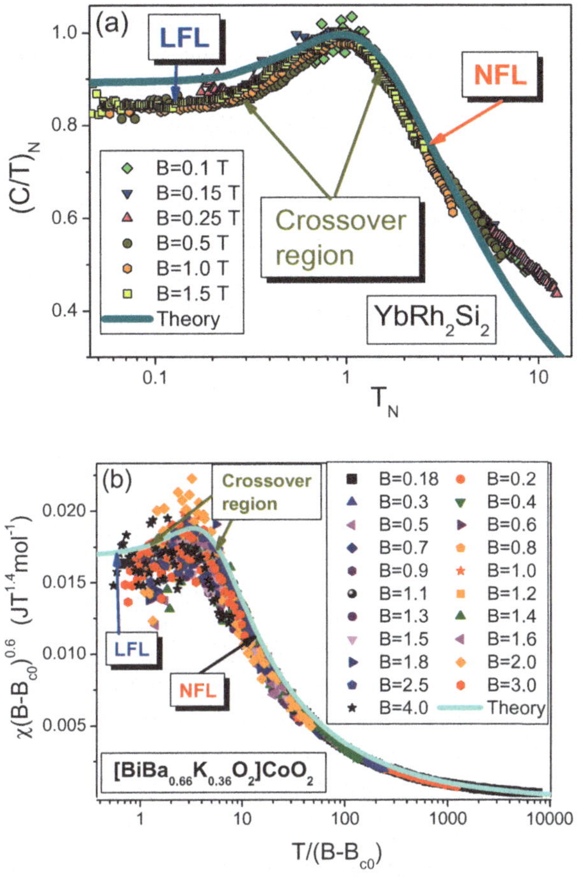

Fig. 13.5 The scaling behavior of the thermodynamic functions. **a** The normalized specific heat $(C/T)_N$ versus normalized temperature T_N. $(C/T)_N$ is extracted from the measurements of the specific heat C/T on $YbRh_2Si_2$ in magnetic fields B [22] listed in the legend. The LFL region, crossover one, and NFL one are depicted by the arrows. The solid curve displays calculations of $(C/T)_N = M^*_N$ based on (8.1) and (13.11) [23]. **b** Scaled susceptibility $\chi(B - B_{c0})^{0.6}$ as a function of scaled temperature $T/(B - B_{c0})$ with $B_{c0} = 0.176$ T for various B values shown in the legend [39]. The LFL region, crossover one, and NFL one are shown by the arrows. The solid curve tracks the results of our calculations based on (8.1) that describe the universal scaling behavior $(C/T)_N = M^*_N \propto \chi(B - B_{c0})^{0.6}$ shown in Fig. 13.5a

13.2.3 Universal Behavior of the Thermopower S_T of Heavy-Fermion Metals

In this section, we demonstrate that the thermopower S_T of such different heavy-fermion (HF) compounds as $YbRh_2Si_2$, β-$YbAlB_4$ and the strongly correlated layered cobalt oxide $[BiBa_{0.66}K_{0.36}O_2]CoO_2$ also exhibits the universal scaling behavior [40].

A study of the thermoelectric power S_T may deliver new insight into the nature of quantum phase transition that defines the features of new state of matter. For example, one may reasonably propose that the thermoelectric power S_T distinguishes two competing scenarios for quantum phase transitions in heavy fermions, namely the spin-density-wave theory and the breakdown of the Kondo effect [42]. Indeed, S_T is sensitive to the derivative of the density of electronic states and the change in the relaxation time at μ [43]. Using the Boltzmann equation, the thermopower S_T can be written as [44]

$$S_T = -\frac{\pi^2 k_B^2 T}{3e}\left[\frac{\partial \ln \sigma(\varepsilon)}{\partial \varepsilon}\right]_{\varepsilon=\mu}, \tag{13.14}$$

where k_B and e are, respectively, the Boltzmann constant and the elementary charge, while σ is the dc electric conductivity of the system, given by

$$\sigma(\varepsilon) = 2e^2\tau(\varepsilon)\int \delta(\mu - \varepsilon(\mathbf{p}))v(\mathbf{p})v(\mathbf{p})\frac{d\mathbf{p}}{(2\pi)^3}, \tag{13.15}$$

where $\mathbf{p}$ is the electron wave vector, τ is the scattering time, and v denotes the velocities of electrons belonging to the bands crossing the Fermi surface at $\varepsilon = \mu$. Thus, we see from (13.15) that the thermoelectric power S_T is sensitive to the derivative of the density of electronic states $N(\varepsilon = \mu)$ and the change in the relaxation time at $\varepsilon = \mu$. On the basis of the Fermi liquid theory description, the term in the brackets on the right-hand side of (13.14) can be simplified, so that one has $S_T \propto TN(\varepsilon = \mu) \propto C \propto M^*T$ at $T \to 0$ [44]. As a result, under general conditions and upon taking into account that charge and heat currents at low temperatures are transported by quasiparticles, the ratio $(S_T/C) \simeq const$ [44]. It is shown in Fig. 13.6 that in the case of $YbRh_2Si_2$ and at $T \geq T_{NL}$, the isotherms $-S_T(B)/T$

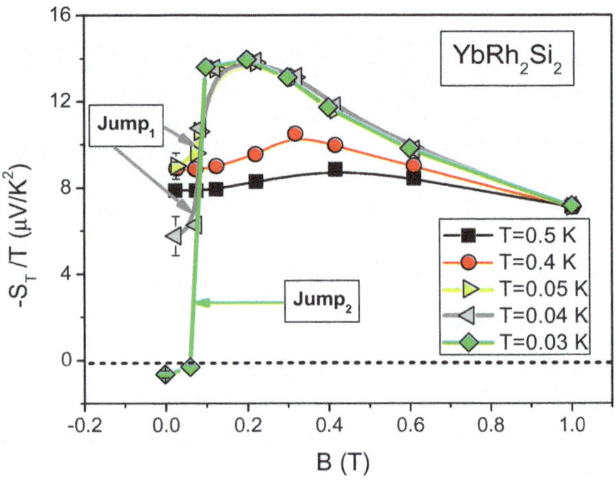

Fig. 13.6 Thermopower isotherm $-S_T(B)/T$ for different temperatures shown in the legend [41]. The labels $Jump_1$ and $Jump_2$ represent the first and second downward jumps in $-S_T(B)/T$ shown by the arrows. The solid lines are guides to the eye

behave like C/T: they exhibit a broad maximum that sharpens and shifts to lower fields upon cooling [41]. Here T_{NL} is the temperature of antiferromagnetic (AF) ordering ($T_{NL} = 70$ mK) at a critical field $B_{c0} = 60$ mT, applied perpendicular to the magnetically hard c axis. Thus, $S_T/T \propto C/T \propto \chi \propto M^*$ over a wide range of T, since in the framework of FC theory, quasiparticles are responsible for the thermodynamic and transport properties. It is worth noting that $S_T/T \propto M^*$ in a low-disorder two-dimensional electron system in silicon, and tends to a diverge at a finite disorder-independent density [45], thus confirming that the charge and heat currents are transported by quasiparticles.

To reveal the universal scaling behavior of the thermopower $S_T/T \propto C/T \propto M^*$, we normalize S_T/T in the same way as in the normalization of C/T: the normalized function $(S_T/T)_N$ is obtained by normalizing (S_T/T) by its maximum value, occurring at $T = T_M$, and the temperature T is scaled by T_M. Taking into account that $S_T/T \propto C/T$ [44], we conclude that $(S_T/T)_N = (C/T)_N = M^*_N$, provided that the system in question is located away from possible phase transitions. This universal function $(C/T)_N = M^*_N$ is displayed in Fig. 13.3b. Figures 13.7a, b report $(S_T/T)_N$ as a function of the normalized magnetic field B_N and T_N, respectively. In Fig. 13.7a, the function $(S_T/T)_N$ is obtained by normalizing (S_T/T) by its maximum occurring at B_M, and the field B is scaled by B_M. As seen from (13.11), the LFL behavior takes place at $B_N > 1$, since $(S_T/T)_N = M^*_N$, and $M^*_N \propto (B - B_{c0})^{-2/3}$ are T-independent, while at $B_N < 1$, M^*_M becomes T-dependent and exhibits the NFL behavior with $M^*_N \propto T_N^{-2/3}$. It is shown in Figs. 13.7a, b that the calculated values of the universal function M^*_N are in good agreement with the corresponding experimental data [41, 46] over the wide range of the normalized magnetic field. Thus, $(S_T/T)_N$ exhibits the universal scaling behavior over a wide range of its scaled variable B_N and T_N. Figure 13.7a also depicts a violation of the scaling behavior for $B \leq B_{c0}$ when the system enters the AF phase. Moreover, as shown in Figs. 13.6 and 13.7a, b, the scaling behavior is violated at $T \leq T_{NL}$ by two downward jumps. As we shall see below, these two jumps reflect the presence of flat band at μ in the single-particle spectrum $\varepsilon(\mathbf{p})$ of heavy electrons in $YbRh_2Si_2$ [38]. In the same way, as it is shown in Fig. 13.7b, the scaling behavior is violated by the superconducting (SC) phase transition, taking place in β-$YbAlB_4$ at $T_c = 80$ mK [46].

We now show that the observed scaling behavior of $(S_T/T)_N$ is universal by analyzing experimental data on the thermopower for $[BiBa_{0.66}K_{0.36}O_2]CoO_2$ [39]. By plotting $(S_T/T)_N$ as a function of T_N in Fig. 13.8, the universal scaling behavior and the three regimes are seen to be in a complete agreement with the reported overall behavior in both Figs. 13.3b and 13.4a, and Figs. 13.7a, b as well.

Before proceeding to the analysis of the jumps observed in measurements of S/T on $YbRh_2Si_2$, some remarks are in order concerning the flattening of the spectra $\varepsilon(\mathbf{p})$ in HF systems, a phenomenon called swelling of the Fermi surface or FC [4]. As indicated in Fig. 13.9, the ground states of systems with flat bands are degenerate, and therefore the occupation numbers $n_0(\mathbf{p})$ of single-particle states belonging to a flat band are given by a continuous function on the interval [0, 1], in contrast to

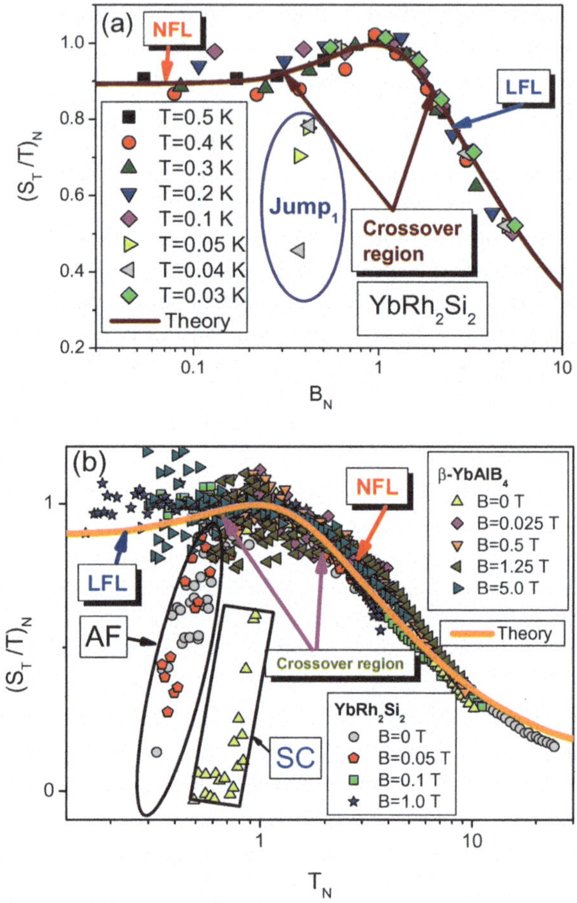

Fig. 13.7 **a** Normalized isotherm $(S_T(B)/T)_N$ versus normalized magnetic field B_N for different temperatures shown in the legend. The LFL behavior takes place at $B_N > 1$. **b** Temperature dependence of the normalized thermopower $(S_T/T)_N$ under several magnetic fields shown in the legend. The experimental data are extracted from measurements on $YbRh_2Si_2$ [41] and on β-$YbAlB_4$ [46]. As it is explained in the text, the data, taken at the AF phase [41] and at the superconducting one (SC) [46] and confined by both the ellipse and the rectangle, respectively, violate the scaling behavior. The solid curves in (**a**) and (**b**) represent calculated $(C/T)_N$ displayed in Fig. 13.3b

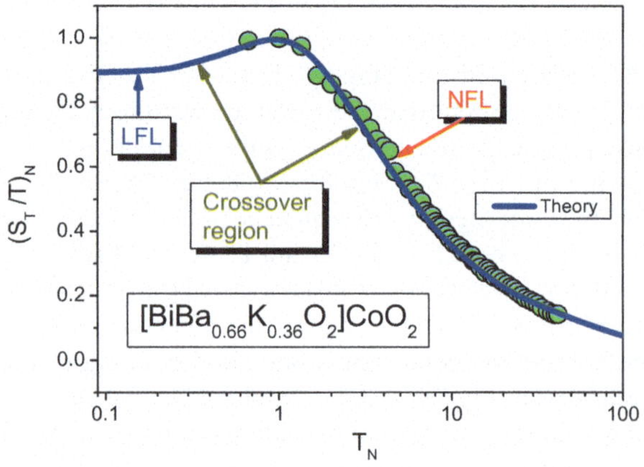

Fig. 13.8 Temperature dependence of $(S_T(T)/T)_N$ at magnetic field $B = 0$, extracted from measurements on $[BiBa_{0.66}K_{0.36}O_2]CoO_2$ [39], is displayed versus T_N. The solid line is the same as one depicted in Fig. 13.3b

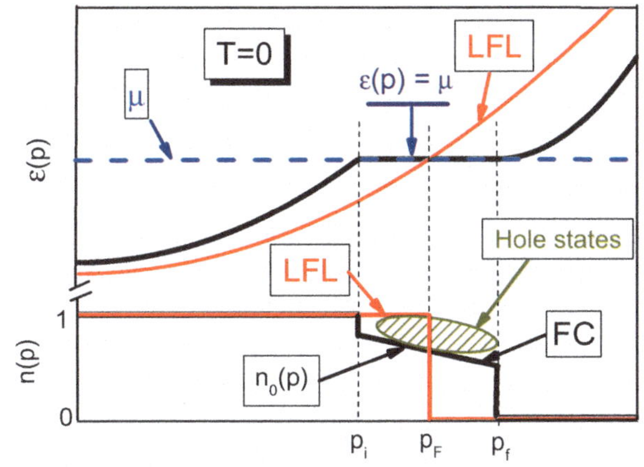

Fig. 13.9 Single-particle energy $\varepsilon(\mathbf{p})$ and distribution function $n(\mathbf{p})$ at $T = 0$. The arrow shows the chemical potential μ. The vertical lines show the area $p_i < p < p_f$ occupied by FC with $0 < n_0(p) < 1$ and $\varepsilon(\mathbf{p}) = \mu$. The Fermi momentum p_F satisfies the condition $p_i < p_F < p_f$ and corresponds to the LFL region, indicated by the arrows, which emerges when the FC state is eliminated. The arrow depicts hole states induced by the FC

the FL restriction to occupation numbers 0 and 1. This property leads to an entropy excess

$$S_0 = -\sum n_0(\mathbf{p}) \ln n_0(\mathbf{p}) + (1 - n_0(\mathbf{p})) \ln(1 - n_0(\mathbf{p})), \tag{13.16}$$

that does not contribute to the specific heat $C(T)$. The entropy excess S_0 contradicts the Nernst theorem. To circumvent violation of the Nernst theorem, FC must be completely eliminated at $T \to 0$. This can happen by virtue of some phase transition, e.g., the AF transition that becomes of first order at some tricritical point occurring at $T = T_{tr}$. Such a first-order phase transition provides for eradication of the flat portion in the spectrum $\varepsilon(\mathbf{p})$. As a consequence, both the density of states N and the hole states, shown by the arrow in Fig. 13.9, vanish discontinuously, while the occupation numbers $n_0(\mathbf{p})$ and the spectrum $\varepsilon(\mathbf{p})$ revert to those LFL state, as indicated by the arrows in Fig. 13.7a. Simultaneously, the Fermi sphere undergoes an abrupt change on the interval from the Fermi momentum p_f to p_F, so as to nullify both the swelling of the Fermi surface and the entropy excess S_0. As a result, the thermopower experiences Jump$_2$, for the entropy abruptly diminishes. We note that the abrupt change is observed as the change of the low-T Hall coefficient [4]. It is shown in Fig. 13.6 that at $T = 0.03$ K, S/T abruptly changes its sign (the second jump—Jump$_2$), for the hole states vanish. The positive sign of S/T of $YbRh_2Si_2$ without the hole states is in agreement with the positive thermopower of its nonmagnetic counterpart $LuRh_2Si_2$, lacking the $4f$ hole states at μ [41]. Contrary, at $T_{NL} > T > T_{cr}$ the AF phase transition is of the second order and the entropy is a continuous function at the border of the phase transition. Therefore, at this second-order phase transition both the occupation numbers and the spectrum do not change, and keep their FC-like shape, while the system with FC is destroyed, converting into HF liquid. This destruction generates the first jump $Jump_1$, shown in Fig. 13.6. Therefore, as the FC state is decayed, its contribution ρ_0^{FC13} to the residual resistivity ρ_0 vanishes, resulting in the change of the scattering time $\tau(\varepsilon = \mu)$. We recall that in the presence of FC, the residual resistivity consists of two terms $\rho_0 = \rho_0^{FC13} + \rho_0^{imp}$, where the residual resistivity ρ_0^{FC13} is formed by the flat band generated by FC, while the resistivity ρ_0^{imp} is formed by impurities [47, 48]. As a result, the thermopower experiences the first jump $Jump_1$, as seen from (13.14) and (13.15). We also conclude that the second downward jump under decreasing B is deeper than the first one, since it is caused by elimination of both ρ_0^{FC13} and the hole states. This is consistent with the experimental observations, as shown in Fig. 13.6.

Thus, we have revealed and explained the universal scaling behavior of the thermopower S_T/T in such different HF compounds as $YbRh_2Si_2$, β-$YbAlB_4$, and $[BiBa_{0.66}K_{0.36}O_2]CoO_2$. Our calculations are in good agreement with experimental observations, and demonstrate that S_T/T exhibits the universal scaling behavior that characterizes and singles out the new state of matter.

13.3 Schematic $T - B$ Phase Diagram

Now we are in a position to construct the $T - B$ phase diagram of the archetypical HF metal $YbRh_2Si_2$. Figure 13.10 displays our constructed schematic $T - B$ phase diagram for $YbRh_2Si_2$. In Fig. 13.10, the NFL region, formed by the FC state, is characterized by the entropy excess S_0 given by (13.16), and labeled by FC. As shown by the solid curve denoted by $T_{NL}(B)$, at $B < B_{c0}$ and $T < T_{NL}(B)$ the system is in its AF state, and exhibits the LFL behavior. The tricritical point T_{cr} at which the AF phase transition becomes of the first order, is indicated by the arrow. At that transition the thermopower experiences the jump $Jump_2$ shown in Fig. 13.6, and changes its sign, becoming $S/T > 0$, for the hole states shown in Fig. 13.9 vanish at $T < T_{cr}$. At $T > T_{cr}$ the AF transition of the second order, at which the thermopower experiences $Jump_1$, see Fig. 13.6. While in the NFL region $S/T > 0$, as it is shown in the phase diagram 13.10. Clearly, on the basis of the phase diagram 13.10, outside the area of the AF phase transition, the behavior of $S_N = M_N^*$, considered as a function of the dimensionless variable T_N or B_N, is almost universal. Indeed, as shown in Figs. 13.5a, b, 13.7a, b, and 14.6, all the data, extracted from measurements on $YbRh_2Si_2$, β-$YbAlB_4$, and $[BiBa_{0.66}K_{0.36}O_2]CoO_2$, collapse on the single scaling curve shown in Fig. 13.5a. As shown in Figs. 13.5a and 13.7b, at $T_N < 1$, $(S/T)_N$ tends to become constant, implying that S/T exhibits LFL behavior. However, at $T_N \simeq 1$ the system enters the narrow crossover region, while at growing temperatures, NFL behavior prevails.

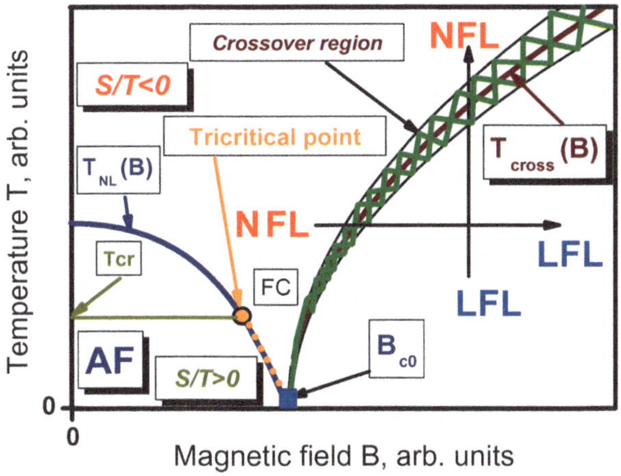

Fig. 13.10 Schematic $T - B$ phase diagram of $YbRh_2Si_2$. The vertical and horizontal arrows, crossing the transition region depicted by the thick lines, show the LFL-NFL and NFL-LFL transitions at fixed B and T, respectively. The hatched area around the solid curve $T_{cross}(B)$ represents the crossover between the NFL and LFL domains. The NFL region is labeled by FC. As shown by the solid curve, at $B < B_{c0}$ the system is in its AF state, and exhibits the LFL behavior. The line of AF phase transitions is denoted by $T_{NL}(B)$. The tricritical point, indicated by the arrow, is at $T = T_{cr}$. At $T < T_{cr}$ the AF phase transition becomes of the first order

13.4 Summary

We have revealed and explained the universal behavior of the thermopower S/T in such different HF compounds as $YbRh_2Si_2$, β-$YbAlB_4$, and $[BiBa_{0.66}K_{0.36}O_2]CoO_2$. Our calculations are in good agreement with experimental observations, and demonstrate that the advocated universal scaling behavior of S/T does take place. This behavior does not depend on the specific properties of the considered HF compounds, and coincides with that of the normalized effective mass $M^*_N = (C/T)_N$, thus representing the scaling behavior as being intrinsic to HF compounds. We have also shown that destruction of the flattening of the single-particle spectrum profoundly affects S/T, leading to the two jumps and the change of sign of the thermopower occurring at the antiferromagnetic phase transition. Thus, despite their drastic microscopic diversity, HF compounds exhibit the uniform scaling behavior. The quantum physics of different strongly correlated Fermi systems is universal and emerges regardless of their underlying microscopic details. The observed behavior resembles the uniform collective behavior exhibited by such different states of matter as the gaseous, liquid, and solid states of matter. Therefore, this uniform behavior, formed by the unique topological FCQPT, allows us to view it as the main manifestation of the new state of matter exhibited by HF compounds [13].

References

1. V.A. Khodel, V.R. Shaginyan, V.V. Khodel, Phys. Rep. **249**, 1 (1994)
2. V.R. Shaginyan, M.Y. Amusia, A.Z. Msezane, K.G. Popov, Phys. Rep. **492**, 31 (2010). (arXiv:1006.2658)
3. V.R. Shaginyan, JETP Lett. **79**, 286 (2004)
4. M.Y. Amusia, K.G. Popov, V.R. Shaginyan, W.A. Stephanowich, *Theory of Heavy-Fermion Compounds*. Springer Series in Solid-State Sciences, vol. 182 (Springer, Berlin, 2015)
5. I.Y. Pomeranchuk, Zh Eksp, Teor. Fiz. **35**, 524 (1958)
6. A. Casey, H. Patel, J. Nyeki, B.P. Cowan, J. Saunders, Low Temp. Phys. **113**, 293 (1998)
7. M. Neumann, J. Nyéki, B. Cowan, J. Saunders, Science **317**, 1356 (2007)
8. V.R. Shaginyan, A.Z. Msezane, K.G. Popov, V.A. Stephanovich, Phys. Rev. Lett. **100**, 096406 (2008)
9. P. Hohenberg, W. Kohn, Phys. Rev. **136**, B864 (1964)
10. W. Kohn, L.J. Sham, Phys. Rev. **140**, A1133 (1965)
11. V.A. Khodel, V.R. Shaginyan, JETP Lett. **51**, 553 (1990)
12. G.E. Volovik, JETP Lett. **53**, 222 (1991)
13. V.R. Shaginyan, V.A. Stephanovich, A.Z. Msezane, P. Schuck, J.W. Clark, M.Y. Amusia, G.S. Japaridze, K.G. Popov, E.V. Kirichenko, J. Low Temp. Phys. **189**, 410 (2017)
14. M.V. Zverev, V.A. Khodel, V.R. Shaginyan, M. Baldo, JETP Lett., 863 (1997)
15. H. v. Löhneysen, J. Phys. Condens. Matter **8**, 9689 (1996)
16. D. Lidsky, J. Shiraishi, Y. Hatsugai, M. Kohmoto, Phys. Rev. B **57**, 1340 (1998)
17. V.Y. Irkhin, A.A. Katanin, M.I. Katsnelson, Phys. Rev. Lett. **89**, 076401 (2002)
18. S.S. Lee, Phys. Rev. D **79**, 086006 (2009)
19. D. Yudin, D. Hirschmeier, H. Hafermann, O. Eriksson, A.I. Lichtenstein, M.I. Katsnelson, Phys. Rev. Lett. **112**, 070403 (2014)
20. G.E. Volovik, *Quantum Analogues: From Phase Transitions to Black Holes and Cosmology*, ed. by W.G. Unruh, R. Schutzhold. Springer Lecture Notes in Physics, vol. 718 (Springer, Orlando, 2007), p. 31
21. S. Link, Forti, A. Stöhr, K. Küster, M. Rösner, D. Hirschmeier, C. Chen, J. Avila, M.C. Asensio, A.A. Zakharov, T.O. Wehling, A.I. Lichtenstein, M.I. Katsnelson, U. Starke, Phys. Rev. B **100**, 121407(R) (2019)
22. N. Oeschler, S. Hartmann, A. Pikul, C. Krellner, C. Geibel, F. Steglich, Phys. B **403**, 1254 (2008)
23. V.R. Shaginyan, M.Y. Amusia, K.G. Popov, Phys. Lett. A **373**, 2281 (2009)
24. M. Brando, L. Pedrero, T. Westerkamp, C. Krellner, P. Gegenwart, C. Geibel, F. Steglich, Phys. Status Solidi B **459**, 285 (2013)
25. D. Takahashi, S. Abe, H. Mizuno, D. Tayurskii, K. Matsumoto, H. Suzuki, Y. Onuki, Phys. Rev. B **67**, 180407(R) (2003)
26. A.W. Rost, S.A. Grigera, J.A.N. Bruin, R.S. Perry, D. Tian, S. Raghu, S.A. Kivelson, A.P. Mackenzie, Proc. Natl. Acad. Sci. USA **108**, 16549 (2011)
27. A.V. Silhanek, N. Harrison, C.D. Batista, A.L.M. Jaime, H. Amitsuka, J.A. Mydosh, Phys. Rev. Lett. **95**, 026403 (2005)
28. J.S. Kim, B. Andraka, G. Fraunberger, G.R. Stewart, Phys. Rev. B **41**, 541 (1990)
29. V.R. Shaginyan, K.G. Popov, V.A. Stephanovich, V.I. Fomichev, E.V. Kirichenko, Europhys. Lett. **93**, 17008 (2011)
30. P. Gegenwart, Y. Tokiwa, T. Westerkamp, F. Weickert, J. Custers, J. Ferstl, C. Krellner, C. Geibel, P. Kerschl, K.H. Müller, F. Steglich, New J. Phys. **8**, 171 (2006)
31. H. v. Löhneysen, A. Rosch, M. Vojta, P. Wölfle, Rev. Mod. Phys. **79**, 1015 (2007)
32. P. Coleman, in *Lectures on the Physics of Highly Correlated Electron Systems VI*, ed. by F. Mancini (American Institute of Physics, New York, 2002), p. 79
33. H.B. Yang, J.D. Rameau, P.D. Johnson, T. Valla, A. Tsvelik, G.D. Gu, Nature **456**, 77 (2008)
34. Y.F. Yang, D. Pines, PNAS **111**, 8398 (2014)

35. P. Wölfle, E. Abrahams, Phys. Rev. B **84**, 041101(R) (2011)
36. Y. Matsumoto, S. Nakatsuji, K. Kuga, Y. Karaki, N. Horie, Y. Shimura, T. Sakakibara, A.H. Nevidomskyy, P. Coleman, Science **331**, 316 (2011)
37. T. Tomita, K. Kuga, Y. Uwatoko, P. Coleman, S. Nakatsuji, Science **349**, 506 (2015)
38. V.R. Shaginyan, A.Z. Msezane, K.G. Popov, J.W. Clark, V.A. Khodel, M.V. Zverev, Phys. Rev. B **93**, 205126 (2016). arXiv:1601.01182
39. P. Limelette, W. Saulquin, H. Muguerra, D. Grebille, Phys. Rev. B **81**, 115113 (2010)
40. V.R. Shaginyan, A.Z. Msezane, G.S. Japaridze, K.G. Popov, J.W. Clark, V.A. Khodel, Front. Phys. **11**, 117103 (2016)
41. S. Hartmann, N. Oeschler, C. Krellner, C. Geibel, S. Paschen, F. Steglich, Phys. Rev. Lett. **104**, 096401 (2010)
42. K.S. Kim, C. Pépin, Phys. Rev. B **83**, 073104 (2011)
43. E.M. Lifshitz, L.P. Pitaevskii, *Statisticheskaya Fizika (Statistical Physics), Part 2*, Course of Theoretical Physics (Pergamon Press, Oxford, 1980)
44. V. Zlatić, R. Monnier, J.K. Freericks, K.W. Becker, Phys. Rev. B **76**, 085122 (2007)
45. A. Mokashi, S. Li, B. Wen, S. Kravchenko, V.T.D.A.A. Shashkin, M.P. Sarachik, Phys. Rev. Lett. **109**, 096405 (2012)
46. Y. Machida, K. Tomokuni, C. Ogura, K. Izawa, K. Kuga, S. Nakatsuji, G. Lapertot, G. Knebel, J.P. Brison, J. Flouquet, Phys. Rev. Lett. **109**, 156405 (2012)
47. V.R. Shaginyan, A.Z. Msezane, K.G. Popov, J.W. Clark, M.V. Zverev, V.A. Khodel, Phys. Rev. B **86**, 085147 (2012)
48. V.R. Shaginyan, A.Z. Msezane, K.G. Popov, J.W. Clark, M.V. Zverev, V.A. Khodel, JETP Lett. **96**, 397 (2012)

Chapter 14
Universal Behavior of the Heavy-Fermion Metal $\beta - \text{YbAlB}_4$

Abstract In this chapter, we reveal and explain a universal scaling behavior of the heavy-fermion metal $\beta - \text{YbAlB}_4$. Informative measurements on the heavy-fermion metal $\beta - \text{YbAlB}_4$ performed with applied magnetic field and pressure as control parameters are analyzed with the goal of establishing a convincing theoretical explanation for the inferred scaling laws and non-Fermi liquid (NFL) behavior, which demonstrate some unexpected features. Most notably, the robustness of the NFL behavior of the thermodynamic properties and of the anomalous $T^{3/2}$ temperature dependence of the electrical resistivity under applied pressure P in zero magnetic field B is at variance with the fragility of the NFL phase under application of a field. We show that a consistent topological basis for this combination of observations, as well as the empirical scaling laws, may be found within fermion-condensation theory in the emergence and destruction of a flat band, and explain that the paramagnetic NFL phase takes place without magnetic criticality, thus not from quantum critical fluctuations. Schematic $T - B$ and $T - P$ phase diagrams are presented to illuminate this scenario.

14.1 Introduction

Measurements on the heavy-fermion metal $\beta - \text{YbAlB}_4$ have been performed under the application of both magnetic field B and hydrostatic pressure P, with results that have received considerable theoretical attention [1–10]. Measurements of the magnetization $M(B)$ at different temperatures T reveal that the magnetic susceptibility $\chi = M/B \propto T^{-1/2}$ demonstrates non-Fermi liquid (NFL) behavior and diverges as $T \to 0$, implying that the quasiparticle effective mass m^* diverges as $m^* \propto B^{-1/2} \propto T^{-1/2}$ at a quantum critical point (QCP) [2, 11, 12]. This kind of quantum criticality is commonly attributed to scattering of electrons off quantum critical fluctuations related to a magnetic instability. However, in a single crystal of $\beta - \text{YbAlB}_4$, the QCP in question is located well away from a possible magnetic instability, making the NFL phase take place without magnetic criticality [2]. Additionally, it is observed that QCP is robust under application of pressure P, in that the divergent T and B dependencies of χ are preserved and accompanied by an

M. Amusia and V. Shaginyan, *Strongly Correlated Fermi Systems*,
Springer Tracts in Modern Physics 283,
https://doi.org/10.1007/978-3-030-50359-8_14

anomalous $T^{3/2}$ dependence of the electrical resistivity ρ [6]. In contrast to resilience of these divergences under pressure, application of even a tiny magnetic field B is sufficient to suppress them, leading to Landau Fermi liquid (LFL) behavior at low temperatures [1, 2]. Thus, among other unusual features, the metal $\beta - YbAlB_4$ presents challenging theoretical problems: how to reconcile the fragility of its NFL behavior under application of a magnetic field, with the robustness of the NFL phase against application of pressure at zero magnetic field. Then, one has to explain that the paramagnetic NFL phase takes place without magnetic criticality, thus not from quantum critical fluctuations [11, 12]. Nonetheless, we will see that $\beta - YbAlB_4$ exhibits universal behavior and belongs to the family of HF compounds forming the new state of matter.

14.2 Universal Scaling Behavior

To solve these problems in the framework of a topological scenario based on the appearance of fermion condensate (FC), we begin by studying the universal behavior of the thermodynamic functions of this HF compound, regarded as a homogeneous fluid HF [13–15], see Chap. 7. The Landau functional $E(n)$, representing the energy of the ground state, depends on the momentum distribution of the quasiparticles $n_\sigma(\mathbf{p})$, see Chap. 3. Near the topological fermion-condensation quantum phase transition (FCQPT), the effective mass m^* is determined by the Landau equation [13, 14, 16]

$$\frac{1}{m^*(T,B)} = \frac{1}{m^*(T=0,B=0)} + \frac{1}{p_F^2}\sum_{\sigma_1}\int \frac{\mathbf{p}_F\mathbf{p}_1}{p_F} F_{\sigma,\sigma_1}(\mathbf{p}_F,\mathbf{p}_1)\frac{\partial \delta n_{\sigma_1}(T,B,\mathbf{p}_1)}{\partial \mathbf{p}_1}\frac{d\mathbf{p}_1}{(2\pi)^3}. \tag{14.1}$$

It is written here in terms of the deviation $\delta n_\sigma(\mathbf{p}) \equiv n_\sigma(\mathbf{p},T,B) - n_\sigma(\mathbf{p},T=0,B=0)$ of the distribution of quasiparticles of its value without a field under zero pressure. Landau interaction $F(\mathbf{p}_1,\mathbf{p}_2) = \delta^2 E/\delta n(\mathbf{p}_1)\delta n(\mathbf{p}_2)$ serves to bring the system to a point FCQPT, where $m^* \to \infty$ at $T = 0$. When this happens, the Fermi surface topology changes, and the effective mass m^* takes on strong temperature and field dependences, while the proportionality $C/T \sim \chi \sim m^*(T,B)$ are saved. Here C is the specific heat and χ is the magnetic susceptibility.

Upon approaching FCQPT, $m^*(T=0,B=0) \to \infty$, (14.1) becomes homogeneous, that is, $m^*(T=0,B) \propto B^{-z}$ and $m^*(T,B=0) \propto T^{-z}$, and z depends on the analytic properties of F, see Sect. 21.1. On the ordered side of FCQPT with $T = 0$, the single-particle spectrum $\varepsilon(\mathbf{p})$ becomes flat in a certain interval $p_i < p_F < p_f$ surrounding the Fermi surface at the point p_F, which coincides there with the chemical potential μ,

$$\varepsilon(\mathbf{p}) = \mu. \tag{14.2}$$

In FCQPT, the flat band interval is compressed, so $p_i \top_F \to p_f$ and $\varepsilon(\mathbf{p})$ are the inflection points in p_F with $\varepsilon(\mathbf{p} \simeq \mathbf{p_F}) - \mu \simeq (p - p_F)^3$. Another case of inflection occurs in the case of the nonanalytic Landau interaction F, so

$$\begin{aligned} \varepsilon - \mu &\simeq -(p_F - p)^2, p < p_F \\ \varepsilon - \mu &\simeq \ (p - p_F)^2, p > p_F \end{aligned} \tag{14.3}$$

in which the effective mass diverges as $m^*(T \to 0) \propto T^{-1/2}$. Such features of ε can be used to identify solutions of (14.1) corresponding to various experimental situations. In particular, experimental results obtained for $\beta - \mathrm{YbAlB_4}$ show that near QCP at $B \simeq 0$ the magnetization obeys [1–10] $M(B) \propto B^{-1/2}$. This behavior corresponds to the spectrum $\varepsilon(\mathbf{p})$ defined by (14.3) with $(p_f - p_i)/p_F \ll 1$. For finite values of B and T near FCQPT, the solutions of (14.1), which determine the dependences of T and B on $m^*(T, B)$, can be quite approximated by a simple universal interpolation function, see Chap. 7. Interpolation occurs between the LFL ($m^* \propto a + bT^2$) and NFL ($m^* \propto T^{-1/2}$) modes, separated by the intersection region in which m^* reaches its maximum value m^*_N at T_M and represents the behavior of universal scaling (see Chap. 7 and Sect. 7.1):

$$m^*_N(T_N) = \frac{m^*(T, B)}{m^*_M} = \frac{1 + c_2}{1 + c_1} \frac{1 + c_1 T_N^2}{1 + c_2 T_N^{5/2}}. \tag{14.4}$$

Here c_1 and c_2 are fitting parameters, $T_N = T/T_M$ is the normalized temperature, and

$$m^*_M \propto B^{-1/2}, \tag{14.5}$$

while

$$T_M \propto B^{1/2} \text{ and } T_M \propto B. \tag{14.6}$$

It follows from (14.4), (14.5), and (14.6) that the effective mass exhibits the universal scaling behavior

$$m^*(T, B) = c_3 \frac{1}{\sqrt{B}} m^*_N(T/B), \tag{14.7}$$

with c_3 being a constant [13–15]. Equations (14.4), (14.5), (14.6), and (14.7) will be used together with (14.1) to account for experimental observations on $\beta - \mathrm{YbAlB_4}$. Note that the scaling behavior refers to the temperatures $T \lesssim T_f$, where T_f is the temperature at which the influence of the CCP becomes insignificant, see Chap. 4. Based on (14.7), we conclude that the magnetization of M, as described in the topological configuration of fermionic condensation, really demonstrates the empirical behavior of scaling, defined as

$$M(T, B) = \int \chi(T, B_1) \mathrm{d}B_1 \propto \int \frac{m^*_N(T/B_1)}{\sqrt{B_1}} \mathrm{d}B_1. \tag{14.8}$$

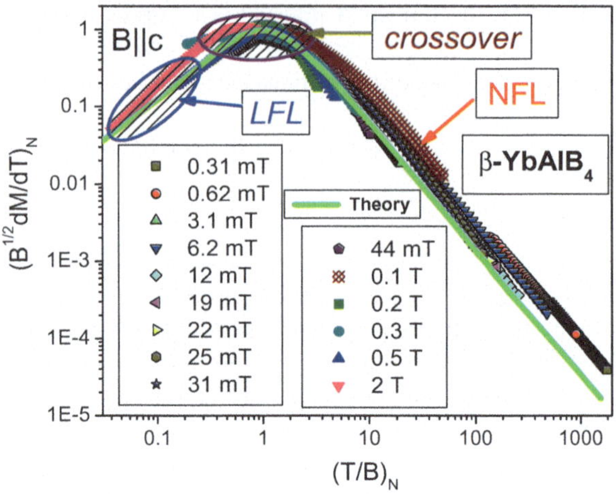

Fig. 14.1 Scaling behavior of dimensionless normalized magnetization $(B^{1/2}\mathrm{d}M(T,B)/\mathrm{d}T)_N$ versus dimensionless normalized $(T/B)_N$ at magnetic field values B given in the legend. Data are extracted from the measurements [6]. Regions of LFL behavior, crossover, and NFL behavior are indicated by arrows. The theoretical prediction is represented by a single scaling function

At $T < B$, the system exhibits the behavior of LFL with $M(B) \propto B^{-1/2}$, while for $T > B$ the system entered the NFL domain and $M(T) \propto T^{-1/2}$. Moreover, $\mathrm{d}M(T,B)/\mathrm{d}T$ again demonstrates the observed scaling behavior, with $\mathrm{d}M(T,B)/\mathrm{d}T \propto T$ for $T < B$ and $\mathrm{d}M(T,B)/\mathrm{d}T \propto T^{-3/2}$ for $T > B$. Thus, our analytical results correspond to the experiment [2, 4, 5] without fitting parameters and empirical functions.

In support of the mentioned scaling behavior analysis, Fig. 14.1 shows our calculations of the dimensionless normalized magnetization $(B^{1/2}\mathrm{d}M(T,B)/\mathrm{d}T)_N$ in comparison with the dimensionless normalized ratio $(T/B)_N$. Normalization is done by dividing $B^{1/2}\mathrm{d}M(T,B)/\mathrm{d}T$ and T/B, respectively, by the maximum value of $(B^{1/2}\mathrm{d}M(T,B)/\mathrm{d}T)_M$ and by the value of $(B/T)_M$ the maximum is reached. It can be seen that the calculated scaling function tracks data for four decades of the normalized value $(B^{1/2}\mathrm{d}M(T,B)/\mathrm{d}T)_N$, while the ratio T/B_N itself has been changing for five decades. From (14.8) it also follows that $(B^{1/2}\mathrm{d}M(T,B)/\mathrm{d}T)_N$ shows the universal scaling behavior depending on $(B/T)_N$. Figure 14.2 illustrates the scaling of $(B^{1/2}\mathrm{d}M(T,B)/\mathrm{d}T)_N$ of the archetypal HF metal $YbRh_2Si_2$. The solid curve representing the theoretical calculations is taken from Fig. 14.1. Thus, we find that the scaling behavior of $\beta - YbAlB_4$ extracted from the measurements [17, 18] and shown in Fig. 14.1 is universal, as Fig. 14.2 shows the same types of the behavior like the LFL, crossover, and NFL types when applying a magnetic field in a wide range of applied pressure. Note that $\beta - YbAlB_4$ exhibits the same universal behavior as 1D QSL does, see Chap. 9 and Fig. 9.1.

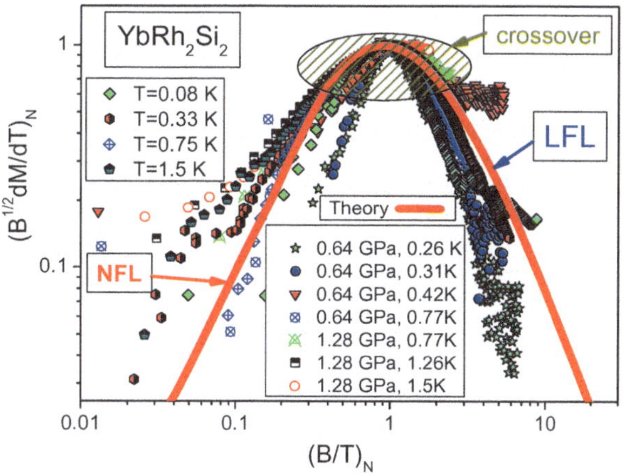

Fig. 14.2 Scaling behavior of the archetypal HF metal $YbRh_2Si_2$. Data for $(B^{1/2}\mathrm{d}M(T,B)/\mathrm{d}T)_N$ versus $(B/T)_N$ are extracted from measurements of $\mathrm{d}M/\mathrm{d}T$ versus B at fixed temperatures [17, 18]. The both solid curve representing the theoretical calculation and the notations are adapted from that of Fig. 14.1. Applied pressures and temperatures are shown in the legends

14.3 The Kadowaki-Woods Ratio

By applying magnetic fields $B > B_{c2} \simeq 30\,\mathrm{mT}$ and at sufficiently low temperatures, one can bring the LFL state having a resistivity of the form $\rho(T) = \rho_0 + AT^2$. The measurements of the T^2 dependence of the coefficient A under the application of magnetic field B provides information about its dependence on B [1]. It turned out that being proportional to the quasiparticle-quasiparticle scattering cross section, $A(B)$ obeys $A \propto (m^*(B))^2$ [19, 20]. According to (14.5), this means that

$$A(B) \simeq A_0 + \frac{D}{B}, \tag{14.9}$$

where A_0 and D are fitting parameters [13, 14]. We rewrite (14.9) in terms of the reduced variable A/A_0, which leads to

$$\frac{A(B)}{A_0} \simeq 1 + \frac{D_1}{B}, \tag{14.10}$$

where $D_1 = D/A_0$ is a constant, which reduces $A(B)$ to a function of B. Figure 14.3 demonstrates the fit of $A(B)$ to the experimental data [1]. The theoretical dependence (14.10) is in good agreement with the experiment in a significant region of B. This agreement suggests that the physics behind the field-induced reentry to the LFL behavior is the same for all classes of HF metals, see Sect. 21.2 and Fig. 21.2. It is important to note here that deviations of the theoretical curve from experimental points at $B > 2.5$ T are due to scaling violation at the QCP [5]. Figure 14.4 displays our calculations $\chi(B) \propto m^*$ and $C/T = \gamma(B) \propto m^*$ with experimental measurements [5]. Turning to (14.5), we see that the behavior of $A(B) \propto (m^*)^2$ is in good agreement with the experimental facts shown in this figure. We check (14.10) and

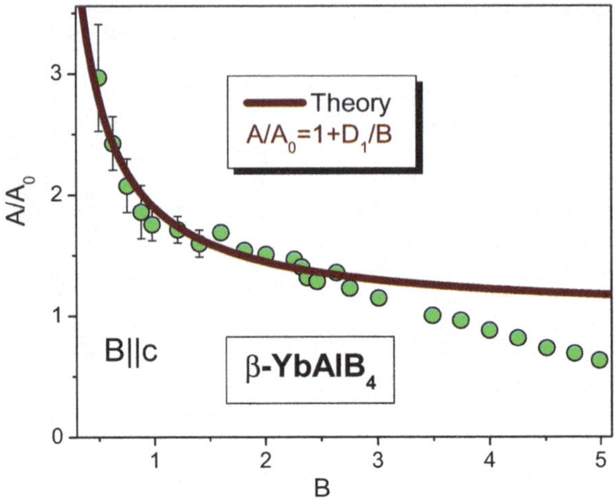

Fig. 14.3 Experimental data for normalized coefficient $A(B)/A_0$ as represented by (14.10), plotted as a function of magnetic field B (solid circles). Measured values of $A(B)$ are taken from [1], with D_1 the only fitting parameter. The solid curve is the theoretical prediction

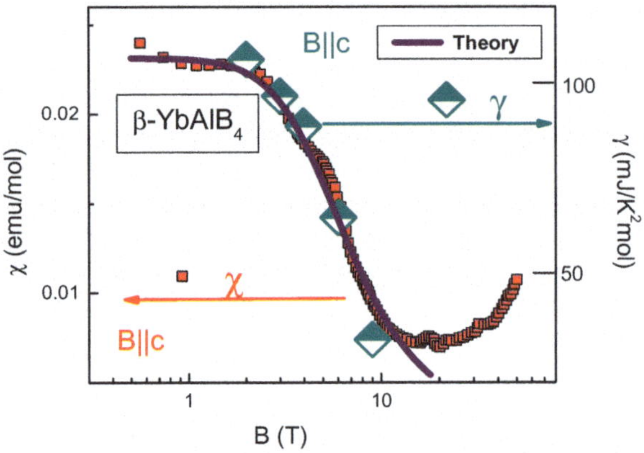

Fig. 14.4 Measurements [5] of magnetic susceptibility $\chi = \mathrm{d}M/\mathrm{d}B = a_1 m_N^*$ (left axis, square data points) and electronic specific heat coefficient $C/T = \gamma = a_2 m_N^*$ (right axis, stars), plotted versus magnetic field B. Solid curve tracing scaling behavior of m_N^*: theoretical results from present study with fitting parameters a_1 and a_2

conclude that in the case of $\beta - YbAlB_4$ the ratio $A/\gamma^2 \propto A/\chi^2 \simeq$ const is preserved, as it is in the other compounds with heavy fermions [1, 3, 13, 21]. Thus, the Kadowaki-Woods ratio is conserved.

14.4 The Schematic Phase Diagrams of HF Compounds

The results of the above analysis of the scaling behavior of this HF system, based on the topological scenario, make it possible to construct the schematic $T - B$ phase diagram of $\beta - YbAlB_4$, shown in Fig. 14.5, with a magnetic field B as a control parameter. For $B = 0$, the system acquires a flat band satisfying (14.2), which implies the presence of the fermion condensate in a highly degenerate state of a substance that becomes susceptible to a transition to the superconducting state [13, 22]. This

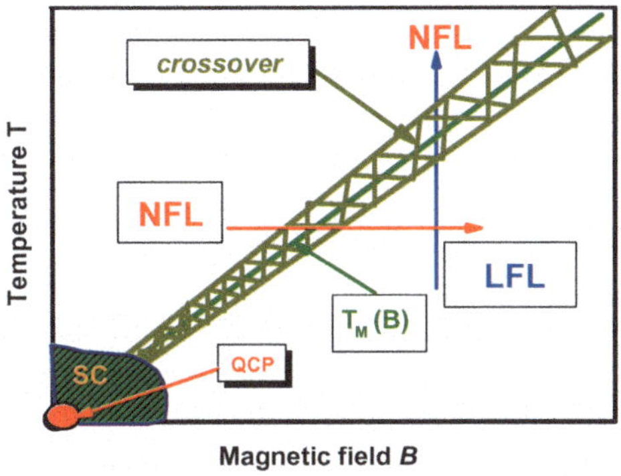

Fig. 14.5 Schematic $T - B$ phase diagram. Vertical and horizontal arrows highlight LFL-NFL and NFL-LFL transitions at fixed B and T, respectively. Hatched area separates the NFL phase from the weakly polarized LFL phase and identifies the transition region. Dashed line in hatched area represents the function $T_M \propto B$ (see (14.6)). The QCP, located at the origin and indicated by the arrow, is the quantum critical point at which the effective mass m^* diverges. It is surrounded by the superconducting phase labeled SC

mode of the NFL fermion condensate takes place at elevated temperatures and fixed magnetic field. QCP shown by the arrow in Fig. 14.5 is located at the origin of the phase diagram, since the application of any magnetic field destroys the flat band and puts the system in the LFL state, provided that the superconducting state is not in the game [13, 14, 23]. The shaded area in the drawing indicates the intersection area that separates the NFL behavior from the LFL one, which is also depicted in Fig. 14.1.

It is important to note that the heavy-fermionic metal $\beta - YbAlB_4$ is actually a superconductor on the ordered side of the topological FCQPT. In that case the superconducting phase transition eliminates the strong degeneracy related to FCQPT, see Sect. 3.5 and Chap. 4. When analyzing the NFL behavior $\rho(T)$ on the disordered side of this transition, it should be kept in mind that a few bands intersect the Fermi surface at the same time, so that the HF band never covers the entire Fermi surface. Accordingly, it turns out that the main contribution to the conductivity is made by quasiparticles not belonging to the HF band, see Chaps. 17 and 22. Therefore, the resistivity takes the form $\rho(T) = m^*_{rmnorm}\gamma(T)$, where $m^*_{\rm norm}$ is the average effective mass of normal quasiparticles, and $\gamma(T)$ describes their attenuation. The main contribution to $\gamma(T)$ can be taken as [24–27] $\gamma \propto T^2 m^*(m^*_{norm})^2$. Based on (14.3) and (14.6), we get [27] $\rho(T) \propto T^{3/2}$. On the other hand, we can expect that at $T \to 0$ the flat band (14.2) comes into play, leading the behavior of $\rho(T) \propto A_1 T$ with the coefficient A_1 proportional to the range of flat bands $(p_f - p_i)/p_F$. However, this behavior is not observed, since this region of the phase diagram is captured by superconductivity, as already indicated in Fig. 14.5. The low-T dependence of $\rho(T, P = 0) \propto T^{3/2}$ found experimentally [6] for the normal state of beta $-$ $YbAlB_4$ is consistent with this analysis. When the pressure P rises to the critical value P_c, a transition to the Landau-like behavior of $\rho(T) = \rho_0 + A_2 T^2$ occurs. Assuming that $P \propto x$, where x is the doping or the density number [18], and we observe that this behavior is very similar to the NFL behavior $\rho(T) \propto T^{1.5\pm0.1}$ found in measurements of the resistivity in electron-doped high-T_c superconductors $La_{2-x}Ce_xCuO_4$ [28, 29]. In this case, the effective mass $m^*(x)$ diverges as $x \to x_c$ or $P \to P_c$, see Chap. 2, [13, 14, 27–29],

$$(m^*(x))^2 \propto A \simeq \left(a_1 + \frac{a_2}{x/x_c - 1}\right)^2, \tag{14.11}$$

where a_1 and a_2 are fitting constants, and x_c is the critical doping, and the NFL behavior changes to LFL behavior. At this point the FC state eliminates at $x = x_c$ and the system gets to the disordered side of the FCQPT.

In Fig. 14.6, we display the schematic $T - x$ phase diagram exhibited by $\beta - \text{YbAlB}_4$ when the system is tuned by pressure P or by number density x. At $P/P_c < 1$ (or $x/x_c < 1$) the system is located on the ordered side of topological phase transition FCQPT and demonstrates NFL behavior at $T \lesssim T_f$. Thus, the NFL behavior induced by the FC that persists at $P < P_c$ is robust under application of pressure $P/P_c < 1$ [14, 23]. We note that such behavior is also observed in quasicrystals [15, 30]. At low temperatures the FC state possessing a flat band, highlighted in the figure, is strongly degenerate. This degeneracy stimulates the onset of certain phase transitions and is thereby lifted before reaching $T = 0$, as required by the Nernst theorem, see Chaps. 4 and 5.

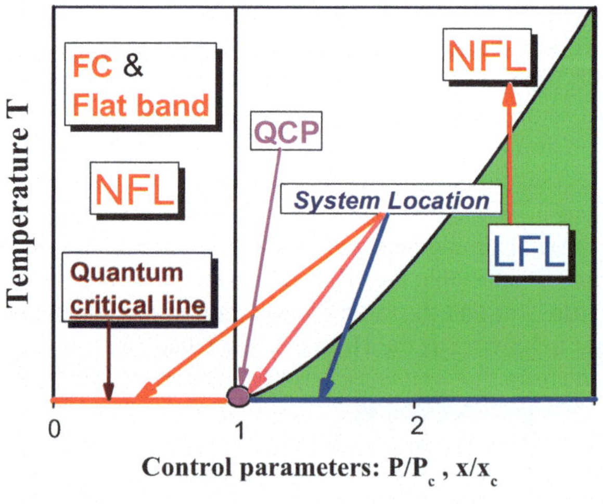

Fig. 14.6 Schematic $T - x$ phase diagram of HF system exhibiting a fermion condensate. Pressure P/P_c and number-density index x/x_c are taken as control parameters, with x_c the critical doping. At $P/P_c > 1$ and sufficiently low temperatures, the system is located in the LFL state (shadowed area). Moving past the QCP point to $P/Pc < 1$ (and forming quantum critical line) into the NFL region, the system develops a flat band that is the signature of the fermion condensation. The upward vertical arrow tracks the system moving in the LFL-NFL direction along T at fixed control parameters. Not shown is the low-temperature stable phase satisfying the Nernst theorem (superconducting in the case of $\beta - \text{YbAlB}_4$) that must exist for P/P_c or x/x_c below unity

Figure 14.6 shows the $T - x$ schematic phase diagram of $\beta - \mathrm{YbAlB_4}$, when the system is tuned by pressure P or number density x. For $P/P_c < 1$ (or $x/x_c < 1$), the system is located on the ordered side of the topological FCQPT and demonstrates the NFL behavior at $T \lesssim T_f$. Thus, the NFL behavior, which persists at $P < P_c$, is stable when pressure is applied at $P/P_c < 1$ [14, 23]. Note that this behavior is also exhibited by quasicrystals [15, 30]. At low temperatures the FC state, generating flat bands and highlighted in Fig. 14.6, is strongly degenerate. This degeneracy stimulates the onset of certain phase transitions and thereby is removed until $T = 0$ is reached, as required by the Nernst theorem that states that the entropy vanishes at $T \to 0$, see Chap. 4, [22, 31]. At elevating pressure (shown by the arrows in Fig. 14.6), the system enters the region $P/P_c > 1$, and situates over there before the onset of the topological FCQPT. It exhibits LFL behavior at low temperatures, as it is depicted by colored area in Fig. 14.6. The temperature range of this region shrinks when $P/P_c \to 1$, and m^* diverges in accordance with (14.11). These findings are in a good agreement with the experimental facts [6].

14.5 Summary

To summarize, we have analyzed the thermodynamic properties of the heavy-fermion metal $\beta - \mathrm{YbAlB_4}$ and explained their enigmatic scaling behavior within the topological scenario in which FC phase plays an essential role. We have explained why the observed NFL behavior is extremely sensitive to magnetic field, and how the thermodynamic properties and anomalous $T^{3/2}$ dependence of the electrical resistivity remains intact under the application of a pressure. We conclude that the observed behavior of $\beta - \mathrm{YbAlB_4}$ can be viewed as the fingerprints of the new state of matter, formed by the topological FCQPT, see also Chaps. 4, 7 and Sect. 4.1.3.

References

1. S. Nakatsuji, K. Kuga, Y. Machida, T. Tayama, T. Sakakibara, Y. Karaki, H. Ishimoto, S. Yonezawa, Y. Maeno, E. Pearson, G.G. Lonzarich, L. Balicas, H. Lee, Z. Fisk, Nat. Phys. **4**, 603 (2008)
2. Y. Matsumoto, S. Nakatsuji, K. Kuga, Y. Karaki, N. Horie, Y. Shimura, T. Sakakibara, A.H. Nevidomskyy, P. Coleman, Science **331**, 316 (2011)
3. Y. Matsumoto, K. Kuga, T. Tomita, Y. Karaki, S. Nakatsuji, Phys. Rev. B **84**, 125126 (2011)
4. Y. Matsumoto, S. Nakatsuji, K. Kuga, Y. Karaki, Y. Shimura, T. Sakakibara, A.H. Nevidomskyy, P. Coleman, J. Phys.: Conf. Ser. **391**, 012041 (2012)
5. Y. Matsumoto, K. Kuga, Y. Karaki, Y. Shimura, T. Sakakibara, M. Tokunaga, K. Kindo, S. Nakatsuji, J. Phys. Soc. Jpn. **84**, 024710 (2015)
6. T. Tomita, K. Kuga, Y. Uwatoko, P. Coleman, S. Nakatsuji, Science **349**, 506 (2015)
7. A.H. Nevidomskyy, P. Coleman, Phys. Rev. Lett. **102**, 077202 (2009)
8. A. Ramires, P. Coleman, A.H. Nevidomskyy, A.M. Tsvelik, Phys. Rev. Lett. **109**, 176404 (2012)

9. S. Watanabe, K. Miyake, J. Phys. Soc. Jpn. **82**, 083704 (2013)
10. S. Watanabe, K. Miyake, J. Phys. Soc. Jpn. **83**, 103708 (2014)
11. V.R. Shaginyan, A.Z. Msezane, K.G. Popov, J.W. Clark, V.A. Khodel, M.V. Zverev, Phys. Rev. B **93**, 205126 (2016)
12. V.R. Shaginyan, A.Z. Msezane, G.S. Japaridze, K.G. Popov, V.A. Khodel, Front. Phys. **11**, 117103 (2016)
13. V.R. Shaginyan, M.Y. Amusia, A.Z. Msezane, K.G. Popov, Phys. Rep. **492**, 31 (2010)
14. M.Y. Amusia, K.G. Popov, V.R. Shaginyan, V.A. Stephanovich, *Theory of Heavy-Fermion Compounds*, Springer Series in Solid-State Sciences, vol. 182 (Cham, Heidelberg, 2014)
15. V.R. Shaginyan, A.Z. Msezane, K.G. Popov, G.S. Japaridze, V.A. Khodel, Phys. Rev. B **87**, 245122 (2013). arXiv:1601.01182
16. L.D. Landau, Sov. Phys. JETP **3**, 920 (1956)
17. Y. Tokiwa, T. Radu, C. Geibel, F. Steglich, P. Gegenwart, Phys. Rev. Lett. **102**, 066401 (2009)
18. Y. Tokiwa, P. Gegenwart, C. Geibel, F. Steglich, J. Phys. Soc. Jpn. **78**, 123708 (2009)
19. V.A. Khodel, P. Schuck, Z. Phys. B: Condens. Matter **104**, 505 (1997)
20. P. Gegenwart, J. Custers, C. Geibel, K. Neumaier, T. Tayama, K. Tenya, O. Trovarelli, F. Steglich, Phys. Rev. Lett. **89**, 056402 (2002)
21. K. Kadowaki, S.B. Woods, Solid State Commun. **58**, 507 (1986)
22. V.A. Khodel, J.W. Clark, M.V. Zverev, Phys. Rev. B **78**, 075120 (2008)
23. V.R. Shaginyan, A.Z. Msezane, K.G. Popov, G.S. Japaridze, V.A. Khodel, Europhys. Lett. **106**, 37001 (2014)
24. V.A. Khodel, M.V. Zverev, JETP Lett. **85**, 404 (2007)
25. V.R. Shaginyan, A.Z. Msezane, K.G. Popov, J.W. Clark, M.V. Zverev, V.A. Khodel, Phys. Rev. B **86**, 085147 (2012)
26. V.A. Khodel, V.R. Shaginyan, P. Schuck, JETP Lett. **63**, 752 (1996)
27. V.A. Khodel, J.W. Clark, K.G. Popov, V.R. Shaginyan, JETP Lett. **101**, 413 (2015)
28. N.P. Armitage, P. Furnier, R.L. Greene, Rev. Mod. Phys. **82**, 2421 (2010)
29. K. Jin, N.P. Butch, K. Kirshenbaum, J. Paglione, R.L. Greene, Nature **476**, 73 (2011)
30. K. Deguchi, S. Matsukawa, N.K. Sato, T. Hattori, K. Ishida, H. Takakura, T. Ishimasa, Nat. Mater. **11**, 1013 (2012)
31. J.W. Clark, M.V. Zverev, V.A. Khodel, Ann. Phys. **327**, 3063 (2012)

Chapter 15
The Universal Behavior of the Archetypical Heavy-Fermion Metals $YbRh_2Si_2$

Abstract Basing on the theory of fermion condensation, we analyze the behavior of the archetypical heavy-fermion metals $YbRh_2Si_2$. We show that the low-temperature thermodynamic and transport properties are formed by quasiparticles, while the dependence of the effective mass on temperature, number density, magnetic fields, etc gives rise to the non-Fermi liquid behavior. Our theoretical study of the heat capacity, magnetization, energy scales, the longitudinal magnetoresistance and magnetic entropy are in good agreement with the remarkable facts collected on the heavy-fermion metal $YbRh_2Si_2$.

15.1 Introduction

The Landau theory of Fermi liquids has a long history and remarkable results in describing the properties of electron liquid in ordinary metals and Fermi liquids of ^{3}He type. The theory is based on the Landau paradigm that elementary excitations determine the behavior of the system at low temperatures. These low-temperature excitations are represented by quasiparticles. They have a certain effective mass M^*, which is approximately independent of temperature T, number density x, and magnetic field strength B and is a parameter of the theory [1]. The discovery of strongly correlated Fermi systems, exhibiting the non-Fermi liquid (NFL) behavior, has opened tremendous challenges in the modern condensed matter physics [2–7]. Facts collected on heavy-fermion (HF) compounds demonstrate that the effective mass strongly depends on T, x, B, etc, while M^* itself can reach very high values or even diverge [4, 5]. Such a behavior contradicts the traditional Landau quasiparticles paradigm so that it does not apply to them.

It exists almost a common view that quantum criticality, describing the collective fluctuations of matter undergoing a second-order phase transition at zero temperature, suppresses quasiparticles and thus generates the NFL behavior, depending on the initial ground state, either magnetic or superconductive [2–6]. Earlier, the theory of the topological fermion-condensation quantum phase transition (FCQPT) preserving quasiparticles and closely related to the unlimited growth of M^* had been suggested [7–10]. Further studies show that it is capable to present an adequate theoretical

M. Amusia and V. Shaginyan, *Strongly Correlated Fermi Systems*, Springer Tracts in Modern Physics 283,
https://doi.org/10.1007/978-3-030-50359-8_15

explanation of vast majority of experimental results in different HF compounds [7, 10–13]. In contrast to the Landau paradigm based on the assumption that M^* is a constant, in FCQPT approach M^* strongly depends on T, x, B, etc. Therefore, in accord with numerous experimental observations the extended quasiparticles paradigm is to be introduced. The main point here is that the well-defined quasiparticles determine as before the thermodynamic and transport properties of strongly correlated Fermi systems, while M^* becomes a function of T, x, B, etc [12–15]. The FCQPT theory had been already successfully applied to describe the thermodynamic properties of such extremely strongly correlated systems as complicated HF compounds [7, 10, 13–18].

In this chapter, we analyze the non-Fermi liquid behavior of HF metal $YbRh_2Si_2$ using the fermion condensation theory. The obtained results are illustrated with calculations of the thermodynamic and transport functions of $YbRh_2Si_2$. Possible energy scales in these functions are discussed. We demonstrate that our calculations of the heat capacity C/T, magnetization M, energy scales, longitudinal magnetoresistance (LMR), magnetic entropy $S(B)$, etc. are in good agreement with impressive data collected on the HF metal $YbRh_2Si_2$ [19–21].

15.2 Scaling Behavior of the Effective Mass

To avoid difficulties associated with the anisotropy generated by the crystal lattice of solids, we study the universal behavior of heavy-fermion metals using the model of the homogeneous heavy electron (fermion) liquid. The model is quite meaningful because we consider the universal behavior exhibited by these materials at low temperatures, as behavior related to power-law divergences of quantities such as the effective mass, heat capacity, magnetization, etc. These divergences and scaling behavior of the effective mass, or the critical exponents that characterize them, are determined by energy and momentum transfers that are small compared to the Debye characteristic temperature and momenta of the order of the reciprocal lattice cell length a^{-1}. Therefore quasiparticles are influenced by the crystal lattice averaged over big distances compared to the length a. Thus, we can substitute the well-known jelly model for the lattice as it is usually done, for example, in the fluctuation theory of second-order phase transitions.

The schematic phase diagram of HF metal is reported in Fig. 15.1. Magnetic field B is taken as the control parameter. In fact, the control parameter can also be pressure P or doping (the number density) x etc as well. At $B = B_{c0}$, FC takes place leading to a strongly degenerated state, where B_{c0} is a critical magnetic field, such that at $B > B_{c0}$ the system is driven toward its Landau Fermi liquid (LFL) regime. In our model B_{c0} is a parameter. The FC state can be captured by the superconducting (SC), ferromagnetic (FM), antiferromagnetic (AFM) etc. states lifting the degeneracy [7, 10, 12, 13]. Below we consider the HF metal $YbRh_2Si_2$. In that case, $B_{c0} \simeq 0.06\,\mathrm{T}$ $(B \perp c)$ and at $T = 0$ and $B < B_{c0}$ the AFM state takes place [21]. At elevated temperatures and fixed magnetic field the NFL regime occurs, while rising B again

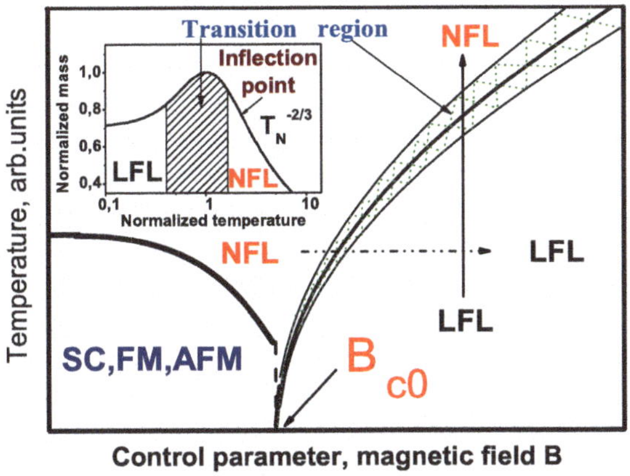

Fig. 15.1 Schematic phase diagram of HF metals. B_{c0} is magnetic field at which the effective mass diverges. SC, FM, AFM denote the superconducting (SC), ferromagnetic (FM), and antiferromagnetic (AFM) states, respectively. At $B < B_{c0}$ the system can be in SC, FM, or AFM states. The vertical arrow shows the transition from the LFL regime to the NFL one at fixed B along T with M^* depending on T. The dash-dot horizontal arrow illustrates the system moving from NFL regime to LFL one along B at fixed T. The inset shows a schematic plot of the normalized effective mass versus the normalized temperature. Transition regime, where M^*_N reaches its maximum value M^*_M at $T = T_M$, is shown by the hatched area both in the main panel and in the inset. The arrows mark the position of inflection point in M^*_N and the transition region

drives the system from NFL region to LFL as shown by the dash-dot horizontal arrow in Fig. 15.1. Below we consider the transition region when the system moves from NFL regime to LFL along the horizontal arrow and it moves from LFL regime to NFL along the vertical arrow as shown in Fig. 15.1. The inset to Fig. 15.1 demonstrates the behavior of the normalized effective mass $M^*_N = M^*/M^*_M$ versus normalized temperature $T_N = T/T_M$, where M^*_M is the maximum value that M^* reaches at $T = T_M$. The $T^{-2/3}$ regime is marked as NFL, since the effective mass depends strongly on temperature. The temperature region $T \simeq T_M$ signifies the crossover between the LFL regime with almost constant effective mass and NFL behavior, given by $T^{-2/3}$ dependence. Thus, temperatures $T \sim T_M$ can be regarded as the crossover region between LFL and NFL regimes.

As it is shown in Chap. 7, when the system is near FCQPT, it turns out that the solution of (7.1) $M^*(T)$ can be well approximated by a simple universal interpolating function. The interpolation occurs between the LFL ($M^* \simeq M^* + a_1 T^2$) and NFL ($M^* \propto T^{-2/3}$) regimes thus describing the above crossover. Introducing the dimensionless variable $y = T_N = T/T_M$, we obtain the desired expression

$$M^*_N(y) \approx c_0 \frac{1 + c_1 y^2}{1 + c_2 y^{8/3}}. \tag{15.1}$$

Here $M^*_N = M^*/M^*_M$ is the normalized effective mass, $c_0 = (1 + c_2)/(1 + c_1)$, c_1, and c_2 are fitting parameters, parameterizing the Landau amplitude.

It follows from (15.1) that in contrast to the Landau paradigm of quasiparticles the effective mass strongly depends on T and B. As we will see it is this dependence that forms the NFL behavior. It follows also from (15.1) that a scaling behavior of M^* near FCQPT point is determined by the absence of appropriate external physical scales to measure the effective mass and temperature. At fixed magnetic fields, the characteristic scales of temperature and the function $M^*(T, B)$ are defined by both T_M and M^*_M, respectively. At fixed temperatures, the characteristic scales are $(B_M - B_{c0})$ and M^*_M. It follows from (7.1), (7.4), and (15.1) that at fixed magnetic fields, $T_M \to 0$, and $M^*_M \to \infty$, and the width of the transition region shrinks to zero as $B \to B_{c0}$ when these are measured in the external scales. In the same way, it follows from (7.1), (7.4), and (15.1) that at fixed temperatures, $(B_M - B_{c0}) \to 0$, and $M^*_M \to \infty$, and the width of the transition region shrinks to zero as $T_{mm} \to 0$. In some cases temperature or magnetic field dependencies of the effective mass or of other observable like the longitudinal magnetoresistance do not have "peculiar points" like maximum. The normalization is to be performed in the other points like the inflection point shown in the inset to Fig. 15.1. Such a normalization is possible since it is established on the internal scales.

15.3 Non-Fermi Liquid Behavior in $YbRh_2Si_2$

In what follows, we compute the effective mass and employ (15.1) for estimations of considered values. Choice of the amplitude is dictated by the fact that the system has to be in the FCQPT point, which means that the first two p-derivatives of the single-particle spectrum $\varepsilon(\mathbf{p})$ should equal zero. Since the first derivative is proportional to the reciprocal quasiparticle effective mass $1/M^*$, its zero (where $1/M^* = 0$ and the effective mass diverges) just signifies FCQPT. Zeros of two subsequent derivatives mean that the spectrum $\varepsilon(\mathbf{p})$ has an inflection point at Fermi momentum p_F so that the lowest term of its Taylor expansion is proportional to $(p - p_F)^3$ [15]. After solving (7.1), the obtained spectrum had been used to calculate the entropy $S(B, T)$, which, in turn, had been recalculated to the effective mass $M^*(T, B)$ by virtue of well-known LFL relation $M^*(B, T) = S(B, T)/T$. We note that our calculations confirm the validity of (15.1).

15.3.1 Heat Capacity and the Sommerfeld Coefficient

Measurements of $C/T \propto M^*$ on $YbRh_2Si_2$ in different magnetic fields B up to 1.5 T [19] allow us to identify the scaling behavior of the effective mass M^* and observe the different regimes of M^* behavior such as the LFL regime, transition region from LFL to NFL regimes, and the NFL regime itself. A maximum structure in $C/T \propto M^*_M$ at T_M appears under the application of magnetic field B and T_M shifts to higher T as B is increased. The value of $C/T = \gamma_0$ is saturated toward lower temperatures

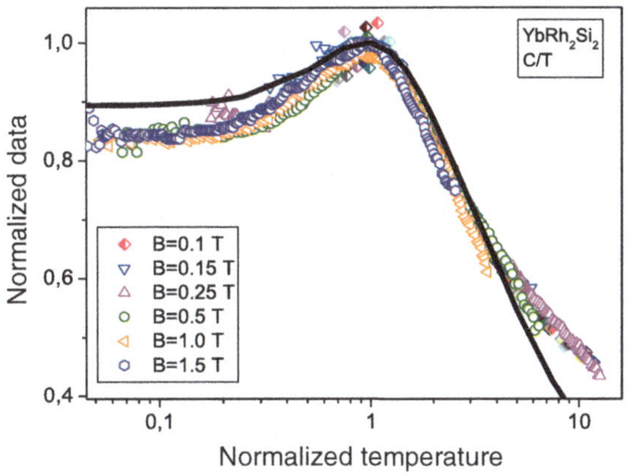

Fig. 15.2 The normalized effective mass M_N^* extracted from the measurements of the specific heat C/T on $YbRh_2Si_2$ in magnetic fields B is shown in the left down corner [19]. Our calculations are depicted by the solid curve tracing the scaling behavior of M_N^*

decreasing at elevated magnetic field, where γ_0 is the Sommerfeld coefficient [19]. It follows from Fig. 15.1 that the transition region corresponds to the temperatures where the vertical arrow in the main panel of Fig. 15.1 crosses the hatched area. The width of the region, being proportional to $T_M \propto (B - B_{c0})$ shrinks, T_M moves to zero temperature and $\gamma_0 \propto M^*$ increases as $B \to B_{c0}$. These observations are in accord with the facts [19].

To obtain the normalized effective mass M_N^*, the maximum structure in C/T was used to normalize C/T, and T was normalized by T_M. In Fig. 15.2, the obtained M_N^* as a function of normalized temperature T_N is shown by geometrical figures; our calculations carried out as described above are presented by the solid line. Figure 15.2 reveals the scaling behavior of the normalized experimental curves—the curves at different magnetic fields B merge into a single one in terms of the normalized variable $y = T/T_M$. As shown in Fig. 15.2, the normalized mass M_N^* extracted from the measurements is not a constant, as would be for an LFL, and shows the scaling behavior given by (15.1) over three decades in normalized temperature. The two regimes (the LFL regime and NFL one) separated by the transition region, as depicted by the hatched area in the inset to Fig. 15.1, are clearly shown in Fig. 15.2 illuminating good agreement between the theory and experiment.

15.3.2 Average Magnetization

Consider now an average magnetization $\widetilde{M} \equiv M + B\chi$ as a function of magnetic field B at fixed temperature $T = T_f$, where χ is the magnetic susceptibility and M is the magnetization,

$$M(B, T) = \int_0^B \chi(b, T)\mathrm{d}b, \tag{15.2}$$

where the magnetic susceptibility is given by [1]

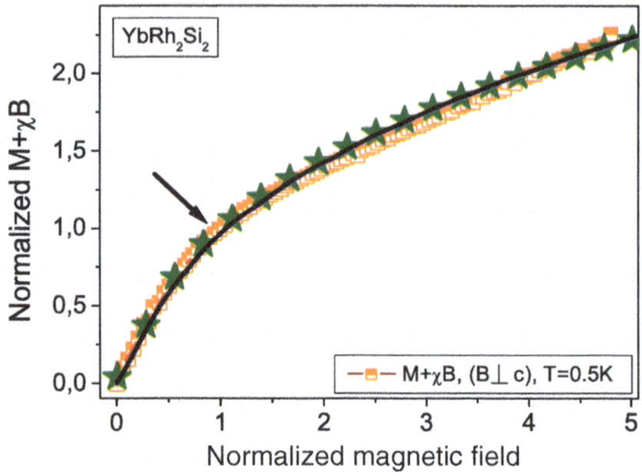

Fig. 15.3 The field dependence of the normalized "average" magnetization $\widetilde{M} \equiv M + B\chi$ is shown by squares and extracted from measurements collected on $YbRu_2Si_2$ [20]. The kink (shown by the arrow) is clearly seen at the normalized field $B_N = B/B_k \simeq 1$. The solid curve and stars (see text) represent our calculations

$$\chi(B,T) = \frac{\beta M^*(B,T)}{1+F_0^a}. \tag{15.3}$$

Here, β is a constant and F_0^a is the Landau amplitude related to the exchange interaction [1]. In the case of strongly correlated systems $F_0^a \geq -0.9$ [22]. Therefore, as seen from (15.3), due to the normalization the coefficients β and $(1 + F_0^a)$ drops out from the result, and $\chi \propto M^*$. To obtain $\widetilde{M}$, we calculate M^* as a function of B at fixed T_{mm}. The obtained curves of $\widetilde{M}$ exhibit energy scales separated by kinks at $B = B_k$. As shown in Fig. 15.3, the kink is a crossover point from the fast to slow growth of $\widetilde{M}$ at rising magnetic field. We use B_k and $\widetilde{M}(B_k)$ to normalize B and $\widetilde{M}$, respectively. We note that the kinks are considered in Chap. 21, see Sect. 21.5.

The normalized $\widetilde{M}$ versus the normalized field $B_N = B/B_K$ are shown in Fig. 15.3. Our calculations are depicted by the solid line. The stars trace out our calculations of $\widetilde{M}$ with $M^*(y)$ extracted from the data C/T shown in Fig. 15.2. The calculation procedure deserves a remark here. Namely, in that case M^* depends on $y = T/T_M$ with T_M is given by (7.4). On the other hand, we can consider $y = T/(B - B_{c0})/$, as it is shown in Chap. 7, and take the data C/T as a function of the scaled variable y.

It is shown in Fig. 15.3 that our calculations are in good agreement with the facts, and all the data exhibit the kink (shown by arrow) at $B_N \simeq 1$ taking place as soon as the system enters the transition region corresponding to the magnetic fields where the horizontal dash-dot arrow in the main panel of Fig. 15.1 crosses the hatched area. Indeed, as shown in Fig. 15.3, at lower magnetic fields $\widetilde{M}$ is a linear function of B since M^* is approximately independent of B. It follows from (15.1) that at elevated magnetic fields M^* becomes a diminishing function of B and generates the kink in $\widetilde{M}(B)$ separating the energy scales observed in [20]. Then, it is seen from (7.6) that the magnetic field $B_k \simeq B_M$ at which the kink appears shifts to lower B as T_{mm} is decreased.

15.3.3 Longitudinal Magnetoresistance

Consider the longitudinal magnetoresistance (LMR) $\rho(B, T) = \rho_0 + AT^2$ as a function of B at fixed T_{mm}. In that case, the classical contribution to LMR due to orbital motion of carriers induced by the Lorentz force is small, while the Kadowaki-Woods relation [23, 24], $K = A/\gamma_0^2 \propto A/\chi^2 = const$, allows us to employ M^* to construct the coefficient A [25], since $\gamma_0 \propto \chi \propto M^*$. As a result, $\rho(B, T) - \rho_0 \propto (M^*)^2$. Figure 15.4 reports the normalized magnetoresistance

$$R_N^\rho(y) = \frac{\rho(y) - \rho_0}{\rho_{inf}} \propto (M_N^*(y))^2 \tag{15.4}$$

versus normalized magnetic field $y = B/B_{inf}$ at different temperatures, shown in the legend. Here ρ_{inf} and B_{inf} are LMR and magnetic field, respectively, taken at the inflection point marked by the arrow in Fig. 15.4.

Since the magnetic field dependence of both the calculated M^* and LMR does not have "peculiar points" like maximums, the normalization have been performed in the corresponding inflection points. To determine the inflection point precisely, we first differentiate $\rho(B, T)$ over B, find the extremum of derivative and normalize the values of the function and the argument by their values in the inflection point. Then, both theoretical (shown by the solid line) and experimental (marked by the geometrical figures) curves have been normalized by their inflection points, which also reveal the universal behavior: The curves at different temperatures merge into single one in terms of the scaled variable y and show the scaling behavior over three decades in the normalized magnetic field.

The transition region at which LMR starts to decrease is shown in the inset to Fig. 15.1 by the hatched area. Obviously, as seen from (7.6), the width of the transition region being proportional to $B_M \simeq B_{inf}$ decreases as the temperature T_{mm} is lowered. In the same way, the inflection point of LMR, generated by the inflection point of M^*

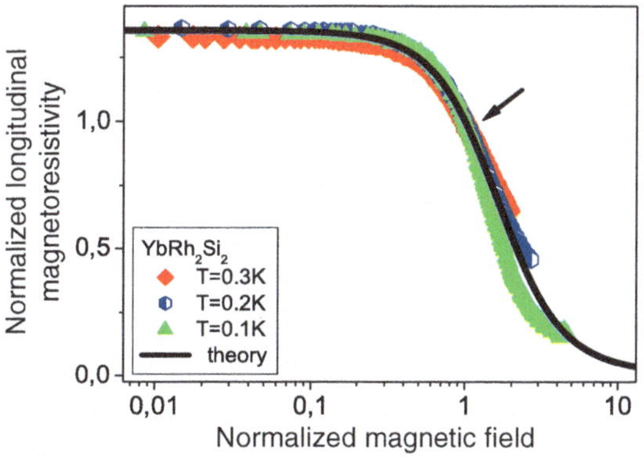

Fig. 15.4 Magnetic field dependence of the normalized (in the inflection point shown by the arrow, see text for details) magnetoresistance R_N^ρ versus normalized magnetic field. R_N^ρ was extracted from LMR of $YbRh_2Si_2$ at different temperatures [20] listed in the legend. The solid line represents our calculations

shown in the inset to Fig. 15.1 by the arrow, shifts to lower B as T_{mm} is decreased. All these observations are in excellent agreement with the facts [20].

15.3.4 Magnetic Entropy

The evolution of the derivative of magnetic entropy $\mathrm{d}S(B,T)/\mathrm{d}B$ as a function of magnetic field B at fixed temperature T_{mm} is of great importance since it allows us to study the scaling behavior of the derivative of the effective mass $T\mathrm{d}M^*(B,T)/\mathrm{d}B \propto \mathrm{d}S(B,T)/\mathrm{d}B$. Scaling properties of the effective mass $M^*(B,T)$ can be analyzed via LMR.

As seen from (15.1) and (7.6), at $y \leq 1$ the derivative $-\mathrm{d}M_N(y)/\mathrm{d}y \propto y$ with $y = (B - B_{c0})/(B_{inf} - B_{c0}) \propto (B - B_{c0})/T_f$. We recall that the effective mass as a function of B does not have the maximum. At elevated y the derivative $-\mathrm{d}M_N(y)/\mathrm{d}y$ possesses a maximum at the inflection point and then becomes a diminishing function of y. Upon using the variable $y = (B - B_{c0})/T_f$, we conclude that at decreasing temperatures, the leading edge of the function $-\mathrm{d}S/\mathrm{d}B \propto -T\mathrm{d}M^*/\mathrm{d}B$ becomes steeper and its maximum at $(B_{inf} - B_{c0}) \propto T_f$ is higher. These observations are in quantitative agreement with striking measurements of the magnetization difference divided by temperature increment, $-\Delta M/\Delta T$, as a function of magnetic field at fixed temperatures T_{mm} collected on $YbRh_2Si_2$ [21]. We note that according to the well-know thermodynamic equality $\mathrm{d}M/\mathrm{d}T = \mathrm{d}S/\mathrm{d}B$, that is one has the approximate relation $\Delta M/\Delta T \simeq \mathrm{d}S/\mathrm{d}B$.

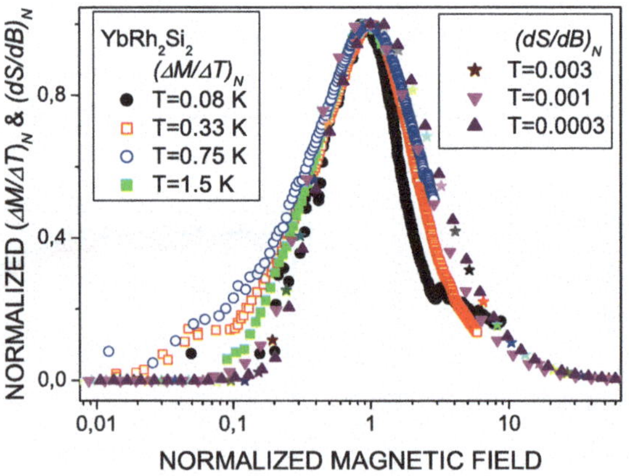

Fig. 15.5 Normalized magnetization difference divided by temperature increment $(\Delta M/\Delta T)_N$ versus normalized magnetic field at fixed temperatures (listed in the legend in the upper left corner) is extracted from the facts collected on $YbRh_2Si_2$ [21]. Our calculations of the normalized derivative $(\mathrm{d}S/\mathrm{d}B)_N \simeq (\Delta M/\Delta T)_N$ versus normalized magnetic field are given at fixed dimensionless temperatures T/μ (listed in the legend in the upper right corner). All the data are marked by the geometrical figures depicted in the legends

To carry out a quantitative analysis of the scaling behavior of $-\mathrm{d}M^*(B,T)/\mathrm{d}B$, we calculate as described above the entropy $S(B,T)$ as a function of B at fixed dimensionless temperatures T_{mm}/μ shown in the upper right corner of Fig. 15.5. This figure reports the normalized $(\mathrm{d}S/\mathrm{d}B)_N$ as a function of the normalized magnetic field. The function $(\mathrm{d}S/\mathrm{d}B)_N$ is obtained by normalizing $(-\mathrm{d}S/\mathrm{d}B)$ by its maximum taking place at B_M, and the field B is scaled by B_M. The measurements of $-\Delta M/\Delta T$ are normalized in the same way and are depicted in Fig. 15.5 as $(\Delta M/\Delta T)_N$ versus normalized field. It is shown in Fig. 15.5 that our calculations are in excellent agreement with the facts and both the experimental functions $(\Delta M/\Delta T)_N$ and the calculated $(\mathrm{d}S/\mathrm{d}B)_N$ ones show the scaling behavior over three decades in the normalized magnetic field.

15.4 Summary

We have analyzed the non-Fermi liquid behavior of $YbRh_2Si_2$ using the theory of the fermion condensation, and showed that extended quasiparticles paradigm is strongly valid, while the dependence of the effective mass on temperature and applied magnetic fields, etc results in the NFL behavior. The obtained results are illustrated with calculations of the thermodynamic and transport functions of $YbRh_2Si_2$. Possible energy scales in these functions are discussed. We have demonstrated that our theoretical study of the heat capacity, magnetization, energy scales, longitudinal magnetoresistance, and magnetic entropy are in good agreement with the impressive data collected on the HF metal $YbRh_2Si_2$. These data show that $YbRh_2Si_2$ exhibits the universal behavior, and enters as the archetypal HF metal into the family of HF compounds that form the new state of matter.

References

1. E.M. Lifshitz, L.P. Pitaevskii, *Statistical Physics, Part 2* (Butterworth-Heinemann, Oxford, 1999)
2. T. Senthil, M. Vojta, S. Sachdev, Phys. Rev. B **69**, 035111 (2004)
3. P. Coleman, A.J. Schofield, Nature **433**, 226 (2005)
4. H.V. Löhneysen, A. Rosch, M. Vojta, P. Wölfle, Rev. Mod. Phys. **79**, 1015 (2007)
5. P. Gegenwart, Q. Si, F. Steglich, Nat. Phys. **4**, 186 (2008)
6. S. Sachdev, Nat. Phys. **4**, 173 (2008)
7. M.Y. Amusia, K.G. Popov, V.R. Shaginyan, W.A. Stephanowich, *Theory of Heavy-Fermion Compounds*. Springer Series in Solid-State Sciences, vol. 182 (Springer, Berlin, 2015)
8. V.A. Khodel, V.R. Shaginyan, JETP Lett. **51**, 553 (1990)
9. M.Y. Amusia, V.R. Shaginyan, Phys. Rev. B **63**, 224507 (2001)
10. V.R. Shaginyan, M.Y. Amusia, A.Z. Msezane, K.G. Popov, Phys. Rep. **492**, 31 (2010)
11. G.E. Volovik, Springer Lecture Notes in Physics, vol. 718 (2007), p. 31
12. V.R. Shaginyan, M.Y. Amusia, K.G. Popov, Phys.-Uspekhi **50**, 563 (2007)
13. V.A. Khodel, J.W. Clark, M.V. Zverev, Phys. Rev. B **78**, 075120 (2008)
14. J. Dukelsky et al., Z. Phys. B: Condens. Matter **102**, 245 (1997)

15. J.W. Clark, V.A. Khodel, M.V. Zverev, Phys. Rev. B **71**, 012401 (2005)
16. V.R. Shaginyan et al., Europhys. Lett. **76**, 898 (2006)
17. V.R. Shaginyan, K.G. Popov, V.A. Stephanovich, Europhys. Lett. **79**, 47001 (2007)
18. V.R. Shaginyan, A.Z. Msezane, K.G. Popov, V.A. Stephanovich, Phys. Rev. Lett. **100**, 096406 (2008)
19. N. Oeschler et al., Physica B **403**, 1254 (2008)
20. P. Gegenwart et al., Science **315**, 969 (2007)
21. Y. Tokiwa et al., Phys. Rev. Lett. **102**, 066401 (2009)
22. M. Pfitzner, P. Wölfle, Phys. Rev. B **33**, 2003 (1986); D. Wollhardt, P. Wölfle, P.W. Anderson, Phys. Rev. B **35**, 6703 (1987)
23. K. Kadowaki, S.B. Woods, Solid State Commun. **58**, 507 (1986)
24. A. Khodel, P. Schuck, Z. Phys. B: Condens. Matter **104**, 505 (1997)
25. V.R. Shaginyan, M.Y. Amusia, A.Z. Msezane, K.G. Popov, V.A. Stephanovich, Phys. Lett. A **373**, 986 (2009)

Chapter 16
Heavy-Fermion Compounds as the New State of Matter

Abstract Such strongly correlated Fermi systems as high-temperature superconductors (HTSC) and heavy-fermion (HF) metals exhibit extraordinary properties. They are so unusual that the traditional Landau paradigm of quasiparticles does not apply. In HTSC, measurements have been performed, demonstrating a puzzling magnetic field-induced transition from non-Fermi liquid to Landau Fermi liquid behavior. We employ the theory of the fermion condensation which is able to describe this transition. We show, that in spite of different microscopic nature of HTSC and HF metals, the behavior of HTSC is similar to that observed in both HF metals, thus confirming that HTSC and HF metals belong to the new state of matter.

16.1 Introduction

The non-Fermi liquid (NFL) behavior of many classes of strongly correlated fermion systems pose one of the prominent challenges in modern condensed matter physics. Many common experimental features of such seemingly different systems as heavy-fermion (HF) metals and high-temperature superconductors (HTSC) suggest that there is a hidden fundamental law of nature, which remains to be recognized. The key word here is quantum criticality, taking place in quantum critical point (QCP).

Heavy-fermion metals provide important examples of strongly correlated Fermi systems [1–6]. The second class of substances to test whether or not the Landau Fermi liquid (LFL) theory is fulfilled in them are HTSC. In these substances, all quantum critical points are almost inaccessible to experimental observations since they are "*hidden in superconductivity*" or more precisely, the superconductive gap opened at the Fermi level changes the physical properties of the corresponding quantum phase transition.

There is a common wisdom that the physical properties of above systems are related to zero temperature quantum fluctuations, suppressing quasiparticles and thus generating their NFL properties [1, 2], depending on their initial ground state, either magnetic or superconductive. On the other hand, it was shown that the electronic system of HF metals demonstrates the universal low-temperature behavior irrespectively of their magnetic ground state [5–7]. The NFL behavior has been discovered

M. Amusia and V. Shaginyan, *Strongly Correlated Fermi Systems*,
Springer Tracts in Modern Physics 283,
https://doi.org/10.1007/978-3-030-50359-8_16

experimentally in 2D ^{3}He [8], and the theoretical explanation has been given to it [9], revealing the similarity in physical properties of 2D ^{3}He and HF metals. We note here that ^{3}He consists of neutral atoms interacting via van der Waals forces, while the mass of He atom is 3 orders of magnitude larger than that of an electron, making ^{3}He to have drastically different microscopic properties than those of HF metals. Therefore it is of crucial importance to check whether this behavior can be observed in other Fermi systems like HTSC. Recently, precise measurements on HTSC $Tl_2Ba_2CuO_{6+x}$ of magnetic field-induced transition from NFL to LFL behavior became available, see the phase diagram 16.1 [10]. This transition takes place under the application of magnetic field $B \geq B_{c0}$, where B_{c0} is the critical field at which the magnetic field-induced QCP takes place.

Here we pay attention to experimental facts that to study the aforementioned transition experimentally, the strong magnetic fields of $B \geq B_{c2}$ are required so that earlier such investigation was technically inaccessible. Here B_{c2} is the critical magnetic field destroying the superconductivity. Basing on recent experimental facts [10, 11], we show that in spite of very different microscopic nature of HTSC, HF metals and 2D ^{3}He, their behavior belong to universal behavior of strongly correlated Fermi systems. We employ the fermion condensation theory, based on fermion-condensation quantum phase transition (FCQPT) [5, 6, 12–15] which is able to demonstrate that the physics underlying the field-induced reentrance of LFL behavior, is the same for both HTSC, HF metals and 2D ^{3}He. We demonstrate that there is at least one quantum phase transition inside the superconducting dome, and this transition is FCQPT. We also show that there is a relationship between the critical fields B_{c2} and B_{c0} so that $B_{c2} \gtrsim B_{c0}$.

16.2 General Properties of Heavy-Fermion Metals

We have shown earlier (see, e.g., Chap. 13) that without loss of generality, to study the abovementioned universal behavior, it is sufficient to use the simplest possible model of a homogeneous heavy electron (fermion) liquid. This permits not only to better reveal the physical nature of observed effects, but to avoid unnecessary complications related to microscopic features, like crystalline structure, defects and impurities, etc of specific substances.

We consider HF liquid at $T = 0$ characterized by the effective mass M^*. Upon applying well-known Landau equation, we can relate M^* to the bare electron mass M [16, 17]

$$\frac{M^*}{M} = \frac{1}{1 - N_0 F^1(x)/3}. \tag{16.1}$$

Here, N_0 is the density of states of a free electron gas, $x = p_F^3/3\pi^2$ is the number density, p_F is Fermi momentum, and $F^1(x)$ is the p-wave component of Landau interaction amplitude F. When at some critical point $x = x_c$, $F^1(x)$ achieves certain threshold value, the denominator in (16.1) tends to zero so that the effective mass diverges at $T = 0$ and the system undergoes FCQPT. The leading term of this divergence is of the form

$$\frac{M^*(x)}{M} = \alpha_1 + \frac{\alpha_2}{x - x_c}, \tag{16.2}$$

where α_1 and α_2 are constants. At $x < x_c$ the FC takes place. The essence of this phenomenon is that at $x < x_c$ the effective mass (16.2) becomes negative signifying that the corresponding state is physically meaningless. To avoid such a state, the system reconstructs its quasiparticle occupation number $n(\mathbf{p})$ and topological structure so as to minimize its ground state energy E [12–14, 18, 19]

$$\frac{\delta E}{\delta n(\mathbf{p})} = \mu, \tag{16.3}$$

here μ is a chemical potential. The main result of such reconstruction is that instead of Fermi step, we have $0 < n(p) < 1$ in certain range of momenta $p_i \leq p \leq p_f$. Accordingly, in the above momenta interval, the quasiparticle spectrum is $\varepsilon(p) = \mu$, see Fig. 16.2 for details of its modification. Due to the above peculiarities of the $n(\mathbf{p})$ function, at $T = 0$ FC state is characterized by the superconducting order parameter $\kappa(\mathbf{p}) = \sqrt{n(\mathbf{p})(1 - n(\mathbf{p}))}$. This means that if the electron system with FC has pairing interaction with coupling constant λ, it exhibits superconductivity with finite value of the superconducting gap Δ, since $\Delta \propto \lambda$ in a weak coupling limit. As it is shown in Chap. 6, this linear dependence is also a peculiarity of FC state and substitutes well-known BCS relation $\Delta \propto \exp(-1/\lambda)$, see, e.g., [20], for the systems with FC [12, 13, 15, 21].

Let us consider the action of external magnetic field on HF liquid in FC phase. Assume now that λ is infinitely small. Any infinitesimal magnetic field $B \neq 0$ (better to say, $B \geq B_{c0}$) destroys both superconductivity and FC state, splitting it into the Landau levels. The simple qualitative arguments can be used to guess what happens to FC state in this case. On one side, the energy gain from FC state destruction is $\Delta E_B \propto B^2$ (see above) and tends to zero as $B \to 0$. On the other side, $n(p)$ in the interval $p_i \leq p \leq p_f$ gives a finite energy gain as compared to the ground state energy of a normal Fermi liquid [15]. It turns out that the state with largest possible energy gain is formed by a multi-connected Fermi surface, so that the smooth function $n(p)$ is replaced in the interval $p_i \leq p \leq p_f$ by the set of rectangular blocks of unit height, formed from Heavyside step functions [19, 22, 23]. In this state the system demonstrates LFL behavior, see Sect. 6.3.1, while the effective mass strongly depends on magnetic field,

$$M^*(B) \propto \frac{1}{\sqrt{B - B_{c0}}}. \tag{16.4}$$

Here B_{c0} is the critical magnetic field driving corresponding QCP toward $T = 0$. In some cases, for example, in HF metal $CeRu_2Si_2$, $B_{c0} = 0$, see e.g., [24]. In our model B_{c0} is to be taken as a parameter.

At elevated temperatures, the system transits from the LFL to NFL regime exhibiting the low-temperature universal behavior independent of its magnetic ground state, composition, dimensionality, (2D or 3D) and even nature of constituent Fermi particles which may be either electrons or ^{3}He atoms [7, 9]. To check, whether the quasiparticles are present in the systems in the transition regime, we use the results of measurements of heat capacity C, entropy S and magnetic susceptibility χ. If these results can be fitted by the well-known relations from Fermi liquid theory $C/T = \gamma_0 \propto S/T \propto \chi \propto M^*$, then quasiparticles define the system properties in the transition regime.

The temperature and magnetic field dependencies of the effective mass $M^*(T, B)$ as system approaches FCQPT are analyzed in Chap. 7. It follows from Fig. 16.1 that in contrast to the standard paradigm of quasiparticles the effective mass strongly depends on temperature, revealing three different regimes at growing temperature. At the lowest temperatures we have the LFL regime. Then the system enters the transition regime: $M^*_N(T_N)$ grows, reaching its maximum $M^*_N = 1$ at $T = T_M$, $(T_N = 1)$, with subsequent diminishing. Near temperatures $T_N \geq 1$ the last "traces" of LFL regime disappear and the NFL state takes place, manifesting itself in decreasing of M^*_N as $T_N^{-2/3}$ and then as

$$M^*_N(T_N) \propto \frac{1}{\sqrt{T_N}}. \tag{16.5}$$

These regimes are reported in the inset to Fig. 16.1. As it follows from Fig. 16.1, M^* reaches the maximum M^*_M at some temperature T_M. Since there is no external

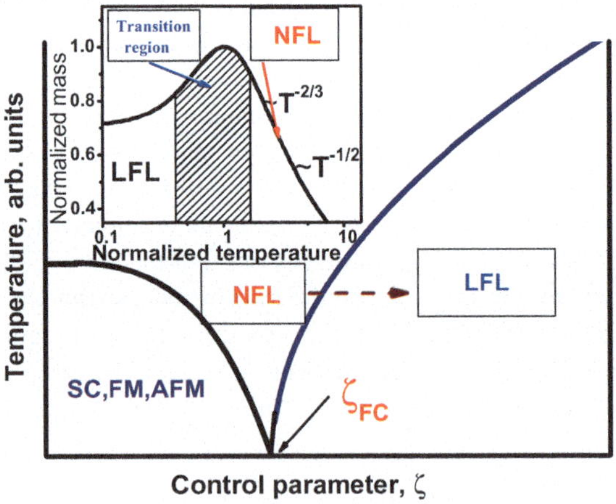

Fig. 16.1 Schematic phase diagram of HF metal. Control parameter ζ represents doping x, magnetic field B, pressure P, etc. ζ_{FC} denotes the point of effective mass divergence. The horizontal arrow shows LFL-NFL transitions at fixed ζ. At $\zeta < \zeta_{FC}$ the system can be in a superconducting (SC), ferromagnetic (FM) or antiferromagnetic (AFM) states. The inset shows a schematic plot of the normalized effective mass versus the normalized temperature. Transition regime, where M^*_N reaches its maximum, is shown by the hatched area

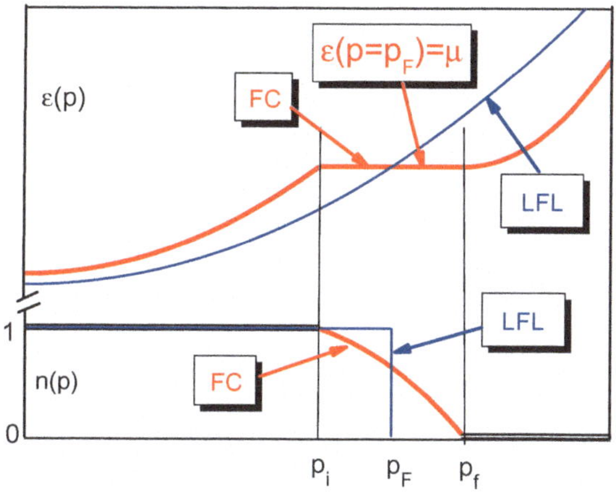

Fig. 16.2 Schematic plot of the quasiparticle occupation number $n(p)$ and spectrum $\varepsilon(p)$ in the LFL and FC states at $T = 0$. In the LFL state $n(p)$ has the jump at $p = p_F$, while $\varepsilon(p)$ corresponds to $M^* > 0$. In the FC state, function $n(p)$ obeys the relations $n(p \leq p_i) = 1, n(p_i < p < p_f) < 1$ and $n(p \geq p_f) = 0$, while $\varepsilon(p_i < p < p_f) = \mu$. Fermi momentum p_F satisfies the condition $p_i < p_F < p_f$

physical scales near FCQPT point, the normalization of both M^* and T by internal parameters M^*_M and T_M immediately reveals the common physical nature of the above thermodynamic functions which we use to extract the effective mass. The normalized effective mass extracted from measurements on the HF metals $YbRh_2(Si_{0.95}Ge_{0.05})_2$ [25, 26], $CeRu_2Si_2$ [24], $CePd_{1-x}Rh_x$ [27], and 2D ^3He [8] along with our theoretical solid curve (also shown in the inset) is reported in Fig. 16.3. It is seen that above normalization of experimental data yields the merging of multiple curves into single one, thus demonstrating a universal scaling behavior [5–7, 9, 28]. It is also seen that the universal behavior of the effective mass given by our theoretical curve agrees well with experimental data (Fig. 16.2).

It is shown in Fig. 16.3 that at $T/T_M = T_N \leq 1$ the T-dependence of the effective mass is weak. This means that the T_M point can be regarded as a crossover between LFL and NFL regimes. Since magnetic field enters the Landau equation as $\mu_B B/T$, we have

$$T^*(B) = a_1 + a_2 B \simeq T_M \sim \mu_B(B - B_{c0}), \tag{16.6}$$

where $T^*(B)$ is the crossover temperature, μ_B is Bohr magneton, a_1 and a_2 are constants. The crossover does not represent a really a phase transition. It necessarily is broad, very much depending on the criteria for determination of the point of such a crossover, as it is seen from the inset to Fig. 16.3. As usually, the temperature $T^*(B)$ is extracted from the field dependence of charge transport, for example, from the resistivity $\rho(T) = \rho_0 + A(B)T^2$ with ρ_0 is the temperature independent part and $A(B)$ is the LFL coefficient. The crossover takes place at temperatures where the resistance starts to deviate from the LFL T^2 behavior, see, e.g., [10]. To verify (16.5) and illus-

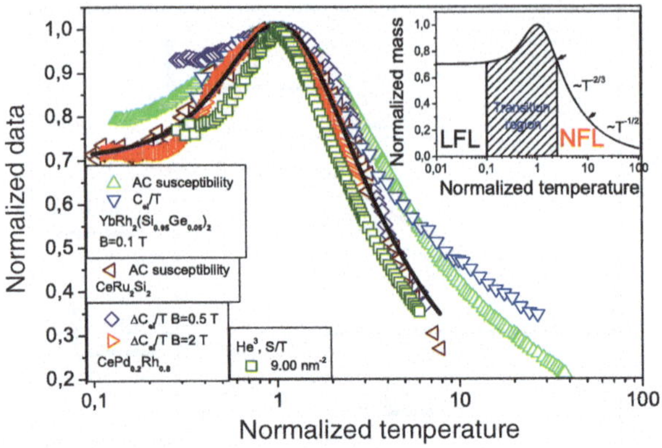

Fig. 16.3 The universal behavior of $M^*_N(T_N)$, extracted from measurements of different thermodynamic quantities, as shown in the legend. The AC susceptibility, $\chi_{AC}(T, B)$, is taken for $YbRh_2(Si_{0.95}Ge_{0.05})_2$ and $CeRu_2Si_2$ [24, 25], the heat capacity divided by temperature, C/T, is taken for $YbRh_2(Si_{0.95}Ge_{0.05})_2$ and $CePd_{0.2}Rh_{0.8}$ [26, 27] and entropy divided by temperature, S/T, for 2D ^{3}He is taken from [8]. The solid curve depicts the theoretical universal behavior of M^*_N determined by (7.8). The inset shows normalized effective mass $M^*_N(T) = M^*(T)/M^*_M$ (M^*_M is the maximal value of the effective mass at $T = T_M$) versus the normalized temperature $T_N = T/T_M$. The hatched area outlines the transition regime. Several regions are shown as explained in the text

trate the transition from the LFL behavior to NFL, we use measurements of $\chi_{AC}(T)$ in $CeRu_2Si_2$ at magnetic field $B = 0.02$ mT at which this HF metal demonstrates the NFL behavior down to lowest temperatures [24]. Indeed, in this case we expect that LFL regime starts to form at temperatures lower than $T_M \sim \mu_B B \sim 0.01$ mK as it follows from (16.6). It is shown in Fig. 16.4 that (16.5) gives good description of the facts in the extremely wide range of temperatures: the susceptibility χ_{AC} as a function of T is not a constant upon cooling, as would be for a Fermi liquid, but shows a $1/\sqrt{T}$ divergence over more than three decades in temperature. The inset of Fig. 16.4 exhibits a fit for M^*_N extracted from measurements of $\chi_{AC}(T)$ at different magnetic fields, clearly indicating the change from LFL behavior at $T_N < 1$ to NFL one at $T_N > 1$, while the function given by (7.8) represents a good approximation for M^*_N.

16.3 Common Field-Induced Quantum Critical Point

Let us now consider the $B - T$ phase diagram of the HTSC substance $Tl_2Ba_2CuO_{6+x}$ shown in Fig. 16.5. The substance is a superconductor with T_c from 15 K to 93 K, depending upon the oxygen content [10]. In Fig. 16.5 open squares and solid circles show the experimental values of the crossover temperature from the LFL to NFL regimes [10]. The solid line shows our fit (16.6) with $B_{c0} = 6$ T that is in good agreement with $B_{c0} = 5.8$ T obtained from the field dependence of the

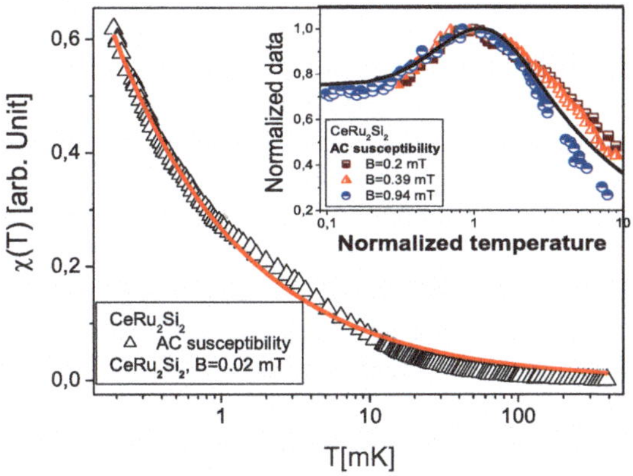

Fig. 16.4 Temperature dependence of the AC susceptibility χ_{AC} for $CeRu_2Si_2$. The solid curve is a fit for the data shown by the triangles at $B = 0.02\,\mathrm{mT}$ [24] and represented by the function $\chi(T) = a/\sqrt{T}$ given by (16.5) with a being a fitting parameter. Inset shows the normalized effective mass versus normalized temperature T_N extracted from χ_{AC} measured at different fields as indicated in the inset [24]. The solid curve traces the universal behavior of $M^*_N(T_N)$ determined by (7.8). Parameters c_1 and c_2 are adjusted to fit the average behavior of the normalized effective mass M^*_N

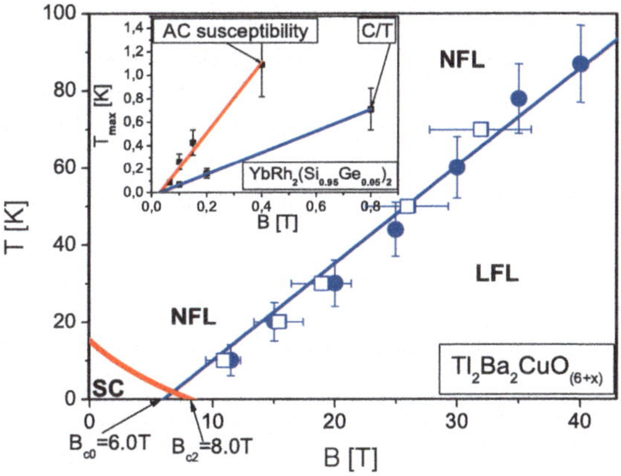

Fig. 16.5 $B - T$ phase diagram of superconductor $Tl_2Ba_2CuO_{6+x}$. The crossover (from LFL to NFL regime) line $T^*(B)$ is given by (16.6). Open squares and solid circles are experimental values [10]. Thick red line represents the boundary between the superconducting and normal phases. Arrows near the bottom left corner indicate the critical magnetic field B_{c2} destroying the superconductivity and the critical field B_{c0}. Inset reports the peak temperatures $T_{\max}(B)$, extracted from measurements of C/T and χ_{AC} on $YbRh_2(Si_{0.95}Ge_{0.05})_2$ [25, 26] and approximated by straight lines (16.6). The lines intersect at $B \simeq 0.03$ T

charge transport [10]. As it is shown in Fig. 16.5, the linear behavior agrees well with experimental data [10]. The peak temperatures T_{max} shown in the inset to Fig. 16.5, depict the maxima of $C(T)/T$ and $\chi_{AC}(T)$ measured on $YbRh_2(Si_{0.95}Ge_{0.05})_2$ [25, 26]. As it follows from (16.6), T_{max} shifts to higher values with increase of the applied magnetic field. It is seen that both functions can be represented by straight lines intersecting at $B \simeq 0.03$ T. This observation is in good agreement with experiments [25, 26].

It is seen from Fig. 16.5 that the critical field $B_{c2} = 8$ T destroying the superconductivity is close to $B_{c0} = 6$ T. Let us show that this is more than a simple coincidence, and $B_{c2} \gtrsim B_{c0}$. Indeed, at $B > B_{c0}$ and low temperatures $T < T^*(B)$, the system is in its LFL state. The superconductivity is then destroyed since the superconducting gap is exponentially small as we have seen above. At the same time, there is FC state at $B < B_{c0}$ and this low-field phase has large prerequisites toward superconductivity as in this case the gap is a linear function of the coupling constant. We note that this is exactly the case in $CeCoIn_5$ where $B_{c0} \simeq B_{c2} \simeq 5$ T [29], while the application of pressure makes $B_{c2} > B_{c0}$ [30]. On the other hand, if the superconducting coupling constant is rather weak, then antiferromagnetic order wins a competition. As a result, $B_{c2} = 0$, while B_{c0} can be finite as in $YbRh_2Si_2$ and $YbRh_2(Si_{0.95}Ge_{0.05})_2$ [25, 31], see also Chap. 15.

Upon comparing the phase diagram of $CeCoIn_5$ with that of $Tl_2Ba_2CuO_{6+x}$, it is possible to conclude that they are similar in many respects. Further, we note that the superconducting boundary line $B_{c2}(T)$ at lowering temperatures acquires a step, i.e., the corresponding phase transition becomes first order [32, 33]. This permits us to speculate that the same may be true for $Tl_2Ba_2CuO_{6+x}$. We expect that in the NFL state the tunneling conductivity is an asymmetrical function of the applied voltage, while it becomes symmetrical at the application of elevated magnetic fields [34] when $Tl_2Ba_2CuO_{6+x}$ transits to the LFL behavior, see Chaps. 18 and 19.

Now we consider the field-induced reentrance of LFL behaviorin$Tl_2Ba_2CuO_{6+x}$ at $B \geq B_{c2}$. The LFL regime is characterized by the temperature dependence of the resistivity, $\rho(T) = \rho_0 + A(B)T^2$, as also seen above. The A coefficient, being proportional to the quasiparticle-quasiparticle scattering cross section, is found to be $A \propto (M^*(B))^2$ [15, 31]. With respect to (16.4), this implies that

$$A(B) \simeq A_0 + \frac{D}{B - B_{c0}}, \tag{16.7}$$

where A_0 and D are parameters. It is pertinent to note that Kadowaki-Woods ratio [35], $K = A/\gamma_0^2$, is constant within the FC theory as it follows from (16.4) and (16.7). It follows from (16.7) that it is impossible to observe the relatively high values of $A(B)$ since in our case $B_{c2} > B_{c0}$. We note that (16.7) is applicable when the superconductivity is destroyed by the application of magnetic field, otherwise the effective mass is finite being given by the relation $M^* \propto 1/\Delta_1$, where Δ_1 is the maximum value of the superconducting gap [21]. Therefore, as was mentioned above, in high T_c superconductors QCP is poorly accessible to experimental observations being "*hidden in superconductivity*". Nonetheless, experimental facts [10] demonstrate that it is possible to study CQP by exploring "shadows" of this QCP.

Figure 16.6 reports the fit of our theoretical dependence (16.7) to the experimental data for two different classes of substances: HF metal $YbRh_2Si_2$ (left panel) and HTSC $Tl_2Ba_2CuO_{6+x}$ (right panel). The different scale of fields is clearly seen as well as good coincidence with theoretical dependence (16.7). This means that the physics underlying the field-induced reentrance of LFL behavior is the same for both classes of substances. To further corroborate this point, we rewrite (16.7) in reduced variables A/A_0 and B/B_{c0}. Such rewriting immediately reveals the universal nature of the behavior of these two substances—both of them are driven to common QCP related to FC and induced by the application of magnetic field. As a result, (16.7) takes the form

$$\frac{A(B)}{A_0} \simeq 1 + \frac{D_N}{B/B_{c0} - 1}, \tag{16.8}$$

where $D_N = D/(A_0 B_{c0})$ is a constant. Fromm (16.8), it is seen that upon applying the scaling for both coefficients $A(B)$ for $Tl_2Ba_2CuO_{6+x}$ and $A(B)$ for $YbRh_2Si_2$ are reduced to a function depending on the single variable B/B_{c0} thus demonstrat-

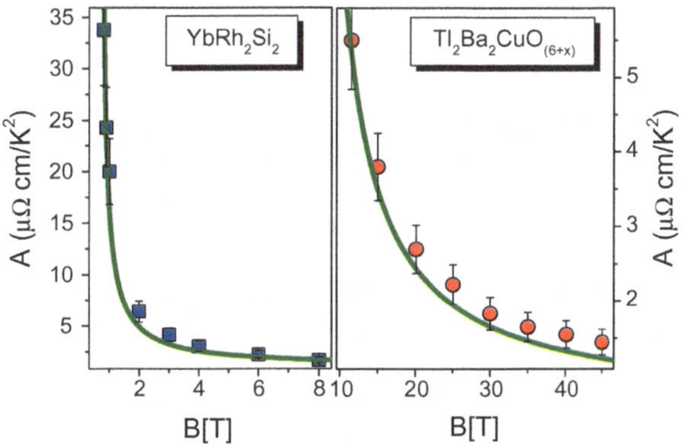

Fig. 16.6 The charge transport coefficient $A(B)$ as a function of magnetic field B obtained in measurements on $YbRh_2Si_2$ [31] and $Tl_2Ba_2CuO_{6+x}$ [10]. The different field scales are clearly seen

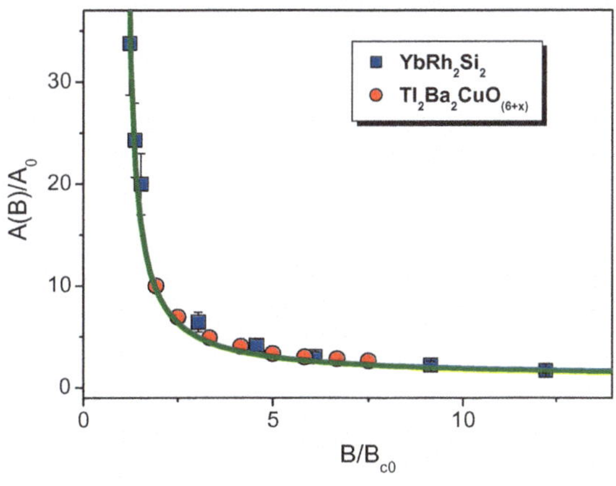

Fig. 16.7 Normalized coefficient $A(B)/A_0 \simeq 1 + D_N/(y-1)$ as a function of normalized magnetic field $y = B/B_{c0}$ shown by squares for $YbRh_2Si_2$ and by circles for $Tl_2Ba_2CuO_{6+x}$. D_N is the only fitting parameter

ing universal behavior. To support (16.8), we replot both dependencies in reduced variables A/A_0 and B/B_{c0} on Fig. 16.7. Such replotting immediately reveals the universal nature of the behavior of these two substances. It is shown in Fig. 16.7 that close to magnetic QCP there is no external physical scales. Therefore the normalization by internal scales A_0 and B_{c0} immediately reveals the common physical nature of these substances behavior.

16.4 Summary

Being based both on theoretical study and experimental facts, our consideration of very different strongly correlated Fermi systems such as high-temperature superconductors and heavy-fermion compounds clearly demonstrate their generic family resemblance. It follows from our study that there is at least one quantum phase transition inside the superconducting dome, and this transition is the fermion-condensation quantum phase transition. Both high-temperature superconductors and heavy-fermion compounds belong to the new state of matter.

References

1. P. Coleman, A.J. Schofield, Quantum criticality. Nature **433**, 226–229 (2005)
2. H.V. Löhneysen, A. Rosch, M. Vojta, P. Wölfle, Fermi-liquid instabilities at magnetic quantum phase transitions. Rev. Mod. Phys. **79**, 1015–1076 (2007)
3. P. Gegenwart, Q. Si, F. Steglich, Quantum criticality in heavy-fermion metals. Nat. Phys. **4**, 186–197 (2008)
4. S. Sachdev, Quantum magnetism and criticality. Nat. Phys. **4**, 173–185 (2008)
5. V.R. Shaginyan, M.Y. Amusia, A.Z. Msezane, K.G. Popov, Phys. Rep. **492**, 31 (2010)
6. M.Y. Amusia, K.G. Popov, V.R. Shaginyan, W.A. Stephanowich, *Theory of Heavy-Fermion Compounds*. Springer Series in Solid-State Sciences, vol. 182 (Springer, Berlin, 2015)
7. V.R. Shaginyan, K.G. Popov, V.A. Stephanovich, Universal low-temperature behavior of the $CePd_{1-x}Rh_x$ ferromagnet. Europhys. Lett. **79**, 47001-p1–47001-p6 (2007)
8. M. Neumann, J. Nyéki, J. Saunders, Bilayer 3He: a simple two-dimensional heavy-fermion system with quantum criticality. Science **317**, 1356–1359 (2007)
9. V.R. Shaginyan, A.Z. Msezane, K.G. Popov, V.A. Stephanovich, Universal behavior of two-dimensional ^{3}He at low temperatures. Phys. Rev. Lett. **100**, 096406-1–096406-4 (2008)
10. T. Shibauchi, L. Krusin-Elbaum, M. Hasegawa, Y. Kasahara, R. Okazaki, Y. Matsuda, Field-induced quantum critical route to a Fermi liquid in high-temperature superconductors. Proc. Natl. Acad. Sci. USA **105**, 7120–7126 (2008)
11. A.P. Mackenzie et al., Normal-state magnetotransport in superconducting $Tl_2Ba_2CuO_{6+\delta}$ to millikelvin temperatures. Phys. Rev. B **53**, 5848–5855 (1996)
12. V.A. Khodel, V.R. Shaginyan, Superfluidity in system with fermion condensate. JETP Lett. **51**, 553–555 (1990)
13. V.A. Khodel, V.R. Shaginyan, V.V. Khodel, New approach in the microscopic Fermi systems theory. Phys. Rep. **249**, 1–134 (1994)
14. G.E. Volovik, Quantum phase transitions from topology in momentum space. Lecture Notes in Physics, vol. 718 (2007), pp. 31–120

15. V.R. Shaginyan, M.Y. Amusia, K.G. Popov, Universal behavior of strongly correlated Fermi systems. Phys.-Uspekhi **50**, 563–593 (2007)
16. L.D. Landau, Fermi liquid theory. Sov. Phys. JETP **3**, 920–929 (1956); E.M. Lifshitz, L.P. Pitaevskii, *Statistical Physics*, Part 2 (Butterworth-Heinemann, Oxford, 1999)
17. M. Pfitzner, P. Wölfle, Quasiparticle interaction in a nearly localized Fermi liquid: application to ^{3}He and heavy-fermion systems. Phys. Rev. B **33**, 2003–2006 (1986)
18. V.A. Khodel, Two scenarios of the quantum critical points. JETP Lett. **86**, 832–838 (2007)
19. V.A. Khodel, J.W. Clark, M.V. Zverev, Topology of the Fermi surface beyond the quantum critical point. Phys. Rev. B **78**, 075120–1–075120-18 (2008)
20. J. Bardeen, L. Cooper, J.R. Shrieffer, Theory of superconductivity. Phys. Rev. **108**, 1175–1204 (1957)
21. M.Y. Amusia, V.R. Shaginyan, Quasiparticle picture of high-temperature superconductors in the frame of a Fermi liquid with the fermion condensate. Phys. Rev. B **63**, 224507-1–224507-6 (2001)
22. M.V. Zverev, M.J. Baldo, The multi-connected momentum distribution and fermion condensation. J. Phys. Condens. Matter **11**, 2059–2071 (1999)
23. Y.G. Pogorelov, V.R. Shaginyan, Transition from non-fermi liquid behavior to Landau-Fermi liquid behavior induced by magnetic fields. JETP Lett. **76**, 532–536 (2002)
24. D. Takahashi et al., AC susceptibility and static magnetization measurements of $CeRu_2Si_2$ at small magnetic fields and ultralow temperatures. Phys. Rev. B **67**, 180407-1–180407-5 (2003)
25. P. Gegenwart et.al., Ferromagnetic quantum critical fluctuations in $YbRh_2$ $(Si_{0.95}Ge_{0.05})_2$. Phys. Rev. Lett. **94**, 076402-1–076402-4 (2005)
26. J. Custers et al., The break-up of heavy electrons at a quantum critical point. Nature **424**, 524–526 (2003)
27. A.P. Pikul et al., Non-Fermi-liquid behaviour close to the disappearance of ferromagnetism in $CePd_{1-x}$ Rh_x. J. Phys. Condens. Matter **18**, L535–L542 (2006)
28. J.W. Clark, V.A. Khodel, M.V. Zverev, Anomalous low-temperature behavior of strongly correlated Fermi systems. Phys. Rev. B **71**, 012401-1–012401-5 (2005)
29. J. Paglione et al., Field-induced quantum critical point in $CeCoIn_5$. Phys. Rev. Lett. **91**, 246405-1–246405-4 (2003)
30. F. Ronning et al., Pressure study of quantum criticality in $CeCoIn_5$. Phys. Rev. B **73**, 064519-1–064519-5 (2006)
31. P. Gegenwart et al., Magnetic-field induced quantum critical point in $YbRh_2$ Si_2. Phys. Rev. Lett. **89**, 056402-1–056402-5 (2002)
32. A. Bianchi et al., First-order superconducting phase transition in $CeCoIn_5$. Phys. Rev. Lett. **89**, 137002-1–137002-5 (2002)
33. V.R. Shaginyan, A.Z. Msezane, V.A. Stephanovich, E.V. Kirichenko, Quasiparticles and quantum phase transition in universal low-temperature properties of heavy-fermion metals. Europhys. Lett. **76**, 898–904 (2006)
34. V.R. Shaginyan, K.G. Popov, Asymmetric tunneling, Andreev reflection and dynamic conductance spectra in strongly correlated metals. Phys. Lett. A **361**, 406–412 (2007)
35. K. Kadowaki, S.B. Woods, Universal relationship of the resistivity and specific heat in heavy-fermion compounds. Solid State Commun. **58**, 507–509 (1986)

Chapter 17
Quasi-classical Physics Within Quantum Criticality in HF Compounds

Abstract In this chapter, we explore how the fermion condensation paves the road for quasi-classical physics in HF compounds. This means simply that systems with FC admit partly the quasi-classical description of their thermodynamic and transport properties. This, in turn, simplifies a lot not only of their description but permits to gain more insights both in the puzzling NFL physics of HF compounds and of the physics of FC itself. The quasi-classical physics starts to be applicable near FCQPT, at which FC generates flat bands and quantum criticality, and makes the density of electron states in strongly correlated metals diverge. As we shall see, due to the formation of flat bands the strongly correlated metals exhibit the classical properties of elemental ones like copper, silver, aluminum, etc, since the strongly correlated metals demonstrate the quasi-classical behavior at low temperatures. We also show that due to the fermion-condensation HF metals exhibits the universal behavior of their resistivity $\rho(T) \propto T$ that is specific to the new state of matter.

17.1 Second Wind of the Dulong-Petit Law at a Quantum-Critical Point

We show that in the systems with quantum criticality like 2D ^{3}He, the group velocity of transverse zero sound depends strongly on temperature. It is this dependence that grants the Dulong–Petit law a "second wind." In 2D liquid ^{3}He, the specific heat becomes temperature-independent at some characteristic temperature of a few mK. In the same way, the heat capacity of the HF metal $YbRh_2Si_2$ contains the temperature-independent term.

Almost two hundred years ago, Pierre-Louis Dulong and Alexis-Thérèse Petit [1] discovered experimentally that the specific heat $C(T)$ of a crystal is close to constant, being independent of the temperature T. This behavior, attributed to lattice vibrations—i.e., phonons—is known as the DP law. Later, Ludwig Boltzmann [2] reproduced the results of Dulong and Petit quantitatively in terms of the equipartition principle. However, subsequent measurements at low temperatures demonstrated that $C(T)$ drops rapidly at $T \to 0$, in sharp contrast to Boltzmann's theory. In 1912, Peter Debye [3] developed a quantum theory for evaluation of the phonon part of

M. Amusia and V. Shaginyan, *Strongly Correlated Fermi Systems*, Springer Tracts in Modern Physics 283,
https://doi.org/10.1007/978-3-030-50359-8_17

the specific heat of solids, correctly explaining the empirical behavior $C(T) \sim T^3$ of the lattice component as $T \to 0$. In the Debye theory, the T-independence of $C(T)$ is recovered at $T \geq T_D$, where T_D is a critical temperature corresponding to the saturation of the phonon spectrum. With the creation of the Landau theory of quantum liquids [4], that predicted a linear-in-T dependence of $C(T)$ for the specific heat that is contributed by itinerant fermions, our understanding of the low-temperature thermodynamic properties of solids and liquids seemed to be firmly established. However, measurements [5, 6] of the specific heat of two-dimensional (2D) ^{3}He, realized as ^{3}He films absorbed on graphite preplated with a ^{4}He bilayer, reveal behavior essentially deviating from the established general view that calls for a new understanding of the low-temperature thermodynamics of strongly correlated many-fermion systems [7].

Owing to its status as a fundamental example of strongly interacting many-fermion systems, liquid ^{3}He remains a valuable touchstone for low-temperature condensed matter physics. Interest to ^{3}He physics has been driven by the observation of non-Fermi liquid (NFL) behavior of dense ^{3}He films at the lowest temperatures $T \simeq 1$ mK reached experimentally [5, 6, 8–12]. In particular, measurements of the specific heat $C(T)$ in the 2D ^{3}He system show the presence of a term β tending to a finite value as $T \to 0$. Such behavior contrasts sharply with that of its counterpart, three-dimensional (3D) liquid ^{3}He. Note, that here we do not consider superfluid phases of ^{3}He.

In looking for the origin of the anomalous contribution β remaining in $C(T)$ at the lowest temperatures attained, it is instructive to examine the schematic low-T phase diagram of 2D liquid ^{3}He shown in Fig. 17.1. The essential features of this picture are

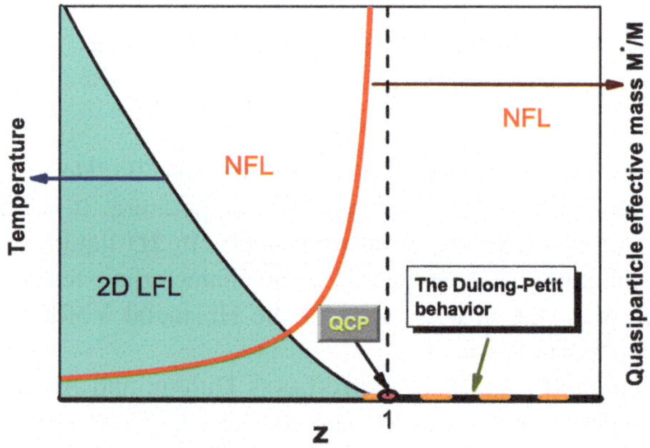

Fig. 17.1 Phase diagram of the 2D liquid ^{3}He system. The region defined by $z = \rho/\rho_{FC} < 1$ is divided into LFL and NFL domains separated by a solid line. The dependence $M^*(z) \propto (1-z)^{-1}$ is shown by the solid line approaching the dashed asymptote, thus depicting the divergence of the effective mass at the quantum-critical point ($z = 1, T = 0$) indicated by the arrow. In the region $z \gtrsim 1$, the fermion condensate (FC) sets in and Dulong-Petit behavior of the specific heat is realized for the strongly correlated quantum many-fermion system (as represented by the dash horizontal line at $T = 0$)

documented in the cited experiments on ^{3}He films. The effective coupling parameter is represented by $z = \rho/\rho_\infty$, where ρ is the number density of the system and ρ_∞ is the critical density at which a quantum-critical point (QCP) occurs. This QCP is associated with a divergence of the effective mass $M^*(z)$, portrayed in Fig. 17.1 by the curve (in red on line) that approaches the dashed asymptote at $z = 1$. The $T - z$ phase plane is divided into regions of 2D LFL and NFL behavior. The part of the diagram where $z < 1$ consists of a FL region at lower T and a NFL region at higher T, separated by a solid curve. The regime where $z \gtrsim 1$ belongs to a NFL state with specific heat taking a finite value $\beta(\rho)$ at very low temperatures. The physical source of this excess heat capacity has not been established with certainty, although it is supposed that the β anomaly is related to peculiarities of the substrate on which the ^{3}He film is placed.

As indicated above, the most challenging feature of the NFL behavior of liquid ^{3}He films involves the specific heat $C(T)$. According to the Landau theory, $C(T)$ varies linearly with T, and at low film densities the experimental behavior of the specific heat of 2D liquid ^{3}He is in agreement with LFL theory. However, for relatively dense ^{3}He films, this agreement is found to hold only at sufficiently *high* temperatures. If T is lowered down to millikelvin region, the function $C(T)$ ceases to fall toward zero and becomes flat [5, 6, 12].

The common explanation [5, 6, 13] of the observed $C(T)$ flattening imputes the phenomenon to disorder associated with the substrate that supports the ^{3}He film. More specifically, it is considered that there exists weak heterogeneity of the substrate (namely, steps and edges on its surface), such that quasiparticles, being delocalized from it, give rise to the low-temperature feature β of the heat capacity [6].

Even if we disregard certain unjustified assumptions [13], there remains the disparate fact that the emergent constant term in $C(T)$ is of comparable order for different substrates [5, 6, 12]. Furthermore, the explanation posed in [13] implies that the departure of $C(T)$ from LFL predictions shrinks as the film density increases, since the effects of disorder are most prominent in weakly interacting systems. On the contrary, the anomaly in $C(T)$ makes it emerge in the density region where the effective mass M^* is greatly enhanced [6, 12] and the 2D liquid ^{3}He system becomes strongly correlated. This reasoning compels us to consider that the NFL behavior of $C(T)$ is an *intrinsic* feature of 2D liquid ^{3}He, which is associated with the divergence of the effective mass rather than with disorder. The flattening of the curve $C(T)$ as seen in ^{3}He films is by no means a unique phenomenon. Indeed, as expressed in the DP law, the specific heat $C(T)$ of solids remains independent of T as long as T exceeds the Debye temperature Ω_D, which is determined by the saturation of the phonon spectrum of the crystal lattice. Normally, the value of Ω_D is sufficiently high so that the DP law belongs to classical physics. However, we will show that the DP behavior of $C(T)$ can also appear at extremely low temperatures in strongly correlated Fermi systems, with zero sound playing the role of phonons.

To clarify the details of this phenomenon and calculate the specific heat $C(T)$, we evaluate the part F_B of the free energy F associated with the collective spectrum $\omega(k) = ck$, based on the standard formula

$$F_B = \int_0^\infty \frac{d\omega}{\pi} \frac{1}{e^{\omega/T} - 1} \int \mathrm{Im}\left[\ln D^{-1}(k, \omega)\right] d\upsilon, \tag{17.1}$$

where $D(k, \omega)$ is the boson propagator and $d\upsilon$ is an element of momentum space. Upon integration by parts this formula recasts into

$$F_B = T \int_0^\infty \frac{d\omega}{\pi} \ln\left(1 - e^{-\omega/T}\right) \int \mathrm{Im}\left(\frac{\partial D^{-1}(k, \omega)/\partial\omega}{D^{-1}(k, \omega)}\right) d\upsilon. \tag{17.2}$$

If the damping of the collective branch is negligible as in the case addressed here, then $D^{-1}(k, \omega) \simeq (\omega - ck)$ and $\partial D^{-1}(k, \omega)/\partial\omega \simeq 1$, while $\mathrm{Im} D^{-1}(k, \omega) \simeq \delta(\omega - ck)$, and we arrive at the textbook formula

$$F_B = T \int \ln\left(1 - e^{-ck/T}\right) \theta(\Omega_0 - ck) d\upsilon, \tag{17.3}$$

where Ω_0 is the characteristic frequency of zero sound and c is the velocity of zero sound. At $T \gg \Omega_0$, the factor $\ln(1 - e^{-\omega/T})$ reduces to $\ln(\omega/T)$, yielding the result

$$F_B(T) \propto T \ln(\Omega_0/T), \tag{17.4}$$

which, upon the double differentiation, leads to the DP law $C(T) = \mathrm{const}$. At first sight, this law has nothing to do with the situation in 2D liquid ^{3}He. Its Fermi energy ε_F^0 is around 1 K at densities where the Sommerfeld ratio $C(T)/T$ soars upward as $T \to 0$, while Ω_0 must be lower than $T \simeq 1$ mK. Indeed, in any conventional Fermi liquid, including 3D liquid ^{3}He, there is no collective degree of freedom, whose spectrum is saturated at such low ratios as Ω_0/ε_F^0.

This conclusion remains valid *assuming* 2D liquid ^{3}He is an ordinary Fermi liquid. However, as it is seen from Fig. 17.1, if the QCP is reached at $T \to 0$ and at some critical density ρ_∞ where the effective mass $M^*(\rho_\infty)$ diverges, as it does in the present case [6, 8, 9, 12, 14], the situation changes dramatically. This is demonstrated explicitly by the results of standard LFL calculations of the velocity c_t of transverse zero sound, which satisfies the equation [15, 16]

$$\frac{c_t}{2v_F} \ln \frac{c_t + v_F}{c_t - v_F} - 1 = \frac{F_1 - 6}{3F_1(c_t^2/v_F^2 - 1)}, \tag{17.5}$$

where $v_F = p_F/M^*$ is the Fermi velocity and $F_1 = p_F M^* f_1/\pi^2$ is a dimensionless version of the first Landau harmonic f_1 [15, 17]. The divergence of the effective mass M^* at the QCP implies that at the critical density determined by $f_1 p_F M/\pi^2 = 3$ [17], one has

$$c_t^2(\rho_\infty) \simeq \frac{p_F^2}{5M^*(\rho_\infty)M} \to 0, \tag{17.6}$$

whereas the sound velocity c_s remains finite in this limit [15, 16, 18].

We see then that in this case the effective mass M^* diverges, the group velocity c_t vanishes as $\sqrt{1/M^*}$. Flattening of the single-particle spectrum $\varepsilon(p)$ prevails as long as $|p - p_F|/p_F < M/M^*$, implying that the transverse mode softens only for rather small wave numbers $k \sim p_F M/M^*$. Unfortunately, the associated numerical prefactor cannot be established, rendering the estimation of $\Omega_0 \sim (p_F^2/M)\sqrt{M/M^*}$ to be uncertain. Nevertheless, one cannot exclude a significant enhancement of the Sommerfeld ratio $C(T)/T$ at $T \simeq 1$ mK due to softening of the transverse zero sound in the precritical density region.

At $T \to 0$ and densities exceeding ρ_∞, the system undergoes a cascade of topological phase transitions in which the Fermi surface acquires additional sheets [19–22]. As indicated in Fig. 17.1, LFL theory continues to hold with quasiparticle momentum distribution $n(p)$ satisfying $n^2 = n$, until a larger critical density ρ_{FC} is reached where a new phase transition, known as FCQPT, takes place [18, 22–26]. Beyond the point of FCQPT, the single-particle spectrum $\varepsilon(p)$ acquires a flat portion. The range L of momentum space adjacent to the Fermi surface, where the FC resides, depends on the difference between the effective coupling constant and its critical value. As will be seen, L is a new dimensional parameter that serves to determine the key quantity Ω_0.

At finite T, the dispersion of the FC spectrum $\varepsilon(p)$, located in the vicinity of the chemical potential, acquires a nonzero value proportional to temperature [18, 22, 27]:

$$\varepsilon(p, T) = T \ln \frac{1 - n_*(p)}{n_*(p)}, \quad p_i < p < p_f, \tag{17.7}$$

where $0 < n_*(p) < 1$ is the FC momentum distribution and p_i and p_f are the lower and upper boundaries of the FC domain in momentum space. Consequently, in the whole FC region, the FC group velocity, given by

$$v(p, T) = \frac{\partial \varepsilon(p)}{\partial p} = -T \frac{\partial n_*(p)/\partial p}{n_*(p)(1 - n_*(p))}, \quad p_i < p < p_f, \tag{17.8}$$

is proportional to T. Significantly, in the density interval $\rho_\infty < \rho < \rho_{FC}$ the formula (17.7) describes correctly the single-particle spectrum $\varepsilon(p, T)$ in case the temperature T exceeds a very low transition temperature [22]. The FC itself contributes a T-independent term to the entropy S. Hence, its contribution to the specific heat $C(T) = T\mathrm{d}S/\mathrm{d}T$ is zero. Accordingly, we focus on the zero-sound contribution to $C(T)$ in systems having a FC.

Due to the fundamental difference between the FC single-particle spectrum and that of the remainder of the Fermi liquid, a system having FC is, in fact, a two-component object. FC subsystem possesses its own set of zero-sound modes, whose wave numbers are relatively small, not exceeding $L = (p_f - p_i) > 0$. The mode of prime interest for our analysis is that of transverse zero sound. As may be seen by comparison of expressions (17.6) and (17.8), its velocity c_t *depends on temperature* so as to vanish like $\sqrt{T}$ as $T \to 0$.

To verify the latter property explicitly, we observe first that for systems with a rather small proportion of FC, evaluation of the spectrum of collective excitations may be performed by employing the familiar LFL kinetic equation [15, 28]

$$(\omega - \mathbf{kv})\,\delta n(\mathbf{p}) = -\mathbf{kn}\frac{\partial n(p)}{\partial p}\int \mathscr{F}(\mathbf{p},\mathbf{p}_1)\delta n(\mathbf{p}_1)\mathrm{d}\upsilon_1. \tag{17.9}$$

Focusing on transverse zero sound in 2D liquid ^{3}He, one needs to retain only the term in the Landau interaction $\mathscr{F}$ proportional to the first harmonic f_1. To proceed further we make the usual identification $(c_t - \cos\theta)\delta n(\mathbf{p}) = (\partial n(p)/\partial p)\,\phi(\mathbf{n})$, where $\cos\theta = \mathbf{kv}/kv$. Equation (17.9) then becomes

$$\phi(\theta) = -f_1 p_F \cos\theta \int \cos\chi \frac{\partial n(p_1)/\partial p_1}{c_t - v(T)\cos\theta_1}\phi(\theta_1)\frac{\mathrm{d}p_1 d\theta_1}{(2\pi)^2}, \tag{17.10}$$

where $\cos\chi = \cos\theta\cos\theta_1 + \sin\theta\sin\theta_1$, while $v(T)$ is given by (17.8). The solution describing transverse zero sound is $\phi(\mathbf{n}) \sim \sin\theta\cos\theta$.

We see immediately that $c_t \gg v(T) \sim T$. Therefore, for the transverse sound in question there is no Landau damping. In such a situation, we come to the simple result

$$c_t^2 = -\frac{p_F}{5M}\int \frac{\partial n(p)}{\partial p} v(p,T)\mathrm{d}p \tag{17.11}$$

upon keeping just the leading relevant term $v(T)\cos\theta/c_t^2$ of the expansion of $(c_t - v(T)\cos\theta)^{-1}$ and executing straightforward transformations. Factoring out an average value of the group velocity $v(p,T) \propto T/p_F$, we arrive at the stated above relation

$$c_t(k) \simeq \sqrt{\frac{T}{M}} \tag{17.12}$$

valid for wave numbers k not exceeding the FC range L. Transverse sound can propagate in the other, non-condensed subsystem of 2D liquid ^{3}He consisting of quasiparticles with normal dispersion [15, 16, 28]. However, its group velocity is T-independent, so the corresponding contribution to the free energy is irrelevant.

As noted above, the characteristic wave number of the soft transverse zero-sound mode is given by the FC range $L(\rho) = p_f - p_i$, treated here as an input parameter. The key quantity Ω_0 is therefore estimated as $\Omega_0 \simeq k_{\max} c_t$, where $k_{\max}$ is the maximum value of the zero-sound momentum at which zero sound still exists. In our case, the zero sound is associated directly with the FC. Hence, $k_{\max} \simeq \sqrt{Lp_F}$ and we have

$$\Omega_0 \simeq k_{\max} c_t \simeq \sqrt{\frac{TLp_F}{M}}. \tag{17.13}$$

As long as the inequality $Lp_F/M < T$ holds (or, equivalently, the relation $T/\varepsilon_F^0 > L/p_F$ is valid), the ratio Ω_0/T is small, and we have the DP result $C(T) = \text{const}$.

Then, in spite of the low temperature, C behaves as if the system was situated in the classical limit rather than at the QCP. Such a behavior is determined by the fact that the system contains a macroscopic subsystem with heavy quasiparticles. As the temperature ultimately goes down to zero at the fixed density ρ, the inequality $Lp_F/M < T$ eventually fails, and the quantum regime is restored and the dominant contribution to C comes from the "normal" fermions. In other words, there exists an extremely low temperature T_0 below which the usual LFL behavior of zero sound is recovered.

The constant term in $C(T)$ can be evaluated in closed form in terms of the FC range L. Upon inserting $\omega_t(k) = c_t k$ into (17.3) and integrating, the T-independent term in the specific heat becomes

$$\frac{C}{N} \simeq \frac{Lp_F}{8\pi\rho}, \tag{17.14}$$

where N is the number of atoms in the investigated film. The FC range parameter L also enters the result obtained for the spin susceptibility χ that is derived similarly to (17.14). The FC component of χ is given by [22, 29]

$$\chi_*(T) \simeq \chi_C(T)\frac{L}{p_F}, \tag{17.15}$$

where $\chi_C(T) = \mu_B^2\rho/T$.

Equations (17.14) and (17.15) jointly establish an unambiguous relation within our model between the T-independent term in the specific heat $C(T)$ and the Curie component of the spin susceptibility $\chi(T)$ that has also been observed experimentally [8, 9]. This relation can be tested using existing experimental data [6]. The T-independent specific heat C/N exists in the density region around $\rho = 9.5$ nm^{-2}. Being referred to one particle, it is readily evaluated from $\beta \simeq 0.25$ mJ/K. One finds $C/N \simeq 0.01$, yielding $L/p_F \simeq 0.05$. On the other hand, the data for the spin susceptibility given in Fig. 2(B) of [6] supports a Curie-like component at $\rho = 9.25$ nm^{-2}. The value of the corresponding numerical factor extracted from the data, which according to (17.15) is to be identified with the ratio L/p_F, is approximately 0.07. Given the uncertainties involved, we conclude that our model is consistent with the experimental data reported in [6].

Thus, in 2D liquid ^{3}He and HF metals, which are located in the vicinity of the quantum critical point associated with divergent quasiparticle effective mass, the group velocity depends strongly on temperature and vanishes at diminishing temperatures. The contribution to the specific heat coming from the boson part of the free energy follows the DP law. Accordingly, the specific heat becomes independent of temperature at some characteristic value of it. At sufficiently low temperature, the usual LFL behavior of zero sound is restored. The other properties of ^{3}He at quantum criticality are considered in Sect. 21.4. Note that the heat capacity C of the HF metal $YbRh_2Si_2$ contains the temperature-independent term C_0 as well [30].

17.2 Transport Properties Related to The Quasi-classical Behavior

In this section, we show that near FCQPT, the quasi-classical approach remains applicable to the description of the resistivity ρ of strongly correlated metals due to the presence of a transverse zero-sound collective mode, reminiscent of the phonon mode in solids. These phonon-like soft and weakly damped branches of the transverse zero-sound mode are found to have extremely low effective Debye temperatures. Their contributions to the collision integral are shown to drive electron transport in the vicinity of the critical point toward a classical regime. A T-linear resistivity occurs due to a mechanism analogous to that affecting the resistivity in conventional metals at room temperature, giving rise to a quasi-classical regime of transport at extremely low temperatures in HF metals.

More than a decade of comprehensive studies [31, 32], the domination of non-Fermi liquid (NFL) behavior in strongly correlated Fermi systems is no longer a big surprise. However, various features of NFL phenomena still await satisfactory explanation, especially the puzzling observations pointing to characteristic quasi-classical behavior in a quantum-critical regime. For example, at extremely low temperatures around 1 mK, the results of experimental measurements of the specific heat of 2D ^{3}He, as observed in dense ^{3}He films are described by the classical formula $C(T) = \beta + \gamma T$, where β and γ are constants, see Sect. 17.1. Also, in contrast to LFL theory, the low-temperature resistivity $\rho(T)$ of many high-T_c compounds and certain HF metals varies linearly with T [33–35]. Hence, these systems behave as if a major contribution to the collision term comes from the electron-phonon interaction, in spite of the fact that the phonon Debye temperature T_D exceeds measurement temperatures by a factor of $10^2 - 10^3$.

In Sect. 17.1, we have attributed the presence of the classical term β in the specific heat $C(T)$ of 2D liquid ^{3}He to softening of the transverse zero-sound mode (TZSM), occurring near QCP [6] where the density of states $N(0)$, proportional to the effective mass M^*, diverges. Here we shall address the impact of the TZSM on transport properties in the QCP regime. The TZSM exists only in those correlated Fermi systems where the effective mass M^* exceeds the bare mass M by a factor more than 3. This requirement is always met while approaching the QCP. In conventional Fermi liquids, the Fermi surface consists of a single sheet, so the TZSM has a single branch with velocity c_t exceeding the Fermi velocity v_F. Consequently, emission and absorption of sound quanta by electrons are forbidden, and the role of the TZSM in kinetics is inessential. However, in heavy-fermion metals, it is usual for several bands to cross the Fermi surface simultaneously, thereby generating several zero-sound branches. For all branches but one the sound velocities are less than the largest Fermi velocity. Hence, the aforementioned ban is lifted, and these branches of the TZSM spectrum experience damping, in a situation similar to that for zero-spin sound. In the latter instance, Landau damping is so strong that the mode cannot propagate through the liquid [15, 36]. It will be seen, however, that this is not the case for TZSM damping due to this mode softening close to the QCP. Due to the softening effect, the contribution of the damped TZSM to the collision integral has the same form as the electron-phonon interaction at room temperature. On the other hand, we will also find that in heavy-fermion metals, similar to the case of liquid ^{3}He films, softening of the TZSM acts to lower the characteristic temperature Ω_t that plays the role of the Debye temperature, setting the stage for the existence of a quasi-classical transport regime at extremely low temperatures.

In the canonical picture of quantum phase transitions, the QCP has been identified with an end point of the line T_N of a second-order phase transition, associated with violation of some Pomeranchuk stability condition. This violation is associated with divergence of the energy derivative $\partial\Sigma(p,\varepsilon)/\partial\varepsilon$ of the self-energy and consequent vanishing of the quasiparticle weight $z = (1 - \partial\Sigma(p,\varepsilon)/\partial\varepsilon)^{-1}$ in single-particle states at the Fermi surface, thus triggering [31, 32, 37] divergence of the effective mass M^* defined by $M/M^* = z(1 + \left(\partial\Sigma(\mathbf{p},\varepsilon)/\partial\varepsilon_p^0\right)|_{p=p_F}$.

A number of important experimental studies, that were performed, fail to support the canonical view of the QCP. In 2D liquid ^{3}He, experimental data [6, 12] have not identified any phase transition that can be associated with the point of the divergence of the effective mass. It has been acknowledged [34, 38] that a similar situation also prevails for the QCPs of heavy-fermion metals. In essence, the point where the density of states diverges is *separated by an intervening NFL phase* from points, where lines of some second-order phase transition terminate. Furthermore, these transitions possess unusual properties such as hidden order parameters. Therefore, within the standard collective scenario, they can hardly qualify as triggers of the observed rearrangements.

We are, therefore, forced to interchange the reason and the consequence in connection to the canonical scenario [39]. Following Sect. 17.1, we attribute the QCP to vanishing of the Fermi velocity v_F at a critical density ρ_∞, which occurs if $1 + \left(\partial\Sigma(\mathbf{p},\varepsilon)/\partial\varepsilon_p^0\right)|_{p=p_F} = 0$. Accordingly, in this scenario for the QCP, it is the *momentum-dependent part* of the mass operator that plays the decisive role.

It is commonly accepted in the theory devoted to the QCP physics that switching on the interaction between particles never produces a significant momentum dependence in the effective interaction function f, and hence the option we propose and develop is irrelevant. However, this assertion is incorrect. The natural measure of the strength of momentum-dependent forces in the medium is provided by the dimensionless first harmonic $F_1 = f_1 p_F M^*/\pi^2$ of the Landau interaction function $f(\mathbf{p}_1, \mathbf{p}_2)$. In a system, such as 3D liquid ^{3}He, where the correlations are of moderate strength, the value $F_1 \geq 6.25$ extracted from specific heat data is already rather large. The data on 2D liquid ^{3}He are yet more damaging to the claim of minimal momentum dependence, since the effective mass *diverges* in dense films [6, 8, 12]. In the case of QCP phenomena occurring in strongly correlated systems of *ionic* crystals, one should have in mind that the electron effective mass is greatly enhanced due to electron-phonon interactions that subserve polaron effects [40, 41].

The change of sign of v_F at QCP results not only in a divergent density of states, but also in a *rearrangement of the Landau state* beyond the QCP. As a rule, however, such a rearrangement already occurs *before* the system reaches QCP. This may be understood from simple arguments based on the Taylor expansion of the group velocity $v(p) = \partial\varepsilon(p)/\partial p$, which has the form

$$v(p) = v_F(\rho) + v_1(\rho)\frac{p - p_F}{p_F} + \frac{1}{2}v_2\frac{(p - p_F)^2}{p_F^2} \tag{17.16}$$

in the vicinity of the QCP. We assert that the last coefficient v_2 is positive, to ensure that the spectrum

$$\varepsilon(p) = v_F(\rho)(p - p_F) + \frac{1}{2p_F}v_1(\rho)(p - p_F)^2 + \frac{1}{6p_F^2}v_2(p - p_F)^3 \tag{17.17}$$

derived from (17.16) exhibits proper behavior even far from the Fermi surface.

By its definition, the QCP is situated at a density ρ_∞ where $v_F(\rho)$ vanishes. The QCP must in fact correspond to an extremum of the function $v(p, \rho_\infty)$, which vanishes at first at $p = p_F$. Thus, the simultaneous vanishing of the coefficient $v_1(\rho_\infty) = (\mathrm{d}v(p, \rho_\infty)/\mathrm{d}p)|_{p=p_F}$ of the second term in the Taylor series is *crucial* to the QCP occurrence. In general, v_1 does not meet this additional requirement. However, in relevant cases, its finite value remains extremely small, making it possible to tune the QCP by imposing an external magnetic field.

When $v_1 \neq 0$ in (17.7), this equation unavoidably acquires two additional *real* roots at a critical density ρ_t where $v_F(\rho_t) = 3v_1^2/(8v_2)$, namely,

$$p_{1,2} - p_F = -p_F\frac{3v_1}{2v_2}\left(1 \pm \sqrt{1 - \frac{8v_F(\rho)v_2}{3v_1^2}}\right). \tag{17.18}$$

Clearly, this transition, identified as a topological phase transition (TPT), see Chap. 4, takes place *already* on the disordered side of the QCP regime, where $v_F(\rho)$ is still positive. Accordingly, as it is discussed in Chap. 4, a new hole pocket opens and the Fermi surface gains two additional sheets, the new $T = 0$ quasiparticle momentum distribution being given by $n(p) = 1$ for $p < p_1$ and $p_2 < p < p_F$, and zero elsewhere. The emergence of new small pockets of the Fermi surface is an integral feature of the QCP phenomenon, irrespectively to whether the strongly correlated Fermi system is 2D liquid ^{3}He, a high-T_c superconductor, or a heavy-fermion metal.

At the TPT point ρ_t, the density of states, given by

$$N(T) = \int \frac{\partial n(\mathbf{p}, T)}{\partial\varepsilon(\mathbf{p})}d\upsilon \tag{17.19}$$

with $d\upsilon$ denoting an element of momentum space, is also divergent. Inserting the spectrum (17.17), straightforward calculation yields $N(T \to 0) \propto T^{-1/2}$, in contrast to the behavior $N(T \to 0, \rho_\infty) \propto T^{-2/3}$ obtained in the case where $v_1(\rho_\infty) = 0$.

Having tracked the initial evolution of the Fermi surface topology in the QCP region, our analysis turns next to the salient features of the TZSM spectrum in systems with a multi-connected (i.e., multi-sheet) Fermi surface. We first examine how the TZSM softens in 3D systems with a singly connected Fermi surface, where (17.23) has the well-known form [15]

$$1 = \frac{F_1}{6}\left[1 - 3(s^2 - 1)\left(\frac{s}{2}\ln\frac{s+1}{s-1} - 1\right)\right] \tag{17.20}$$

with $s = c_t/v_F$ and $F_1 = f_1 p_F M^*/\pi^2$. The TZSM is seen to propagate only if $F_1 > 6$, i.e., $M^* > 3M$. Near the QCP where $M^*(\rho) \to \infty$, one has $v_F/c_t \to 0$, and (17.20) simplifies to

$$1 = \frac{F_1}{15}\frac{v_F^2}{c_t^2}, \tag{17.21}$$

which implies

$$c_t(\rho \to \rho_\infty) \to \sqrt{\frac{p_F v_F(\rho)}{M}} \propto \frac{p_F}{M}\sqrt{\frac{M}{M^*(\rho)}} \to 0, \tag{17.22}$$

an analogous formula being obtained for a 2D system.

To simplify analysis of TZSM damping in the systems with *multi-connected* Fermi surface, we restrict consideration to the case of two electron bands. TPT is assumed to occur at one of the bands, so that its Fermi velocity, denoted again by v_F, tends to zero, while the Fermi velocity v_o of the other band remains unchanged in the critical density region. The model dispersion relation for the complex sound velocity $c = c_R + ic_I$ becomes

$$1 = \frac{F_1}{6}\left[1 - 3\left(\frac{c_t^2}{v_F^2} - 1\right)\left(\frac{c_t}{2v_F}\ln\frac{c_t + v_F}{c_t - v_F} - 1\right)\right] +$$

$$+ \frac{F_1}{6}\frac{v_F}{v_o}\left[1 - 3\left(\frac{c_t^2}{v_o^2} - 1\right)\left(\frac{c_t}{2v_o}\ln\frac{c_t + v_o}{c_t - v_o} - 1\right)\right]. \tag{17.23}$$

It can easily be verified that the contribution of the second term to the real part of the right-hand side of (17.23) is small compared to that of the first term, since $v_F/v_o \to 0$ toward the QCP. On the other hand, noting that

$$\ln\left[(c_R + ic_I + v_o)/(c_R + ic_I - v_o)\right] \simeq -i\pi,$$

the corresponding contribution $i\pi F_1 v_F c_R/(4v_o^2)$ to the imaginary part of the right-hand side cannot be ignored, else $c_I = 0$. Thus, (17.23) assumes the simplified form

$$1 = \frac{F_1}{15}\frac{v_F^2}{(c_R + ic_I)^2} - i\frac{\pi}{4v_o^2}F_1 v_F c_R \tag{17.24}$$

analogous to (17.21). Its solution obeys

$$c_R \propto \sqrt{\frac{M}{M^*(\rho)}}, \quad c_I \propto \frac{M}{M^*(\rho)}. \tag{17.25}$$

Importantly, we see then that the ratio $c_I/c_R \propto \sqrt{M/M^*(\rho)}$ is *suppressed* in the QCP regime, which allows us to analyze the contribution of the TZSM to the collision term entering the resistivity similar to the familiar case of the electron-phonon interaction. By contrast, the group velocities of the damped branches of *longitudinal* zero sound are found to be insensitive to variation of the effective mass in the QCP region [15, 36]. No quenching by a small parameter M/M^* arises, so these modes cannot propagate in the Fermi liquid.

It is now clear that approaching QCP, the effective Debye temperature $\Omega_t = \omega(k_{\max}) = c_R k_{\max}$ *goes down to zero*, independently of the $k_{\max}$ value, characterizing the cutoff of the TZSM spectrum. Thus, the necessary condition $\Omega_t < T$ for quasi-classical behavior emergence is always met. However, another condition must also be satisfied, if a well-pronounced classical domain at extremely low temperature exists. Indeed, let us take into account that the Boson contribution

$$F_B = T \int \ln\left(1 - e^{-ck/T}\right) \theta(k - k_{\max}) d\upsilon \tag{17.26}$$

to the free energy is proportional to some power of $k_{\max}$, depending on the dimensionality of the problem. The same is true for the corresponding contributions to kinetic phenomena. Therefore, the extra condition needed is that $k_{\max}$ should not be too small. We identify $k_{\max}$ with the new characteristic momentum arising beyond the point of the TPT, namely, with the distance $d = p_2 - p_1$ between the new sheets of the Fermi surface. Indeed, for momenta p situated at distances from the Fermi surface significantly bigger than d, the single-particle spectrum $\varepsilon(p)$ is no longer flat. The collective spectrum $\omega(k)$ determined from the corresponding Landau kinetic equation is no longer soft, and consequently the imaginary part c_I of c becomes of the same order as c_R, preventing propagation of the TZSM.

This option is illustrated by the key property of resistivity in electron systems of solids. The kernel of the electron-TZSM collision integral underlying the resistivity $\rho(T)$ contains terms $n_o(\mathbf{p}+\mathbf{k})(1-n_o(\mathbf{p}))N(\mathbf{k}) - n_o(\mathbf{p})(1-n_o(\mathbf{p}+\mathbf{k}))(1+N(\mathbf{k}))$ and $n_o(\mathbf{p}+\mathbf{k})(1-n_o(\mathbf{p}))(1+N(\mathbf{k})) - n_o(\mathbf{p})(1-n_o(\mathbf{p}+\mathbf{k}))N(\mathbf{k})$, in which $N(\mathbf{k})$ denotes a nonequilibrium TZSM momentum distribution and $n_o(\mathbf{p})$ is a nonequilibrium electron momentum distribution of the band that can absorb and emit the TZSM. The explicit linearized electron-phonon-like form of the corresponding component of the collision integral is [42, 43]

$$I_{e,ph} \propto \int w(\mathbf{p},\mathbf{k})\omega(k)\frac{\partial N_0(\omega)}{\partial\omega}(\delta n_i - \delta n_f)\delta(\varepsilon_i - \varepsilon_f)d\upsilon. \tag{17.27}$$

In this expression, w is the collision probability, $N_0(\omega) = [\exp(\omega(\mathbf{k})/T) - 1]^{-1}$ is the equilibrium TZSM momentum distribution, and $\delta n_{i,f}$ stands for the deviations of the real momentum distributions in the electron band labeled o to distinguish it from its nonequilibrium counterpart, with $n_i = n(\mathbf{p})$ and $n_f = n(\mathbf{p} + \mathbf{k})$. In the classical case of the electron-phonon interaction, at $T_D < T$, one has $\partial N_0(\omega)/\partial\omega \propto -T/\omega^2$ while all the other factors are T-independent, resulting [42, 43] in linear variation of the resistivity $\rho(T)$ with T.

Based on the analogy we have established between the roles of phonons and the TZSM, the resistivity of the strongly correlated electron system must obey the FL law $\rho(T) \propto T^2$ only at $T < \Omega_t$. In the opposite case $\Omega_t < T$, the resistivity exhibits a *linear* dependence on T. Imposition of a magnetic field cannot kill the soft mode of transverse zero sound as long as the flattening of the single-particle spectrum responsible for strong depression of the effective Debye temperature Ω_t persists.

These results and conclusions are in qualitative agreement with experimental data [34] on the low-temperature resistivity of the doped HF metal $YbRh_2(Si_{0.95}Ge_{0.05})_2$. The data indicate that the linear-in-T dependence of the resistivity is robust, down to temperatures as low as 20 mK. Remarkably, the linearity of $\rho(T)$ continues to hold in external magnetic fields up to $B \simeq 2T$, far in excess of the critical value $B_c \simeq 0.3T$, at which this compound undergoes some phase transition with a hidden order parameter [34]. A linear T dependence of $\rho(T)$ is present as well in the Hertz-Millis spin-density-wave (SDW) scenario for the QCP in 2D electron systems [44, 45]. However, critical spin fluctuations die out at $B > B_c$, since the SDW transition is suppressed. Thus, the observed behavior of $\rho(T)$ contradicts the SDW scenario. As to our scenario, it is in fact compatible with the observed behavior, since the TZSM spectrum is less sensitive than the structure of critical spin fluctuations to the magnitude of the magnetic field. A linear T dependence of $\rho(T)$ can also emerge, if the light carriers are scattered by heavy bipolarons [40, 46]. However, there is no evidence for the presence of these quasiparticles in heavy-fermion metals.

Let us now briefly turn to the analysis of the soft TZSM contribution to the thermopower. In the classical situation, the phonon-drag thermopower $S_d(T)$ associated with nonequilibrium phonons is known [42, 43, 47] to account for a substantial part of the Seebeck coefficient $S(T)$. The same is true for the class of quantum-critical systems considered here, except that the domain, where the drag term $S_d(T)$ contributes appreciably, extends down to extremely low temperatures. Significantly, at $T \to 0$ the drag contribution increases as T^3, whereas at $T > \Omega_t$ it falls off as T^{-1}, producing a bell-like shape [42, 43] of $S_d(T)$ with a sharp maximum at $T \simeq \Omega_t$. Since other contributions to S are rather smooth, this remarkable feature of S_d appears to be responsible for the change of sign of the full Seebeck coefficient $S(T)$ at extremely low T observed experimentally [48] in the heavy-fermion metal $YbRh_2Si_2$, as well as the irregular behavior of $S(T \to 0)$ found in several heavy-fermion compounds [49]. The TZSM scenario proposed here predicts that the Seebeck coefficient in $YbRh_2(Si_{0.95}Ge_{0.05})_2$ exhibits the same anomalous behavior at magnetic fields substantially exceeding the corresponding critical value B_c.

The conditions promoting the formation of soft damped collective modes play the same role in kinetic phenomena as phonons. Thus, a damped soft branch belonging to the transverse zero-sound mode emerges when several bands cross the Fermi surface simultaneously, with one of the bands subject to divergence of the effective mass of carriers. As a result, there are prerequisites for lowering the corresponding Debye temperature Ω_t, which makes the inequality $\Omega_t < T$ valid near the quantum critical point, where the quasi-classical regime sets in at extremely low temperatures. As we shall see in Sect. 17.3, such a behavior illuminates unexpectedly close relationships existing between HF compounds and ordinary metals [50].

17.3 Quasi-Classical Physics and T-Linear Resistivity

As it was demonstrated above, the Debye temperature T_D can be extremely low in the case of HF compounds. As a result, at $T > T_D$ the resistivity $\rho(T)$ varies linearly with T, since the mechanism, forming the dependence $\rho(T)$, is the same as the electron-phonon mechanism that prevails at high temperatures in ordinary metals. Thus, in the region of T-linear resistivity, the electron-phonon scattering yields almost material independence of the lifetime τ_q of quasiparticles that is expressed as the ratio of Planck $\hbar$ and Boltzmann constants k_B, $T\tau_q \sim \hbar/k_B$. As an example, we analyze the resistivity of the HF metal $Sr_3Ru_2O_7$.

Discoveries of surprising universality in the properties of both strongly correlated and ordinary metals provide unique opportunities to check and expand our understanding of quantum criticality in strongly correlated compounds. When exploring at different temperatures T the linear in temperature resistivity of these utterly different metals, universality of their fundamental physical properties has been revealed [51]. On the one hand, at low T the linear T-resistivity,

$$\rho(T) = \rho_0 + AT, \tag{17.28}$$

has been observed in many strongly correlated compounds like high-temperature superconductors and HF metals located near their QCP, and therefore exhibiting quantum criticality. Here ρ_0 is the residual resistivity and A is a T-independent coefficient. Explanations based on quantum criticality for the T-linear resistivity have been given in the literature, see, e.g., [52–55] and references therein. On the other hand, at room temperatures the T-linear resistivity is exhibited by conventional metals such as Al, Ag, or Cu. In case of a simple metal with a single Fermi surface pocket, the resistivity is expressed as $e^2 n\rho \simeq p_F/(\tau_q v_F)$ [28], where e is the electronic charge, τ_q is the quasiparticles lifetime, n is the carrier concentration, and p_F and v_F are the Fermi momentum and the Fermi velocity, respectively. Writing the lifetime (or inverse scattering rate) τ_q of quasiparticles in the form [56, 57]

$$\frac{\hbar}{\tau_q} \simeq a_1 + \frac{k_B T}{a_2}, \tag{17.29}$$

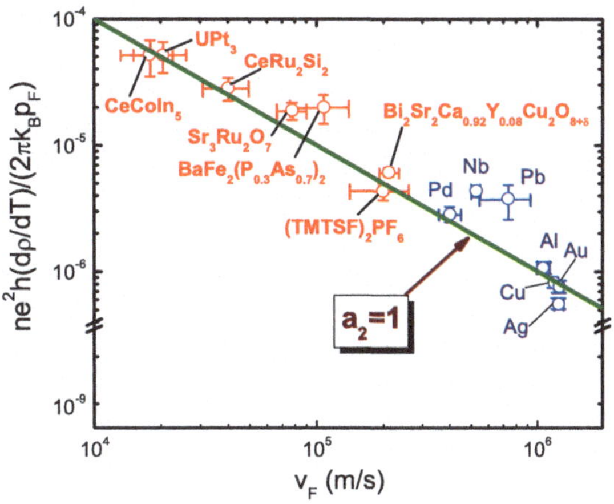

Fig. 17.2 Experimental measurements of $e^2n\hbar/p_Fk_B\partial\rho/\partial T$ (the left-hand side of (17.30)) versus $1/v_F$ [51]. Ordinary metals are shown with blue symbols and strongly correlated metals are depicted with red ones. The line shown by the arrow represents the case $a_2 = 1$

we obtain

$$\frac{e^2n\hbar}{p_Fk_B}\frac{\partial\rho}{\partial T} = \frac{1}{a_2v_F}, \tag{17.30}$$

where $\hbar = h/2\pi$, h is Planck's constant, k_B is Boltzmann's constant, and a_1 and a_2 are T-independent dimensionless parameters. Figure 17.2 reports experimental measurements of $e^2n\hbar/p_Fk_B\partial\rho/\partial T$, that it is the left-hand side of (17.30), versus $1/v_F$ [51]. One can see that the scattering rate per Kelvin is approximately constant across a wide range of materials [51], and (17.30) is in good agreement with the experimental facts. It is worth noting that elemental metals like copper, silver, aluminum, etc. are not strongly correlated. In contrast, as we have seen above, due to the formation of flat bands the strongly correlated metals exhibit the classical properties of elemental ones, for the strongly correlated metals demonstrate the quasi-classical behavior at low temperatures.

An important point for the theory is that experimental data confirm (17.30) for both strongly correlated and normal metals under the condition that latter demonstrate linear T-dependence of their resistivity [51]. Moreover, as it is seen from Fig. 17.2, the analysis of the data from the literature for the majority of compounds with the linear dependence of $\rho(T)$ shows that the coefficient a_2 is always close to unity, $0.7 \leq a_2 \leq 2.7$, despite huge distinction in the absolute value of ρ, T, and Fermi velocities v_F that are varying by two orders of magnitude [51]. As a result, it follows from (17.29) that the T-linear scattering rate is of universal dependence, $1/(\tau_q T) \sim k_B/\hbar$, valid for different systems displaying the T-linear dependence. Indeed, on the one hand, this dependence is demonstrated by ordinary metals at $T \geq T_D$, where it is due to electron-phonon mechanism. On the other hand, it occurs in strongly correlated metals which are assumed to be fundamentally different from the ordinary ones. Therefore, the linear dependence at their quantum criticality and at temperatures of a few Kelvin is assumed to come from excitations of electronic origin rather than from phonons [51]. We note that in some of the cuprates the scattering rate has a momentum and doping dependence omitted in (17.30) [58–60]. Nonetheless, the fundamental picture outlined by (17.30) is strongly supported by measurements of the resistivity on $Sr_3Ru_2O_7$ for a wide range of temperatures: At $T \geq 100$ K, the resistivity becomes again T-linear at all applied magnetic fields, as it does at low temperatures and at the critical field $B_c \simeq 7.9$ T, but with the coefficient A lower than that seen at low temperatures [51]. Thus, the same strongly correlated compound exhibits the same behavior of the resistivity at both quantum critical regime and high-temperature one, allowing us to expect that the same physics governs the T-linear resistivity in spite of possible peculiarities of some compounds.

To explain the constancy of the T-linear scattering rate $1/(\tau_q T)$, it is necessary to recollect the nature and consequences of flattening of single-particle excitation spectra $\varepsilon(\mathbf{p})$ ("flat bands") in strongly correlated Fermi systems [18, 23, 29, 61] (see [27, 62, 63] for reviews). At $T = 0$, the ground state of a system with a flat band is degenerate, and the occupation numbers $n_0(\mathbf{p})$ of single-particle states belonging to the flat band are continuous functions of momentum $\mathbf{p}$, in contrast to discrete standard LFL values 0 and 1, as it is seen from Fig. 17.3. Such behavior of $n_0(\mathbf{p})$ leads to a temperature-independent entropy term

$$S_0 = -\sum_{\mathbf{p}}[n_0(\mathbf{p}) \ln n_0(\mathbf{p}) + (1 - n_0(\mathbf{p})) \ln(1 - n_0(\mathbf{p}))]. \tag{17.31}$$

Unlike the corresponding LFL entropy, which vanishes linearly as $T \to 0$, the term S_0 produces the NFL behavior that includes T-independent thermal expansion coefficient [27, 57, 64, 65]. T-independent behavior is observed in measurements on $CeCoIn_5$ [66–68] and $YbRh_2(Si_{0.95}Ge_{0.05})_2$ [69], while measurements on $Sr_3Ru_2O_7$ indicate the same behavior [70, 71]. In the theory of fermion condensation, the degeneracy of the NFL ground state is removed at any finite temperature, with the flat band acquiring a small dispersion [18, 27]

$$\varepsilon(\mathbf{p}) = \mu + T \ln \frac{1 - n_0(\mathbf{p})}{n_0(\mathbf{p})} \tag{17.32}$$

proportional to T with μ being the chemical potential. The occupation numbers n_0 of FC remain unchanged at relatively low temperatures and, accordingly, so does the entropy S_0. Due to the fundamental difference between the FC single-particle spectrum and that of the remainder of the Fermi liquid, a system having FC is, in fact, a two-component system. The range L of momentum space adjacent to the Fermi surface where FC resides is given by $L \simeq (p_f - p_i)$, as seen from Fig. 17.3.

In strongly correlated metals at high temperatures, a light electronic band, characterized by electrons with small effective mass, coexists with f or d-electron narrow bands, placed below the Fermi surface. At lower temperatures, when the quantum criticality is formed, a hybridization between this light band and f or d-electron bands results in its splitting into new flat bands, while some of the bands remain light, thus representing LFL states [72].

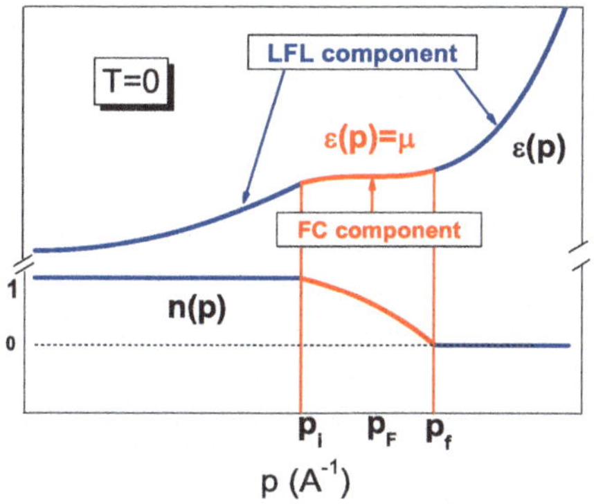

Fig. 17.3 Schematic plot of two-component electron liquid at $T = 0$ with FC. Due to the presence of FC, the system is separated into two components. The first component is a normal liquid with the quasiparticle distribution function $n_0(p < p_i) = 1$, and $n_0(p > p_f) = 0$. The second one is FC with $0 < n_0(p_i < p < p_f) < 1$ and the single-particle spectrum $\varepsilon(p_i < p < p_f) = \mu$. The Fermi momentum p_F satisfies the condition $p_i < p_F < p_f$

A flat band can also be formed by a van Hove singularity (vHs) [73–80]. We assume that at least one of these flat bands crosses the Fermi level and represents the FC subsystem that is shown in Fig. 17.3. Remarkably, the FC subsystem possesses its own set of zero-sound modes. The mode of interest for our analysis is that of transverse zero sound with its T-dependent sound velocity $c_t \simeq \sqrt{T/M}$ and the Debye temperature [81]

$$T_D \simeq c_t k_{max} \simeq \beta\sqrt{T T_F}. \tag{17.33}$$

Here, β is a numeric factor, M is the effective mass of electron generated by vHs or by the hybridization, T_F is the Fermi temperature, while M^* is the effective mass formed finally by some interaction, e.g., by the Coulomb interaction, that leads to flat bands [72]. The characteristic wave number $k_{\max}$ of the soft transverse zero-sound mode is estimated as $k_{\max} \sim p_F$, since we assume that the main contribution forming the flat band comes from vHs or from the hybridization. Note that the numerical factor β cannot be established and is considered as a fitting parameter, rendering T_D value (17.33) to be uncertain. Estimating $T_F \sim 10$ K and taking $\beta \sim 0.3$, and noting that the quasi-classical regime takes place at $T > T_D \simeq \beta\sqrt{T T_F}$, we obtain that $T_D \sim 1$ K and expect that strongly correlated Fermi systems can exhibit a quasi-classical behavior at their quantum criticality [39, 81] with the low-temperature coefficient A entering (17.28) $A = A_{LT}$. In case of HF metals with few bands crossing Fermi level and populated by LFL and HF quasiparticles, the transverse zero sound makes the resistivity possess the T-linear dependence at the quantum criticality, as the normal sound (or phonons) does in the case of ordinary metals [39]. It is quite natural to assume that the scattering of sound in these materials is almost material-independent, so that electron-phonon processes both in the low-temperature limit at the quantum criticality and in the high-temperature limit of ordinary metals have the same T-linear scattering rate that can be expressed as

$$\frac{1}{\tau_q T} \sim \frac{k_B}{\hbar}. \tag{17.34}$$

Thus, in case of the same material, the coefficient $A = A_{HT}$, defining the classical linear T-dependence generated by the common sound (or phonons) at high temperatures, coincides with that of low-temperature coefficient A_{LT}, $A_{HT} \simeq A_{LT}$. As we shall see, this observation is in accordance with measurements on $Sr_3Ru_2O_7$ [51]. It is worth noting that the transverse zero-sound contribution to the heat capacity C follows the Dulong–Petit law, making C possess a T-independent term C_0 at $T \gtrsim T_D$, as it does in case of ordinary metals [81]. It is obvious that the zero sound contributes to the heat transport as the normal sound does in case of ordinary metals and its presence can violate the Wiedemann-Franz low. A detailed consideration of the properties of the transfers zero sound is presented in Sects. 17.1 and 17.2.

There is another mechanism contributing to the T-linear dependence at the quantum criticality that we name the second mechanism, in contrast to the first one described above and related to the transverse zero sound. We turn to consideration of the next contribution to the resistivity ρ in the range of quantum criticality, at which the dispersion of the flat band is governed by (17.32). It follows from (17.32) that the temperature dependence of $M^*(T)$ of the FC quasiparticles is given by

$$M^*(T) \sim \frac{\eta p_F^2}{4T}, \tag{17.35}$$

where $\eta = L/p_F$ [27, 62, 63]. Thus, the effective mass of FC quasiparticles diverges at low temperatures, while their group velocity, and hence their current vanishes and the main contribution to the resistivity is provided by light quasiparticles bands. Nonetheless, the FC quasiparticles still play a key role in determining the behavior of both the T-dependent resistivity and ρ_0. The resistivity has the conventional dependence [28]

$$\rho(T) \propto M_L^* \gamma \tag{17.36}$$

on the effective mass and the damping of the normal quasiparticles. Based on the fact that all the quasiparticles have the same lifetime, one can show that in playing its key role, the FC makes all quasiparticles belonging to light and flat bands to have the same unique width γ and lifetime τ_q given by (17.29) [57, 82]. As a result, the first term a_0 on the right-hand side of (17.29) forms an irregular residual resistivity ρ_0^c, while the second one forms the T-dependent part of the resistivity. The term "residual resistivity" ordinarily refers to impurity scattering. In the present case, the irregular residual resistivity ρ_0^c is, instead, determined by the onset of a flat band and has no relation to scattering of quasiparticles by impurities [83]. The two mechanisms described above contribute to the coefficient A on the right-hand side of (17.28) and it can be represented as $A \simeq A_{LT} + A_{FC}$, where A_{LT} and A_{FC} are formed by the zero sound and by FC, respectively. Coefficients A_{LT} and A_{FC} can be identified and distinguished experimentally, because A_{LT} is accompanied by the temperature-independent heat capacity C_0, while A_{FC} is escorted by the emergence of ρ_0^c.

As we have seen above, the presence of flat bands generates the characteristic behavior of the resistivity. Besides, it has a strong influence on the systems properties by creating the term S_0, making the spin susceptibility of these systems exhibit the Curie–Weiss law, as it is observed in the HF metal $CeCoIn_5$ [29]. The term S_0 serves as a stimulator of phase transitions that could lift the degeneracy and make S_0 vanish in accordance with the Nernst theorem. As we shall see, in case of $Sr_3Ru_2O_7$, the nematic transition emerges. If a flat band is absent, the T-dependence of the resistivity is defined by the dependence of the term γ, entering (17.36), on the effective mass $M^*(T)$ of heavy electrons, while the spin susceptibility is determined by $M^*(T)$ [27].

To illustrate the emergence of the both mechanisms contributing to the linear T-dependence of the resistivity, we now consider the HF compound $Sr_3Ru_2O_7$. To achieve a consistent picture of the quantum critical regime underlying the quasi-classical region in $Sr_3Ru_2O_7$, we have to construct its $T - B$ phase diagram. We employ the model [73–80] based on vHs that induces a peak in the single-particle density of states (DOS) and yields a field-induced flat band [84]. At fields in the range $B_{c1} < B < B_{c2}$, the vHs is moved through the Fermi energy and the DOS peak turns out to be at or near the Fermi energy. A key point in this scenario is that within the range $B_{c1} < B < B_{c2}$, a repulsive interaction (e.g., Coulomb) is sufficient to induce FC and the formation of a flat band with the corresponding DOS singularity is locked to the Fermi energy [27, 62, 63, 84]. Now, it is seen from (17.32) that finite temperatures, while removing the degeneracy of the FC spectrum, do not change the excess entropy S_0, threatening the violation of the Nernst theorem. To avoid such a singularity, the FC state must be altered at $T \to 0$, so that S_0 is to be somehow removed before zero temperature is reached. This can take place by means of a specific phase transitions. In case of $Sr_3Ru_2O_7$, this mechanism is naturally identified with the electronic nematic transition [73–75].

The schematic $T - B$ phase diagram of $Sr_3Ru_2O_7$ based on the proposed scenario is presented in Fig. 17.4. Its main feature is the magnetic field-induced quantum critical domain created by quantum critical points that are situated at B_{c1} and B_{c2}, generating FC and associated flat band. It is seen that in contrast to the typical phase diagram of a HF metal [27], the domain occupied by the ordered phase in Fig. 17.4 is approximately symmetric with respect to the magnetic field $B_c \simeq (B_{c2} + B_{c1})/2 \simeq 7.9$ T [77]. The emergent FC and quantum critical points are considered to be hidden or concealed in a phase transition. The area occupied by this phase transition is indicated by horizontal lines and restricted by the thick boundary lines. At the critical temperature T_c, where new ordered phase sets in, the entropy is a continuous function. Therefore, the top of the domain, occupied by the new phase, is a line of second-order phase transitions. As T is lowered, some temperatures $T_{\rm tr}^1$ and $T_{\rm tr}^2$ are reached, at which the entropy of the ordered phase becomes larger than that of the adjacent disordered phase, due to the remnant entropy S_0 from the highly entropic flat-band state. Therefore, under the influence of the magnetic field, the system undergoes a

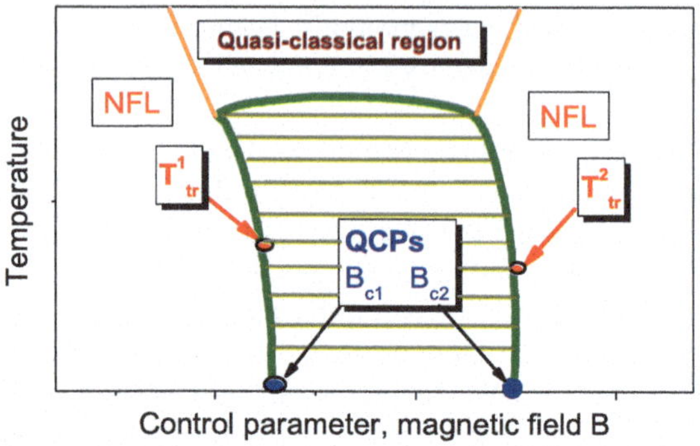

Fig. 17.4 Schematic phase diagram of the metal $Sr_3Ru_2O_7$. The quantum critical points (QCPs) situated at the critical magnetic fields B_{c1} and B_{c2} are indicated by arrows. The ordered phase bounded by the thick curve and marked by horizontal lines emerges to remove the entropy excess given by (17.31). Two arrows label the tricritical points T^1_{tr} and T^2_{tr} where the lines of the second-order phase transitions change to the first order. Quasi-classical region is confined by two lines at the top of the figure and by the top line of the ordered phase

first-order phase transition upon crossing a sidewall boundary at $T = T^1_{tr}$ or $T = T^2_{tr}$, since entropy cannot be equalized there. It follows then that the line of the second-order phase transitions is changed to lines of the first-order transitions at tricritical points indicated by arrows in Fig. 17.4. It is seen from Fig. 17.4 that the sidewall boundary lines are not strictly vertical, due to the stated behavior of the entropy at the boundary and as a consequence of the magnetic Clausius-Clapeyron relation (as discussed in [75, 76]). Quasi-classical region is located above the top of the second-order phase transition and restricted by two lines shown in Fig. 17.4. Therefore, the T-linear dependence is located in the same region, and is represented by AT dependence with $A \simeq A_{LT} + A_{FC}$. We predict that in this region the heat capacity C contains the temperature-independent term C_0 as that of the HF metal $YbRh_2Si_2$ does [30], while jumps of the residual resistivity, represented by ρ_0^c in $Sr_3Ru_2O_7$ [73], are generated by the second mechanism.

The coefficients A_{FC}, A_{LT}, and A_{HF} can be extracted from the results of measurements of the resistivity $\rho(T)$ shown in the left and right panels of Fig. 17.5 [51, 77]. For clarity, the left panel shows only a part of the data on $\rho(T)$ measured from 0.1 K to 18 K at $B = B_c$. This part exhibits the T-linear dependence between 1.4 K and 18 K and between 0.1 K and 1 K [77]. The coefficient $A \simeq A_{LT} + A_{FC} \simeq 1.1$ $\mu\Omega$cm/K between 18 K and 1.4 K. Since $T_D \sim 1$ K, we expect that between 1 K and 0.1 K the coefficient A is formed by the second mechanism and $A_{FC} \simeq 0.25$ $\mu\Omega$cm/K. The right panel reports the measurements of $\rho(T)$ for $T > T_c$ up to 400 K [51]. The dash line shows the extrapolation of the low-temperature linear resistivity at $T < 20$ K and B_c with $A \simeq 1.1$ $\mu\Omega$cm/K, and the solid line shows the extrapolation of the high-temperature linear resistivity at $T > 100$ K with $A_{HT} \simeq 0.8$ $\mu\Omega$cm/K [51]. The obtained values of A allow us to estimate the coefficients A_{LT} and A_{FC}. Due to our assumption that $A_{LT} \simeq A_{HT}$, we have $A - A_{LH} \simeq A_{FC} \simeq 0.3$ $\mu\Omega$cm/K.

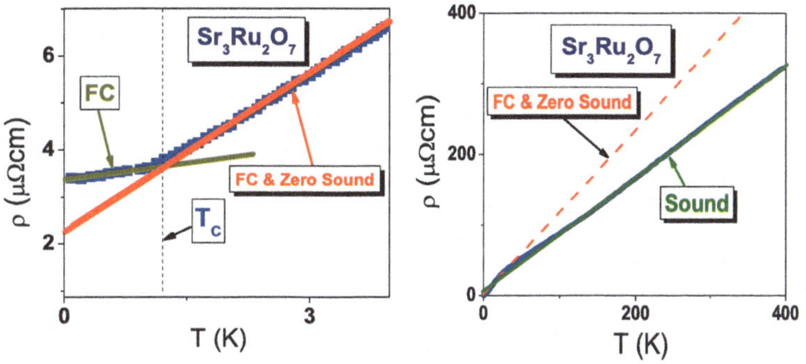

Fig. 17.5 Resistivity versus temperature. The left panel: The resistivity $\rho(T)$ for $Sr_3Ru_2O_7$ at the critical field $B_c = 7.9$ T [77]. Two straight lines display the T-linear dependence of the resistivity exhibiting a kink at $T = T_c$. At $T > T_c$ the T-linear resistivity is formed by zero-sound and FC contributions, while at $T < T_c$ it comes from the FC contribution. The right panel: The resistivity at B_c over an extended temperature ranges up to 400 K [51]. The dashed line shows the extrapolation of the low-T-linear resistivity at $T > T_c$, and the solid line shows the extrapolation of the high-T-linear resistivity formed at $T > 100$ K by the common sound (phonons)

This value is in good agreement with $A_{FC} \simeq 0.25$ $\mu\Omega$cm/K. As a result, we conclude that for $Sr_3Ru_2O_7$, where the measurements are precise, the scattering rate is given by (17.34) and does not depend on T, provided that $T \geq T_D$. The relatively small term A_{FC} is omitted. On the other hand, at $T < T_D$ $A_{HT}/A_{FC} \simeq 3$, and the constancy of the lifetime τ_q is violated, while the resistivity exhibits the T-linear dependence. It is seen from the left panel of Fig. 17.5 that the transition from the resistivity, characterized by the coefficient A_{LT}, to that with A_{FC} occurs as a kink at $T_c = 1.2$ K representing both the entry into the ordered phase and a transition region, where the resistivity alters its slope. We expect that the constancy of the scattering rate can also fail in such HF metals as $YbRh_2Si_2$ and the quasicrystal $Au_{51}Al_{34}Yb_{15}$ that exhibits the NFL behavior related to the presence of the fermion condensation [85, 86]. This universal behavior confirms that HF compounds form the new state of matter generated by the topological FCQPT that induces flat bands and the fermion condensation, see Chap. 4 and Sect. 4.1.3.

References

1. A.I. Petit, P.L. Dulong, Annales de Chimie ed de Physique **10**, 395 (1819)
2. L. Boltzmann, Wiener Berichte **74**, 553 (1877)
3. P. Debye, Annalen der Physik (Leipzig) **39**, 789 (1912)
4. L.D. Landau, Zh. Eksp. Teor. Fiz. **30**, 1058 (1956)
5. D.S. Greywall, Phys. Rev. B **41**, 1842 (1990)
6. M. Neumann, J. Nyéki, B. Cowan, J. Saunders, Science **317**, 1356 (2007)
7. V.A. Khodel, J.W. Clark, V.R. Shaginyan, M.V. Zverev, JETP Lett. **92**, 585 (2010)
8. C. Bäuerle, Y.M. Bunkov, A. Chen, S. Fisher, H. Godfrin, J. Low Temp. **110**, 333 (1998)
9. C. Bäuerle, Y.M. Bunkov, A.S. Chen, S.N. Fisher, H. Godfrin, J. Low Temp. **110**, 345 (1998)

10. M. Morishita, K. Ishida, K. Yawata, H. Nagatani, H. Fukuyama, J. Low Temp. **110**, 351 (1998)
11. M. Morishita, H. Nagatani, H. Fukuyama, J. Low Temp. **113**, 299 (1998)
12. A. Casey, H. Patel, J. Cowan, B.P. Saunders, Phys. Rev. Lett. **90**, 115301 (2003)
13. A. Golov, F. Pobell, Europhys. Lett. **38**, 353 (1997)
14. V.R. Shaginyan, A.Z. Msezane, K.G. Popov, V.A. Stephanovich, Phys. Rev. Lett. **100**, 096406 (2008)
15. I.M. Khalatnikov, *An Introduction to the Theory of Superfluidity* (Benjamin, New York, 1965)
16. A.A. Abrikosov, I.M. Halatnikov, Soviet Phys. Uspekhi **1**, 68 (1958)
17. M. Pfitzner, P. Wölfle, Phys. Rev. B **33**, 2003 (1986)
18. P. Nozières, J. Phys. I France **2**, 443 (1992)
19. M.V. Zverev, M. Baldo, JETP **87**, 1129 (1998)
20. M.V. Zverev, M. Baldo, J. Phys. Condens. Matter **11**, 2059 (1999)
21. S.A. Artamonov, Y.G. Pogorelov, V.R. Shaginyan, JETP Lett. **68**, 942 (1998)
22. V.A. Khodel, J.W. Clark, M.V. Zverev, Phys. Rev. B **78**, 075120 (2008)
23. V.A. Khodel, V.R. Shaginyan, JETP Lett. **51**, 553 (1990)
24. G.E. Volovik, JETP Lett. **53**, 222 (1991)
25. V.A. Khodel, V.R. Shaginyan, V.V. Khodel, Phys. Rep. **249**, 1 (1994)
26. V.R. Shaginyan, M.Y. Amusia, K.G. Popov, Phys. Usp. **50**, 563 (2007)
27. V.R. Shaginyan, M.Y. Amusia, A.Z. Msezane, K.G. Popov, Phys. Rep. **492**, 31 (2010). (arXiv:1006.2658)
28. A.A. Abrikosov, L.P. Gorkov, I.E. Dzyaloshinski, *Methods of Quantum Field Theory in Statistical Physics* (Dover, New York, 1975)
29. V.A. Khodel, M.V. Zverev, V.M. Yakovenko, Phys. Rev. Lett. **95**, 236402 (2005)
30. P. Wölfle, E. Abrahams, Phys. Rev. B **84**, 041101(R) (2011)
31. H.V. Löhneysen, A. Rosch, M. Vojta, P. Wölfle, Rev. Mod. Phys. **79**, 1015 (2007)
32. P. Gegenwart, Q. Si, F. Steglich, Nat. Phys. **4**, 186 (2008)
33. P. Gegenwart, J. Custers, T. Tayama, K. Tenya, C. Geibel, O. Trovarelli, F. Steglich, K. Neumaier, Acta Phys. Pol. **34**, 323 (2003)
34. J. Custers, P. Gegenwart, S. Geibel, F. Steglich, P. Coleman, S. Paschen, Phys. Rev. Lett. **104**, 186402 (2010)
35. L. Taillefer, Ann. Rev. Cond. Matt. Phys. **1**, 51 (2010)
36. D. Pines, P. Noziéres, *Theory of Quantum Liquids* (Benjamin, New York, Amsterdam, 1966)
37. D.M. Maslov, A.V. Chubukov, Phys. Rev. B **81**, 045110 (2010)
38. S.L. Bud'ko, E. Morosan, P.C. Canfield, Phys. Rev. B **69**, 014415 (2004)
39. J.W. Clark, V.A. Khodel, M.V. Zverev, Phys. Lett. A **377**, 647 (2013)
40. A.S. Alexandrov, N. Mott, *Polarons and Bipolarons* (World Scientific, Singapore, 1996)
41. A.S. Alexandrov, P.P. Kornilovitch, Phys. Rev. Lett. **82**, 807 (1999)
42. E.M. Lifshitz, L.P. Pitaevskii, *Physical Kinetics* (Pergamon Press, Oxford, 1981)
43. C. Kittel, *Introduction to Solid State Physics* (John Wiley & Sons, New York, 1996)
44. J.A. Hertz, Phys. Rev. B **14**, 1165 (1976)
45. A.J. Millis, Phys. Rev. B **48**, 7183 (1993)
46. A.S. Alexandrov, Physica C **274**, 237 (1997)
47. R.D. Barnard, *Thermoelectricity in Metals and Alloys* (Taylor and Francis, London, 1972)
48. S. Hartmann, N. Oeschler, C. Krellner, C. Geibel, S. Paschen, F. Steglich, Phys. Rev. Lett. **104**, 096401 (2010)
49. K. Behnia, D. Jaccard, J. Flouquet, J. Phys. Condens. Matter **16**, 5187 (2004)
50. V.R. Shaginyan, K.G. Popov, V.A. Khodel, Phys. Rev. B **88**, 115103 (2013)
51. A.N. Bruin, H. Sakai, R.S. Perry, A.P. Mackenzie, Science **880**, 339 (2013)
52. C.M. Varma, P.B. Littlewood, S. Schmittrink, E. Abrahams, A.E. Ruckenstein, Phys. Rev. Lett. **63**, 1996 (1989)
53. M.E. Simon, C.M. Varma, Phys. Rev. Lett. **89**, 247003 (2002)
54. P. Phillips, Phil. Trans. R. Soc. A **369**, 1572 (2011)
55. P. Phillips, Phil. Trans. R. Soc. A **369**, 1574 (2011)

56. P. Aynajian, E. Neto, A. Gyenis, R.E. Baumbach, J.D. Thompson, Z. Fisk, E.D. Bauer, A. Yazdani, Nature **486**, 201 (2012)
57. V.R. Shaginyan, A.Z. Msezane, K.G. Popov, J.W. Clark, M.V. Zverev, V.A. Khodel, Phys. Rev. B **86**, 085147 (2012)
58. D.C. Peets, D.G. Hawthorn, K.M. Shen, Y.J. Kim, D.S. Ellis, H. Zhang, S. Komiya, Y. Ando, G.A. Sawatzky, R. Liang, D.A. Bonn, W.N. Hardy, Phys. Rev. Lett. **103**, 087402 (2009)
59. M.M.J. French, J.G. Analytis, A. Carrington, L. Balicas, N.E. Hussey, New J. Phys. **11**, 055057 (2009)
60. J.W. Alldredge, J. Lee, K. McElroy, M. Wang, K. Fujita, Y. Kohsaka, C. Taylor, H. Eisaki, S. Uchida, P.J. Hirschfeld, J.C. Davis, Nat. Phys. **4**, 319 (2008)
61. G.E. Volovik, in *Quantum Analogues: From Phase Transitions to Black Holes and Cosmology*, ed. by W.G. Unruh, R. Schutzhold. Springer Lecture Notes in Physics, vol. 718 (Springer, Orlando, 2007), p. 31
62. V.R. Shaginyan, Phys. At. Nuclei **74**, 1107 (2011)
63. V.A. Khodel, J.W. Clark, M.V. Zverev, Phys. At. Nuclei **74**, 1237 (2011)
64. M.V. Zverev, V.A. Khodel, V.R. Shaginyan, M. Baldo, JETP Lett. **65**, 863 (1997)
65. M.Y. Amusia, A.Z. Msezane, V.R. Shaginyan, Phys. Lett. A **320**, 459 (2004)
66. N. Oeschler, P. Gegenwart, M. Lang, R. Movshovich, J.L. Sarrao, J.D. Thompson, F. Steglich, Phys. Rev. Lett. **91**, 076402 (2003)
67. J.G. Donath, F. Steglich, E.D. Bauer, J.L. Sarrao, P. Gegenwart, Phys. Rev. Lett **100**, 136401 (2008)
68. S. Zaum, K. Grube, R. Schäfer, E.D. Bauer, J.D. Thompson, H.V. Löhneysen, Phys. Rev. Lett. **106**, 087003 (2011)
69. R. Kúchler, N. Oeschler, P. Gegenwart, T. Cichorek, K. Neumaier, O. Tegus, C. Geibel, J.A. Mydosh, F. Steglich, L. Zhu, Q. Si, Phys. Rev. Lett. **91**, 066405 (2003)
70. P. Gegenwart, F. Weickert, M. Garst, R.S. Perry, Y. Maeno, Phys. Rev. Lett. **96**, 136402 (2006)
71. C. Stingl, R.S. Perry, Y. Maeno, P. Gegenwart, Phys. Rev. Lett. **107**, 026404 (2011)
72. V.A. Khodel, J.W. Clark, H. Li, M.V. Zverev, Phys. Rev. Lett. **98**, 216404 (2007)
73. S.A. Grigera, P. Gegenwart, R.A. Borzi, F. Weickert, A.J. Schofield, R.S. Perry, T. Tayama, T. Sakakibara, Y. Maeno, A.G. Green, A.J. Millis, A.P. Mackenzie, Science **306**, 1154 (2004)
74. R.A. Borzi, S.A. Grigera, J. Farrell, R.S. Perry, S.J.S. Lister, S.L. Lee, D.A. Tennant, Y. Maeno, A.P. Mackenzie, Science **315**, 214 (2007)
75. A.W. Rost, R.S. Perry, J.F. Mercure, A.P. Mackenzie, S.A. Grigera, Science **325**, 1360 (2009)
76. S. Raghu, A. Paramekanti, E.A. Kim, R.A. Borzi, S.A. Grigera, A.P. Mackenzie, S.A. Kivelson, Phys. Rev. B **79**, 214402 (2009)
77. A.W. Rost, S.A. Grigera, J.A.N. Bruin, R.S. Perry, D. Tian, S. Raghu, S.A. Kivelson, A.P. Mackenzie, Proc. Natl. Acad. Sci. USA **108**, 16549 (2011)
78. B. Binz, M. Sigrist, Europhys. Lett. **65**, 816 (2004)
79. A.M. Berridge, S.A. Grigera, B.D. Simons, A.G. Green, Phys. Rev. B **81**, 054429 (2010)
80. A.P. Mackenzie, J.A.N. Bruin, R.A. Borzi, A.W. Rost, S.A. Grigera, Physica C **481**, 207 (2012)
81. V.A. Khodel, J.W. Clark, V.R. Shaginyan, M.V. Zverev, JETP Lett. **92**, 532 (2010)
82. V.R. Shaginyan, A.Z. Msezane, K.G. Popov, J.W. Clark, M.V. Zverev, V.A. Khodel, JETP Lett. **96**, 397 (2012)
83. M.Y. Amusia, K.G. Popov, V.R. Shaginyan, W.A. Stephanowich, *Theory of Heavy-Fermion Compounds*. Springer Series in Solid-State Sciences, vol. 182 (Springer, Berlin, 2015)
84. V.R. Shaginyan, A.Z. Msezane, K.G. Popov, J.W. Clark, M.V. Zverev, V.A. Khodel, Phys. Lett. A **377**, 2800 (2013)
85. K. Deguchi, S. Matsukawa, N.K. Sato, T. Hattori, K. Ishida, H. Takakura, T. Ishimasa, Nat. Mater. **11**, 1013 (2012)
86. V.R. Shaginyan, A.Z. Msezane, K.G. Popov, G.S. Japaridze, V.A. Khodel, Phys. Rev. B **87**, 245122 (2013)

Chapter 18
Asymmetric Conductivity of Strongly Correlated Compounds

Abstract In this chapter, we show that since the FC solution for distribution function $n_0(\mathbf{p})$ generates NFL behavior and violates the particle-hole symmetry inherent in LFL, this dramatically changes transport properties of HF metals, particularly, the differential conductivity becomes asymmetric. As it is demonstrated in Sect. 4.1, Fermi quasiparticles can behave as a Bose one. Such a state is viewed as possessing the supersymmetry (SUSY) that interchanges bosons and fermions eliminating the difference between them. In the case of asymmetrical conductivity, it is the emerging SUSY that violates the time invariance symmetry. Thus, restoring one important symmetry, the FC state violates another essential symmetry. As is shown in Sect. 6.3, the LFL behavior can be restored under the application of tiny magnetic field, as it takes place in the case of graphene, see Chap. 19. Therefore, we expect that in magnetic fields SUSY is violated and the asymmetric part of the differential conductivity is suppressed. Scanning tunneling microscopy and point-contact spectroscopy closely related to the Andreev reflection are sensitive to both the density of states and the probability of the population of quasiparticle states determined by the function $n(\mathbf{p}, T)$. The above experimental techniques are ideal tools for studying specific features of the NFL behavior of HF metals and high-T_c superconductors.

18.1 Normal State

Direct experimental studies of quantum phase transitions in HTSC and HF metals are of great importance for understanding the underlying physical mechanisms responsible for their anomalous properties. However, such studies of HF metals and HTSC are difficult because the corresponding critical points are usually concealed by the proximity to other phase transitions, commonly antiferromagnetic (AF) and/or superconducting (SC). Most of the experiments on HF metals and HTSCs explore their thermodynamic properties. However, it is equally important to determine other properties of these strongly correlated systems, notably quasiparticle occupation numbers $n(p, T)$ as a function of momentum p and temperature T. These quantities are not linked directly to the density of states (DOS) $N(\varepsilon \simeq \mu)$ determined by the quasiparticle energy ε or to the behavior of the effective mass M^*. Scanning

M. Amusia and V. Shaginyan, *Strongly Correlated Fermi Systems*,
Springer Tracts in Modern Physics 283,
https://doi.org/10.1007/978-3-030-50359-8_18

tunneling microscopy [1, 2] and point-contact spectroscopy [3], being sensitive to both the density of states and quasiparticle occupation numbers, are ideal tools for exploring the effects of C and T symmetry violation. When C and T symmetries are not conserved, the differential tunneling conductivity and dynamic conductance are no longer symmetric functions of the applied voltage V.

Indeed, if under the application of bias voltage V, the current of electrons with charge $-e$, traveling from HF to a common (i.e., "non-HF") metal, changes the sign of a charge carrier to $+e$, then current character and direction alters. Namely, now the carriers are holes with charge $+e$ traveling from the common to HF metal. Turning this around, one can obtain the same current of electrons provided that V is changed to $-V$. The resulting asymmetric differential conductivity $\Delta\sigma_d(V) = \sigma_d(V) - \sigma_d(-V)$ becomes nonzero. On the other hand, if time t is changed to $-t$ (but charge is kept intact), the current changes its direction only. The same result can be achieved by $V \to -V$, and we conclude that T symmetry is broken, provided that $\Delta\sigma_d(V) \neq 0$. Thus, we detect $\Delta\sigma_d(V) \neq 0$ signals under both C and T symmetries violation. At the same time, the change of both $e \to -e$ and $t \to -t$ returns the system to its initial state so that CT symmetry is conserved. Note that the parity symmetry P is conserved and the well-known CPT symmetry is not broken in the considered case. On the other hand, the time-reversal invariance and particle-hole symmetry remain intact in normal Fermi systems, the differential tunneling conductivity and dynamic conductance are symmetric functions of V. Thus, a conductivity asymmetry is not observed in conventional metals at low temperatures.

To determine the tunneling conductivity, we first calculate the tunneling current $I(V)$ through the point contact between two metals. This is done using the method of Harrison [1, 2], based on the observation that $I(V)$ is proportional to the particle transition probability. Bardeen considered the probability P_{12} of a particle (say an electron) making a transition from a state 1 on one side of the tunneling layer to a state 2 on the other side. This quantity has the behavior $P_{12} \sim |t_{12}|^2 N_2(0) n_1(1-n_2)$ in terms of the density of states $N_2(0)$ (at $\varepsilon = 0$) in state 2. The electron occupation number is $n_{1,2}$ in these states and a transition matrix element is t_{12}. The total tunneling current I is then proportional to the difference between the current from 1 to 2 and that from 2 to 1, with the result taking the form

$$\begin{aligned} I \sim P_{12} - P_{21} \sim |t_{12}|^2 N_1(0) N_2(0) \times \\ \left[n_1(1-n_2) - n_2(1-n_1)\right] = \\ |t_{12}|^2 N_1(0) N_2(0)(n_1 - n_2). \end{aligned} \tag{18.1}$$

Harrison applied the WKB approximation to calculate the matrix element [1, 2], $t_{12} = t(N_1(0)N_2(0))^{-1/2}$, where t denotes the resulting transition amplitude. Multiplication of expression (18.1) by 2 to account for the electron spin and integration over the energy ε leads to the expression for total (or net) tunneling current:

$$I(V) = 2|t|^2 \int [n_F(\varepsilon - \mu - V) - n_F(\varepsilon - \mu)]\, d\varepsilon. \tag{18.2}$$

Here $n_F(\varepsilon)$ is the electron occupation number for a metal in the absence of FC, and we have adopted the units $e = m = \hbar = 1$, where e and m are the electron charge and mass, respectively. Since temperature is low, $n_F(\varepsilon)$ can be approximated by the step function $\theta(\varepsilon - \mu)$, where μ is the chemical potential.

It follows from (18.2) that quasiparticles with single-particle energies ε in the range $\mu \le \varepsilon \le \mu + V$ contribute to the current, while $I(V) = c_1 V$ and $\sigma_{\rm d}(V) \equiv {\rm d}I/{\rm d}V = c_1$, with $c_1 =$ const. Thus, in the framework of LFL theory, the differential tunneling conductivity $\sigma_{\rm d}(V)$, being a constant, is a symmetric function of the voltage V, i.e., $\sigma_{\rm d}(V) = \sigma_{\rm d}(-V)$. In fact, the symmetry of $\sigma_{\rm d}(V)$ holds provided C and T symmetries are observed, as is customary for LFL theory. The symmetry of $\sigma_{\rm d}(V)$ is therefore quite obvious and common in case of contact of two ordinary metals (without FC), regardless whether they are a normal or superconducting state. We note that a more rigorous consideration of the densities of states N_1 and N_2 entering (18.1) for $\varepsilon \simeq \mu$ requires their inclusion in the integrand of (18.2), see, e.g., [4]. For example, see (7) of [4], where this refinement has been carried out for a system of a magnetic adatom and scanning tunneling microscope tip. However, this complication does not break the C symmetry in LFL case. On the other hand, if the system hosts FC, the presence of the density-of-states factors in the integrand of (18.2) acts to promote the asymmetry of tunneling spectra, for the density of states strongly depends on $\varepsilon \simeq \mu$, see Figs. 18.1 and 18.6. Indeed, the situation becomes quite different in the case of a strongly correlated Fermi system in the vicinity of FCQPT that generates a flat band and violates the C symmetry [5, 6]. We note that as we have seen above, the violation of C symmetry entails the violation of T symmetry. Panel (a) of Fig. 18.1 illustrates the resulting low-temperature single-particle energy spectrum $\varepsilon(\mathbf{k}, T)$. Panel (b), which displays the momentum dependence of the occupation numbers $n(\mathbf{k}, T)$ in such a system, shows that the flat band induced by the FCQPT, as we have seen above, in fact, violates T symmetry as well. The violation of C symmetry is reflected in the asymmetry of the regions occupied by particles (labeled p) and holes (labeled h) [5, 6]. We note that a system in its superconducting state and located near FCQPT exhibits asymmetrical tunneling conductivity, for the C symmetry remains broken in both the superconducting and the normal states.

The tunneling current I flowing through a point contact of two ordinary metals is proportional to the applied voltage V and to the square of the modulus of the quantum mechanical transition interaction t. This is to be multiplied by the quantity $N_1(0)N_2(0)(n_1(\mathbf{p}, T) - n_2(\mathbf{p}, T))$ [2], where $N_1(0)$ and $N_2(0)$ are the densities of states of the metals 1 and 2 and $n_1(\mathbf{p}, T))$ and $n_2(\mathbf{p}, T))$ are, respectively, the distribution functions of these metals. We note that at low temperatures the density of states of ordinary metals is constant at the Fermi level. On the other hand, in the semiclassical approximation, the wave function that determines the interaction t is proportional to $(N_1(0)N_2(0))^{-1/2}$. Therefore, the density of states cancels down in the final result and the tunnel current becomes independent of $N_1(0)N_2(0)$. Because the distribution $n(p, T \to 0) \to \theta(p_F - p)$ as $T \to 0$, where $\theta(p_F - p)$ is the step function, it can be verified that the differential tunnel conductivity $\sigma_d(V) = {\rm d}I/{\rm d}V$ is a symmetric or even function of V in the LFL theory. Actually, the symmetry of $\sigma_{\rm d}(V)$ is obeyed if there is the (quasi) particle-hole symmetry, which is always

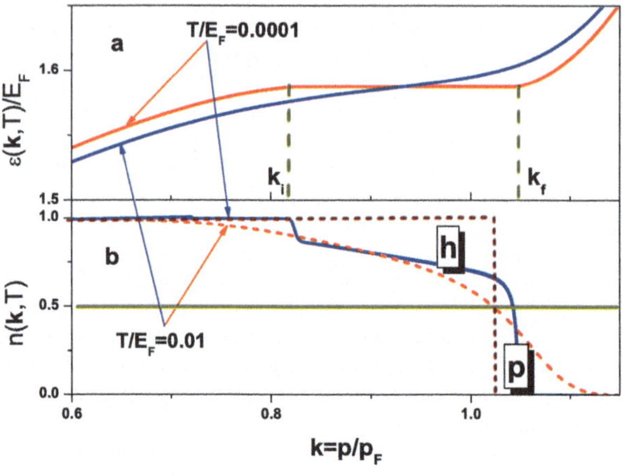

Fig. 18.1 The single-particle energy $\varepsilon(\mathbf{k}, T)$ (**a**) and the distribution function $n(\mathbf{k}, T)$ (**b**) at finite temperatures as functions of the dimensionless variable $k = p/p_F$. The arrows show temperature measured in the units of E_F. At $T = 0.0001E_F$, the vertical line shows the position of the Fermi level E_F, at which $n(\mathbf{k}, T) = 0.5$ as depicted by the horizontal line. At $T \to 0$, the single-particle energy $\varepsilon(\mathbf{k}, T)$ becomes more flat in the range $(p_f - p_i)$ so that the distribution function $n(\mathbf{k}, T)$ becomes more asymmetrical with respect to the Fermi level E_F, generating the particle-hole asymmetry related to the NFL behavior. To illuminate the asymmetry at $T = 0.01E_F$, the area bounded by the short dash lines and occupied by holes is marked by "**h**," and the area bounded by the lines and occupied by quasiparticles is marked by "**p**"

present in the LFL theory. Hence, the fact that $\sigma_d(V)$ is symmetric is obvious and natural in the case of metal-metal contacts for ordinary metals that are in the normal or superconducting state.

To study the tunneling current at low temperatures in the case of HF metals and common ones, we use the simple (18.2), for that simplification does not change availability of (18.7) and (18.10), which we use to analyze the asymmetrical part of the conductivity. Here we normalize the transition interaction to unity, $|t|^2 = 1$. Since the temperatures are low, we can approximate the distribution function $n(\varepsilon)$ by the step function $\theta(\mu - \varepsilon)$. Then, (18.2) yields $I(V) = a_1 V$ so that the differential conductivity $\sigma_d(V) = \mathrm{d}I/\mathrm{d}V = a_1 = const$ is a symmetric function of the applied voltage V. To examine quantitatively the behavior of the asymmetric part of the conductivity $\sigma_d(V)$, we differentiate both sides of (18.2) with respect to V. The result is the following equation for $\sigma_d(V)$:

$$\sigma_d = \frac{1}{T}\int n[\varepsilon(z) - V, T][1 - n(\varepsilon(z) - V, T)]\frac{\partial \varepsilon}{\partial z}\mathrm{d}z. \qquad (18.3)$$

We use the dimensionless momentum $z = p/p_F$ instead of ε in the integrand of (18.3), since for strongly correlated electron liquid n is no longer a function of ε. This is because for the latter case the energy-momentum dependence is no more linear (see panel a of Fig. 18.1), so that the proper dependence is upon momentum. Namely, the variable ε in the interval $(p_f - p_i)$ is equal to μ so that the quasiparticle distribution function varies within this interval. It is seen from (18.3) that the violation of the particle-hole symmetry makes $\sigma_d(V)$ to be asymmetric as a function of the applied voltage V [7–10].

The single-particle energy $\varepsilon(\mathbf{k}, T)$ shown in Fig. 18.1a and the corresponding $n(\mathbf{k}, T)$ shown in the panel b evolve from the FC state characterized by $n_0(\mathbf{k}, T = 0)$ determined by the equation

$$\frac{\delta E}{\delta n(\mathbf{p})} = \varepsilon(\mathbf{p}) = \mu; \; p_i \leq p \leq p_f. \tag{18.4}$$

The momenta $p_i = k_i p_F$ and $p_f = k_f p_F$ are shown in Fig. 18.1a. It is seen from Fig. 18.1a that at elevated temperatures the dispersion $\varepsilon(\mathbf{k}, T)$ becomes steeper, since the effective mass $M^*(T)$ diminishes, as it is seen from (4.13). At the Fermi level $\varepsilon(p, T) = \mu$, then from (2.5) the distribution function $n(p, T) = 1/2$. The vertical line in Fig. 18.1, crossing the distribution function at the Fermi level, illustrates the asymmetry of the distribution function with respect to the Fermi level at $T = 0.0001 E_F$. It is clearly seen that the FC state strongly violates the particle-hole symmetry at decreasing temperatures. As a result, at low temperatures the asymmetric part of the differential conductivity becomes larger. This state is viewed as the state possessing SUSY that interchanges bosons and fermions, and eliminates the difference between them, see Sect. 4.1. Thus, the asymmetrical conductivity is induced by the emerging SUSY, accompanying by the violation of the time invariance symmetry. Under the application of magnetic fields the system transits to the LFL state that strongly supports the particle-hole symmetry and violates SUSY. Therefore, the application of magnetic fields restoring the symmetry suppresses the asymmetric part of the differential conductivity.

Fairly simple transformations of (18.3) generate the following form for the asymmetric part of the differential conductivity

$$\Delta\sigma_d(V) = \frac{1}{2}[\sigma_d(V) - \sigma_d(-V)]. \tag{18.5}$$

Its explicit form yields

$$\Delta\sigma_d(V) = \frac{1}{2}\int \frac{\alpha(1-\alpha^2)}{\left[n(z,T) + \alpha[1 - n(z,T)]\right]^2}$$
$$\times \frac{\partial n(z,T)}{\partial z} \frac{1 - 2n(z,T)}{\left[\alpha n(z,T) + [1 - n(z,T)]\right]^2} dz, \; \alpha = \exp(-V/T). \tag{18.6}$$

Asymmetric tunneling conductivity can be observed in measurements that involve metals, whose electron system is located near FCQPT or behind it. High-T_c superconductors and HF compounds like $YbRh_2(Si_{0.95}Ge_{0.05})_2$, $CeCoIn_5$, $YbCu_{5-x}Al_x$, or $YbRh_2Si_2$ are among such metals. The measurements should be done when the HF metal is in either superconducting or normal state. If the metal is in its normal state, the measurements of $\Delta\sigma_d(V)$ can be done in a magnetic field $B > B_{c0}$ at temperatures $T^*(B) < T \leq T_f$ or in a zero magnetic field at temperatures higher than the corresponding critical temperature, when the electron system is in the paramagnetic state and its properties are determined by the entropy S_0. Very often the experimentally measurable quantity is conductance (the characteristic of a specific sample) rather than the conductivity, which is conductance multiplied by length and divided by cross-sectional area of a sample. Since in our theoretical considerations, the above length and cross-sectional area are kept constants, here we do not make a difference between conductance and conductivity. However, in the plots, related to experiment, we mention conductance as the quantity which has been actually measured.

We now derive an approximate expression to analyze the asymmetric part of the differential conductivity. It follows from (18.6) that for small V, the asymmetric part behaves as $\Delta\sigma_d(V) \propto V$. Note that the asymmetric part of the tunneling conductivity is an odd function of V, $\Delta\sigma_d(-V) = -\Delta\sigma_d(V)$. The natural unit for measuring voltage is $2T$ as this quantity determines the characteristic energy for FC, as shown by (4.15). The asymmetric part should be proportional to the size $(p_f - p_i)/p_F$ of the region occupied by FC:

$$\Delta\sigma_d(V) \simeq c_1 \frac{V}{2T} \frac{p_f - p_i}{p_F} \simeq c \frac{V}{2T} \frac{S_0}{x_{FC}}, \tag{18.7}$$

where $S_0/x_{FC} \sim (p_f - p_i)/p_F$ is the temperature-independent part of the entropy and c_1 is a constant of the order of unity. From (18.7), we see that when $V \simeq 2T$ and FC occupies a sizable part of the Fermi volume, $(p_f - p_i)/p_F \simeq 1$, the asymmetric part becomes comparable to the differential tunneling conductivity.

Figure 18.3a shows the results of calculations of the asymmetric part $\Delta\sigma_d(V)$ of the conductivity $\sigma_d(V)$ obtained from (18.6) [8]. In this case, $(p_f - p_i)/p_F \simeq 0.1$. Figure 18.3a also shows that the asymmetric part $\Delta\sigma_d(V)$ of the conductivity (conductance) is a linear function of V for small voltages. The asymmetry of $n(\mathbf{k}, T)$ diminishes at elevated temperatures, the asymmetric part decreases with increasing temperature, which agrees with the behavior of the experimental curves in the panel b of Fig. 18.3. It is seen from Fig. 18.3a that (18.6) is in accordance with calculation based on (18.6): At low voltage, the asymmetric conductance exhibits a linear dependence as a function of the voltage.

Recent measurements of the differential conductivity in $CeCoIn_5$ carried out by the point-contact spectroscopy technique [11] have clearly revealed its asymmetry in the superconducting ($T_c = 2.3$ K) and normal states. Figure 18.2 shows the results of these measurements. It is seen from Fig. 18.3b that $\Delta\sigma_d(V)$ is nearly constant when the HF metal is in the superconducting state, experiencing no substantial variation near T_c, see also Fig. 18.7. Then it monotonically decreases as the temperature

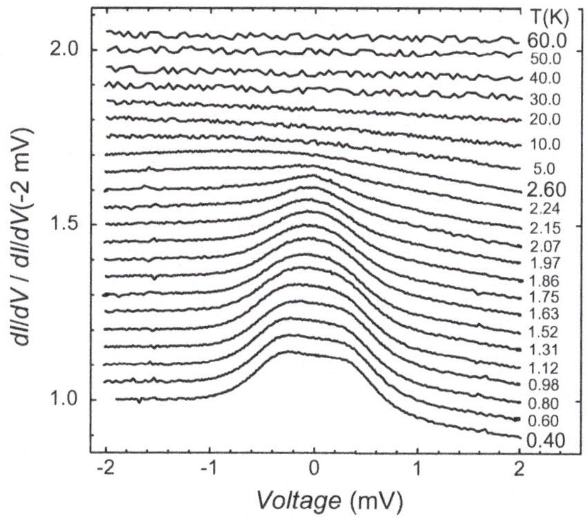

Fig. 18.2 Measured differential conductivity $\sigma_d(V)$ of point contacts Au/CeCoIn$_5$. The curves $\sigma_d(V)$ are displaced along the vertical axis by 0.05. The conductivity is normalized to its value at $V = -2$ mV. The asymmetry becomes noticeable at $T < 45$ K and increases as the temperature decreases [11]

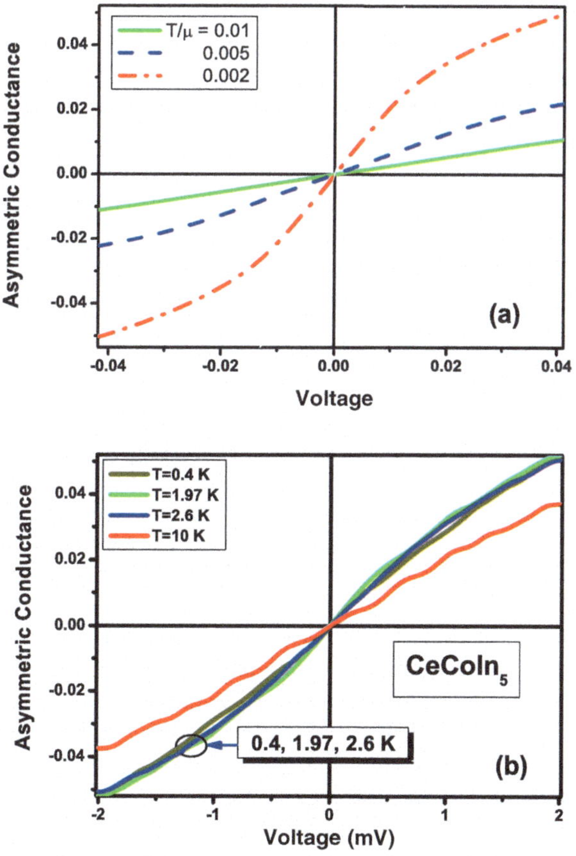

Fig. 18.3 Asymmetric conductance in CeCoIn$_5$. Panel a presents the asymmetric conductance $\Delta\sigma_d(V)$ as a function of V/μ for three normalized temperatures T/μ. Panel b shows the asymmetric conductance extracted from the data in Fig. 18.2. At $T \le T_c$ ($T_c = 2.3$ K is the superconducting phase transition temperature) the conductance (and conductivity) becomes temperature-independent

increases. Such a behavior of $\Delta\sigma_d(V)$ is related to the constancy of the effective mass at $T \leq T_c$, as it is discussed in Sect. 6.1.2. We pay attention to the fact that in the superconducting phase of $CeCoIn_5$ with $T \leq T_c = 2.3$ K the asymmetric conductivity does not depend on temperature, see (18.10). Moreover, this independence continues at $T \geq T_c$ K. Such a behavior suggests that there is the pseudogap in $CeCoIn_5$, resembling that of high-T_c superconductors. If we assume that there exists a pseudogap, then we expect that $\Delta\sigma_d(V)$ remains approximately constant in the pseudogap state. It is seen from Fig. 18.3b that $\Delta\sigma_d(V)$ is constant at $T \leq 2.6$ K, which is notably higher than T_c. This observation shows the existence of the pseudogap state in the HF compound $CeCoIn_5$. Thus, the study of the asymmetrical conductivity can help in revealing the pseudogap state.

18.1.1 Suppression of the Asymmetrical Differential Resistance in $YbCu_{5-x}Al_x$ *in Magnetic Fields*

Now consider the asymmetric part of the differential conductivity $\Delta\sigma_d(V)$ under the application of magnetic field B, see also Chap. 19. Obviously, the differential conductivity being a scalar should not depend on the current I direction. Thus, the nonzero value of $\Delta\sigma_d(V)$ manifests the violation of the particle-hole symmetry on a macroscopic scale. As we have seen in Chap. 7, at sufficiently low temperatures $T < T^*(B)$, the application of a magnetic field $B > B_{c0}$ leads to restoration of the LFL behavior eliminating the particle-hole asymmetry, and therefore the asymmetric part of the differential conductivity disappears [8, 9]. This prediction coincides with the experimental data on the differential resistance $\mathrm{d}V/\mathrm{d}I(V)$ under the application of magnetic fields in $YbCu_{5-x}Al_x$ [12]. Representing the differential resistance as the sum of its symmetric $\mathrm{d}V/\mathrm{d}I^s(V)$ and asymmetric parts $\mathrm{d}V/\mathrm{d}I^{as}(V)$,

$$\frac{\mathrm{d}V}{\mathrm{d}I(V)} = \frac{\mathrm{d}V}{\mathrm{d}I^s(V)} + \frac{\mathrm{d}V}{\mathrm{d}I^{as}(V)},$$

we obtain the equation

$$\Delta\sigma_d(V) \simeq -\frac{\mathrm{d}V/\mathrm{d}I^{as}(V)}{[\mathrm{d}V/\mathrm{d}I^s(V)]^2}. \tag{18.8}$$

To derive (18.8), we have assumed that $\mathrm{d}V/\mathrm{d}I^s(V) \gg \mathrm{d}V/\mathrm{d}I^{as}(V)$. Figure 18.4 [12] shows the temperature evolution of the symmetric $\mathrm{d}V/\mathrm{d}I^s(V)$ (a) and the asymmetric (b) $\mathrm{d}V/\mathrm{d}I^{as}(V)$ parts at zero value of the applied magnetic field. Also, the symmetric part does not show a decrease in $\rho(T)$, while the asymmetric one decreases with growth of temperatures [12]. It is seen from Fig. 18.4 that the behavior of the asymmetric part of the differential resistance given by (18.7) and (18.8) is in accord with the experimental data. It is seen from Figs. 18.4 and 18.5 that the asymmetric part shows a linear behavior as a function of the voltage below about 1 mV [12], just as it was predicted in [8].

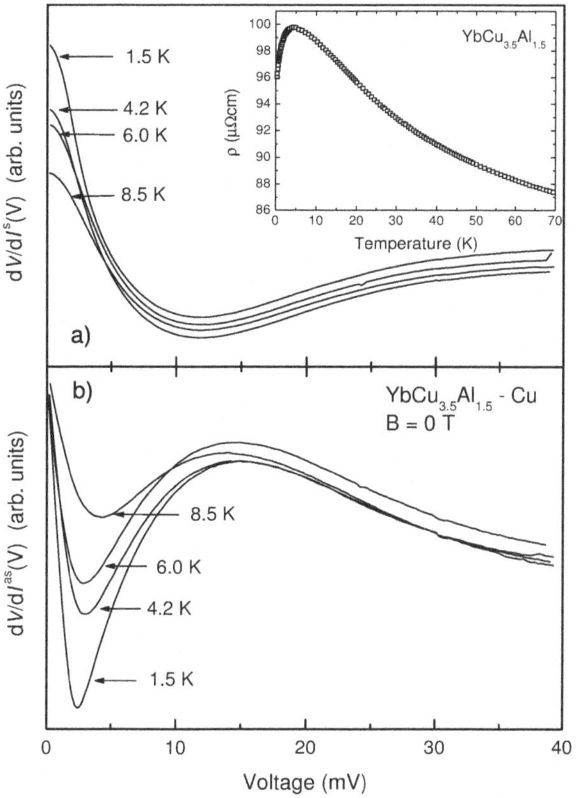

Fig. 18.4 Characteristic temperature behavior of (**a**) symmetric $\mathrm{d}V/\mathrm{d}I^s(V)$ and (**b**) asymmetric $\mathrm{d}V/\mathrm{d}I^{as}(V)$ parts of $\mathrm{d}V/\mathrm{d}I(V)$ for heterocontact $YbCu_{3.5}Al_{1.5}$ − Cu at $B = 0$ and different temperatures shown by the arrows. The inset shows the bulk resistivity $\rho(T)$ of $YbCu_{3.5}Al_{1.5}$ [12]

One can see from Fig. 18.5 [12] that with increase of magnetic fields the asymmetric part is suppressed. Thus, the application of magnetic fields destroys the NFL behavior and recovers both the LFL state and the particle-hole symmetry. We conclude that the particle-hole symmetry is macroscopically broken in the absence of applied magnetic fields, while the application of magnetic fields restores both the particle-hole symmetry and the LFL state. We note that the violation of the particle-hole symmetry destroys the T-symmetry, that is, the symmetry of physical laws under a time-reversal transformation.

18.2 Superconducting State

Tunnel conductivity may remain asymmetric as a high-T_c superconductor or a HF metal passes into the superconducting state from the normal one. The reason is that the function $n_0(\mathbf{p})$ again determines the differential conductivity. As we have seen in Sect. 6.1, $n_0(\mathbf{p})$ is inessential distorted by the pairing interaction. This is because the

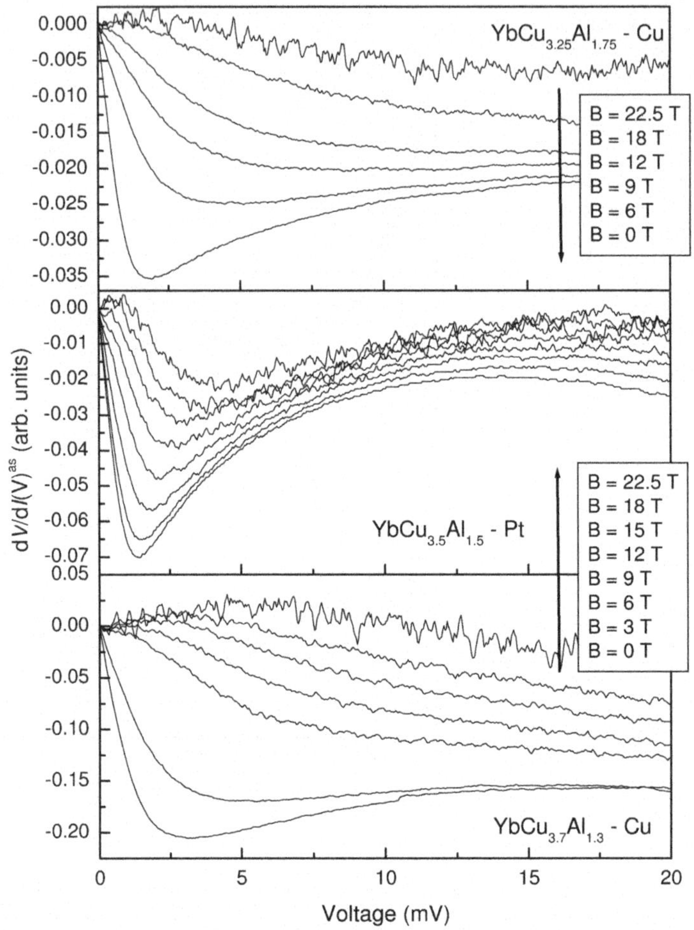

Fig. 18.5 Asymmetric part $dV/dI^{as}(V)$ of the differential conductivity versus magnetic fields. The values of the field are displayed in the legends for heterocontacts with different $x = 1.3, 1.5$, and 1.75 at 1.5 K [12]

latter interaction is weaker than the Landau one, which shapes the distribution function $n_0(\mathbf{p})$. Hence, the asymmetric part of the conductivity remains almost unchanged for $T \le T_c$, which agrees with the experimental results, see Fig. 18.2. In calculating the conductivity using the results of tunneling microscopy measurements, we should keep in mind that the density of states of a superconductor

$$N_S(E) = N(\varepsilon - \mu)\frac{E}{\sqrt{E^2 - \Delta^2}} \tag{18.9}$$

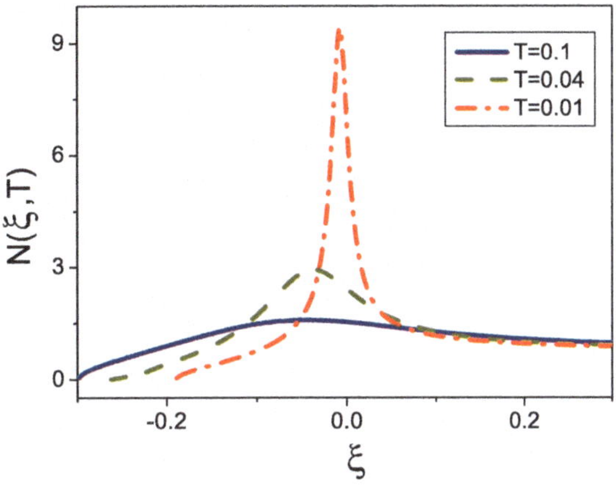

Fig. 18.6 Density of states $N(\xi, T)$ as a function of $\xi = (\varepsilon - \mu)/\mu$. $N(\xi, T)$ is calculated for three values of the temperature T, normalized to μ

determines the conductivity, which is zero for $E \leq |\Delta|$, see, e.g., [13–15]. Here, E is the quasiparticle energy given by (6.5), and $\varepsilon - \mu = \sqrt{E^2 - \Delta^2}$. Equation (18.9) implies that the tunnel conductivity may be asymmetric, if the density of states in the normal state $N(\varepsilon)$ is asymmetric with respect to the Fermi level [16], which is the case for strongly correlated Fermi systems with FC. Our calculations of the above density of states corroborate this conclusion.

Figure 18.6 depicts the results of the calculated density of states $N(\xi, T)$. It is seen that $N(\xi, T)$ is strongly asymmetric with respect to the Fermi level. If the system is in the superconducting state, the normalized temperature listed in the legend can be related to Δ_1. With $\Delta_1 \simeq 2T_c$, we find that $2T/\mu \simeq \Delta_1/\mu$. Since $N(\xi, T)$ is asymmetric, the first derivative $\partial N(\xi, T)/\partial \xi$ is finite at the Fermi level, and the function $N(\xi, T)$ can be written as $N(\xi, T) \simeq a_0 + a_1 \xi$ for small values of ξ. The coefficient a_0 does not contribute to the asymmetric part. Obviously, the value of $\Delta \sigma_{\rm d}(V)$ is determined by the coefficient $a_1 \propto M^*(\xi = 0)$. In turn, $M^*(\xi = 0)$ is determined by (6.8). As a result, (18.7) becomes

$$\Delta \sigma_{\rm d}(V) \sim c_1 \frac{V}{|\Delta|} \frac{S_0}{x_{FC}}, \tag{18.10}$$

because $(p_f - p_i)/p_F \simeq S_0/x_{FC}$, the energy E is replaced by the voltage V, and $\xi = \sqrt{V^2 - \Delta^2}$. The entropy S_0 here refers to the normal state of a heavy-fermion metal.

In fact, (18.10) coincides with (18.7) if we have in mind the fact that the characteristic energy of the superconducting state is determined by (6.9) and is temperature-independent. In studies of the universal behavior of the asymmetric conductivity, (18.10) proved to be more convenient than (18.9). It follows from (18.7) and (18.10) that the measurements of the transport properties, for instance, the asymmetric part of the conductivity, allow to determine the thermodynamic properties of the normal phase that are related to the entropy S_0. Equation (18.10) clearly shows that the

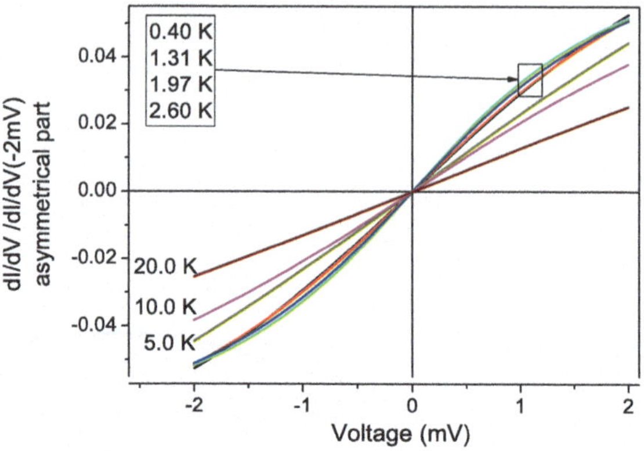

Fig. 18.7 Temperature dependence of the asymmetric parts $\Delta\sigma_d(V)$ of the conductance spectra extracted from measurements on $CeCoIn_5$ [11]. The temperatures are in the box and shown by the arrow for $T \leq 2.60$ K. Otherwise, the higher temperatures are given by numbers near the corresponding curves

asymmetric part of the differential tunneling conductivity becomes comparable to the differential tunneling conductivity at $V \sim 2|\Delta|$ if FC occupies a substantial part of the Fermi volume, $(p_f - p_i)/p_F \simeq 1$. In the case of the d-wave symmetry of the gap, the right-hand side of (18.10) must be averaged over the gap angular distribution $\Delta(\phi)$. This simple procedure amounts to redefining the gap size or the constant c_1. As a result, (18.10) can also be applied when $V < \Delta_1$, where Δ_1 is the maximum size of the d-wave gap [9]. For the Andreev reflection [1–3], where the current is finite for any small V, (18.10) also holds for $V < \Delta_1$ in the case of the s-wave gap.

It is seen from Fig. 18.7 that the asymmetric part $\Delta\sigma_d(V)$ of the conductivity remains the same up to temperatures of about T_c and persists up to temperatures well above T_c. At small voltages, the asymmetric part is a linear function of V and starts to diminish at $T \geq T_c$. It follows from Fig. 18.7 that the above description of the asymmetric part based on (18.7) and (18.10) coincides with the experimental data for $CeCoIn_5$.

Low-temperature tunneling microscopy and spectroscopy measurements have been used in [17] to detect an inhomogeneity in the electron density distribution in $Bi_2Sr_2CaCu_2O_{8+x}$. This inhomogeneity manifests itself as spatial variations in the local density of states in the low-energy part of the spectrum and in the size of the superconducting gap. The inhomogeneity observed in the integrated local density of states is not caused by impurities but is inherent in the system. This observation helps to relate the integrated local density of states to the concentration x of local oxygen impurities.

Spatial variations in the differential tunneling conductivity spectrum are reported in Fig. 18.8. Clearly, the latter conductivity is highly asymmetric in the superconducting state of $Bi_2Sr_2CaCu_2O_{8+x}$. The differential tunneling conductivity shown in Fig. 18.8 may be interpreted as measured at different values of $\Delta_1(x)$ but at the same

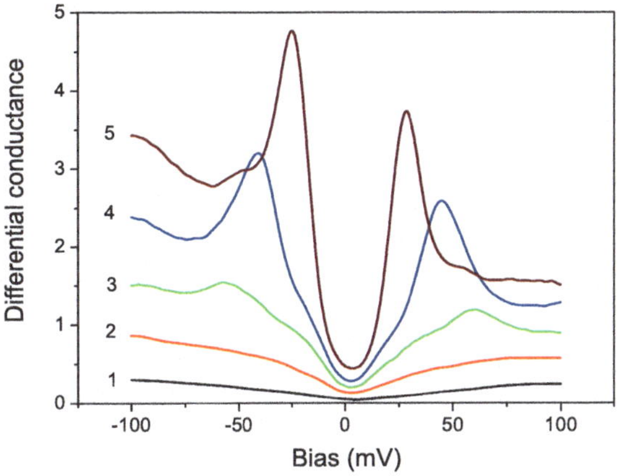

Fig. 18.8 Spatial variation of the differential tunneling conductance spectra of the $Bi_2Sr_2CaCu_2O_{8+x}$. Lines 1 and 2 belong to regions where the integrated local density of states is very low. Low differential conductivity and the absence of a gap show that we are dealing with an insulator. Line 3 corresponds to a large gap (65 meV) with mildly pronounced peaks. The integrated value of the local density of states for curve 3 is small, but is higher than that for lines 1 and 2. Line 4 corresponds to a gap of about 40 meV, which is close to the average value. Line 5 corresponds to the maximal integrated local density of states and the smallest gap of about 25 meV. Also, it has two sharp coherent peaks [17]

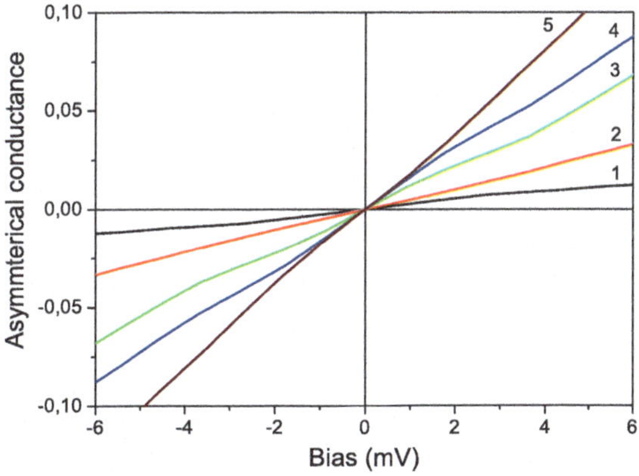

Fig. 18.9 The asymmetric part $\Delta\sigma_d(V)$ of the differential tunneling conductance in the high-T_c superconductor $Bi_2Sr_2CaCu_2O_{8+x}$. The corresponding values are extracted from the data in Fig. 18.8, and are presented as a function of the voltage V (mV). The lines numbering is consistent with that in Fig. 18.8

temperature, which allows studying the $\Delta\sigma_d(V)$ dependence on $\Delta_1(x)$. Figure 18.9 shows the asymmetric conductivity diagrams obtained from the data in Fig. 18.8. Clearly, for small voltages, $\Delta\sigma_d(V)$ is a linear function of V consistent with (18.10) and the slope of the respective straight lines $\Delta\sigma_d(V)$ is inversely proportional to the gap size Δ_1.

Figure 18.10 reports the variation of the asymmetric part $\Delta\sigma_d(V)$ with the temperature increase. The measurements have been performed on YBa_2 Cu_3O_{7-x} / $La_{0.7}Ca_{0.3}MnO_3$ with $T_c \simeq 30$ K [18]. It is seen that at $T < T_c \simeq 30$ K, the asym-

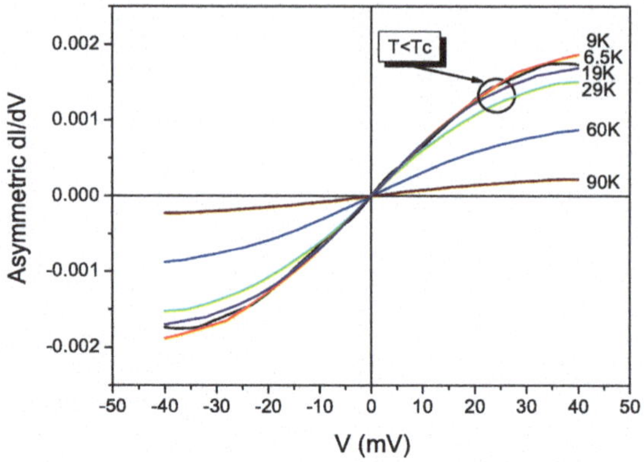

Fig. 18.10 Temperature dependence of the asymmetric part $\Delta\sigma_d(V)$ of the conductivity spectra. The data are obtained in measurements on $YBa_2Cu_3O_{7-x}/La_{0.7}Ca_{0.3}MnO_3$ by the contact spectroscopy method; the critical temperature $T_c \simeq 30$ K [18]

metric part $\Delta\sigma_d(V)$ depends on temperature only weakly in the region of the linear V-dependence. Such behavior agrees with (18.10). At $T > T_c$, the slope of the straight parts of $\Delta\sigma_d(V)$ dependence decreases with increase in temperature. This behavior is described by (18.7). We conclude that the description of the universal behavior of $\Delta\sigma_d(V)$ based on the FCQPT is in good agreement with the results of the experiments presented in Figs. 18.3, 18.4, 18.5, 18.7, 18.9, and 18.10 and is valid for both high-T_c superconductors and HF metals.

18.3 Relation to the Baryon Asymmetry in the Early Universe

The demonstrated above particle-hole symmetry violation in the NFL state of HF compounds has its large-scale counterpart in the asymmetry between matter and antimatter in the early Universe [19]. In this case, the FCQPT concept delivers underlying physical mechanism for both abovementioned processes, which means that the FC phenomenon is rather general and not seldom in Nature. As the details of matter-antimatter (baryon) asymmetry is discussed in detail in [6], here we make some general remarks regarding this question.

As it is well known (see, e.g., [20–22]), the relation between particles and antiparticles in the Universe is governed by so-called CP (or more generally CPT) symmetry, which is the result of successive action of charge conjugation (C), transforming particle into antiparticle and parity (P), which reverses the directions of spatial coordinates. One more (T) symmetry results in time reversal, which reverses the time arrow. Simply speaking, the CPT symmetry is responsible for spin and charge conjugation, transforming particles into antiparticles. It is widely believed that at the initial stages of Universe, called Big Bang, creation the number of particles and antiparticles (or baryons and antibaryons, i.e., the rest mass carrying elementary particles with nonzero rest mass, consisting of only either quarks or antiquarks, respectively,

[22]) was the same, while at the later stages, when the Universe began to cool down, this symmetry disappeared, giving rise to the current state where we have large clusters of visible matter (baryons) and the dark matter in their vicinity. The analogy between baryon-antibaryon and (quasi) particle-hole symmetry breaking can be obtained if we match baryon to a hole and antibaryon to a quasiparticle in a Fermi liquid. The standard observation in many ordinary metals is the symmetric character of their conductivity, which is a direct consequence of LFL theory, admitting complete particle-hole symmetry. As we have seen above, latter symmetry breaks down if we are going beyond LFL, which is the case for HF compounds.

Indeed, experimental observations of low-temperature electric conductivity in high-T_c superconductors [1] and in HF metals like $CeCoIn_5$ and $YbCu_{5-x}Al_x$ [11, 12] show that it is clearly asymmetric. Latter asymmetry vanishes as the temperature or magnetic field increases. We have demonstrated above that the asymmetry cannot be explained in the framework of LFL theory since its particle-hole symmetry unavoidably leads to the step function for the fermion distribution function at low temperatures, which, in turn, results in a symmetric conductivity. To explain this asymmetry, the FCQPT notion has been invoked [19]. We note that although fundamental microscopic interaction in FCQPT theory is fully symmetric with respect to quasiparticles and holes, at low temperatures it causes the spontaneous symmetry breaking [5, 8]. Asymmetry is due to the simple fact that, contrary to LFL, in strongly correlated fermion systems, the single-particle energy $\varepsilon(\mathbf{p})$ is temperature- and magnetic-field-dependent. Thus, $n(\mathbf{p}, T)$ is not the step function in the low-temperature limit [23, 24]. The FCQPT approach of [19] is based on the observation that some condensed matter systems like HF compounds have topologically protected gapless and dispersionless fermions forming flat bands [25–27], which promote the FCQPT. This quantum phase transition, in turn, breaks the particle-hole symmetry generating, among others, the observable asymmetric conductivity.

The same notion can be used to explain the baryon-antibaryon asymmetry in the Universe, which does not require any artificial extension of the standard models of cosmology. Namely, it is suggested that the initial state of the Universe was completely symmetric with the baryon number and CP conserved at the end of the inflation when the particle production started [19]. The observed asymmetry has been explained by suggesting that after the initial inflation, as the Universe cooled down in approximately 10 orders of magnitude, it came close to FCQPT similar to the situation in HF metals. At that time an excess of matter over the antimatter in the Universe had been generated by FC phenomenon. As the Universe cools down further, it has all NFL properties, dictated by FCQPT. We assume that this model describes the particle-antiparticle content of the Universe. At finite temperatures, baryon-antibaryon asymmetry emerges as an inherent property of the system located in FC state. The asymmetry results from the distortion of the Fermi surface, or, in other words, because of the deviation of the distribution function $n(\mathbf{p})$ from the step function at low temperatures. At temperature lowering, the system approaches the quantum-critical line which increases the asymmetry. Details of such increase depend on the model interparticle potential and other parameters. This interesting and important analogy is worth further extensive studies, which, in our opinion, can shed light not only on yet unsolved problems in condensed matter physics, but also on those in cosmology and particle physics.

18.4 Conclusion

In this chapter, we have attracted attention to the important experimental technique represented by scanning tunneling microscopy, which has the advantage of being sensitive both to the density of states and quasiparticle occupation numbers. The reason for this dual sensitivity is that this technique is well suited to studying effects related to the violation of particle-hole symmetry and time-reversal invariance. Their violation leads to asymmetry of the differential tunneling conductivity and resistance to the applied voltage V or current I. Based on the experimental results, we have demonstrated that the asymmetric part of both the conductivity and the resistivity vanishes under the application of a magnetic field, as it has been predicted within the frameworks of the fermion-condensation theory [9, 10]. The theory gives excellent explanations to experimental facts related to the observation of the asymmetry, see also Chap. 19, thus laying firm grounds to studying HF compounds as the new state of matter.

References

1. G. Deutscher, Rev. Mod. Phys. **77**, 109 (2005)
2. A.M. Zagoskin, *Quantum Theory of Many-Body Systems* (Springer-Verlag Inc., New York, 1998)
3. A.F. Andreev, Zh Eksp, Teor. Fiz. **46**, 1823 (1964)
4. A. Schiller, S. Hershfield, Phys. Rev. B **61**, 9036 (2000)
5. V.R. Shaginyan, M.Y. Amusia, A.Z. Msezane, K.G. Popov, Phys. Rep. **492**, 31 (2010). (arXiv:1006.2658)
6. M.Y. Amusia, K.G. Popov, V.R. Shaginyan, W.A. Stephanowich, *Theory of Heavy-Fermion Compounds*. Springer Series in Solid-State Sciences, vol. 182 (Springer, Berlin, 2015)
7. V.R. Shaginyan, M.Y. Amusia, K.G. Popov, Phys. Usp. **50**, 563 (2007)
8. V.R. Shaginyan, K.G. Popov, Phys. Lett. A **361**, 406 (2007)
9. V.R. Shaginyan, JETP Lett. **81**, 222 (2005)
10. V.R. Shaginyan, K.G. Popov, V.A. Stephanovich, E.V. Kirichenko, J. Alloys Comp. **442**, 29 (2007)
11. W.K. Park, L.H. Greene, J.L. Sarrao, J.D. Thompson, Phys. Rev. B **72**, 052509 (2005)
12. G. Pristáš, M. Reiffers, E. Bauer, A.G.M. Jansen, D.K. Maude, Phys. Rev. B **78**, 235108 (2008)
13. L.N. Oliveira, E.K.U. Gross, W. Kohn, Phys. Rev. Lett. **60**, 2430 (1988)
14. J. Bardeen, L.N. Cooper, J.R. Schrieffer, Phys. Rev. **108**, 1175 (1957)
15. D.R. Tilley, J. Tilley, *Superfluidity and Superconductivity* (Hilger, Bristol, 1985)
16. P.W. Anderson, N.P. Ong (2006)
17. S.H. Pan, J.P. O'Neal, R.L. Badzey, C. Chamon, H. Ding, J.R. Engelbrecht, Z. Wang, H. Eisaki, S. Uchida, A.K. Gupta, K.W. Ng, E.W. Hudson, K.M. Lang, J.C. Davis, Nature **413**, 282 (2001)

18. S. Piano, F. Bobba, A.D. Santis, F. Giubileo, A. Scarfato, A.M. Cucolo, J. Phys. Conf. Ser. **43**, 1123 (2006)
19. V.R. Shaginyan, G.S. Japaridze, M.Y. Amusia, A.Z. Msezane, K.G. Popov, Europhys. Lett. **94**, 69001 (2011)
20. M. Sozzi, *Discrete symmetries and CP violation: From experiment to theory* (Oxford University Press, Oxford, 2008)
21. G.C. Branco, L. Lavoura, J.P. Silva, *CP violation* (Clarendon Press, Oxford, 1999)
22. D.J. Griffiths, *Introduction to elementary particles* (Wiley, New York, 1987)
23. V.A. Khodel, V.R. Shaginyan, JETP Lett. **51**, 553 (1990)
24. V.A. Khodel, V.R. Shaginyan, V.V. Khodel, Phys. Rep. **249**, 1 (1994)
25. G.E. Volovik (2010). http://arxiv.org/abs/cond-mat/:1012.0905v3
26. G.E. Volovik, J. Low Temp. Phys. **110**, 23 (1998)
27. G.E. Volovik, *The Universe in a Helium Droplet* (Clarendon Press, Oxford, 2003)

Chapter 19
Asymmetric Conductivity, Pseudogap and Violations of Time and Charge Symmetries

Abstract In this Chapter we show that tunneling differential conductivity (or resistivity) is a sensitive tool to experimentally test the non-Fermi liquid (NFL) behavior of strongly correlated Fermi systems. In the case of common metals the Landau Fermi liquid (LFL) theory demonstrates that the differential conductivity is a symmetric function of bias voltage V. This is because the particle-hole symmetry is conserved in LFL state. When a strongly correlated Fermi system approaches the topological fermion condensation quantum phase transition, its LFL properties disappear so that the particle-hole symmetry breaks making the differential tunneling conductivity an asymmetric function of V. This asymmetry can be observed when a strongly correlated metal is in its normal, superconducting or pseudogap states. We show that the asymmetric part of the dynamic conductance does not depend on temperature provided that the metal is in its superconducting or pseudogap states. In normal state the asymmetric part diminishes at rising temperatures. Under the application of magnetic field the metal transits to the LFL state and the differential tunneling conductivity becomes a symmetric function of V. These findings are in good agreement with experimental observations on graphene.

19.1 Introduction

The unusual properties of strongly correlated Fermi systems observed in high-T_c superconductors (HTSC) and heavy-fermion (HF) metals are due to a quantum phase transition in its quantum critical point (QCP) at $T = 0$. The above quantum phase transition (QPT) is driven by chemical composition, pressure, number density x of electrons (holes), magnetic field B, etc. It is commonly accepted that QPT is the main cause of the NFL behavior exhibited by strongly correlated Fermi systems. Experimental observations of strongly correlated Fermi systems explore mainly their thermodynamic, relaxation and transport properties. It is highly desirable to probe the other properties of the Fermi systems, which are not directly linked to the density of states (DOS) or to the behavior of the effective mass M^*. Scanning tunneling microscopy and point contact spectroscopy based on Andreev reflection (AR) [1] are sensitive to quasiparticle occupation numbers, and are ideal techniques to study the

M. Amusia and V. Shaginyan, *Strongly Correlated Fermi Systems*,
Springer Tracts in Modern Physics 283,
https://doi.org/10.1007/978-3-030-50359-8_19

effects of particle-hole symmetry violation, making the differential tunneling conductivity and dynamic conductance to be asymmetric function with respect of the sign of both the applied voltage V and the supported electrical current I. The asymmetric part of the conductivity (or the conductance) $\sigma_{\rm asym}(V) \equiv I'(V) - I'(-V)$ ($I' \equiv dI/dV$) can be observed when strongly correlated Fermi system with heavy electrons are either normal or superconducting [2–5]. This asymmetry does not occur in conventional metals, especially at low temperatures. The reason is that accordingly to the LFL theory, the particle-hole symmetry is conserved and both the differential tunneling conductivity and dynamic conductance are symmetric functions of V [6–8]. The theory of fermion condensation (FC) taking place at the topological fermion condensation quantum phase transition (FCQPT) [9–11] allows one to explain the NFL behavior of many strongly correlated Fermi systems like HF metals, HTSC, quantum spin liquids, quasicrystals, one and two dimensional Fermi systems [7, 8]. Moreover, it has been predicted within the framework of the FC theory, that the differential tunneling conductivity, which is commonly symmetric function of V, becomes noticeably asymmetric in the case of HF metals. The reason for that is that their electronic system is located near FCQPT similar to the situation with the archetypical HF metal $YbRh_2Si_2$. The asymmetry can be observed in HF metals regardless of their normal or superconducting state. It has also been predicted that under the application of magnetic field B that drives a HF metal to its LFL state $\sigma_{\rm asym}(V) \to 0$ [2, 4, 5]. In fact, the application of magnetic field B restores the LFL properties so that the conductivity becomes a symmetric function of V due to the restoration of a particle-hole symmetry. More generally, in this case the time (T) reversal symmetry and particle-antiparticle (C) invariance are restored.

The experimental observation of flat band and superconducting (SC) state in graphene [12] attract strong attention to the band flattening, for it can lead to the bulk room-T superconductivity in graphite. Both the band flattening and the emergence of SC state are in accordance with experimental and theoretical observations suggesting that the flattening indeed raises T_c, and can be accompanied by asymmetrical conductivity that can induce the time-reversal symmetry breaking [13].

19.2 Asymmetric Conductivity and the NFL Behavior

In this Chapter, based on experimental data, we show that particle-hole symmetry breaks when HF metal and/or HTSC are located near FCQPT and exhibit the NFL behavior in their superconducting, pseudogap and normal states. As a result, the asymmetric part $\sigma_{\rm asym}(V)$ of differential tunneling conductivity becomes nonzero. Since the application of magnetic field destroys the NFL behavior, it restores the above symmetries, nullifying $\sigma_{\rm asym}(V)$. This nullification can be extracted from experimental observations [12, 14, 15] and has been predicted theoretically in [2, 4, 5]. Thus, the existence of the asymmetric part $\sigma_{\rm asym}(V)$ is manifestation of the NFL behavior that can be destroyed by the application of magnetic field. Therefore the measurements of $\sigma_{asym}(V)$ can be viewed as a powerful tool to investigate the NFL behavior of strongly correlated Fermi systems, see also Chap. 18.

Direct experimental studies of quantum phase transitions in HTSC and HF metals are of great importance for understanding the underlying physical mechanisms responsible for their anomalous properties. However, such studies of HF metals and HTSC are difficult because the corresponding critical points are usually concealed by the proximity to other phase transitions, commonly antiferromagnetic (AF) and/or superconducting (SC). Recently, unusual properties of tunneling conductivity in the presence of a magnetic field were observed in a graphene sample having a flat band [12], as well as in HTSC's and the HF metal $YbRh_2Si_2$ [14, 15]. Measuring and analyzing these properties will shed light on the nature of the quantum phase transitions occurring in these substances.

Most of the experiments on HF metals and HTSC's explore their thermodynamic properties. However, it is equally important to determine other properties of these strongly correlated systems, notably quasiparticle occupation numbers $n(p, T)$ as a function of momentum p and temperature T. These quantities are not linked directly to the density of states (DOS) $N(\varepsilon = 0)$ determined by the quasiparticle energy ε or to the behavior of the effective mass M^*. Scanning tunneling microscopy [6] and point contact spectroscopy [1], being sensitive to both the density of states and quasiparticle occupation numbers, are ideal tools for exploring the effects of C and T symmetry violation. When C and T symmetries are not conserved, the differential tunneling conductivity and dynamic conductance are no longer symmetric functions of the applied voltage V.

Indeed, if under the application of bias voltage V, the current of electrons with charge $-e$, traveling from HF to a common (i.e. "non-HF") metal changes the sign of a charge carrier to $+e$, then current character and direction alters. Namely, now the carriers are holes with charge $+e$ traveling from the common to HF metal. Turning this around, one can obtain the same current of electrons provided that V is changed to $-V$. The resulting asymmetric differential conductivity $\Delta\sigma_d(V) = \sigma_d(V) - \sigma_d(-V)$ becomes nonzero, as it is seen from Fig. 19.2. On the other hand, if time t is changed to $-t$ (but charge is kept intact), the current changes its direction only. The same result can be achieved by $V \to -V$, and we conclude that T symmetry is broken, provided that $\Delta\sigma_d(V) \neq 0$. Thus, we detect the presence of $\Delta\sigma_d(V) \neq 0$ signals due to both C and T symmetries violation. At the same time, the change of both $e \to -e$ and $t \to -t$ returns the system to its initial state so that CT symmetry is conserved. Note that the parity symmetry P is conserved and the well-known CPT symmetry is not broken in the considered case. On the other hand, since the time-reversal invariance and particle-hole symmetry remain valid in normal Fermi systems, the differential tunneling conductivity and dynamic conductance are symmetric functions of V. Thus, a conductivity asymmetry is not observed in conventional metals at low temperatures.

19.3 Schematic Phase Diagram

We briefly recollect the main ideas of FC theory, for details see Chap. 4. As soon as the number density x of the liquid approaches some threshold value x_{FC}, the electronic quasiparticle dispersion disappears in some part of its spectrum, forming so-called flat bands [7, 8]. This dispersionless part is usually situated between, say, momentum p_i (standing for initial momentum) and p_f (final momentum). It is clear, that in this dispersionless part of the spectrum, the effective mass M^* diverges. Beyond $x_{\rm FC}$, the system undergoes FCQPT with FC formation, so that the step quasiparticle distribution function $n_F(p, T \to 0) = \theta(p - p_F)$ does not deliver the minimum to the Landau functional $E[n(\mathbf{p})]$ and the quasiparticle distribution is determined by the equation for the minimum of the functional [9]

$$\frac{\delta E[n(\mathbf{p})]}{\delta n(\mathbf{p}, T = 0)} = \varepsilon(\mathbf{p}) = \mu; \; p_i \le p \le p_f. \tag{19.1}$$

At $T = 0$ (19.1) determines the quasiparticle distribution function $n_0(\mathbf{p})$ minimizing the ground state energy $E \equiv E[n(\mathbf{p})]$. Figure 19.1 reports a typical $n_0(\mathbf{p})$ at dimensionless temperature $T/E_F = 0.0001$ with E_F being the Fermi energy. It is seen from Fig. 19.1 that the particle-hole symmetry is violated by FC since $n_0(\mathbf{p})$ does not evolve from the Fermi–Dirac distribution function (which is a step at $T = 0$, shown by dash-dotted line), it is expected that the conductivity possesses an asymmetric part.

The $CeCoIn_5$ schematic $T - B$ phase diagram is shown in Fig. 19.2. The key feature here is a magnetic field-induced QCP at B_{c0} hidden inside the SC state [16].

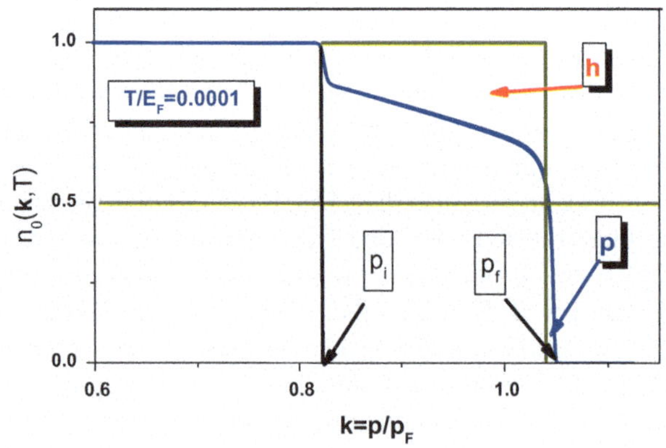

Fig. 19.1 The quasiparticle distribution function $n_0(p)$. Since $n_0(p)$ is a solution of (19.1) it implies that $n_0(p < p_i) = 1$, $0 < n_0(p_i < p < p_f) < 1$ and $n_0(p > p_f) = 0$. Quasiparticles located at $p_i \le p \le p_f$ form fermion condensate similar to the case of Bose condensation. The Fermi momentum p_F obeys the condition $p_i < p_F < p_f$. Momenta p_i and p_f and the quasiparticles (p) and quasiholes (h) areas are shown by the arrows

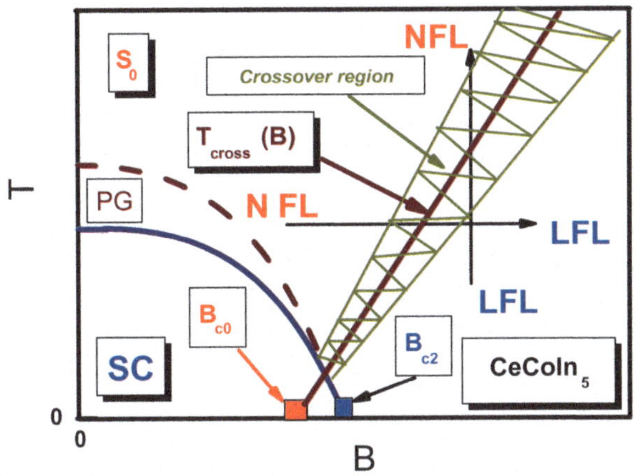

Fig. 19.2 Schematic $T - B$ phase diagram of $CeCoIn_5$. The vertical and horizontal arrows, crossing the transition region depicted by the thick lines, show the LFL-NFL and NFL-LFL transitions at fixed B and T, respectively. As shown by the solid curve, at $B < B_{c2}$ the system is in its SC state. The points B_{c2} and B_{c0} denote, respectively, the critical field eliminating the SC state and the quantum critical point hidden beneath the "SC dome" where the flat band could exist at $B \leq B_{c0}$. Pseudogap state is labeled by "PG". The hatched area with the solid line $T_{\rm cross}(B)$ represents the crossover, separating the domains of NFL and LFL behaviors. A part of the crossover shown by $T_{\rm cross}(B)$ is hidden inside the SC state. The NFL state is characterized by the entropy excess S_0 that leads to $\sigma_{\rm asym}(V)$, see (19.3) and (19.4), and Fig. 19.4, panels (a), (b) and Fig. 19.5

As it is seen from Fig. 19.2, $B_{c2} > B_{c0}$ [17, 18] and the LFL behavior persists at $T \leq T_{\rm cross}$ until the SC state emerges, eliminating the temperature independent part of the normal (rather then superconducting) state entropy $S_0 \sim \sum_p [n_0(p)(1 - n_0(p))]$ [7]. The hatched area, including line $T_{\rm cross}(B)$, describes crossover, which separates the domain of NFL behavior from that of LFL one.

19.4 Heavy Fermion Compounds and Asymmetric Conductivity

Here we briefly consider the tunneling current. For details see Sect. 18.1. The tunneling current I between two ordinary metals is proportional to the driving voltage V and to the squared modulus of the quantum mechanical transition amplitude t multiplied by the expression $N_1(0)N_2(0)(n_1(p,T) - n_2(p,T))$ [6], where $N(0)$ is DOS of the corresponding metal. On the other hand, the wave function calculated in the WKB approximation is proportional to $(N_1(0)N_2(0))^{-1/2}$ so that the density of states cancels down and the tunneling current becomes independent of $N_1(0)N_2(0)$ [6]. Consider the tunneling current at low temperatures that in the case of ordinary metals is given by

$$I(V) = 2|t|^2 \int [n_F(\varepsilon - V) - n_F(\varepsilon)]\, d\varepsilon, \tag{19.2}$$

for the density of states is constant at the Fermi level, and be included into the factor t. In the case of HF metals the density of states magnify the asymmetry; for simplicity we use (19.2), for details see Sect. 18.1. At low temperatures we approximate $n_F(\varepsilon)$ by the step function $\theta(\varepsilon - \mu)$, μ is the chemical potential. It follows from (19.2) that quasiparticles with the energy ε, $\mu - V \leq \varepsilon \leq \mu$, contribute to the current. It follows from (19.2) that $I(V) = a_1 V$ and $\sigma_d(V) = dI/dV = a_1$, $a_1 = const$ and within the framework of the LFL theory the differential tunneling conductivity $\sigma_d(V)$ is a symmetric function of the voltage V and $\sigma_{asym} = 0$. Obviously, σ_{asym} becomes finite provided that the step function n_F has to be substituted by n_0, and therefore particle-hole symmetry is violated. As a result, we obtain the expression for $\sigma_{asym}(V)$ when the system in its normal (nonsuperconducting) state [4, 5, 7]

$$\sigma_{asym}(V) \simeq c\left(\frac{V}{2T}\right)\frac{p_f - p_i}{p_F} \simeq c\frac{V}{2T}\frac{S_0}{x}, \tag{19.3}$$

and another expression if the system in its superconducting state

$$\sigma_{asym}(V) \simeq c\frac{V}{\Delta_1}\frac{p_f - p_i}{p_F} \simeq c\frac{V}{\Delta_1}\frac{S_0}{x}. \tag{19.4}$$

Here, $x = p_F^3/(3\pi^2)$ is the number density of the heavy electron liquid, c is a constant that approximately $c \simeq 1$. We note that S_0, characterizing the normal state, can be estimated measuring the asymmetric part of tunneling conductance of the superconducting state. The scale $2T$ entering (19.3) is replaced by the scale Δ_1 in (19.4), where Δ_1 is the maximum value of the superconducting gap. In the same way, as (19.3) is valid up to $V \sim 2T$, (19.4) is valid up to $V \sim \Delta_1$ [4, 5, 7].

The asymmetric conductivity $\sigma_{asym}(V)$ can be observed when both HTSC and HF metals are shifted from their normal to superconducting phase since $n_0(\mathbf{p})$ is responsible for the asymmetric part of their measured differential conductivity. The function $n_0(\mathbf{p})$ is not appreciably disturbed by the superconductive pairing interaction which is relatively weak as compared to the Landau interaction forming the distribution function $n_0(\mathbf{p})$ [4, 5, 8, 19]. Therefore, the asymmetric conductance remains approximately the same below T_c. This result is in good agreement with experimental facts as it is seen from Fig. 19.3. To illustrate the ability of (19.3) and (19.4) to describe experimental data, we consider a temperature dependence of the asymmetric parts $\sigma_{asym}(V)$ of point contact spectra on $YBa_2Cu_3O_{7-x}/La_{0.7}Ca_{0.3}MnO_3$ bilayers with $T_c = 30$ K [20], showing that $\sigma_{asym}(V)$ remains constant up to temperatures of T_c and persists up to temperatures well above T_c, see Fig. 19.3 [4, 5, 7]. It is also seen from Fig. 19.3 that $\sigma_{asym}(V)$ starts to diminish at $T \geq T_c$. These observations are in good agreement with the behavior described by (19.3) and (19.4), and are strong evidence for the physics of FC [7–10].

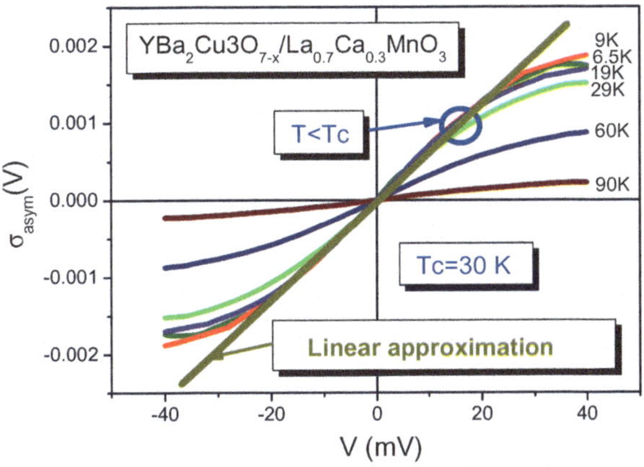

Fig. 19.3 Temperature dependence of the asymmetric parts $\sigma_{\rm asym}(V)$ extracted from measurements on $YBa_2Cu_3O_{7-x}/La_{0.7}Ca_{0.3}MnO_3$ bilayers of the differential conductivity at different temperatures shown in the figure [20]. The solid line shows the approximate constancy of the asymmetric part at $T < T_c$, as it follows from (19.4)

Both the point contact spectroscopy and the tunneling conductivity have been used to investigate the HF metal $CeCoIn_5$ in its superconducting, PG and normal states [19, 21]. The asymmetric part $\sigma_{\rm asym}(V)$ shown in Fig. 19.4 (a) (the point contact spectroscopy [19]) and (b) (the tunneling conductivity [21]), have been extracted from experimental data [19, 21], and $\sigma_{\rm asym}(V)$ has been observed in $CeCoIn_5$ both in its superconducting, PG and the NFL normal states. It is seen from Fig. 19.4a that $\sigma_{\rm asym}(V)$ takes place in the NFL normal states at 10 K, while in the PG state at 2.6 K $\sigma_{\rm asym}(V)$ coincides with that of the SC states. As it is seen from Fig. 19.4b, $\sigma_{\rm asym}(V)$ exhibits the same behavior. The asymmetry becomes more pronounced with decreasing temperature down to the PG state, and remains almost the same at least down to $\simeq 0.3$ K. As a result, we conclude that the physics of the PG state is defined by the presence of FC.

At sufficiently low temperatures the application of magnetic field B restores the LFL behavior making the asymmetric part $\sigma_{\rm asym}(V)$ of the tunneling conductivity vanish [2, 4, 5, 7]. In order to observe the finite asymmetry of the conductivity, the measurements have to be carried out when the corresponding HF metal demonstrates the NFL behavior with particle-hole asymmetry. This asymmetry is typical for HF metals located near FCQPT. Thus, we conclude that the existence of $\sigma_{\rm asym}(V)$ is the typical feature of the NFL behavior. It is seen from the phase diagram of Fig. 19.2 that under the application of magnetic field the HF metal transits from its NFL to LFL part so that the asymmetric part $\sigma_{\rm asym}(V)$ should vanish. It has been predicted within the framework of the FC theory that $\sigma_{\rm asym}(V)$ becomes finite in the archetypical HF metal $YbRh_2Si_2$, and disappears if the metal is moved to the LFL state by magnetic field [4, 5]. Such a suppression of $\sigma_{\rm asym}(V)$ has been observed in the HF metal $YbCu_{5-x}Al_x$ (for $x = 1.5$) in high magnetic fields of 20 T [7, 22]. The asymmetric conductivity [14] and its suppression [15] have been observed in $YbRh_2Si_2$ as it is reported in Fig. 19.5. In that case magnetic fields B is applied parallel to the magnetically hard c axis and the suppression takes place at $B \geq 7$ T.

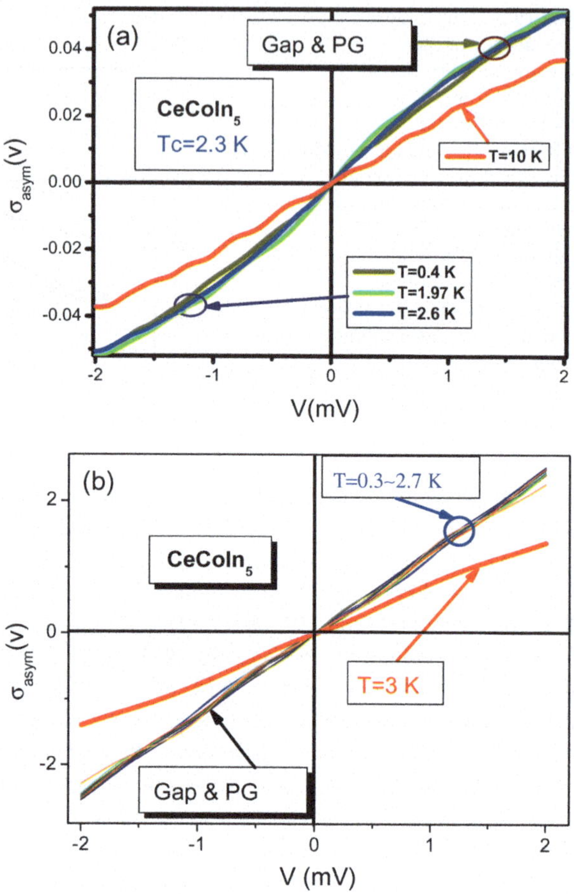

Fig. 19.4 The asymmetric part $\sigma_{\text{asym}}(V)$ in $CeCoIn_5$. Panel (a). The asymmetric part is extracted from the experimental data [19]. At $T \leq 2.6$ K in the PG state and in the SC state, $\sigma_{\text{asym}}(V)$ becomes temperature independent, while $T_c = 2.3$ K is the superconducting phase transition temperature. Panel (b). The asymmetric part is extracted from the experimental data [21]. At $T \leq 2.7$ K in the PG state and down to 0.3 K $\sigma_{\text{asym}}(V)$ are temperature independent

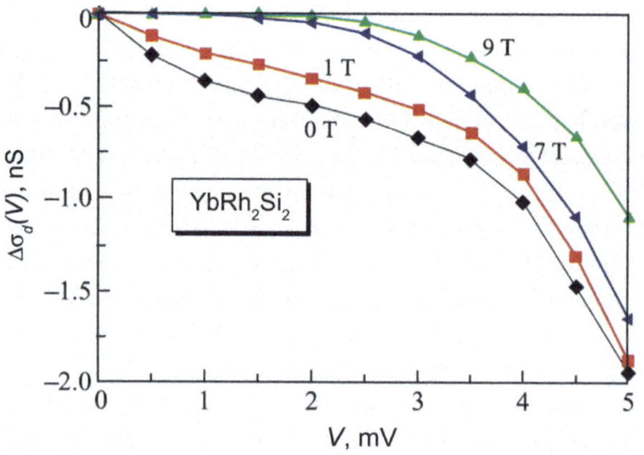

Fig. 19.5 The asymmetric part $\sigma_{\text{asym}}(V)$ at different applied magnetic fields, shown in the legend. $\sigma_{\text{asym}}(V)$ is extracted from the experimental data [15]. At $B \geq 7\text{T}\sigma_{\text{asym}}(V)$ vanishes at $V \leq 2\,\text{mV} \sim 9\,\text{T}$ as it is seen from the phase diagram in Fig. 19.2

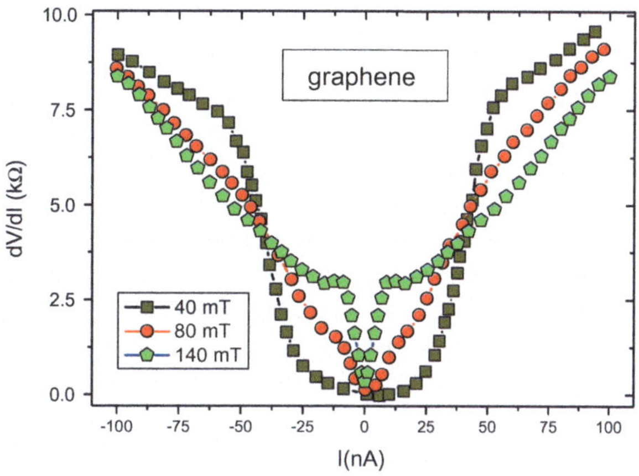

Fig. 19.6 Differential resistance dV/dI as a function of bias current I versus magnetic field B shown in the legend [12]. It is seen that the asymmetry emerges at $B \leq 80$ mT and vanishes at elevated magnetic fields

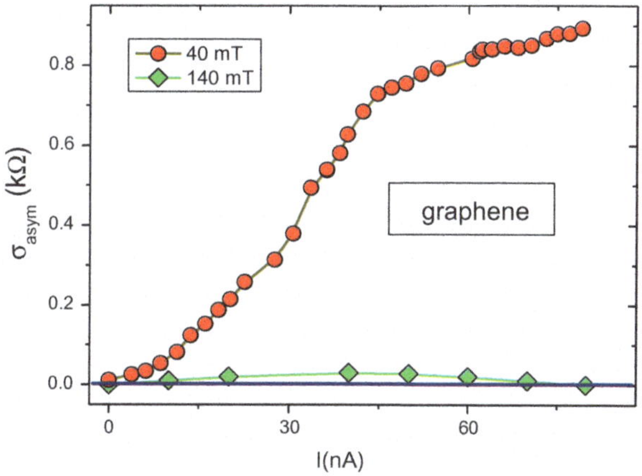

Fig. 19.7 The asymmetric part of the differential resistance $\sigma_{\text{asym}}(k\Omega) = dV/dI(I) - dV/dI(-I)$ is extracted from experimental data [12] exposed in Fig. 19.6

The differential resistance dV/dI as a function of current I can be a hallmark of the NFL behavior since the symmetry properties of dV/dI are similar to those of $\sigma_d(V)$. They are described by (19.3) and (19.4). The differential resistance dV/dI as a function of bias current I at two different magnetic fields B measured on graphene [12] is reported in Fig. 19.6. The asymmetric part of the differential resistance $\sigma_{\text{asym}}(k\Omega) = dV/dI(I) - dV/dI(-I)$ diminishes at elevated magnetic field, and vanishes at $B \simeq 140$ mT, as it is seen from Figs. 19.6 and 19.7. This observation is of great importance, for the graphene has a perfect flat band, making the particle-hole asymmetry generated by FC be obviously exhibited [12]. It is clearly seen from Fig. 19.7, that under the application of tiny magnetic field $B \simeq 140$ mT the asymmetric part $\sigma_{\text{asym}}(k\Omega)$ of the differential resistance vanishes, for the system transits to the LFL state, as it follows from Fig. 19.2.

19.5 Conclusions

The central message of the present Chapter is that if the electronic spectrum of a substance features a dispersionless part, or "flat bands," it is just this aspect that is responsible for measured properties that depart radically from those of familiar condensed-matter systems described by the Landau Fermi liquid theory. This is the case irrespective of varied microscopic details characterizing these substances such as crystal symmetry and defect structure. The explanation for this finding lies in the fact that FC most easier occurs in substances hosting flat bands. Experimental manifestations of FC phenomena are varied, which implies that different experimental techniques are most suitable for detecting and analyzing them. We have shown that one such technique is scanning tunneling microscopy, which has the advantage of being sensitive both to the density of states and quasiparticle occupation numbers. The reason for this dual sensitivity is that this technique is a well suited to studying effects related to the violation of particle-hole symmetry, that is, particle-antiparticle symmetry, and time-reversal invariance. Their violation leads to asymmetry of the differential tunneling conductivity and that of resistance to the applied voltage V or current I. Based on the experimental results, we have demonstrated that the asymmetric part of both the conductivity and the resistivity vanishes under the application of a magnetic field, as predicted in [4, 5]. We have shown that the asymmetric part $\sigma_{\rm asym}$ of the tunneling conductivity/resistivity arises due to the fermion condensation phenomenon in the corresponding substance. This asymmetry is obviously an NFL property. It is observed in HF metals, high-T_c superconductors and graphene. The theories based either on model calculations within Hubbard and Kondo models (see [8] for appropriate references) and ab initio simulations, based actually on the more complicated versions of the same models, are unable to explain the destruction of asymmetric conductivity in sufficiently high magnetic fields. To the best of our knowledge, the presented FC theory is the only one capable to explain the above discussed experimental puzzles. Namely, in accordance with the prediction [2, 4, 5], $\sigma_{\rm asym}$ vanishes under the application of magnetic field as a material transits to the LFL state and the particle-hole symmetry restores. Finally, our study of experimental results strongly suggests that the topological FCQPT is an intrinsic feature of many strongly correlated Fermi systems and can be viewed as the universal agent of their non-Fermi liquid behavior that generates the universal behavior of HF compounds representing the new state of matter.

References

1. A.F. Andreev, Zh Eksp, Teor. Fiz. **46**, 1823 (1964)
2. V.R. Shaginyan, JETP Lett. **81**, 222 (2005)
3. P.W. Anderson, N.P. Ong, J. Phys. Chem. Solids **67**, 1 (2006)
4. V.R. Shaginyan, K.G. Popov, Phys. Lett. A **361**, 406 (2007)
5. V.R. Shaginyan, K.G. Popov, V.A. Stephanovich, E.V. Kirichenko, J. Alloys Comp. **442**, 29 (2007)

6. G. Deutscher, Rev. Mod. Phys. **77**, 109 (2005)
7. V.R. Shaginyan, M.Y. Amusia, A.Z. Msezane, K.G. Popov, Phys. Rep. **492**, 31 (2010). (arXiv:1006.2658)
8. M.Y. Amusia, K.G. Popov, V.R. Shaginyan, W.A. Stephanowich, *Theory of Heavy-Fermion Compounds*. Springer Series in Solid-State Sciences, vol. 182 (Springer, Berlin, 2015)
9. V.A. Khodel, V.R. Shaginyan, JETP Lett. **51**, 553 (1990)
10. G.E. Volovik, JETP Lett. **53**, 222 (1991)
11. V.A. Khodel, V.R. Shaginyan, V.V. Khodel, Phys. Rep. **249**, 1 (1994)
12. Y. Cao, V. Fatemi, S. Fang, K. Watanabe, T. Taniguchi, E. Kaxiras, P. Jarillo-Herrero, Nature **556**, 43 (2018)
13. F. Arnold, J. Nyéki, J. Saunders, JETP Lett. **107**, 577 (2018)
14. S. Ernst, S. Kirchner, C. Krellner, C. Geibel, G. Zwicknagl, F. Steglich, S. Wirth, Nature **474**, 362 (2011)
15. S. Ernst, S. Kirchner, C. Krellner, C. Geibel, G. Zwicknagl, F. Steglich, S. Wirth, Nat. Comm. **9**, 3324 (2018)
16. J. Paglione, M. Tanatar, D. Hawthorn, E. Boaknin, R.W. Hill, F. Ronning, M. Sutherland, L. Taillefer, C. Petrovic, P. Canfield, Phys. Rev. Lett. **91**, 246405 (2003)
17. F. Ronning, C. Capan, E.D. Bauer, J.D. Thompson, J.L. Sarrao, R. Movshovich, Phys. Rev. B **73**, 064519 (2006)
18. F. Ronning, C. Capan, A. Bianchi, R. Movshovich, A. Lacerda, M.F. Hundley, J.D. Thompson, P.G. Pagliuso, J.L. Sarrao, Phys. Rev. B **71**, 104528 (2005)
19. W.K. Park, L.H. Greene, J.L. Sarrao, J.D. Thompson, Phys. Rev. B **72**, 052509 (2005)
20. S. Piano, F. Bobba, A.D. Santis, F. Giubileo, A. Scarfato, A.M. Cucolo, J. Phys. Conf. Ser. **43**, 1123 (2006)
21. S. Wirth, Y. Prots, M. Wedel, S. Ernst, S. Kirchner, Z. Fisk, J.D. Thompson, F. Steglich, Y. Grin, Nat. Comm. **83**, 061009 (2014)
22. G. Pristáš, M. Reiffers, E. Bauer, A.G.M. Jansen, D.K. Maude, Phys. Rev. B **78**, 235108 (2008)

Chapter 20
Violation of the Wiedemann-Franz Law in Strongly Correlated Electron Systems

Abstract The renormalization of the Wiedemann-Franz (WF) ratio in strongly correlated electron systems is analyzed. We demonstrate that the WF law is violated at the quantum-critical point, where the effective mass diverges, and beyond point of fermion condensation (FC), where the single-particle spectrum $\epsilon(p)$ becomes flat. Experimental observations of the much-studied compounds $CeCoIn_5$ and $YbRh_2Si_2$ at vanishing temperatures carefully probe the nature of their magnetic-field-tuned QCPs. The violation of Wiedemann-Franz (WF) law, along with jumps revealed both in the residual resistivity ρ_0 and the Hall resistivity R_H, provides vital clues to the origin of their non-Fermi liquid behavior. The empirical facts point unambiguously to association of the observed QCP with FC forming flat bands.

20.1 Introduction

Wiedemann and Franz [1] discovered the empirical law stating that for different metals at low temperatures the ratio of its thermal conductivity $\kappa(T)$ to its electrical conductivity $\sigma(T)$ is the same, $\kappa(T)/\sigma(T)$ =const. Lorenz showed that the above ratio is proportional to the temperature T, $\kappa(T)/\sigma(T) = LT$, and the proportionality constant L is known as the Lorenz number. In fact, what is really named Wiedemann-Franz (WF) law is indeed is the independence of the Lorenz number L on temperature. However, it was firmly established that the WF law is obeyed both at room temperatures and at low temperatures (several Kelvins), while at intermediate temperatures $L = L(T)$.

Strictly speaking, the Lorenz number is temperature-independent only at low temperatures. Its theoretical value is

$$L_0 = \lim_{T\to 0} = \frac{\kappa(T)}{T\sigma(T)} = \frac{\pi^2}{3}\frac{k_B}{e^2}. \tag{20.1}$$

It had been calculated by Sommerfeld [2] in the model of noninteracting electrons, obeying Fermi-Dirac statistics. The same result is obtained much later in LFL theory and reflects merely the fact that both thermal and electrical conductivities of a metal at

M. Amusia and V. Shaginyan, *Strongly Correlated Fermi Systems*, Springer Tracts in Modern Physics 283,
https://doi.org/10.1007/978-3-030-50359-8_20

low temperatures are determined by Landau quasiparticles. Due to this fact, possible deviations from the WF law can be regarded as a signature of NFL behavior in a considered compound.

Actually, (20.1) is usually referred to as the Wiedemann-Franz (WF) law. It was shown that at $T = 0$ (20.1) remains valid for arbitrarily strong scattering [3], disorder [4], and interactions [5]. This law holds for ordinary metals [6–9] and does not hold for HF metals [10–12] CeNiSn and $CeCoIn_5$ [12], $YbRh_2Si_2$ [13] (see also [14]), the electron-doped material [15] $Pr_{2-x}Ce_xCuO_{4-y}$, and the underdoped compound [16] $YbBa_2Cu_3O_y$. We note here that WF law is also violated in mesoscopic systems, see [17] for review. In CeNiSn, the experimental value of the reduced Lorenz number $L(T)/L_0 \sim 1.5$ varies little at $T < 1$ K. This rules out the phonon contribution to the violation of the WF law. In the electron-doped compound $Pr_{2-x}Ce_xCuO_{4-y}$, the departure of $L(T)$ from L_0 at $T > 0.3$ K is also considerably more then unity and even larger [15] than that in CeNiSn. Other experimental tests of the WF law have been undertaken in the normal state of cuprate superconductors. The phase diagram of these compounds shows evolution from Mott insulator for undoped materials toward metallic Fermi liquid behavior for overdoped cases. Upward shift $L/L_0 \simeq 2 - 3$ was found in underdoped cuprates at the lowest studied temperatures [15, 16, 18]. In strongly overdoped cuprates, the WF law was found to be perfectly correct [19].

The physical mechanism for the WF law violation is usually attributed to the NFL behavior, as it takes place in Luttinger and Laughlin liquids [20–23] or in the case of a marginal Fermi liquid [24]. Yet another possibility for the LFL theory and the WF law (20.1) violation occurs near QCPs where the effective mass M^* of a quasiparticle diverges. This is because at the QCP the Fermi liquid spectrum with finite Fermi velocity $v_F = p_F/M^*$ becomes meaningless as in this case $v_F \to 0$. Here, we analyze the magnetotransport and violation of the WF law in HF compounds like $YbRh_2Si_2$ and $CeCoIn_5$ across a magnetic-field-tuned QCP. Close similarity between the properties of the Hall coefficient R_H and magnetoresistivity ρ at QCP indicates that all manifestations of magnetotransport stem from the same underlying physics. Thus, WF law violation along with the jumps of the Hall coefficient and magnetoresistivity in the zero-temperature limit provide unambiguous evidence for interpreting the QCP in terms of FCQPT forming a flat band in HF compounds [25].

The schematic phase diagram, demonstrating possible regions of the WF law violation, is depicted in Fig. 20.1, with the magnetic field B serving as the control parameter. At $B = 0$, the HF liquid acquires a flat band corresponding to a strongly degenerate state. The NFL regime dominates with increase of temperature and fixed magnetic field. With increasing B, the system is driven from the NFL region to the LFL domain. As it is shown in Fig. 20.1, the system moves from the NFL to the LFL regime along the horizontal arrow, and from the LFL to NFL along the vertical arrow. The magnetic-field-tuned QCP is indicated by the arrow and located at the origin of the phase diagram, since application of magnetic field destroys the flat band and shifts the system into the LFL state. The hatched area denotes the transition region that separates the NFL state from the weakly polarized LFL state and contains the dashed line tracing $T_M(B)$, see Sect. 7.1 and comments below (7.7). This line is defined by the function $T = a_1 \mu_B B$, and the width $W(B)$ of the NFL state is proportional to

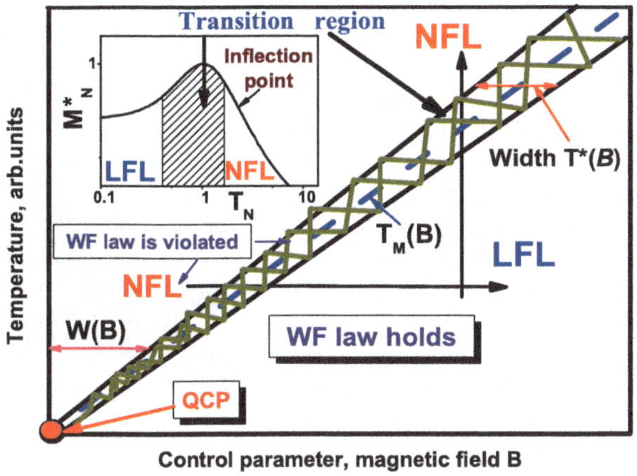

Fig. 20.1 Violation of the Wiedemann-Franz law and $T - B$ phase diagrams of HF metals. Schematic $T - B$ phase diagram of HF liquid with magnetic field as the control parameter. The vertical and horizontal arrows show LFL-NFL and NFL-LFL transitions at fixed B and T, respectively. At $B = 0$, the system is in its NFL state having a flat band down to $T \to 0$. The hatched area separates the NFL and the weakly polarized LFL phase and represents the transition state. The dashed line in the hatched area represents the function $T_M(B)$. The functions $W(B) \propto T$ and $T^*(B) \propto T$ shown by two-headed arrows define the widths of the NFL and the transition states, respectively. The QCP located at the origin denotes the critical point where the effective mass M^* diverges and both $W(B)$ and $T^*(B)$ tend to zero. The areas where WF law holds (LFL state) and is violated (NFL and transition state) are also displayed. The inset shows a schematic plot of the normalized effective mass versus the normalized temperature. The transition regime, where M^*_N reaches its maximum value at $T_N = T/T_M = 1$, is shown as the hatched area in both the main panel and the inset. Arrows indicate the transition region and the inflection point $T_{\rm inf}$ in the M^*_N plot

T. In the same way, it can be shown that the width $T^*(B)$ of the transition region is also proportional to T. The regions, where WF law is violated and/or holds, are also shown.

We now focus on the empirical phase diagrams of HF compounds $YbRh_2Si_2$ (Fig. 20.2 **a**, **b**) and $CeCoIn_5$ (Fig. 20.2 **c**). Panel **a** of Fig. 20.2 is similar to the main panel of Fig. 20.1, but with the difference that this HF compound possesses the AF state. To avert realization of a strongly degenerate ground state, induced by the flat band, the FC must be completely eliminated at $T \to 0$. In a general scenario, this occurs by means of an antiferromagnetic (AF) phase transition with an ordering temperature $T_{NL} = 70$ mK, while application of a magnetic field $B = B_{c0}$ destroys the AF state at $T = 0$ [26]. In other words, the field B_{c0} places the HF metal at the magnetic-field-tuned QCP and nullifies the Nèel temperature $T_N(B_{c0}) = 0$ of the corresponding AF phase transition. Imposition of a magnetic field $B > B_{c0}$ drives the system to the LFL state. Thus, in the case of $YbRh_2Si_2$, the QCP is shifted from the origin to $B = B_{c0}$. In FC theory, the quantity B_{c0} is a parameter determined by the properties of the specific HF compound. In some cases, notably the HF metal $CeRu_2Si_2$, B_{c0} does vanish [27], whereas in $YbRh_2Si_2$, $B_{c0} \simeq 0.06$ T,

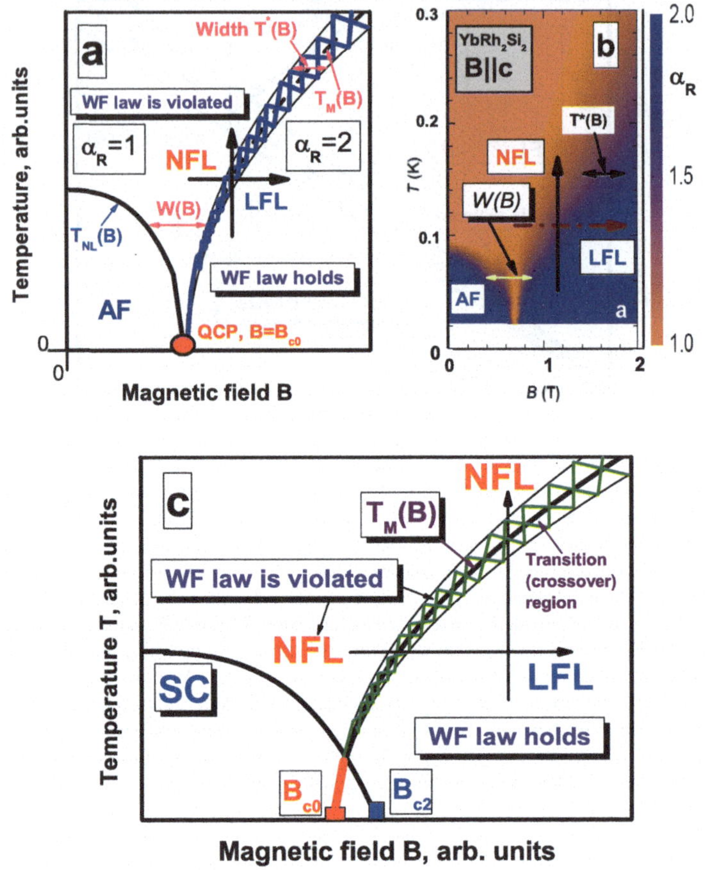

Fig. 20.2 Schematic phase diagrams of the HF compounds $YbRh_2Si_2$ (panels **a**, **b**) and $CeCoIn_5$ (panel **c**). In panel **a**, $T_{NL}(B)$ displays the Nèel temperature as a function of magnetic field B. The QCP is shifted from $B = 0$ to $B = B_{c0}$, see the phase diagram Fig. 20.1. At $B < B_{c0}$, the system is in its AF state. As in Fig. 20.1, the vertical and horizontal arrows show the transitions between the LFL and NFL states, the functions $W(B) \propto T$ and $T^*(B) \propto T$ indicated by the two-sided arrows define the width of the NFL state and of the transition region, respectively. The dashed line in the hatched area represents the function $T_M(B)$. The exponent α_R determines the temperature-dependent part of the resistivity, with α_R taking values 2 and 1, in LFL and NFL states. In the transition regime, the exponent evolves between LFL and NFL values. Panel **b** shows the experimental $T - B$ phase diagram [13, 28]. The evolution of α_R is depicted by color. The NFL behavior takes place at the lowest temperatures at the magnetic-field-tuned QCP, see panel **a**. The transition regime between the NFL state and the field-induced LFL state broadens with rising magnetic fields $B > B_{c0}$ and $T \sim T^*(B)$. As in panel **a**, transitions from LFL to NFL state and from NFL to LFL state are indicated by the corresponding arrows, as are $W(B) \propto T$ and $T^*(B) \propto T$. In panel **c**, as shown by the solid curve, at $B < B_{c2}$ the system is in its superconducting state, with B_{c0} denoting a QCP hidden beneath the SC dome, where the flat band could exist at $B \leq B_{c0}$. The rest of the lines are similar to those from panel **a**. A part of the crossover above B_{c0} marked by thick red line is hidden in the SC state. The areas, where WF law holds (LFL state) and is violated (NFL and crossover region), are displayed in panels **a** and **c**

$B \perp c$ [26]. Panel **b** of Fig. 20.2 portrays the experimental $T - B$ phase diagram in a manner showing the evolution of the exponent $\alpha_R(T, B)$ [13, 28]. At the critical field $B_{c0} \simeq 0.66$ T ($B \parallel c$), the NFL behavior extends down to the lowest temperatures, while $YbRh_2Si_2$ transits from the NFL to LFL behavior under increase of the applied magnetic field. Panel **c** of Fig. 20.2 depicts the $T - B$ phase diagram of compound $CeCoIn_5$. This phase diagram resembles those from the above panels **a** and **b** but with the difference that at low magnetic fields $CeCoIn_5$ is in superconducting (SC) state, while its QCP is hidden beneath the superconducting dome. This implies the difference in the positions of critical fields B_{c0} in $CeCoIn_5$ and $YbRh_2Si_2$. In former compound, this QCP is hidden under superconducting dome, $B_{c0} < B_{c2}$. It means that it lies in the superconducting state, which closes at $B = B_{c2}$, in latter one it corresponds to the QCP, comprising a boundary between NFL and magnetic-field-induced LFL parts of its phase diagram. Above the critical temperature T_c of the SC phase transition, the zero-field resistivity $\rho(T, B = 0)$ varies linearly with T. On the other hand, at $T \to 0$ and magnetic fields $B \geq B_{c2} \simeq 5$ T, the curve $\rho(T, B_{c2})$ is parabolic [10, 29]. Below we will show that WF law holds in LFL regime, while it is violated in NFL and in transition regions of the phase diagram.

20.2 Wiedemann-Franz Law Violations

The violation of the WF law at the QCP in HF metals has been predicted and estimated a few years ago [30, 31] and observed recently [13]. Predictions of LFL theory fail in the vicinity of a QCP, where the effective mass M^* diverges, since the single-particle spectrum possesses a flat band at that point. In a once-standard scenario for such a QCP [32, 33], the divergence of the effective mass is attributed to vanishing of the quasiparticle weight z. However, as already indicated, this scenario has failed [34]. We therefore employ a different scenario for the QCP, in which the departure of the Lorenz number L from the Wiedemann-Franz value is associated with a rearrangement of single-particle degrees of freedom leading to a flat band. Within the quasiparticle paradigm, the relation between the Seebeck thermodynamic coefficient S and the thermal κ and electric σ conductivities has the form [35, 36]

$$\frac{\kappa(T)}{\sigma(T)T} + S^2(T) = \frac{1}{e^2}\frac{I_2(T)}{I_0(T)}. \tag{20.2}$$

Here

$$S(T) = \frac{1}{e}\frac{I_1(T)}{I_0(T)}, \tag{20.3}$$

with

$$I_k(T) = -\int\left(\frac{\epsilon(p)}{T}\right)^k\left(\frac{d\epsilon(p)}{dp}\right)^2\tau(\epsilon, T)\frac{\partial n(p)}{\partial\epsilon(p)}d\upsilon, \tag{20.4}$$

where τ is the collision time and dv is the phase space volume element. Overwhelming contributions to the integrals I_k come from a narrow vicinity $|\epsilon| \sim T$ of the Fermi surface. In case of LFL, the Seebeck coefficient $S(T)$ vanishes linearly with T at $T \to 0$. The group velocity can be factored out from the integrals (20.4). The same is true for the collision time τ, which at $T \to 0$ depends merely on impurity scattering, and one obtains $I_1(T = 0) = 0$ and $I_2(T \to 0)/I_0(T \to 0) = \pi^2/3$. Inserting these results into (20.2), we find that the WF law holds, even if several bands cross the Fermi surface simultaneously [35].

It is worth illustrating the violation of the WF law by a numerical example [31]. This situation is convenient for the demonstration of the impact of the particle-hole symmetry breaking in the systems with FC on the WF law violation. Below we address the model with the interaction function

$$F(k) = g\frac{\exp(-\beta k)}{k}. \tag{20.5}$$

Since inside FC domain the single-particle spectrum $\varepsilon(p, T = 0) = \mu$, while outside it [37]

$$\frac{d\varepsilon(p, T = 0)}{dp} \propto \sqrt{|\varepsilon(p) - \mu|}, \tag{20.6}$$

such behavior results in a marked violation of the WF law. This conclusion is confirmed by results of numerical calculations shown in Fig. 20.3.

The transport integrals I_k have been calculated numerically. We present the results of these calculations on Fig. 20.3. The calculations have been performed for $\beta p_F = 3$ and $\beta p_F = 30$. As it is seen from this figure, at the QCP and beyond it, the integrals $I_0(T = 0)$ and $I_1(T = 0)$ have the same order, implying that in contrast to LFL theory, the $T = 0$ value of the Seebeck coefficient $S(0) = I_1(0)/eI_0(0)$ differs from 0. This is a fingerprint of particle-hole symmetry breaking, inherent in any system with FC. On the other hand, the ratio $L(0)/L_0$ turns out to be even larger than that at the QCP. Furthermore, calculations demonstrate that with increasing β, the value of this ratio increases as well, i.e., the longer the radius of the interaction function (20.5) in the coordinate space, the larger is the departure from the WF law. Thus, taking into account the fact that the reduction of the ratio L/L_0 occurs in the NFL state at the QCP [31], we conclude that the violation of the WF law takes place in the narrow segment of the $T - B$ phase diagram displayed in Figs. 20.1 and 20.2 with the width $W \to 0$ at $T \to 0$. In other words, at $T \to 0$ the ratio L/L_0 becomes abruptly $L/L_0 \sim 0.9$ at $B/B_{c0} = 1$, while $L/L_0 = 1$ at $B/B_{c0} \neq 1$ when the system is in its AF or LFL state, shown in Fig. 20.2. This observation is in a good agreement with experimental data for $YbRh_2Si_2$ [13]. We conclude that at $T \to 0$, the WF law holds in the LFL state, at which the Fermi distribution function is reduced to the step function. The violation of the WF law at $B = B_{c0}$ and at $T \to 0$ seen in $YbRh_2Si_2$ thus suggests that a sharp Fermi surface does exist at $B/B_{c0} \neq 1$ but does not exist only at $B/B_{c0} = 1$, where the flat band emerges, the WF law is violated, and the jump of ρ_0 takes place, as it is shown in Fig. 20.4 by the arrow.

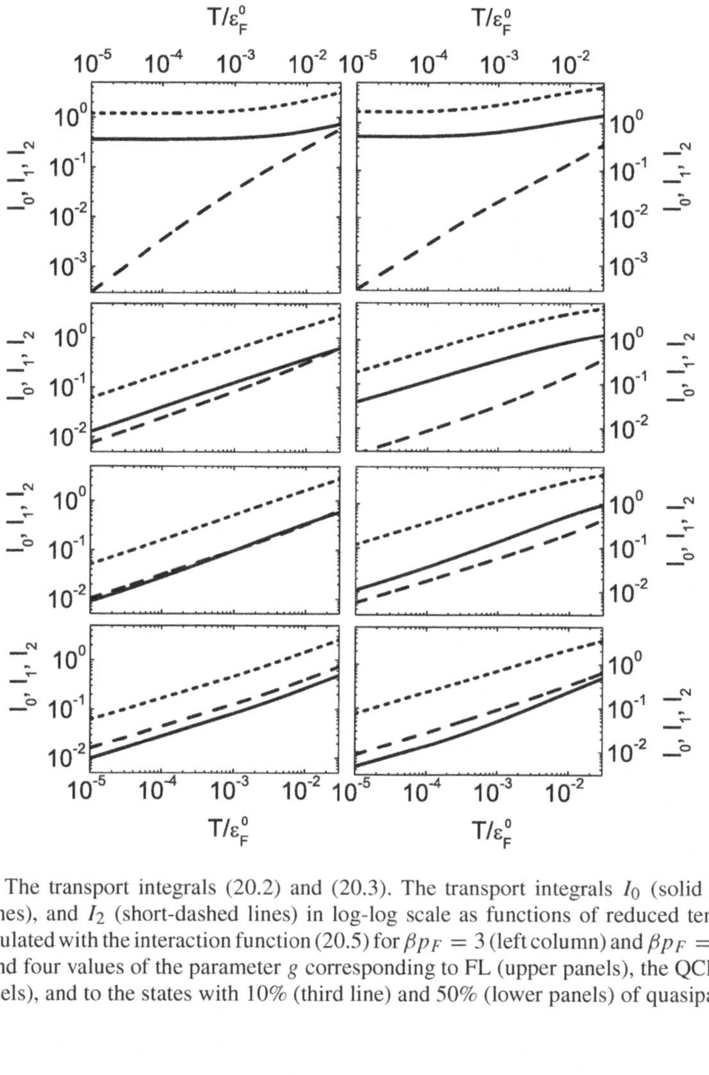

Fig. 20.3 The transport integrals (20.2) and (20.3). The transport integrals I_0 (solid lines), I_1 (dashed lines), and I_2 (short-dashed lines) in log-log scale as functions of reduced temperature T/ϵ_F^0, calculated with the interaction function (20.5) for $\beta p_F = 3$ (left column) and $\beta p_F = 30$ (right column) and four values of the parameter g corresponding to FL (upper panels), the QCP (second line of panels), and to the states with 10% (third line) and 50% (lower panels) of quasiparticles in the FC

The application of a magnetic field to $YbRh_2Si_2$ creates the step-like drop in ρ_0 [38]. As it is seen from Fig. 20.4, when the system transits from the NFL state to the LFL at fixed T and under the application of growing magnetic fields B, the step-like drop in its resistivity $\rho(T, B)$ becomes more pronounced (see the experimental curves for $T = 0.3$, 0.2, and 0.1 K). This behavior comes from the fact that $W \propto T$. As WF law is violated in the transition and NFL regions of the phase diagram, we conclude that at $T = 0$ this law is violated in the narrow region of the ρ jump (see Fig. 20.4 with the arrow, showing the region of the WF law violation), while at higher temperatures this region widens and becomes diffuse. Note that at $T \to 0$ the NFL behavior can be captured by some states which destroy the flat band. For example, a

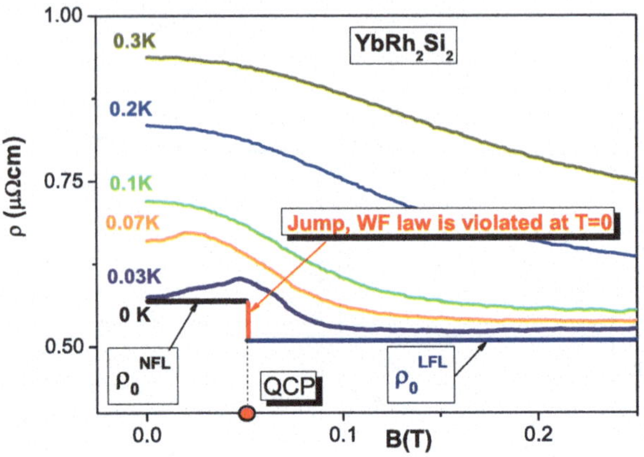

Fig. 20.4 Experimental results [38] for the longitudinal magnetoresistivity $\rho(T, B)$ of $YbRh_2Si_2$ versus B at various temperatures T. The temperatures are indicated in the figure. The maxima of the curves for $T = 0.03$ and 0.07 K correspond to boundary points of the AF ordered state shown in Fig. 20.2 **b**. The solid lines marked with 0 K represent the schematic behavior of the residual resistivity ρ_0 as a function of B. The arrows pointing to the horizontal solid lines identify the residual resistivities ρ_0^{NFL} and ρ_0^{LFL} in $YbRh_2Si_2$. The jump of ρ_0 that occurs at the QCP is identified by an arrow. This is the point where WF law is violated at $T = 0$

short-range magnetic order destroys the NFL behavior down to $T \to 0$ by destroying the flat band. In that case, at low temperatures, the validity of the WF law is restored. This means, in turn, that the experimental observation of deviation of L/L_0 ratio from unity (i.e., WF law violation) is extremely difficult. This observation is in agreement with experimental results, some of which directly point to the WF law violation in the case of the HF metal $YbRh_2Si_2$ [13], while the others give evidence that the WF law holds [14, 39].

The anisotropy of WF law violation near the QCP has been experimentally observed in the HF metal $CeCoIn_5$ [12]. In that paper, the above HF compound has been studied experimentally in external magnetic fields, close to the critical value H_{c2} that suppresses the superconductivity. Under these conditions, the WF law is violated. The violation is anisotropic and cannot be attributed to the scenario of quasiparticle distraction by the fluctuation taking place at QCP. At the same time, close to the QCP, sufficiently large external magnetic fields reveal the anisotropy of the electrical conductivities $\sigma_{ik} \propto < v_i v_k >$ (v_i are the components of the group velocity vector) and thermal conductivities $\kappa_{ik} \propto < \varepsilon(\mathbf{p}) v_i v_k >$ of a substance. This is because the magnetic field does not affect the z-components of the group velocity $\mathbf{v}$ so that the QCP T-dependence of the transport coefficients holds, triggering the violation of the WF relation $L_{zz} = \sigma_{zz}/T\kappa_{zz} = \pi^2 k_B/3e^2$. On the other hand, the magnetic field $\mathbf{B}$ alters substantially the electron motion in the perpendicular direction, yielding considerable increase of the x- and y-components of the group velocity so that the corresponding components L_{ik} do not depart from their WF value.

20.3 Conclusion

The flattening of the single-particle spectrum $\varepsilon(\mathbf{p})$ of strongly correlated electron systems alters considerably their transport properties, especially beyond the FCQPT point due to particle-hole symmetry breaking. In topologically different "iceberg" phases, the WF law is also violated near its QCP. The results of theoretical [31] and experimental investigations demonstrate that the FCQPT scenario with further occurrence of both "iceberg" and FC phases give natural and universal explanation of the NFL changes of the transport properties of HF compounds, in general, and of the WF law violation, in particular. Therefore, as we have demonstrated above, to describe theoretically the violation of the WF law within the FCQPT formalism, it is sufficient to use the well-known LFL formulas for thermal and electrical conductivities, with the substitution of the modified single-particle spectrum into them. Such theory has been advanced in [31, 40]. It is shown that close to the QCP the Lorenz number $L_{\rm QCP}(T=0) = 1.81\,L_0$. This result agrees well with the experimental values [11, 15]. Furthermore, the dependence $L(T)/L_0$ has been calculated for two topologically distinct phases (see Sect. 4.2)—"iceberg" phase and FC phase [31, 40]. Theoretical calculations have shown that in both phases the largest violation from the WF law occurs near QCP [31, 40]. Deep in the "iceberg" phase we have the reentrance of the "classical" WF law in a sense that $L = L_0$ while in the deep FC phase the Lorenz number is temperature-independent at low temperatures, but its value is slightly larger than L_0. This is due to the particle-hole symmetry violation in FC phase, see Chaps. 18 and 19. These violations, being fingerprints of the topological FCQPT and generated flat bands, are important features of the new state of matter.

References

1. G. Wiedemann, R. Franz, Ann. Phys. **89**, 497 (1853)
2. A. Sommerfeld, Naturwissenschaften **15**, 825 (1927)
3. G.V. Chester, A. Tellung, Proc. Phys. Soc. **77**, 1005 (1961)
4. M.J. Kearney, P.N. Butcher, J. Phys. C.: Solid State Phys. **21**, L265 (1988)
5. C. Castellani, C. DiCastro, G. Kotliar, P.A. Lee, G. Strinati, Phys. Rev. Lett. **59**, 477 (1987)
6. H.R. Ott, O. Marti, F. Hulliger, Solid St. Commun. **49**, 1129 (1984)
7. A. Amato, D. Jaccard, E. Walker, J. Sierro, J. Flouquet, J. Magn. Magn. Mater. **63–64**, 300 (1987)
8. S. Kambe, H. Suderow, T. Fukuhara, J. Flouquet, T. Takimoto, J. Low Temp. Phys. **117**, 101 (1999)
9. R.T. Syme, M.J. Kelly, M. Pepper, J. Phys.: Condens. Matter **1**, 3375 (1989)
10. J. Paglione, M.A. Tanatar, D.G. Hawthorn, F. Ronning, R.W. Hill, M. Sutherland, L. Taillefer, C. Petrovic, Phys. Rev. Lett. **97**, 106606 (2006)
11. S. Paschen, B. Wand, G. Sparn, F. Steglich, Y. Echizen, T. Takabatake, Phys. Rev. B **62**, 14912 (2000)
12. M.A. Tanatar, J. Paglione, L. Taillefer, C. Petrovich, Science **316**, 1320 (2007)

13. H. Pfau, S. Hartmann, U. Stockert, P. Sun, S. Lausberg, M. Brando, S. Friedemann, C. ius Krellner, C. Geibel, S. Wirth, S. Kirchner, E. Abrahams, Q. Si, F. Steglich, Nature **484**, 493 (2012)
14. J.P. Reid, M.A. Tanatar, R. Daou, R. Hu, C. Petrovic, L. Taillefer (2013), http://arxiv.org/abs/cond-mat/:1309.6315v1
15. R.W. Hill, C. Proust, L. Taillefer, P. Fournier, R.L. Greene, Nature **414**, 711 (2001)
16. N. Doiron-Leyrand, M. Sutherland, S.Y. Li, L. Taillefer, R. Liang, D.A. Bonn, W.N. Hardy, Phys. Rev. Lett. 207001 (2006)
17. G.B. Lesovik, I.A. Sadovsky, Phys. Uspekhi **54**, 1007 (2011)
18. C. Proust, K. Behnia, R. Bel, D. Maude, S.I. Vedeneev, Phys. Rev. B **72**, 214511 (2005)
19. C. Proust, E. Boaknin, R.W. Hill, L. Taillefer, A.P. Mackenzie, Phys. Rev. Lett. **89**, 147003 (2002)
20. C.L. Kane, M.P.A. Fisher, Phys. Rev. Lett. **76**, 3192 (1996)
21. C.L. Kane, M.P.A. Fisher, Phys. Rev. B **55**, 15832 (1997)
22. L.G.C. Rego, G. Kirczenow, Phys. Rev. B **59**, 13080 (1999)
23. M.R. Li, E. Orignac, Europhys. Lett **60**, 432 (2002)
24. C.M. Varma, P.B. Littlewood, S. Schmittrink, E. Abrahams, A.E. Ruckenstein, Phys. Rev. Lett. **63**, 1996 (1989)
25. M.Y. Amusia, K.G. Popov, V.R. Shaginyan, W.A. Stephanowich, *Theory of Heavy-Fermion Compounds*, Springer Series in Solid-State Sciences, vol. 182 (Springer, Berlin, 2015)
26. P. Gegenwart, J. Custers, C. Geibel, K. Neumaier, K.T.T. Tayama, O. Trovarelli, F. Steglich, Phys. Rev. Lett. **89**, 056402 (2002)
27. D. Takahashi, S. Abe, H. Mizuno, D. Tayurskii, K. Matsumoto, H. Suzuki, Y. Onuki, Phys. Rev. B **67**, 180407(R) (2003)
28. J. Custers, P. Gegenwart, H. Wilhelm, K. Neumaier, Y. Tokiwa, O. Trovarelli, C. Geibel, F. Steglich, C. Pépin, P. Coleman, Nature **424**, 524 (2003)
29. J. Paglione, M. Tanatar, D. Hawthorn, E. Boaknin, R.W. Hill, F. Ronning, M. Sutherland, L. Taillefer, C. Petrovic, P. Canfield, Phys. Rev. Lett. **91**, 246405 (2003)
30. V.R. Shaginyan, M.Y. Amusia, A.Z. Msezane, K.G. Popov, Phys. Rep. **492**, 31 (2010). (arXiv:1006.2658)
31. V.A. Khodel, V.M. Yakovenko, M.V. Zverev, JETP Lett. **86**(86), 772 (2007)
32. P. Coleman, C. Pépin, Q. Si, R. Ramazashvili, J. Phys. Condens. Matter **13**, R723 (2001)
33. P. Coleman, C. Pepin, Phys. B **312–313**, 383 (2002)
34. V.A. Khodel, JETP Lett. **86**, 721 (2007)
35. E.M. Lifshitz, L.P. Pitaevskii, *Physical Kinetics* (Pergamon Press, Oxford, 1981)
36. N.W. Ashkroft, N.D. Mermin, *Solid State Physics* (HRW, Philadelphia, 1976)
37. V.A. Khodel, V.R. Shaginyan, V.V. Khodel, Phys. Rep. **249**, 1 (1994)
38. P. Gegenwart, T. Westerkamp, C. Krellner, Y. Tokiwa, S. Paschen, C. Geibel, F. Steglich, E. Abrahams, Q. Si, Science **315**, 969 (2007)
39. Y. Machida, K. Tomokuni, K. Izawa, G. Lapertot, G. Knebel, J.P. Brison, J. Flouquet, Phys. Rev. Lett. **110**, 236402 (2013)
40. V.A. Khodel, J.W. Clark, V.M. Yakovenko, M.V. Zverev, Phys. B **403**, 1227 (2008)

Chapter 21
Quantum Criticality of Heavy-Fermion Compounds

Abstract In this chapter, we continue to consider the nature of quantum criticality in HF compounds. The quantum criticality induced by the fermion-condensation quantum phase transition extends over a wide range in the $T - B$ phase diagram. As we shall see, the quantum criticality in all such different HF compounds, as high-T_c superconductors, HF metals, compounds with quantum spin liquids, quasicrystals, and 2D quantum liquids, is of the same nature. This challenging similarity between different HF compounds expresses universal physics that transcends the microscopic details of the compounds. This uniform behavior, induced by the universal quantum-critical physics, allows us to view it as the main characteristic of the new state of matter. We construct the $T - B$ phase diagrams and explain main features of experimental facts of low-temperature thermodynamic in terms of FCQPT that leads to the formation of both the flat band and the new state of matter.

21.1 Quantum Criticality of High-Temperature Superconductors and HF Metals

At low temperatures, the normal state (which is Landau Fermi liquid) of high-T_c superconductors and HF metals is recovered by the application of a magnetic field larger than the critical field. In this state, the Wiedemann-Franz and Korringa laws are held and the elementary excitations are Landau quasiparticles. Contrary to what one might expect from the LFL, the effective mass of quasiparticles depends on the magnetic field. We show that the magnetic-field-induced transition from NFL to LFL in high-temperature superconductors is similar to the transition observed in HF metals.

The extended quasiparticles paradigm supports quasiparticles that define the major part of the low-temperature properties of high-T_c superconductors, including their NFL behavior. At sufficiently low temperatures, as soon as the order parameter $\kappa(\mathbf{p})$ is suppressed by magnetic field $B > B_{c2}$, the field-induced LFL emerges, see Chap. 6. It was reported that in the normal state obtained by applying a magnetic field greater than the upper critical field B_{c2}, in hole-doped cuprates at overdoped concentration

M. Amusia and V. Shaginyan, *Strongly Correlated Fermi Systems*,
Springer Tracts in Modern Physics 283,
https://doi.org/10.1007/978-3-030-50359-8_21

($Tl_2Ba_2CuO_{6+\delta}$) [1] and at optimal doping concentration ($Bi_2Sr_2CuO_{6+\delta}$) [2], there are no sizable violations of the Wiedemann-Franz (WF) law. Since the validity of the WF law is a robust signature of LFL, these experimental facts demonstrate that the observed elementary excitations can be considered as Landau quasiparticle. At a constant magnetic field, the low-energy elementary excitations are characterized by $M^*(B)$ and cannot be distinguished from Landau quasiparticles. On the other hand, in contrast to the LFL theory, the effective mass $M^*(B)$ depends on the magnetic field. As it was shown in Sect. 6.3.1, in this case, the magnetic field B plays the role of the control parameter determining the effective mass

$$M^*(B) \propto \frac{1}{\sqrt{B - B_{c0}}}. \tag{21.1}$$

We recall that B_{c0} is the critical magnetic field driving corresponding AF phase transition toward $T = 0$. Since $B_{c0} > B_{c2}$, (21.1) is valid at $B > B_{c2}$. In that case, the effective mass $M^*(B)$ is finite, and the system is driven back to LFL and acquires the LFL behavior induced by the magnetic field.

Equation (21.1) shows that by applying a magnetic field $B > B_c$ the system can be driven back into LFL with the effective mass $M^*(B)$, which is finite and temperature-independent. This means that the low-temperature properties of the considered compounds depend on the effective mass in accordance with the LFL theory. In particular, the resistivity $\rho(T)$ as a function of the temperature behaves as $\rho(T) = \rho_0 + \Delta\rho(T)$ with $\Delta\rho(T) = AT^2$, where the factor A behaves as $A \propto (M^*)^2 \propto 1/(B - B_{c0})^2$. At finite temperatures, the system persists as LFL, but there is the transition region (crossover) from the LFL behavior to the non-Fermi liquid behavior at temperature $T^*(B) \propto \mu_B(B - B_{c0})$. At $T \sim T^*(B)$, the system is located in the transition region, the effective mass starts to depend on the temperature, and the resistivity possesses the non-Fermi liquid behavior with a substantial T-linear term, $\Delta\rho(T) = aT + bT^2$. Here, $T^*(B)$ is the transition temperature. Since magnetic field enters the Landau equation as $\mu_B B/T$, we have

$$T^*(B) \sim \mu_B(B - B_{c0}). \tag{21.2}$$

The transition temperature is not really a phase transition. It is necessarily broad, very much depending on the criteria for determination of the point of such a transition. As it is seen from the $T - B$ phase diagram reported in Fig. 21.1, at raising magnetic field the system enters the LFL regime, and as a result, the T-linear term vanishes. Such a behavior of the resistivity was observed in the cuprate superconductors $Tl_2Ba_2CuO_{6+\delta}$ ($T_c < 15$ K) [4] and $La_{2-x}Ce_xCuO_4$ [5]. For example, at B $= 10$ T, $\Delta\rho(T)$ is a linear function of the temperature between 120 mK and 1.2 K, whereas at B $= 18$ T, the temperature dependence of the resistivity is consistent with $\rho(T) = \rho_0 + AT^2$ over the same temperature range [4].

It has been shown in Sect. 11.1 that in the LFL phase, the nuclear spin-lattice relaxation rate $1/T_1$ is determined by the quasiparticles near the Fermi level, whose population is proportional to M^*T, so that $1/T_1T \propto (M^*)^2$ is a constant [6, 7].

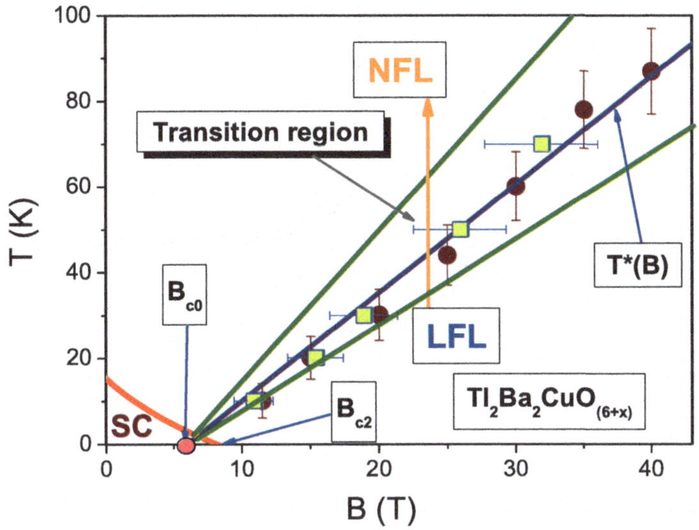

Fig. 21.1 $B-T$ phase diagram of superconductor $Tl_2Ba_2CuO_{6+x}$. The transition region from LFL to NFL regime marked by the line $T^*(B)$ is shown by the arrow. The corresponding transition temperature $T^*(B)$ is given by (21.2), while the squares and the circles are the experimental points [3]. Thick line is a boundary between the superconducting and normal phases. The arrows near the bottom left corner indicate the critical magnetic fields B_{c2} destroying the superconductivity and B_{c0} destroying the AF order

When the superconducting state is removed by the application of a magnetic field, the underlying ground state can be seen as the field-induced LFL with effective mass depending on the magnetic field. As a result, the rate $1/T_1$ follows the $T_1T = constant$ relation, and so the Korringa law is held. Unlike the behavior of LFL, as it follows from (21.1), $1/T_1T \propto (M^*(B))^2$ decreases with increasing the magnetic field at $T < T^*(B)$. Note that at $T > T^*(B)$, we observe that $1/T_1T$ is a decreasing function of the temperature, $1/T_1T \propto M^* \propto 1/T$. These observations are in good agreement with the experimental facts [6]. Since $T^*(B)$ is an increasing function of the magnetic field, the Korringa law remains valid to higher temperatures at higher magnetic fields.

Let us now turn to the $B-T$ phase diagram of a high-T_c superconductor. The corresponding $T-B$ phase diagram of $Tl_2Ba_2CuO_{6+x}$ is shown as an example in Fig. 21.1. The substance is a superconductor with T_c from 15 K to 93 K depending on oxygen content x. In Fig. 21.1, open squares and solid circles show the experimental values of the crossover temperature from the LFL to NFL regimes. The transition region between LFL and NFL regimes is shown by the arrow. The solid line shows our fit using (21.2) with $B_{c0} = 6$ T that is in good agreement with $B_{c0} = 5.8$ T obtained from the field dependence of the charge transport, see Sect. 6.3.2.1. As it is seen from Fig. 21.1, the linear behavior agrees well with experimental data [3]. Now we consider the field-induced reentrance of LFL behavior in $Tl_2Ba_2CuO_{6+x}$

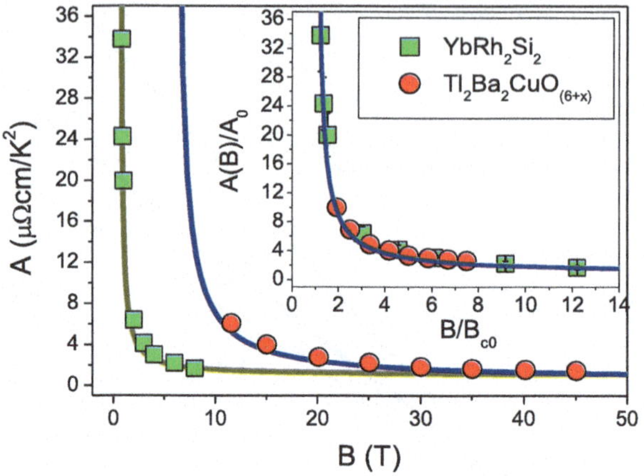

Fig. 21.2 The universal behavior of the charge transport coefficient $A(B)$ in the LFL state induced by the application of magnetic field B. Main panel: $A(B)$ obtained in measurements on $YbRh_2Si_2$ (squares) [8] and $Tl_2Ba_2CuO_{6+x}$ (circles) [3]. Our calculations of $A(B)$ based on (21.3) are depicted by the solid lines. The inset: Normalized coefficient $A(B)/A_0 \simeq 1 + D_N/(y-1)$ as a function of normalized magnetic field $y = B/B_{c0}$ is given by (21.4), and is shown by squares for $YbRh_2Si_2$ and by circles for $Tl_2Ba_2CuO_{6+x}$

at $B \geq B_{c2}$. In that case, the effective mass M^* depends on magnetic field B that becomes the control parameter, while the system is in the LFL regime as it is shown by the dashed horizontal arrow in Fig. 15.1. The LFL regime is characterized by the temperature dependence of the resistivity $\rho(T) = \rho_0 + A(B)T^2$, see also above. The coefficient A, being proportional to the quasiparticle scattering cross-section, is found to be $A \propto (M^*(B))^2$. With respect to $M^* \propto B^{-1/2}$ (see (6.39)), this implies that

$$A(B) \simeq A_0 + \frac{D}{B - B_{c0}}, \tag{21.3}$$

where A_0 and D are parameters. It seen from Fig. 21.1 and follows from (21.3) that it is impossible to observe the relatively high values of $A(B)$ since in our case $B_{c2} > B_{c0}$. Therefore, as it was mentioned above, in high-T_c superconductors, their QCP is poorly accessible to experimental observations being *"hidden under superconducting dome."* Nonetheless, it is possible to study QCP by exploring the quantum criticality.

Figure 21.2 reports the fit of our theoretical dependence (21.3) to the experimental data for two different classes of substances: HF metal $YbRh_2Si_2$ and HTSC $Tl_2Ba_2CuO_{6+x}$. The different scale of fields is clearly seen as well as good coincidence with the theoretical dependence determined by (21.3). This means that the physics underlying the field-induced reentrance of LFL behavior is the same for both classes of substances. To corroborate this point further, let us rewrite (21.3) in reduced variables A/A_0 and B/B_{c0}. This rewriting immediately reveals the universal nature of the behavior of these two substances: Both of them are driven to common

QCP, related to FC and induced by the application of magnetic field. In reduced values, (21.3) takes the form

$$\frac{A(B)}{A_0} \simeq 1 + \frac{D_N}{B/B_{c0} - 1}, \tag{21.4}$$

where $D_N = D/(A_0 B_{c0})$ is the only fitting parameter. It is seen from (21.4) that upon applying the scaling in these variables, the quantities $A(B)$ for $Tl_2Ba_2CuO_{6+x}$ and $YbRh_2Si_2$ are described by a function of single variable B/B_{c0}, thus demonstrating universal behavior. To support (21.4), we replot both dependencies in reduced variables A/A_0 and B/B_{c0} as it is depicted in the inset to Fig. 21.2. Such replotting immediately reveals the universal nature of the behavior of these two quite different substances. Indeed, in the case of $Tl_2Ba_2CuO_{6+x}$ the critical field $B_{c0} \simeq 6$ T, while the value of the critical field of $YbRh_2Si_2$ is $B_{c0} \simeq 0.06$ T. Nonetheless, it is seen from the inset to Fig. 21.2 that close to magnetic QCP there is no external physical scales, so that the normalization by internal scales A_0 and B_{c0} shows straightforwardly the common quantum criticality of HF metals and high-T_c superconductors.

Thus, our theoretical study of high-T_c superconductors and HF metals clearly demonstrates their generic family features. We have shown that the physics underlying the field-induced reentrance of LFL behavior is the same for both high-T_c compounds and HF metals. It follows from our study that there is at least one quantum phase transition inside the superconducting dome, and this transition is FCQPT.

21.2 Quantum Criticality of Quasicrystals

New exotic materials named quasicrystals and characterized by noncrystallographic rotational symmetry and quasiperiodic translational properties have attracted great attention. Studies of quasicrystals may shed light on the most basic notions related to the quantum-critical state observed in HF metals. We show here that the electronic system of some quasicrystals is already located at the FCQPT point without any tuning. Therefore, quasicrystals have the quantum-critical state with NFL phase, which in magnetic fields transforms into the LFL state. Remarkably, the quantum-critical state is robust despite the strong disorder experienced by the electrons. Quasicrystals exhibit the typical scaling of their thermodynamic properties like the magnetic susceptibility, belonging also to the family of HF metals.

When we encounter the unusual behavior of strongly correlated metals, we anticipate to learn more about quantum-critical physics. Such an opportunity is provided by quasicrystals (QCs) [9, 10]. These substances, characterized by the absence of translational symmetry in combination with good ordered atomic arrangement and rotational symmetry, can be viewed as materials located between crystalline and disordered solids. QCs, their crystalline approximants, and related complex metallic phases reveal unusual mechanical, magnetic, electronic transport, and thermodynamic properties. The mentioned crystalline approximants are the arrangements of atoms which within their unit cells closely approximate the local atomic structures in QCs [10]. The aperiodicity of QCs plays an important role in the formation of the properties since the band electronic structure governed by the famous Bloch theorem cannot be well defined [11]. As an example, QCs exhibit a high resistivity although DOS at the Fermi energy is not small [12]. One expects that the transport properties are defined by a small diffusivity of electrons which occupy a new class of states denoted as "critical states," neither being extended nor localized, and making the velocity of charge carriers very low [12]. Associated with these critical states, characterized by an extremely degenerate confined wave function, are the so-called "spiky" DOS [11, 13]. The presence of those predicted DOS is confirmed by experiments revealing that the single-particle spectra of the local DOS demonstrate a spiky DOS, [14, 15]. Clearly, these spiky states are associated with flat bands [16, 17]. On one hand, we expect the properties related to the itinerant states governed by the spiky ones of QCs to coincide with those of HF metals, while, on the other hand, the pseudolocalized states may result in those of amorphous materials. Therefore, the question of how quasicrystalline order influences the electronic properties of QCs, whether they resemble those of HF metals or amorphous materials, is of crucial importance.

Experimental measurements on the gold-aluminum-ytterbium quasicrystal $Au_{51}Al_{34}Yb_{15}$ with a six-dimensional lattice parameter $a_d = 0.7448$ nm have revealed quantum-critical behavior with the unusual exponent $\alpha \simeq 0.51$ defining the divergency of the magnetic susceptibility $\chi \propto T^{-\alpha}$ as temperature $T \to 0$ [18]. The measurements have also exposed that the observed NFL behavior transforms into LFL under the application of a tiny magnetic field H, while it exhibits robustness against hydrostatic pressure. The quasicrystal shows also metallic behavior with the $T-$dependent part $\Delta\rho$ of the resistivity, $\Delta\rho \propto T$, at low temperatures [18].

We start with constructing a model to explain the behavior of the gold-aluminum-ytterbium QC [19]. Taking into account that the spiky states are associated to flat bands [16, 20, 21], which are a typical signature of FCQPT, we have the right to assume that the electronic system of the gold-aluminum-ytterbium QC $Au_{51}Al_{34}Yb_{15}$ is located very close to FCQPT [20]. Thus, $Au_{51}Al_{34}Yb_{15}$ turns out to be located at FCQPT without tuning this substance by pressure, magnetic field, etc. Then, the system exhibits the robustness of its critical behavior against the hydrostatic pressure since this pressure does not change the topological structure of QC leading to the spiky DOS and, correspondingly, flat bands. As we will see, the spiky DOS cannot prevent the field-induced LFL state.

To study the low-temperature thermodynamic and scaling behavior, we use again the model of homogeneous HF liquid [20]. This model avoids the complications associated with the anisotropy of solids and considering both the thermodynamic properties and NFL behavior by treating the effective mass $M^*(T, H)$ as a function of temperature T and magnetic field H. To study the behavior of the effective mass $M^*(T, H)$, we use the Landau equation for the quasiparticle effective mass. The only modification is that in our formalism the effective mass is no longer constant but depends on temperature and magnetic field. For the model of homogeneous HF liquid at finite temperatures and magnetic fields, this equation takes the form [20–23]

$$\frac{1}{M^*_\sigma(T,H)} = \frac{1}{M} + \sum_{\sigma_1}\int \frac{\mathbf{p}_F\mathbf{p}}{p_F^3} F_{\sigma,\sigma_1}(\mathbf{p_F},\mathbf{p}) \frac{\partial n_{\sigma_1}(\mathbf{p},T,H)}{\partial p}\frac{d\mathbf{p}}{(2\pi)^3}. \quad (21.5)$$

The single-particle spectrum is a variational derivative of the system energy $E[n_\sigma(\mathbf{p})]$ with respect to the quasiparticle distribution function or occupation numbers n,

$$\varepsilon_\sigma(\mathbf{p}) = \frac{\delta E[n(\mathbf{p})]}{\delta n_\sigma(\mathbf{p})}. \quad (21.6)$$

In our case, F is fixed by the condition that the system is situated at FCQPT. The sole role of the Landau interaction is to bring the system to FCQPT point, where $M^* \to \infty$ at $T = 0$ and $H = 0$, and the Fermi surface alters its topology so that the effective mass acquires temperature and field dependence [20–22, 24]. Provided that the Landau interaction is an analytical function, at the Fermi surface, the momentum-dependent part of the Landau interaction can be taken in the form of truncated power series $F = aq^2 + bq^3 + cq^4 + ...$, where $\mathbf{q} = \mathbf{p}_1 - \mathbf{p}_2$, a, b, and c are fitting parameters which are defined by the condition that the system is at FCQPT.

A direct inspection of (21.5) shows that at $T = 0$ and $H = 0$, the sum of the first term and the second one on the right side vanishes, since $1/M^*(T \to 0) \to 0$, since the system is located at FCQPT [20, 24]. In case of analytic Landau interaction with respect to the momenta variables, at finite T the right-hand side is proportional $F'(M^*)^2T^2$, where F' is the first derivative of $F(q)$ with respect to q at $q \to 0$. Calculations of the corresponding integrals can be found in [25]. Thus, we have $1/M^* \propto (M^*)^2T^2$, and obtain, see Chap. 7 and [20, 24]

$$M^*(T) \simeq a_T T^{-2/3}. \quad (21.7)$$

At finite temperatures, the application of magnetic field H drives system to the LFL region with

$$M^*(H) \simeq a_H H^{-2/3}. \quad (21.8)$$

On the other hand, an analytic function $F(q)$ can lead to the general topological form of the spectrum $\varepsilon(p) - \mu \propto (p - p_b)^2(p - p_F)$ with $(p_b < p_F)$ and $(p_F - p_b)/p_F \ll 1$ that makes $M^* \propto T^{-1/2}$ and creates a quantum-critical point [26]. As we shall see below, the same critical point is generated by the interaction $F(q)$ represented by an integrable over x nonanalytic function with $q = \sqrt{p_1^2 + p_2^2 - 2xp_1p_2}$ and $F(q \to 0) \to \infty$ [20, 27]. Both cases lead to $M^* \propto T^{-1/2}$, and (7.4) becomes

$$M^*(T) \simeq a_T T^{-1/2}. \quad (21.9)$$

In the same way, we obtain

$$M^*(H) \simeq a_H H^{-1/2}, \quad (21.10)$$

where a_T and a_H are parameters. Taking into account that (21.9) leads to the spiky DOS with the vanishing of spiky structure with increasing temperature T [28], as it is observed in quasicrystals [15, 18], we assume that the general form of $\varepsilon(p)$ produces the behavior of M^*, given by (21.9) and (21.10). This is materialized in quasicrystals, which can be viewed as a generalized form of common crystals [29]. We note that the behavior $1/M^* \propto \chi^{-1} \propto T^{1/2}$ is in good agreement with $\chi^{-1} \propto T^{0.51}$ observed experimentally [18]. Our explanation is consistent with the robustness of the exponent 0.51 against the hydrostatic pressure [18] since the robustness is guaranteed by the unique singular DOS of QCs that survives under the application of pressure [11, 13, 15, 16, 18]. Nonanalytic Landau interaction $F(q)$ can also serve as a good approximation, generating the observed behavior of the effective mass. We speculate that the nonanalytic interaction is generated by the nonconservation of the quasimomentum in QCs, making the Landau interaction $F(q)$ a nonlocal function of momentum q. Such a function can be approximated by a nonanalytic one.

In calculations of low-temperature resistivity, we employ a two-band model, one of which is occupied by heavy quasiparticles, with the effective mass given by (21.9), while the second band possesses LFL quasiparticles with a $T-$independent effective mass [30]. As a result, we find that the quasiparticles have a width of $\gamma \propto T$ and that the $T-$dependent part of the resistivity is $\Delta\rho \propto T$. This observation is in accordance with experimental facts [18]. At finite H and T near FCQPT, the solutions of (21.5) $M^*(T, H)$ can be well approximated by a simple universal interpolating function (7.14), see Chap. 7 for details.

The inset to the left panel of Fig. 21.3 shows the scaling behavior of the normalized effective mass. It is seen from the inset that the total width W of the LFL and the transition region $W \propto T$ vanish as $H \to 0$ since $T_{\rm max} \propto H$. In the same way, the common width of the NFL and the transition region tend to zero as soon as $T \to 0$.

Now we construct the schematic phase diagram of the gold-aluminum-ytterbium QC $Au_{51}Al_{34}Yb_{15}$. The phase diagram is reported in Fig. 21.3, left panel. The magnetic field H is the control parameter that drives the system outward FCQPT that occurs at $H = 0$ and $T = 0$ without tuning since the QC critical state is formed by singular density of states [11, 13, 15, 16, 18]. It follows from (7.14) and seen from the left panel of Fig. 21.3 that at fixed temperatures the increase of H moves the system along the horizontal arrow from NFL state to LFL one. On the contrary, at fixed magnetic field and increasing temperatures, the system transits along the vertical arrow from LFL state to NFL one. The inset to the left panel demonstrates the behavior of the normalized effective mass M^*_N versus normalized temperature T_N following from (21.9). The $T^{-1/2}$ regime is marked as NFL since contrary to the LFL case, where the effective mass is constant, the effective mass depends strongly on temperature. It is seen that the temperature region $T_N \sim 1$ signifies a transition regime between the LFL behavior with almost constant effective mass and the NFL one, given by $T^{-1/2}$ dependence. Thus, temperatures $T \simeq T_{\rm max}$, shown by arrows in the inset and the main panel, marks the transition regime between LFL and NFL states. The common width W of the LFL transition regions $W \propto T$ is shown by the heavy arrow. These theoretical results are in good agreement with the experimental facts [18]. The right panel of Fig. 21.3 illustrates the behavior of the dimension-

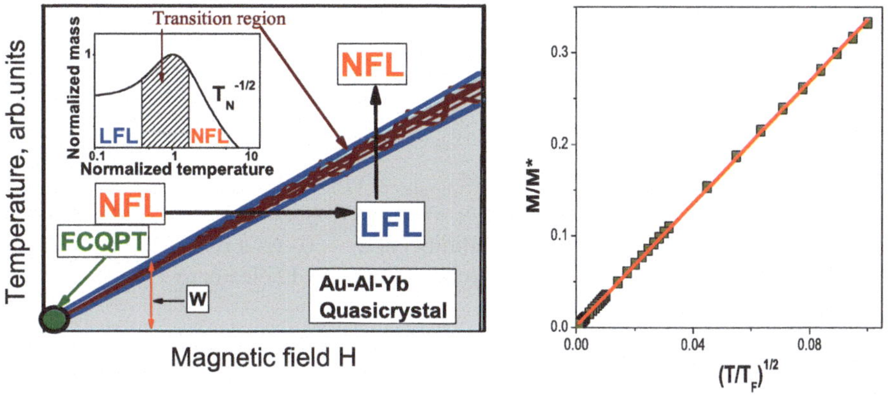

Fig. 21.3 The $T - H$ phase diagram of $Au_{51}Al_{34}Yb_{15}$ with the effective mass $M^*(T) \propto T^{-1/2}$. Left panel: The $T - H$ phase diagram of $Au_{51}Al_{34}Yb_{15}$. Magnetic field H is the control parameter. The vertical and horizontal arrows show LFL-NFL and reverse transitions at fixed H and T, respectively. At $H = 0$ and $T = 0$, the system is at FCQPT shown by the solid circle. The total width of the LFL and the transition regions $W \propto T$ are shown by the double arrows. Inset shows a schematic plot of the normalized effective mass versus the normalized temperature. Transition region, where M_N^* reaches its maximum at $T/T_{\max} = 1$, is marked by the hatched area. The right panel reports the dimensionless inverse effective mass M/M^* versus dimensionless temperature $(T/T_F)^{1/2}$. The line is a linear fit

less inverse effective mass M/M^* versus the dimensionless temperature $(T/T_F)^{1/2}$, where T_F is the Fermi temperature of electron gas. To calculate M/M^*, we use the model Landau functional presented in [20, 27]

$$E[n(p)] = \int \frac{\mathbf{p}^2}{2M} \frac{d\mathbf{p}}{(2\pi)^3} + \frac{1}{2} \int V(\mathbf{p}_1 - \mathbf{p}_2) n(\mathbf{p}_1) n(\mathbf{p}_2) \frac{d\mathbf{p}_1 d\mathbf{p}_2}{(2\pi)^6},$$

with the Landau interaction

$$V(\mathbf{q}) = g_0 \frac{\exp(-\beta_0 \sqrt{\mathbf{q}^2 + \gamma^2})}{\sqrt{\mathbf{q}^2 + \gamma^2}}, \tag{21.11}$$

where the parameters g_0 and β_0 are fixed by the requirement that the system is located at FCQPT. At $\gamma = 0$, the interaction becomes nonanalytic function of q. Note that the other investigated nonanalytic interactions lead to the same behavior of M/M^*, see, e.g., [27].

To demonstrate this, we apply (21.6) to construct $\varepsilon(p)$ using the functional (21.11). Taking into account that $\varepsilon(p \simeq p_F) - \mu \simeq p_F(p - p_F)$ and integrating over the angular variables, we obtain

$$\frac{1}{M^*} = \frac{1}{M} + \frac{\partial}{\partial p} \int [\Phi(p + p_1) - \Phi(|p - p_1|)] \frac{n(p_1, T) p_1 dp_1}{2 p_F^2 \pi^2}. \tag{21.12}$$

Here the derivative on the right-hand side of (21.12) is taken at $p = p_F$ and

$$\int_{|p-p_1|}^{p+p_1} V(z, \gamma = 0) z dz = \Phi(p + p_1) - \Phi(|p - p_1|). \quad (21.13)$$

The derivative $\partial\Phi(|p - p_1|)/\partial p|_{p \to p_F} = (p_F - p_1)/(|p_F - p_1|)\partial\Phi(z)/\partial z$ becomes a discontinuous function at $p_1 \to p_F$, provided that $\partial\Phi(z)/\partial z$ is finite (or integrable if the function tends to infinity) at $z \to 0$. As a result, the right-hand side of (21.12) becomes proportional to M^*T and (21.12) leads to $1/M^* \propto M^*T$, making $M^* \propto T^{-1/2}$.

The analytic Landau interaction (21.11) with $\gamma > 0$ makes $M/M^* \propto T^{0.5}$ at raising temperatures, while at $T \to 0$ the system demonstrates the LFL behavior [20, 26]. This interaction can serve as model one to describe the behavior of the quasicrystal's crystalline approximant $Au_{51}Al_{35}Yb_{14}$, [18]. The approximant $Au_{51}Al_{35}Yb_{14}$ shows the LFL behavior at low temperatures, $\chi^{-1} \propto a + bT^{0.51}$ with the conventional LFL behavior of the resistivity [18]. We interpret this behavior of χ^{-1} through the absence of the unique electronic state of QCs, which results in the shift of the electronic system of the approximant from FCQPT into the LFL region. Such a behavior is achieved by making the interaction (21.11) an analytic function with $\gamma > 0$ as soon as the quasicrystal is transformed into its crystalline approximant. The finite γ, creating the LFL behavior at $T = 0$, makes $T_{\max}$ finite even at $H = 0$. Then, it follows from (7.14) that $1/M^* \propto \chi^{-1} \propto a + bT^{1/2}$ and the approximant is to demonstrate the conventional LFL behavior: $\Delta\rho \propto T^2$. The same result is achieved by transforming the spectrum into $\varepsilon(p) - \mu \propto ([p - p_b]^2 + \gamma^2)(p - p_F)$ [26].

We now investigate the behavior of χ as a function of temperature at fixed magnetic fields. The effective mass $M^*(T, H)$ can be measured in experiments for $M^*(T, H) \propto \chi$ where χ is the AC or DC magnetic susceptibility. If the corresponding measurements are carried out at fixed magnetic field H then, as it follows from (7.14), χ reaches the maximum $\chi_{\max}$ at some temperature $T_{\max}$. Upon normalizing both χ and the specific heat C/T by their peak values at each field H and the corresponding temperatures by $T_{\max}$, we observe from (7.14) that all the curves merge into a single one, thus demonstrating a scaling behavior typical for HF metals [20]. As seen from Fig. 21.4, χ_N extracted from measurements on $Au_{51}Al_{34}Yb_{15}$ [18] shows the scaling behavior given by (7.14) and agrees well with the normalized specific heat $(C/T)_N$ extracted from measurements in magnetic fields H on $YbRh_2Si_2$ [31]. Our calculations shown by the solid curve are in good agreement with χ_N over four orders of magnitude in the normalized temperature.

To affirm the phase diagram in Fig. 21.3, we focus on the LFL, NFL, and the transition LFL-NFL regions exhibited by the QC. In Fig. 21.5a, we report the normalized χ_N in the log-log scale. As seen from Fig. 21.5a, χ_N extracted from the measurements is not a constant, as would be for LFL. Two (NFL and LFL) regions, separated by the transition one, as depicted by the hatched area in the inset in Fig. 21.3, are

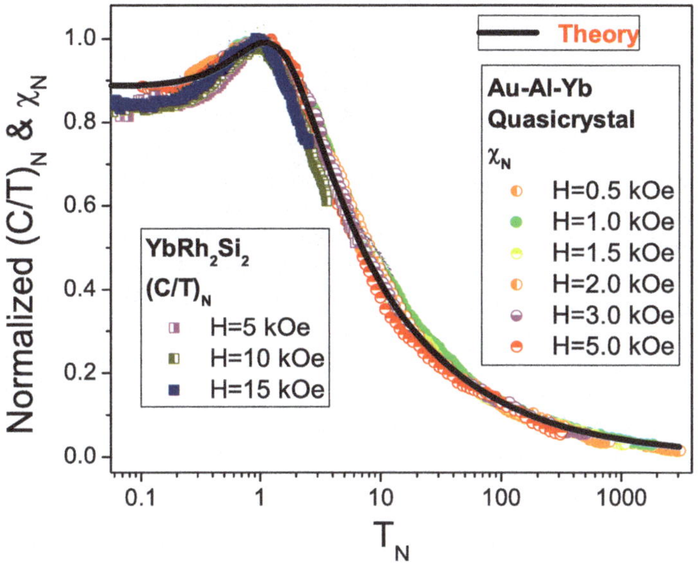

Fig. 21.4 The normalized specific heat $(C/T)_N$ and magnetic susceptibility χ_N extracted from measurements in magnetic fields H (shown in the legends) on $YbRh_2Si_2$ [31] and $Au_{51}Al_{34}Yb_{15}$ [18], respectively. Our calculations are depicted by the solid curve (7.14) tracing the scaling behavior of $(C/T)_N = \chi_N = M^*_N$

clearly seen in Fig. 21.5a, illuminating good agreement between the theory and measurements. The straight lines in Fig. 21.5a outline both the LFL and NFL behaviors of $\chi_N \propto const$ and $\chi_N \propto T_N^{-1/2}$, and are in good agreement with the behavior of M^*_N displayed in the inset of Fig. 21.3. In Fig. 21.5b, the solid squares denote temperatures $T_{\max}(H)$ at which the maxima $\chi_{\max}$ of $\chi(T)$ and, (c), the corresponding values of the maxima $\chi_{\max}(H)$ occur. It is seen that the agreement between the theory and experiment is good in the entire magnetic field domain. It is also seen from Fig. 21.5b that $T_{\max} \propto H$; thus, a tiny magnetic field H destroys the NFL behavior hereby driving the system to the LFL region. This behavior is consistent with the phase diagram displaced in Fig. 21.3: at increasing temperatures, ($T_N \simeq 1$) the LFL state first converts into the transition one and then disrupts into the NFL state, while at given magnetic field H the width $W \propto T$.

Thus, the quasicrystal $Au_{51}Al_{34}Yb_{15}$ exhibits typical scaling behavior of its thermodynamic properties, thus belonging, in fact, to the HF metals family, while the quantum-critical physics of the quasicrystal is universal and emerges regardless its underlying microscopic details.

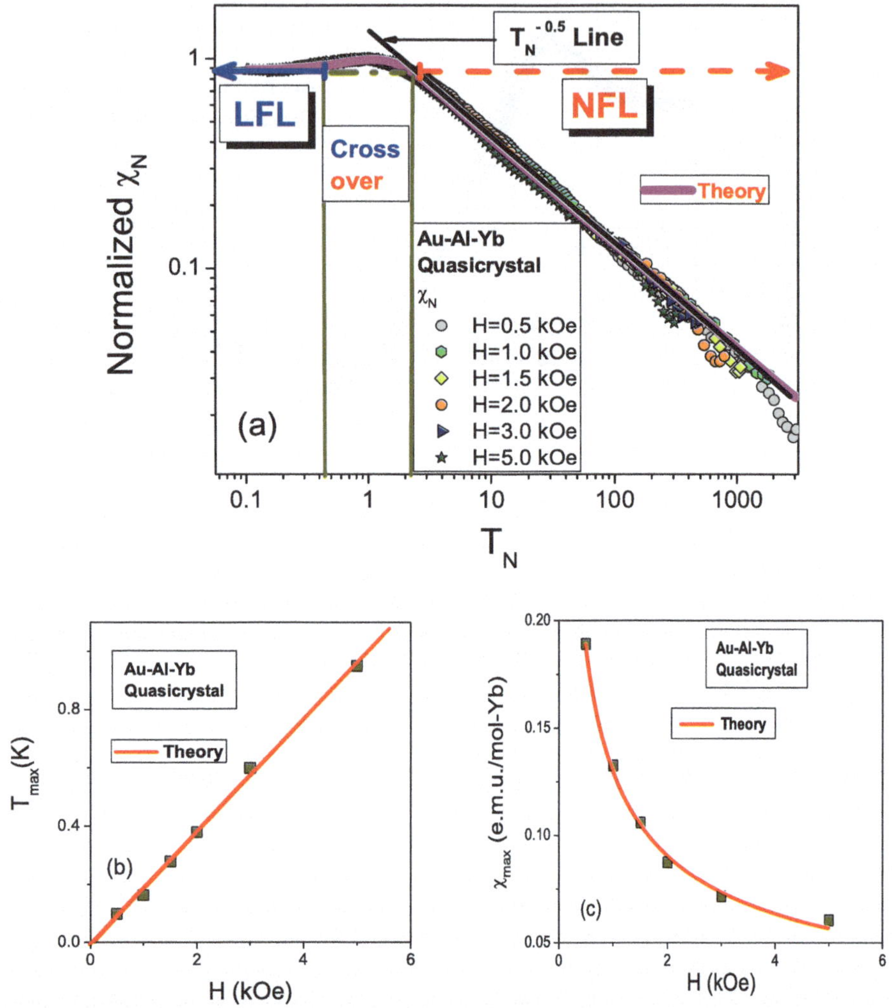

Fig. 21.5 The thermodynamic properties of the magnetic susceptibility. Panel **a** Temperature dependence in log-log scale of the magnetic susceptibility χ_N at different magnetic fields [18] given in the legend. The LFL and NFL regions are marked by the solid and dashed arrows, respectively. The solid line depicts $\chi_N \propto T_N^{-0.5}$ behavior. Panel **b** The temperatures $T_{\rm max}$ where the maxima of χ (see Fig. 21.3) are located. The solid line represents the function $T_{\rm max} = aH$, where a is a fitting parameter. Panel **c** The maxima $\chi_{\rm max}$ versus magnetic field H. The solid curve is approximated by $\chi_{\rm max} = tH^{-1/2}$, see (21.10), t is a fitting parameter

21.3 Quantum Criticality at Metamagnetic Phase Transitions

The nature of field-tuned metamagnetic quantum criticality in HF metals is a significant challenge for condensed matter physics. Here we center our attention on the role of the applied magnetic field B in the formation of quantum criticality within a restricted range of B at low temperatures T. Here we analyze quantum criticality and metamagnetic phase transitions in HF metals.

21.3.1 *Typical Properties of the Metamagnetic Phase Transition in* $Sr_3Ru_2O_7$

Here we consider a coherent picture of both the quantum-critical regime and the metamagnetic phase transition in $Sr_3Ru_2O_7$. We take $Sr_3Ru_2O_7$ as an example of quantum criticality formed at metamagnetic phase transition to outline its main features. In constructing both the field-induced quantum criticality and the corresponding metamagnetic phase transition, we employ the model based on a vHs that induces a peak in the single-particle DOS. At fields in some range $B_{c1} < B < B_{c2}$, the DOS peak turns out to be at or near the Fermi energy. As a result, a relatively weak repulsive interaction (e.g., Coulomb) is sufficient to move the system to FCQPT, or even to induce FC and formation of a flat band, as it takes place in the case of $Sr_3Ru_2O_7$.

To reveal signatures of the quantum criticality, we resume with an analysis of the properties of the C/T electronic specific heat observed in $Sr_3Ru_2O_7$ [33, 34]. The measurements of $C/T \propto M^*$ on $Sr_3Ru_2O_7$ in magnetic fields allow us to uncover the universal scaling behavior of the effective mass M^* that is characteristic of HF metals. As it is seen from Fig. 13.3 displaying the typical behavior of HF metals, the maximum of $C/T \propto M^*$ sharpens and shifts to lower temperatures as the field B approaches the critical value $B = B_m \simeq (B_{c1} + B_{c2})/2$, where the maximum disappears. In contrast to HF metals, the maximum of the function C/T is symmetric with respect to B_m: the maximum appears at $B \to B_m$ and reappears on the high-field side at $B > B_m$. To reveal the scaling behavior of C/T, we normalize the measured C/T and T values (obtaining $(C/T)_N$ and T_N, respectively) by their maxima T_M and $(C/T)_M$, respectively [20]. The spin susceptibility data $\chi(T)$ are normalized in the same way. The behavior of the normalized effective mass extracted from measurements of χ and (C/T) on $CeRu_2Si_2$ [32], $CePd_{0.8}Rh_{0.8}$ [31], and $Sr_3Ru_2O_7$ [33] are presented in Fig. 21.6. The figure displays the main features of scaling behavior of the normalized effective mass M^*_N shown in Fig. 15.1. Namely, at low temperatures $T_N < 1$ the normalized effective mass is in the LFL region, then it enters the transition region, and finally moves to the NFL regime. The solid curve shows the result of our calculation of the scaling behavior. It is seen from Fig. 21.6 that $Sr_3Ru_2O_7$ is located at the metamagnetic transition and HF metals exhibit the same scaling

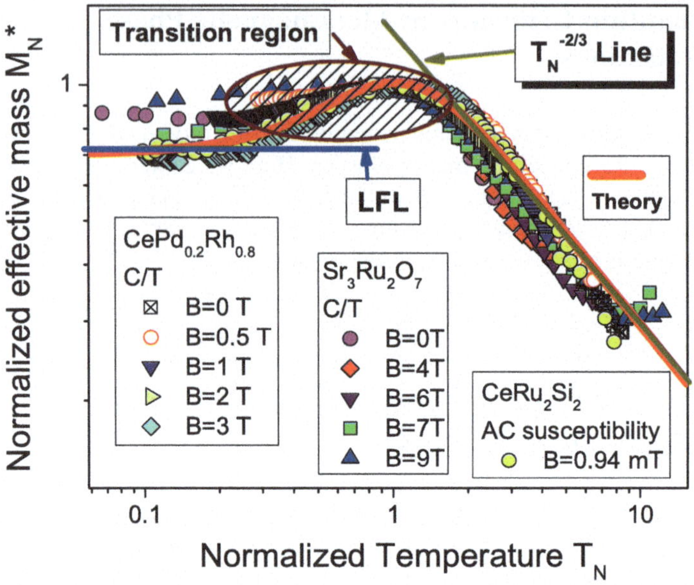

Fig. 21.6 The universal scaling behavior of the normalized effective mass M_N^* versus T_N. M_N^* is extracted from the measurements of χ and C/T (in magnetic fields B shown in the legends) on $CeRu_2Si_2$ [32], $CePd_{1-x}Rh_x$ with $x = 0.80$ [31], and $Sr_3Ru_2O_7$ [33]. The LFL and NFL regimes (latter having $M_N^* \propto T_N^{-2/3}$) are shown by the arrows and straight lines. The transition regime is depicted by the shaded area. The solid curve represents our calculation of the universal behavior of $M_N^*(T_N)$

behavior, as other HF compounds which can be understood within the framework of fermion condensation or flat-band theory.

21.3.2 *Metamagnetic Phase Transition in HF Metals*

As we have seen above, a HF metal can be driven to FCQPT when narrow (flat) bands situated close to the Fermi surface are formed by the application of a critical magnetic field B_m. The emergence of such state is known as metamagnetism that occurs when this transformation comes abruptly at B_m [34, 35].

Thus, the magnetic field B_m is similar to that of B_{c0} that moves a HF metal to its magnetic-field-tuned QCP. In our simple model, both B_{c0} and B_m are taken as parameters. To apply (7.14) when the critical magnetic field is not zero, we have to replace B by $(B - B_m)$. Acting as above, we can extract the normalized effective mass $M_N^*(T_N)$ from data collected on HF metals at their metamagnetic quantum phase transition. In Fig. 21.7, the extracted normalized mass is displayed. $M_N^*(T_N)$ is extracted from measurements of C/T collected on $URu_{1.92}Rh_{0.08}Si_2$ and $CeRu_2Si_2$

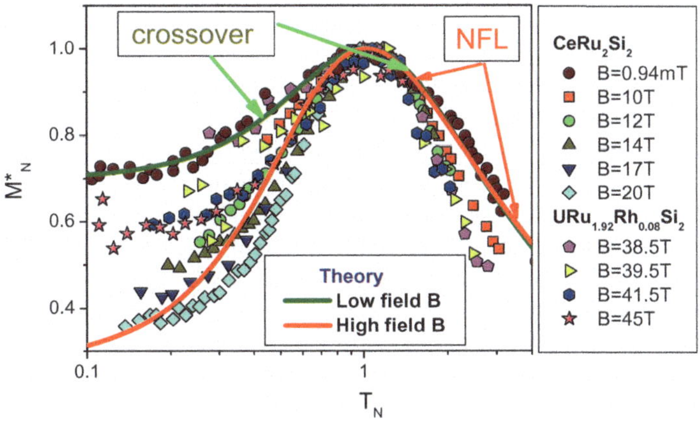

Fig. 21.7 The normalized effective mass versus the normalized temperature at different magnetic fields B shown in the legend. $M_N^*(T_N)$ is extracted from measurements of C/T collected on $URu_{1.92}Rh_{0.08}Si_2$ and $CeRu_2Si_2$ (data for $CeRu_2Si_{1.8}Ge_{0.2}$ see Fig. 13.4) [36, 37]. The two solid curves give the universal behavior of M_N^* at low and high magnetic fields B, (7.14)

(data for $CeRu_2Si_{1.8}Ge_{0.2}$ see Fig. 13.4) at their metamagnetic QCP with $B_m \simeq 35$ T, $B_m \simeq 7$ T, and $B_m \simeq 1.2$ T, respectively [36, 37]. As it is seen from Fig. 21.7, the effective mass $M_N^*(T_N)$ in different HF metals reveals the same form, both in the high and in low magnetic fields as soon as the corresponding bands become flat, that is, the electronic system of HF metals is driven to FCQPT. This observation allows us to check the universal behavior in HF metals, when these are under the application of essentially different magnetic fields. Namely, the magnitude of the applied field ($B \sim 10$ T) at the metamagnetic point is four orders of magnitude larger than that of the field applied to tune $CeRu_2Si_2$ to the LFL behavior ($B \sim 1$ mT). Relatively small values of $M_N^*(T_N)$ observed in $URu_{1.92}Rh_{0.08}Si_2$ and $CeRu_2Si_2$ at the high fields and small temperatures can be explained by taking into account that the narrowband is completely polarized [36]. As a result, at low temperatures, the summation over spin projections "up" and "down" reduces to a single direction producing the coefficient $1/2$ in front of the normalized effective mass. At high temperatures, the polarization vanishes and the summation is restored. As it is seen from Fig. 21.7, these observations are in accord with the experimental facts.

21.4 Universal Behavior of Two-Dimensional ^{3}He at Low Temperatures

On the example of 2D ^{3}He we demonstrate that the main universal features of its temperature T -density x phase diagram^{3}He allows us to propose a simple expression for effective mass $M^*(T, x)$, describing all diverse experimental facts in 2D ^{3}He in unified manner and demonstrating that the universal behavior of $M^*(T, x)$ coincides with that observed in HF metals.

It was discussed in Chap. 7 that the electronic system of HF metals demonstrates the universal low-temperature behavior irrespectively of their magnetic ground state [38]. Therefore, it is of importance to check whether this behavior can be observed in 2D Fermi systems. Fortunately, the measurements on 2D ^{3}He are available [39]. Their results allow one to check the presence of the universal behavior in the system formed by ^{3}He atoms, which are essentially different from electrons [40]. Namely, atoms of 2D ^{3}He are neutral fermions with spin $S = 1/2$. They interact with each other by van der Waals forces with strong hardcore repulsion and a weakly attractive tail. The different character of interparticle interaction along with the fact that the mass of He atom is three order of magnitude larger than that of an electron makes ^{3}He have drastically different microscopic properties than that of 3D HF metals. Because of this difference nobody can be sure that the macroscopic physical properties of both abovementioned fermionic systems would essentially be different from each other.

The bulk liquid ^{3}He is historically the first object, to which a Landau Fermi liquid (LFL) theory had been applied [23]. This substance being intrinsically isotropic Fermi liquid with negligible spin-orbit interaction is an ideal object to test the LFL theory. 2D films of ^{3}He have been fabricated and its thermodynamic properties have been thoroughly investigated [39]. Among these properties are measurements of the entropy S as a function of the number density x versus T. As it is seen from Fig. 21.8, at $x \geq 8.00$ nm^{-2}, the entropy is no more a linear function of temperature T, exhibiting the NFL behavior.

The NFL properties manifest themselves in the power-law behavior of the physical quantities of strongly correlated Fermi systems located close to their QCPs, with exponents, characterizing the power-law behavior, different from those of a Fermi liquid. The main output of theory should be the explanation of these exponents, which are at least dependent on the magnetic character of QCP and dimensionality of the system. On the other hand, the observed behavior of the thermodynamic properties cannot be described entirely by these exponents, as it is seen from Figs. 21.8 and 21.9. The behavior of the entropy $S(T)$ of two-dimensional (2D) ^{3}He [39] shown in Fig. 21.8 is positively different from that described by a simple function $a_1 T^{a_2}$ where a_1 is a constant and a_2 is the exponent. It is seen from Fig. 21.8 that at low densities $x \simeq 7$ nm^{-2} the entropy demonstrates the LFL behavior characterized by a linear function of T with $a_2 = 1$. The behavior becomes quite different at higher densities, where $S(T)$ has an inflection point that shifts to lower temperatures. The effective mass $M^*(x)$ diverges at $x \to x_c$, see Figs. 12.3 and 13.1. Obviously, near the inflection point $S(T)$ cannot be fit by the simple function $a_1 T^{a_2}$, as in the case of HF metals.

In order to show that the behavior of S displayed in Fig. 21.8 and 21.9 is of generic character, let us recollect that in the vicinity of QCP it is helpful to use "internal" scales to measure the effective mass $M^* \propto S/T$ and temperature T [20]. The internal scales of the thermodynamic functions, such as S or C/T, are related to "peculiar points" like the inflection or maximum. Since the entropy has no maxima,

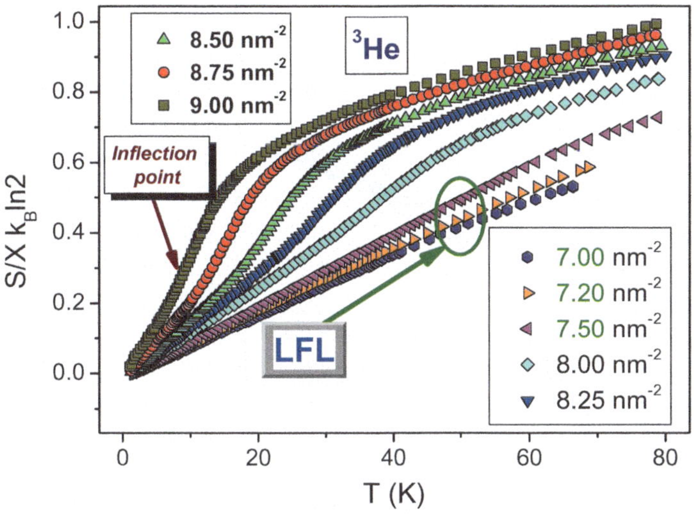

Fig. 21.8 Temperature dependence of the entropy S of 2D ^{3}He at different densities x (shown in the legends) [39]. The inflection point of S at $x = 9.00$ nm^{-2} is shown by the arrow. The LFL behavior at high densities ($S(T) \propto T$) is displayed by the arrow

its normalization is to be performed at the inflection point that takes place at $T = T_{inf}$. Note that T_{inf} is a function of x, and it is seen from Fig. 21.8 that the inflection point moves toward lower temperatures with increase of x. The normalized entropy S_N as a function of the normalized temperature $T_N = T/T_{inf} = y$ is reported in Fig. 21.9. We normalize the entropy by its value at the inflection point $S_N(y) = S(y)/S(1)$. As it is seen from Fig. 21.9, the normalization reveals the scaling behavior of S_N. As we have seen previously, this means that the curves at different temperatures and densities x merge into a single one in terms of the variable y. We have excluded the experimental data taken at $x \leq 8\,\text{nm}^{-2}$ since the corresponding curves do not contain the inflection points. It is seen from Fig. 21.9 that $S_N(y)$ extracted from the measurements is not a linear function of y, as would be for a LFL, and shows the scaling behavior in the normalized temperature T_N.

Thus, our analysis of the experimental measurements shows that the behavior of 2D ^{3}He is very close to that of 3D HF compounds with various ground state magnetic properties. Because of van der Waals interparticle interaction, ^{3}He has an important and specific feature that, generally speaking, cannot be realized in full in HF metals. This feature is that it is possible to change the total density of ^{3}He film. This change allows one to drive 2D film toward its QCP at which the quasiparticle effective mass M^* diverges [39, 41]. This peculiarity permits to plot the experimental temperature-density phase diagram, which in turn can be directly compared with theoretical predictions.

Let us consider HF liquid at $T = 0$ characterized by the effective mass M^* [20, 27, 42–45]. Upon applying the well-known equation, we can relate M^* to the bare

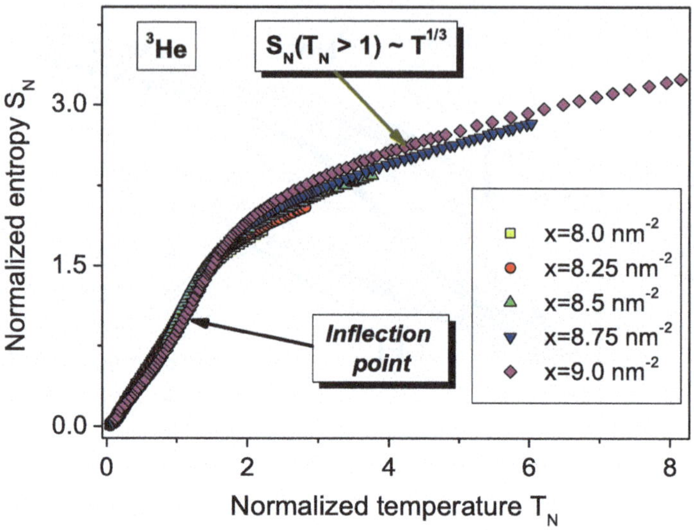

Fig. 21.9 The normalized entropy $S_N(T_N)$, extracted from the experimental data of Fig. 21.8. Densities x are shown in the legend. The arrow shows the inflection point

gas mass M [23, 46] $M^*/M = 1/(1 - N_0F^1(p_F, p_F)/3)$, see Sect. 2.3.1. Here N_0 is the density of states of a free gas, p_F is Fermi momentum, and $F^1(p_F, p_F)$ is the p-wave component of Landau interaction F. Since LFL theory implies that the number density has the form $x = p_F^3/3\pi^2$, we can rewrite the interaction as $F^1(p_F, p_F) = F^1(x)$. When at some critical point $x = x_c$, $F^1(x)$ achieves certain threshold value, the denominator tends to zero so that the effective mass diverges at $T = 0$ and the system undergoes FCQPT. The leading term of this divergence looks as

$$\frac{M^*(x)}{M} = A + \frac{B}{1-z}, \; z = \frac{x}{x_c}. \tag{21.14}$$

Equation (21.14) is valid in both 3D and 2D cases, while the values of factors A and B depend on dimensionality and interparticle interaction [45]. At $x > x_c$, the fermion condensation takes place. However, we consider at first the case $x < x_c$.

When the system approaches FCQPT, the dependence $M^*(T, x)$ is governed by the Landau equation

$$\frac{1}{M^*(T,x)} = \frac{1}{M} + \int \frac{\mathbf{p}_F\mathbf{p}}{p_F^3} F(\mathbf{p_F}, \mathbf{p}) \frac{\partial n(\mathbf{p}, T, x)}{\partial p} \frac{d\mathbf{p}}{(2\pi)^3}, \tag{21.15}$$

where $n(\mathbf{p}, T, x)$ is the distribution function of quasiparticles. The approximate solution of this equation is of the form [38, 40]

$$\frac{M}{M^*(T)} = \frac{M}{M^*(x)} + \beta f(0) \ln\{1 + \exp(-1/\beta)\} + \lambda_1 \beta^2 + \lambda_2 \beta^4 + ..., \quad (21.16)$$

where $\lambda_1 > 0$ and $\lambda_2 < 0$ are constants of the order of unity, $\beta = TM^*(T)/p_F^2$, and $f(0) \sim F^1(x_c)$. It follows from (21.16) that the effective mass $M^*(T)$ as a function of T and x reveals three different regimes at growing temperature. At the lowest temperatures, we have LFL regime with $M^*(T) \simeq M^*(x) + aT^2$ with $a < 0$ since $\lambda_1 > 0$. This observation coincides with experimental situation [39, 41]. The effective mass as a function of T decreases up to a minimum and afterward grows, reaching its maximum $M^*_M(T, x)$ at some temperature $T_{\max}(x)$ with subsequent diminishing as $T^{-2/3}$ [24, 45]. Moreover, the closer is the number density x to its threshold value x_c, the higher is the growth rate. The peak value M^*_M grows also, but the maximum temperature $T_{\max}$ lowers. Near this temperature the last "traces" of LFL regime disappear, manifesting themselves in the divergence of above low-temperature series and substantial growth of $M^*(x)$. The temperature region that starts near above minimum and continuing up to $T_{\max}(x)$ signifies the crossover between LFL regime with almost constant effective mass and NFL behavior, given by $T^{-2/3}$ dependence, as shown in Chap. 7. Thus, the $T_{\max}$ point can be considered as the crossover between LFL and NFL regimes. The latter regime sets up at $T \leq T_{\max}$, when $M^*(x) \to \infty$, giving rise to the effective mass decrease

$$M^*(T) \propto T^{-2/3}. \quad (21.17)$$

It turns out that $M^*(T, x)$ in the entire T and x range can be well approximated by a simple universal interpolating function similar to the case of the application of magnetic field. The interpolation occurs between LFL ($M^* \propto T^2$) and NFL ($M^* \propto T^{-2/3}$) regimes thus describing the above crossover. Introducing the dimensionless variable $y = T/T_{\max}$, we present the desired expression

$$\frac{M^*(T, x)}{M^*_M} = \frac{M^*(y)}{M^*_M} = M^*_N(y) \approx \frac{M^*(x)}{M^*_M} \frac{1 + c_1 y^2}{1 + c_2 y^{8/3}}. \quad (21.18)$$

Here $M^*_N(y)$ is the normalized effective mass, and c_1 and c_2 are parameters obtained from the condition of best fit to experiment. Equation (21.14) shows that the maximum value M^*_M of the effective mass $M^*_M \propto 1/(1 - z)$. On the other hand, it follows from (21.17) that $M^*_M \propto T^{-2/3} \sim T_{\max}^{-2/3}$. Thus, we have

$$T_{\max} \propto (1 - z)^{3/2}. \quad (21.19)$$

Note that the obtained results are in agreement with numerical calculations [24, 40, 45].

$M^*(T)$ can be measured in experiments on strongly correlated Fermi systems. For example, $M^*(T) \propto C(T)/T \propto S(T)/T \propto M_0(T) \propto \chi(T)$ where $C(T)$ is the specific heat, $S(T)$—entropy, $M_0(T)$—magnetization, and $\chi(T)$—AC magnetic susceptibility. If the measurements are performed at fixed x, then, as it follows from

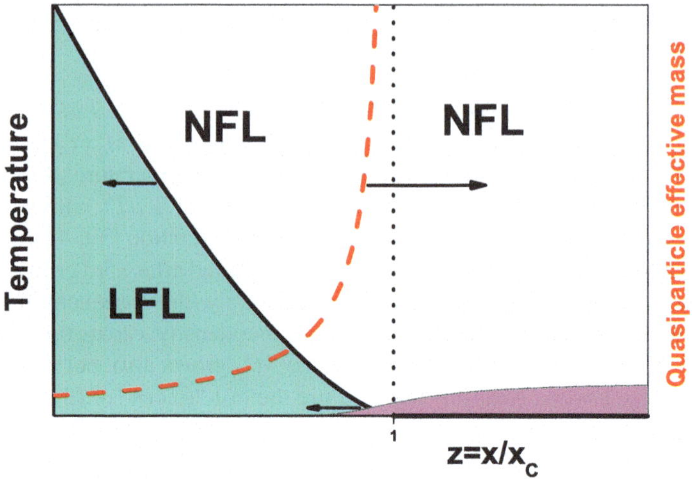

Fig. 21.10 The phase diagram of 2D ^{3}He. The part for $z < 1$ corresponds to HF behavior divided to the LFL and NFL parts by the line $T_{\max}(z) \propto (1-z)^{3/2}$, where $T_{\max}$ is the effective mass maximum temperature. The exponent $3/2 = 1.5$ coming from (21.19) is in good agreement with the experimental value 1.7 ± 0.1 [39]. The dependence $M^*(z) \propto (1-z)^{-1}$ is shown by the dashed line. The regime for $z \geq 1$ consists of LFL piece (the shadowed region, beginning in the intervening phase $z \leq 1$ [39], which is probably due to the quasi-classical behavior of the heat capacity C, see text) and NFL regime at higher temperatures

(21.18), the effective mass reaches the maximum at $T = T_{\max}$. Upon normalizing both, $M^*(T)$ by its peak value at each x and the temperature by $T_{\max}$, we see from (21.18) that all the curves merge into single one demonstrating a scaling behavior.

In Fig. 21.10, we show the phase diagram of 2D ^{3}He in the variables T - z (see (21.14)). For the sake of comparison, the plot of the effective mass versus z is shown by dashed line. The part of the diagram where $z < 1$ corresponds to HF behavior and consists of LFL and NFL parts, divided by the line $T_{\max}(z) \propto (1-z)^{3/2}$. We pay attention here that exponent $3/2 = 1.5$ is exact as compared to that from [39] 1.7 ± 0.1. The agreement between theoretical and experimental exponents suggests that the FCQPT scenario takes place both in 2D ^{3}He and in HF metals. The regime for $z > 1$ consists of low-temperature LFL piece, (shadowed region, beginning in the intervening phase $z \leq 1$ [39]) and NFL regime at higher temperatures. The former LFL piece is assigned to the peculiarities of substrate on which 2D ^{3}He film is placed. Namely, it is related to weak substrate heterogeneity (steps and edges on its surface) so that Landau quasiparticles, being localized (pinned) on it, give rise to LFL behavior [39]. On the other hand, the intervening phase, shown in Fig. 21.10, can be related to the quasi-classical behavior of the heat capacity C that leads to a temperature-independent contribution to C, as it is discussed in Sect. 17.1. This important observation deserves a further investigation that would prompt new theoretical work, supporting the idea that the physics of quantum criticality seen in HF compounds is universal.

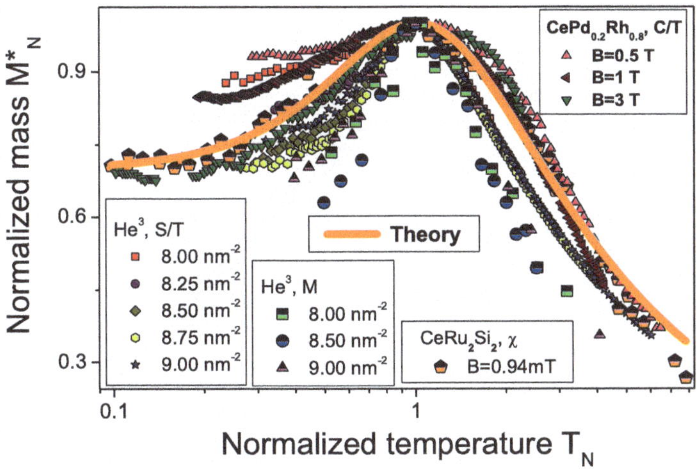

Fig. 21.11 The normalized effective mass M_N^* as a function of the normalized temperature $T/T_{\max}$ at densities shown in the left down corner. M_N^* is extracted from experimental data for $S(T)/T$ in 2D ^{3}He [39] and 3D HF compounds with different magnetic ground states, such as $CeRu_2Si_2$ and $CePd_{1-x}Rh_x$ [31, 32], fitted by the universal function (21.18)

In Fig. 6.5, the experimental values of effective mass $M^*(z)$ obtained by the measurements on ^{3}He monolayer are reported [41]. These measurements, in agreement with those from [39], demonstrate the divergence of the effective mass at $x = x_c$. To show that our FCQPT approach is able to describe the above data, we present the fit of $M^*(z)$ by the rational expression $M^*(z)/M \propto A + B/(1 - z)$ and the reciprocal effective mass by the linear fit $M/M^*(z) \propto A_1 z$. We note here that the linear fit has been used to describe the experimental data for bilayer ^{3}He [39] and we use this function here for the sake of illustration. It is seen from Fig. 6.5 that the data in [39] (^{3}He bilayer) can be equally well approximated by both linear and rational functions, while the data in [41] cannot. For instance, both fitting functions give for the critical density in bilayer $x_c \approx 9.8$ nm^{-2}, while for monolayer [41] these values are different - $x_c = 5.56$ for linear fit and $x_c = 5.15$ for fractional fit. It is seen from Fig. 6.5 that linear fit is unable to properly describe the experiment [41] at small $1 - z$ (i.e., near $x = x_c$), while the fractional fit describes the experiment pretty good. This means that more detailed measurements are necessary in the vicinity $x = x_c$.

Now we apply the universal dependence, given by (7.14) and (7.8), to fit the experiment not only in 2D ^{3}He but in 3D HF metals as well. $M_N^*(y)$ extracted from the entropy measurements on the ^{3}He film [39] at different densities $x < x_c$ smaller than the critical point $x_c = 9.9 \pm 0.1$ nm^{-2} is reported in Fig. 21.11. In the same figure, the data extracted from heat capacity of ferromagnet $CePd_{0.2}Rh_{0.8}$ [31] and AC magnetic susceptibility of paramagnet $CeRu_2Si_2$ [32] are plotted for different magnetic fields. It is seen that the universal behavior of the effective mass given by (21.18) (solid curve in Fig. 21.11) is in accord with experimental data. All substances are located at $x < x_c$, where the system progressively disrupts its LFL behavior at

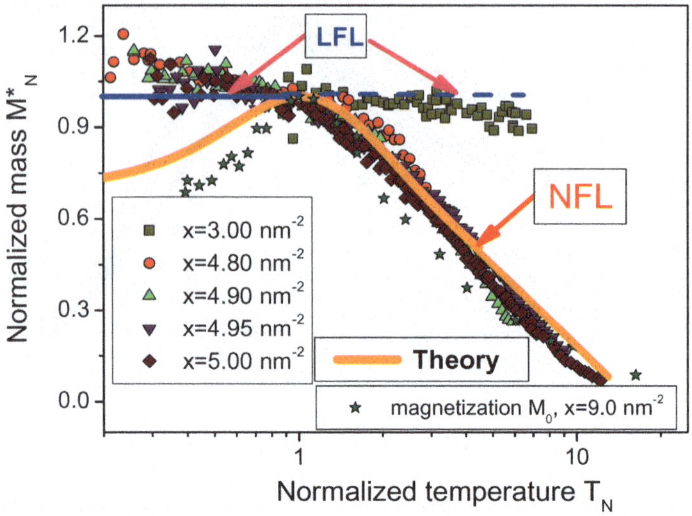

Fig. 21.12 The dependence of $M_N^*(T/T_{\max})$ versus T at densities shown in the left down corner. The behavior M_N^* is extracted from experimental data for $C(T)/T$ in 2D ^{3}He [41] and for the magnetization M_0 in 2D ^{3}He [39]. The solid curve shows the universal function, see Fig. 21.11. Note that at low densities $x < 7.50$ nm^{-2} the system exhibits the LFL behavior at low temperatures. At elevated temperatures and growing densities, the LFL behavior changes to NFL one. At $x = 3.00$ nm^{-2}, the system demonstrates the LFL behavior at relatively high temperatures

elevated temperatures. In that case, the control parameter, driving the system toward its critical point x_c is merely a number density x. It is seen that the behavior of the effective mass $M_N^*(y)$, extracted from $S(T)/T$ in 2D ^{3}He (the entropy $S(T)$ is reported in Fig. 21.8), looks very much like that in 3D HF compounds. As we shall see from Fig. 21.13, the interaction and positions of the maxima of magnetization $M_0(T)$ and $S(T)/T$ in 2D ^{3}He follow nicely the interpolation formula (21.18). We conclude that (21.18) allows us to reduce a four-variable function describing the effective mass to a function of a single variable. Indeed, the effective mass depends on magnetic field, temperature, number density, and the composition so that all these parameters can be merged in the single variable by means of interpolating function like (21.18), see also [38].

The attempt to fit the available experimental data for $C(T)/T$ in ^{3}He [41] by the universal function $M_N^*(y)$ is reported in Fig. 21.12. Here, the data extracted from heat capacity $C(T)/T$ for ^{3}He monolayer [41] and magnetization M_0 for bilayer [39] are reported. It is seen that the effective mass extracted from these thermodynamic quantities can be well described by the universal interpolation formula (7.14). We note the qualitative similarity between the double layer [39] and monolayer [41] of ^{3}He seen from Fig. 21.12.

On the left panel of Fig. 21.13, we show the density dependence of $T_{\max}$, extracted from measurements of the magnetization $M_0(T)$ on ^{3}He bilayer [39]. The peak

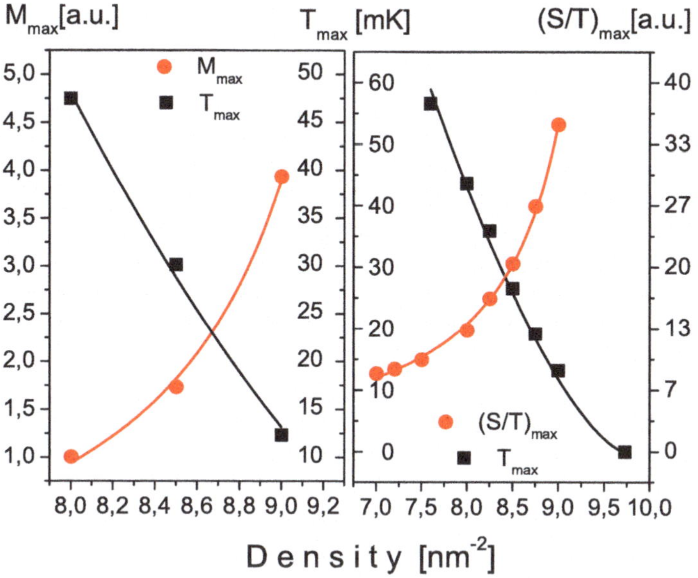

Fig. 21.13 Left panel. The peak temperatures $T_{\max}$ and values $M_{\max}$ extracted from measurements of the magnetization M_0 in ^{3}He [39]. Right panel shows $T_{\max}$ and the peak values $(S/T)_{\max}$ extracted from measurements of $S(T)/T$ in ^{3}He [39]. We approximate $T_{\max} \propto (1-z)^{3/2}$ and $(S/T)_{\max} \propto M_{\max} \propto A/(1-z)$

temperature is fitted by (21.19). At the same figure, we have also presented the maximal magnetization $M_{\max}$. It is seen that $M_{\max}$ is well described by the expression $M_{\max} \propto (S/T)_{\max} \propto (1-z)^{-1}$, see (21.14). The right panel of Fig. 21.13 reports the peak temperature $T_{\max}$ and the maximal entropy $(S/T)_{\max}$ versus the number density x. They are extracted from the measurements of $S(T)/T$ on ^{3}He bilayer [39]. The fact that both the left and right panels have the same behavior shows once more that there are indeed the quasiparticles, determining the thermodynamic behavior of 2D ^{3}He (and also 3D HF compounds [38]) near their QCP.

We conclude that despite absolutely different microscopic nature of 2D ^{3}He and 3D HF metals, their main universal macroscopic features and quantum criticality are the same [40]. As a result, the main features of ^{3}He experimental T-x phase diagram look like those in HF metals and can be well captured utilizing our notion of FCQPT, based on the extended quasiparticles paradigm, and described in detail in Chap. 2. The modification is that in contrast to the Landau quasiparticle effective mass, the ^{3}He effective mass $M^*(T, x)$, as well as that of HF metals, becomes temperature- and density-dependent. We have demonstrated that the universal behavior of $M^*(T, x)$ coincides with that observed in HF metals.

21.5 Scaling Behavior of HF Compounds and Kinks in the Thermodynamic Functions

In this section, we visualize kinks or energy scales in the thermodynamic functions measured on HF compounds such as HF metals, 2D ^{3}He, quasicrystals, etc. The kink is a crossover point from the fast to slow growth of the thermodynamic function of HF compound at raising magnetic field and at fixed temperature and vise versa.

The $T - B$ schematic phase diagram of a HF metal like $YbRh_2Si_2$ is shown in Fig. 21.14. The LFL phase prevails at $T \ll T_M$, followed by the $T^{-2/3}$ regime at $T \gtrsim T_M$, see (21.18) and Sect. 7.1. The latter phase is designated as NFL due to the strong temperature dependence of the effective mass in it. The region $T \simeq T_M$ encompasses the transition between the LFL regime with almost constant effective mass and the NFL behavior. When constructing the phase diagram in Fig. 21.14, we assumed that the AF order prevails, destroying the S_0 term at low temperatures. At $B = B_{c0}$, the HF liquid acquires a flat band, where B_{c0} is a critical magnetic field, such that at $T \to 0$ the application of magnetic field $B \gtrsim B_{c0}$ destroys the AF state, restoring the paramagnetic (PM) state with LFL behavior. In some cases $B_{c0} = 0$ as in the HF metal $CeRu_2Si_2$ (see, e.g., [32]), while in $YbRh_2Si_2$, $B_{c0} \simeq 0.06$ T [8]. Obviously, B_{c0} is defined by the specific system properties; therefore, we consider it as a parameter. At elevated temperature and fixed magnetic field, the NFL regime is dominant. With B increasing, the system is driven from the NFL to the LFL domain. The magnetic-field-tuned FCQPT is indicated by the arrow and is located at $B = B_{c0}$. The transition region separates the NFL state from the weakly polarized

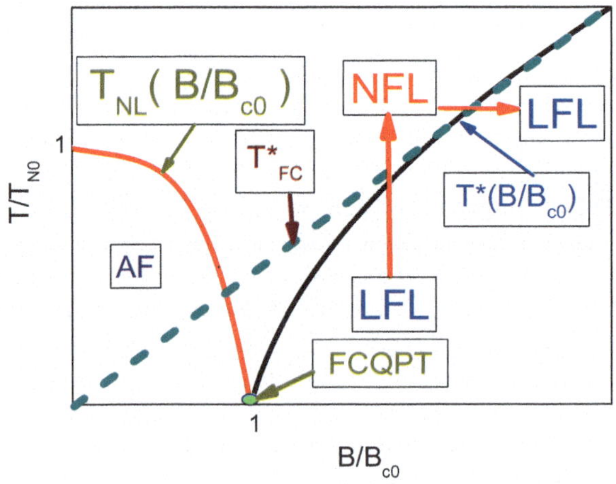

Fig. 21.14 Schematic $T - B$ phase diagram of a HF metal with dimensional magnetic field B/B_{c0} as control parameter. The AF phase boundary line is shown by the arrow and plotted by the solid curve, representing the Néel temperature $T_{NL}(T/T_{NO})$. The solid curve represents the transition temperature $T^*(B/B_{c0})$. The dashed line $T^*_{FC} \propto B\mu_B$ shows the transition temperature provided that the AF state was absent

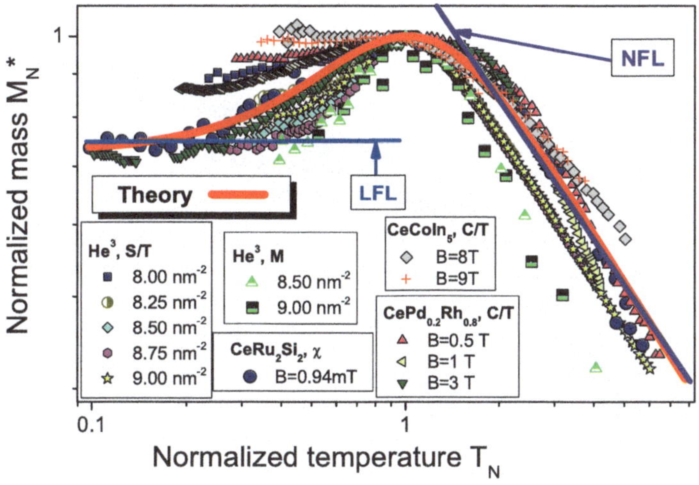

Fig. 21.15 Energy scales in HF metals and 2D ^{3}He. Normalized effective mass M^*_N versus normalized temperature $T_N = T/T_M$. The dependence $M^*_N(T_N)$ is extracted from measurements of $S(T)/T$ and magnetization M of 2D ^{3}He [39]), from AC susceptibility $\chi(T)$ of $CeRu_2Si_2$ [32] and from $C(T)/T$ of both $CePd_{1-x}Rh_x$ [31] and $CeCoIn_5$ [49]. The data are collected for different densities and magnetic fields shown in the legends. The solid curve traces the universal behavior of the normalized effective mass (21.18). Parameters c_1 and c_2 are adjusted for $\chi_N(T_N, B)$ at $B = 0.94$ mT. The LFL and NFL regions are shown by the straight lines, and are marked by the arrows

LFL one. Referring to (21.18), the latter is defined by the function $T^* \propto \mu_B B$ and merges with $T^*_{FC}(B)$ at relatively high temperatures, and $T^* \propto \mu_B(B - B_{c0})$ at lower $T \sim T_{NL}$, with $T_{NL}(B)$ being the Néel temperature. As seen from (21.18), the width of the transition region is proportional to T. The AF phase boundary line is shown by the arrow and depicted by the solid curve. As mentioned above, the dashed line $T^*_{FC}(B) \propto B\mu_B$ represents the transition temperature provided that the AF state was absent. In that case, the FC state is destroyed by any weak magnetic field $B \to 0$ at $T \to 0$ and the dashed line T^*_{FC} crosses the origin, as displayed in Fig. 21.14. At $T \gtrsim T_{NL}(B = 0)$, the transition temperature $T^*_{FC}(B)$ coincides with $T^*(B)$ shown by the solid curve, since the properties of the system are given by its local free energy, describing the PM state of the system. One might say that the system exhibits the universal behavior, and "does not remember" the AF state, emerging at lower temperatures. This conclusion is in good agreement with the data collected on the HF metal $YbRh_2Si_2$ [47], and demonstrates the modification of the topological FCQPT by a phase transition like AF one that takes place to eliminate the residual entropy S_0, $S(T \to 0) = S_0$ [48].

To better visualize kinks or energy scales in the thermodynamic functions measured in HF metals [50] and 2D ^{3}He, we present the normalized effective mass M^*_N extracted from the thermodynamic functions versus normalized temperature and the normalized thermodynamic functions proportional to $T_N M^*_N$ in Figs. 21.15 and 21.16, respectively.

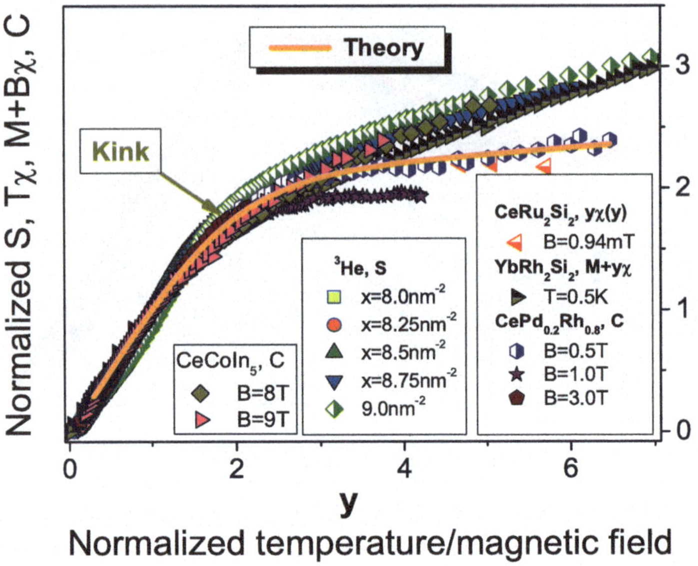

Fig. 21.16 Energy scales in HF metals and 2D ^{3}He. We plot the normalized specific heat $C(y)$ of $CePd_{1-x}Rh_x$ and $CeCoIn_5$ at different magnetic fields B, and the normalized entropy $S(y)$ of ^{3}He at different number densities x, and the normalized $y\chi(y)$ at $B = 0.94$ mT versus normalized temperature $y = T_N$ or magnetic field $y = B_N$. The normalized "average" magnetization $\underline{M} = M + B\chi$ is collected on $YbRu_2Si_2$ [50]. The kink (shown by the arrow) in all the data is clearly seen at the transition region. The solid curve represents $yM_N^*(y)$ with parameters c_1 and c_2 adjusted for the magnetic susceptibility of $CeRu_2Si_2$ at $B = 0.94$ mT

$M_N^*(y)$ extracted from the entropy $S(T)/T$ and magnetization M measurements on the ^{3}He film [39] at different densities x are presented in Fig. 21.15. The data are extracted from the heat capacity C of the ferromagnet $CePd_{0.2}Rh_{0.8}$ [31], $CeCoIn_5$ [49], and the AC magnetic susceptibility of the paramagnet $CeRu_2Si_2$ [32], and are plotted for different magnetic fields. It is seen that the universal behavior of the normalized effective mass given by (21.18) and shown by the solid curve is in accord with the experimental data. Note that the behavior of $M_N^*(y)$, extracted from $S(T)/T$ and magnetization M of 2D ^{3}He look very much like that of 3D HF compounds. The LFL and NFL regimes are shown by the arrows. As it is seen from Fig. 21.15, the transition area is located between the LFL and NFL, and is relatively broad. In Fig. 21.16, we present the normalized data on $C(y)$, $S(y)$, $y\chi(y)$, and $\underline{M} = M(y) + y\chi(y)$ extracted from data collected on $CePd_{1-x}Rh_x$ [31], ^{3}He [39], $CeRu_2Si_2$ [32], $CeCoIn_5$ [49], and $YbRu_2Si_2$ [50], respectively. Note that in the case of $YbRu_2Si_2$, the variable $y = (B - B_{c0})\mu_B/T_M$ can be viewed as effective normalized temperature. Thus, we observe that these quite various HF compounds exhibit the same behavior as that shown in Fig. 21.15. Again, we see that despite absolutely different microscopic nature of these HF compounds, their main universal macroscopic features and quantum criticality are of the same nature, generated by FCQPT.

It is seen from Fig. 21.16 that all the data exhibit the kink shown by arrow and taking place as soon as the system enters the transition region from the LFL state to the NFL one. This region corresponds to the temperatures where the vertical arrow in Fig. 15.1 crosses the hatched area separating the LFL from NFL. It is also seen in Fig. 21.16 that the low-temperature LFL scale of the thermodynamic functions, as a function of y, is characterized by fast growth, and the high-temperature scale related to the NFL region is characterized by slow growth. As a result, we can identify the energy scales near QCP, discovered in [50, 51]: The thermodynamic characteristics exhibit kinks, i.e., crossover points from the fast to slow growth at elevated temperatures, which separate the low-temperature LFL scale and high-temperature scale related to the NFL state.

21.6 New State of Matter

In Chap. 7, we have shown that the theory of the fermion condensation gives firm grounds to the explanation of diverse experimental facts related to temperature, magnetic field, and number density dependencies of the characteristics of HF compounds, revealing their universal scaling behavior. This universal behavior is also inherent to HF compounds with different magnetic ground states. We also analyzed the positions of the maxima of magnetization M and the entropy S in 2D ^{3}He as functions of the number density. These data could be obtained for ^{3}He only, while they were inaccessible for analysis in HF metals and other compounds. As a result, we were able to show the universality of the quantum criticality and the observed scaling behavior. Thus, by bringing the experimental data collected on different strongly correlated Fermi systems to the above form related to the internal scales immediately we reveal their universal behavior. As we have seen above, the topological FCQPT takes place in many compounds, generating the quantum-critical state with the NFL behavior by forming flat bands [20, 35]. We have carried out a systematic theoretic study of the phase diagrams of strongly correlated Fermi systems, including HF metals, the new type of insulators with strongly correlated quantum spin liquid, and quasicrystals. We have demonstrated that these diagrams have universal features. The obtained results are in good agreement with experimental data. We have shown both theoretically and using arguments based entirely on the experimental grounds that the data collected on very different HF compounds, such as HF metals, compounds with quantum spin liquid, quasicrystals, and 2D ^{3}He, have a universal scaling behavior at the quantum criticality domain in spite of their microscopic diversity. Our analysis of HF compounds relies on numerous experimental results and on calculations of the physical properties, related to NFL behavior. It is seen that the FC theory, based on fermion-condensation paradigm, delivers pretty good description of the NFL behavior of different strongly correlated Fermi systems. Moreover, the topological FCQPT can be considered as the universal reason for the NFL behavior observed in various HF metals, liquids, insulators with quantum spin liquids, and quasicrystals. Thus, the quantum-critical physics of different HF compounds can be well explained within the framework of the fermion-condensation theory, being universal, and emerges regardless of the underlying microscopic details of the compounds. This uniform behavior, induced by the universal quantum-critical physics and generated by the topological FCQPT, allows us to view it as the main characteristic of the new state of matter [52–54].

References

1. C. Proust, E. Boaknin, R.W. Hill, L. Taillefer, A.P. Mackenzie, Phys. Rev. Lett. **89**, 147003 (2002)
2. R. Bel, K. Behnia, Y. Nakajima, K. Izawa, Y. Matsuda, H. Shishido, R. Settai, Y. Onuki, Phys. Rev. Lett. **92**, 217002 (2004)
3. T. Shibauchi, L. Krusin-Elbaum, M. Hasegawa, Y. Kasahara, R. Okazaki, Y. Matsuda, Proc. Natl. Acad. Sci. USA **105**, 7120 (2008)
4. A.P. Mackenzie, S.R. Julian, D.C. Sinclair, C.T. Lin, Phys. Rev. B **53**, 5848 (1996)
5. N.P. Butch, K. Jin, K. Kirshenbaum, R.L. Greene, J. Paglione, PNAS **109**, 8440 (2012)
6. G.Q. Zheng, T. Sato, Y. Kitaoka, M. Fujita, K. Yamada, Phys. Rev. Lett. **90**, 197005 (2003)
7. J. Korringa, Physica (Utrecht) **16**, 601 (1950)
8. P. Gegenwart, J. Custers, C. Geibel, K. Neumaier, K.T.T. Tayama, O. Trovarelli, F. Steglich, Phys. Rev. Lett. **89**, 056402 (2002)
9. D. Shechtman, I. Blech, D. Gratias, J.W. Cahn, Phys. Rev. Lett. **53**, 1951 (1984)
10. A.I. Goldman, R.F. Kelton, Rev. Mod. Phys. **65**, 213 (1993)
11. T. Fujiwara, in *Physical Properties of Quasicrystals*, Springer Series in Solid-State Sciences, ed. by S.Z. M (Springer, Berlin, 1999)
12. D. Mayou, G.T. de Laissardière, in *Quasicrystals*, Handbook of Metal Physics, ed. by T. Fujiwara, Y. Ishii (Elsevier Science, Amsterdam, 2008)
13. T. Fujiwara, T. Yokokawa, Phys. Rev. Lett. **66**, 333 (1991)
14. R. Widmer, O. Gröning, P. Ruffieux, P. Gröning, Philos. Mag. **86**, 781 (2006)
15. R. Widmer, P. Gröning, M. Feuerbacher, O. Gröning, Phys. Rev. B **79**, 104202 (2009)
16. T. Fujiwara, S. Yamamoto, G.T. de Laissardière, Phys. Rev. Lett. **71**, 4166 (1993)
17. G.T. de Laissardière, Z. Kristallogr **224**, 123 (2009)
18. K. Deguchi, S. Matsukawa, N.K. Sato, T. Hattori, K. Ishida, H. Takakura, T. Ishimasa, Nat. Mater. **11**, 1013 (2012)
19. V.R. Shaginyan, A.Z. Msezane, K.G. Popov, G.S. Japaridze, V.A. Khodel, Phys. Rev. B **87**, 245122 (2013)
20. V.R. Shaginyan, M.Y. Amusia, A.Z. Msezane, K.G. Popov, Phys. Rep. **492**, 31 (2010). (arXiv:1006.2658)
21. V.R. Shaginyan, Phys. At. Nucl. **74**, 1107 (2011)
22. V.A. Khodel, J.W. Clark, M.V. Zverev, Phys. At. Nucl. **74**, 1237 (2011)
23. L.D. Landau, Zh Eksp, Teor. Fiz. **30**, 1058 (1956)
24. J.W. Clark, V.A. Khodel, M.V. Zverev, Phys. Rev. B **71**, 012401 (2005)
25. E.M. Lifshitz, L. Pitaevskii, *Statistical Physics. Part 2.* (Butterworth-Heinemann, Oxford, 2002)
26. V.A. Khodel, M.V. Zverev, J.W. Clark, JETP Lett. **81**, 315 (2005)
27. V.A. Khodel, V.R. Shaginyan, V.V. Khodel, Phys. Rep. **249**, 1 (1994)
28. J. Dukelsky, V. Khodel, P. Schuck, V. Shaginyan, Z. Phys. **102**, 245 (1997)
29. R. Lifshitz, Z. Phys. **51**, 1156 (2011)
30. V.R. Shaginyan, A.Z. Msezane, K.G. Popov, J.W. Clark, M.V. Zverev, V.A. Khodel, Phys. Rev. B **86**, 085147 (2012)

31. N. Oeschler, S. Hartmann, A. Pikul, C. Krellner, C. Geibel, F. Steglich, Phys. B **403**, 1254 (2008)
32. D. Takahashi, S. Abe, H. Mizuno, D. Tayurskii, K. Matsumoto, H. Suzuki, Y. Onuki, Phys. Rev. B **67**, 180407(R) (2003)
33. A.W. Rost, S.A. Grigera, J.A.N. Bruin, R.S. Perry, D. Tian, S. Raghu, S.A. Kivelson, A.P. Mackenzie, Proc. Natl. Acad. Sci. USA **108**, 16549 (2011)
34. R.A. Borzi, S.A. Grigera, J. Farrell, R.S. Perry, S.J.S. Lister, S.L. Lee, D.A. Tennant, Y. Maeno, A.P. Mackenzie, Science **315**, 214 (2007)
35. M.Y. Amusia, K.G. Popov, V.R. Shaginyan, W.A. Stephanowich, *Theory of Heavy-Fermion Compounds*. Springer Series in Solid-State Sciences, vol. 182 (Springer, Berlin, 2015)
36. A.V. Silhanek, N. Harrison, C.D. Batista, A.L.M. Jaime, H. Amitsuka, J.A. Mydosh, Phys. Rev. Lett. **95**, 026403 (2005)
37. J.S. Kim, B. Andraka, G. Fraunberger, G.R. Stewart, Phys. Rev. B **41**, 541 (1990)
38. V.R. Shaginyan, K.G. Popov, V.A. Stephanovich, Europhys. Lett. **79**, 47001 (2007)
39. M. Neumann, J. Nyéki, B. Cowan, J. Saunders, Science **317**, 1356 (2007)
40. V.R. Shaginyan, A.Z. Msezane, K.G. Popov, V.A. Stephanovich, Phys. Rev. Lett. **100**, 096406 (2008)
41. A. Casey, H. Patel, J. Cowan, B.P. Saunders, Phys. Rev. Lett. **90**, 115301 (2003)
42. V.A. Khodel, V.R. Shaginyan, JETP Lett. **51**, 553 (1990)
43. M.Y. Amusia, V.R. Shaginyan, Phys. Rev. B **63**, 224507 (2001)
44. G.E. Volovik, in *Quantum Analogues: From Phase Transitions to Black Holes and Cosmology*, Springer Lecture Notes in Physics, vol. 718, ed. by W.G. Unruh, R. Schutzhold (Springer, Orlando, 2007), p. 31
45. V.R. Shaginyan, M.Y. Amusia, K.G. Popov, Phys. Usp. **50**, 563 (2007)
46. M. Pfitzner, P. Wölfle, Phys. Rev. B **33**, 2003 (1986)
47. M. Brando, L. Pedrero, T. Westerkamp, C. Krellner, P. Gegenwart, C. Geibel, F. Steglich, Phys. Status Solidi B **459**, 285 (2013)
48. V.R. Shaginyan, A.Z. Msezane, K.G. Popov, G.S. Japaridze, V.A. Khodel, Europhys. Lett. **106**, 37001 (2014)
49. R. Movshovich, M. Jaime, J.D. Thompson, C. Petrovic, Z. Fisk, P.G. Pagliuso, J.L. Sarrao, Phys. Rev. Lett. **91**, 5152 (2001)
50. P. Gegenwart, T. Westerkamp, C. Krellner, Y. Tokiwa, S. Paschen, C. Geibel, F. Steglich, E. Abrahams, Q. Si, Science **315**, 969 (2007)
51. V.R. Shaginyan, M.Y. Amusia, A.Z. Msezane, K.G. Popov, V.A. Stephanovich, Phys. Lett. A **373**, 986 (2009)
52. M.Y. Amusia, V.R. Shaginyan, Contrib. Plasma Phys. **53**, 721 (2013)
53. D. Yudin, D. Hirschmeier, H. Hafermann, O. Eriksson, A.I. Lichtenstein, M.I. Katsnelson, Phys. Rev. Lett. **112**, 070403 (2014)
54. V.R. Shaginyan, V.A. Stephanovich, A.Z. Msezane, P. Schuck, J.W. Clark, M.Y. Amusia, G.S. Japaridze, K.G. Popov, E.V. Kirichenko, J. Low Temp. Phys. **189**, 410 (2017)

Chapter 22
Quantum Criticality, T-linear Resistivity, and Planckian Limit

Abstract In this chapter, we continue to consider the nature of quantum criticality and its relationships with Planckian limit. We explain experimental observations of universal scattering rate related to the linear temperature resistivity exhibited by a large number of both strongly correlated Fermi systems and conventional metals. We show that the observed scattering rate in strongly correlated Fermi systems like heavy-fermion metals and high-T_c superconductors stems from phonon contribution that induces the linear temperature dependence of a resistivity. These phonons are formed by the presence of flat band, resulting from the topological fermion-condensation quantum phase transition (FCQPT). We emphasize that so-called Planckian limit, widely used to explain the above universal scattering rate, may occur accidentally because in conventional metals its experimental manifestations (e.g., scattering rate at room and higher temperatures) are indistinguishable from those generated by the well-known phonons being the classic lattice excitations. To support this point of view, we show that despite the $\rho(T) \propto T$, the universal scattering rate does not occur in all cases. This fact shows that the Planckian limit, quantum-critical behavior, and the T-linear resistivity are not directly related. Our results are in good agreement with experimental data and show convincingly that the fermion-condensation theory can be viewed as the universal agent explaining the very unusual physics of strongly correlated Fermi systems. In its turn, this physics, being universal, reveals that HF compounds represent the new state of matter.

22.1 Introduction

Unusual properties of different classes of strongly correlated Fermi systems still remain largely unexplained due to the lack of universal underlying physical mechanism. These properties are very often attributed to so-called non-Fermi liquid (NFL) behavior. This behavior is widely observed in heavy-fermion (HF) metals, graphene, and high-T_c superconductors (HTSC). Experimental data collected on many of these systems show that at $T = 0$ a portion of their excitation spectrum becomes dispersionless, giving rise to so-called flat bands, see, e.g., [1–4]. The presence of flat band indicates that the system is close to a special quantum-critical point (QCP), in

M. Amusia and V. Shaginyan, *Strongly Correlated Fermi Systems*,
Springer Tracts in Modern Physics 283,
https://doi.org/10.1007/978-3-030-50359-8_22

which the topological class of Fermi surface alters. This QCP is coined as topological fermion-condensation quantum phase transition (FCQPT). [1, 2, 4], leading to flat bands ($\varepsilon(\mathbf{k}) = \mu$, where $\varepsilon(\mathbf{k})$ is quasiparticle energy and μ is a chemical potential) formation. Latter phenomenon is called fermion condensation (FC) and had been predicted long ago [1, 2]. The flat bands are formed by interaction (while a geometric frustration can help the process) and have a special T-dependence. At rising temperatures, a nonzero slope appears in the $\varepsilon(\mathbf{k})$ dispersion law, while the quasiparticle width $\gamma \propto T$ [5–7]. This observation is in accordance with experimental data, see, e.g., [8–10]. Moreover, the FC theory allows one to describe adequately, both qualitatively and quantitatively, the above NFL behavior of strongly correlated Fermi systems [1, 2, 6, 11–13]. Here we analyze HF metals and high-T_c superconductors exhibiting $\rho(T) \propto T$ at $T \to 0$ and, therefore, located near their topological FCQPT [12, 14]. Note that FC theory permits also to consider the systems located relatively far from their FCQPTs, see, e.g., [6, 12].

Experimental findings of linear temperature dependence of the resistivity $\rho(T) \propto T$ collected on HTSC, graphene, HF, and conventional metals have revealed that the scattering rate $1/\tau$ of charge carriers reaches the so-called universal Planckian limit $1/(T\tau) = k_B/\hbar$ (k_B and $\hbar = h/2\pi$ are the Boltzmann and Planck constants, respectively) [15, 16]. Note that above Planckian limit, used to explain the universal scattering rate in the so-called Planckian metals [15, 16], can occur accidentally since its experimental manifestations in other, than Planckian metals, metals may be equally well explained by more conventional physical mechanisms like those related to phonon contribution [13, 14]. For instance, the conventional metals exhibit the universal linear scattering rate at room and higher temperatures, generated by the well-known phonons being the classic lattice excitations [15].

In this chapter, within the framework of the fermion-condensation theory, we show that the quasi-classical physics is still applicable to describe the universal scattering rate $1/\tau$ experimentally observed in strongly correlated metals at their quantum-critical region. This is because flat bands, responsible for quantum criticality, generate transverse zero-sound mode, reminiscent of the phonon mode in solids, with the Debye temperature T_D [13, 14, 17]. At $T \geq T_D$, the mechanism of the T-linear dependence of the resistivity is the same both in conventional metals and strongly correlated metals, and is represented by electron-phonon scattering. Therefore, it is electron-phonon scattering at $T \geq T_D$ that leads to the near material independence of the lifetime τ that is expressed as $1/(\tau T) \sim k_B/\hbar$. As a result, we describe and explain experimental observations of universal scattering rate related to linear temperature resistivity of a large number of both strongly correlated Fermi systems and conventional metals [15, 16]. Thus, the observed scattering rate is explained by the emergence of flat bands formed by the topological FQCPT, rather than by the so-called Planckian limit at which the assumed Planckian scattering rate takes place. The reason is that Planckian limit which occurs in the conventional metals at relatively high temperatures is that at $T > T_D$ the linearity in T resistivity is induced by scattering electrons by phonons, [15].

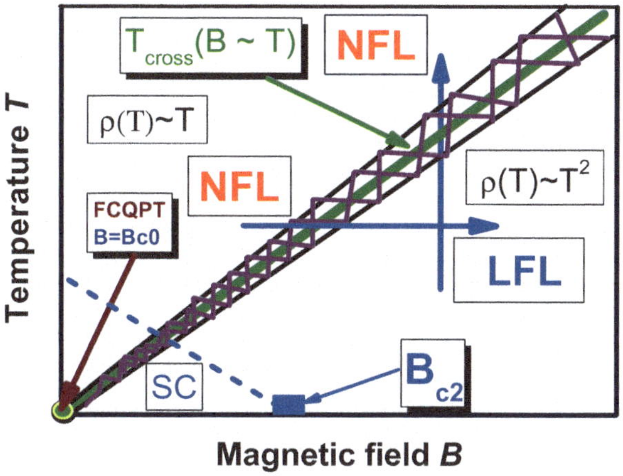

Fig. 22.1 Schematic $T - B$ phase diagram of a strongly correlated Fermi system. The vertical and horizontal arrows crossing the transition region marked by the thick lines depict the LFL-NFL and NFL-LFL transitions at fixed B and T, respectively. At $B < B_{c2}$ (dashed line beginning at solid rectangle), the system is in its possible SC state, with B_{c2} being a critical magnetic field. The hatched area with the solid curve $T_{\rm cross}(B \sim T)$ represents the crossover separating the NFL and LFL domains. A part of the crossover is hidden inside possible SC state. The boxes $\rho(T) \propto T$ and $\rho(T) \propto T^2$ demonstrate the NFL and LFL behaviors of resistivity, respectively

22.2 Phase Diagram

We begin with considering the schematic $T - B$ phase diagram of strongly correlated Fermi system [18] depicted in Fig. 22.1. The magnetic field B plays the role of a control parameter, driving the system toward its QCP represented by FCQPT. The FCQPT occurs at $B = B_{c0}$, yielding new strongly degenerate state at $B = B_{c0}$. To lift this degeneracy, the system forms either superconducting (SC), magnetically ordered (ferromagnetic (FM), antiferromagnetic (AFM), etc.), or nematic states [6]. In the case of $CeCoIn_5$, this state is located at $B_{c0} \simeq 4.9$ T and covered with superconducting "dome" with the critical field $B = B_{c2} \simeq 5$ T [19]. In case of $Sr_3Ru_2O_7$ $B_{c0} \simeq 7.9$ T, while the SC state is absent [15]. It is seen from Fig. 22.1 that at fixed temperature the increase of B drives the system along the horizontal arrow from NFL state to LFL one. On the contrary, at fixed magnetic field and rising temperatures, the system transits along the vertical arrow from LFL state to NFL one. The region shown by the arrow and labeled by $T_{\rm cross}(B \sim T)$ signifies a transition regime between the LFL part with almost constant effective mass and NFL region at which $\rho(T) \propto T$.

At low temperatures, the observed resistivity in HTSC and HF metals located near their QCPs obeys linear T-resistivity *see* Chap. 17,

$$\rho(T) = \rho_0 + AT. \tag{22.1}$$

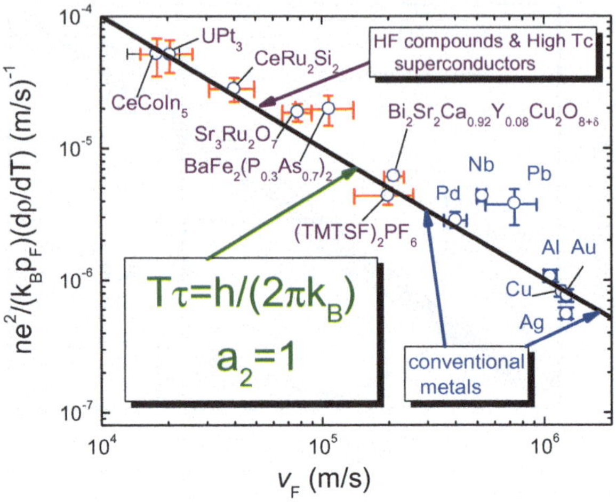

Fig. 22.2 Scattering rates per kelvin of different strongly correlated metals like HF metals, HTSC, organic, and conventional metals [15]. All these metals exhibit $\rho(T) \propto T$ and demonstrate two orders of magnitude variations in their Fermi velocities v_F. The parameter $a_2 \simeq 1$ gives the best fit shown by the solid line, and corresponds to the scattering rate $\tau T = h/(2\pi k_B)$, with $h = 2\pi\hbar$, see (22.3) and (22.4). The region occupied by the conventional metals is displayed by the two arrows. The arrow shows the region of strongly correlated metals, including organic ones. Note that at low temperatures $T \ll T_D$, the scattering rate per kelvin of a conventional metal is order of magnitude lower, and does not correspond to the Planckian limit

This law demonstrates their quantum criticality and the fact that they form the new state of matter [14, 23]. Here ρ_0 is the residual resistivity and A is a T-independent coefficient. Explanations based on quantum criticality for the T-linear resistivity have been given in the literature, see, e.g., [14, 17, 24, 25] and references therein. On the other hand, at room temperature, the T-linear resistivity is exhibited by conventional metals such as Al, Ag, or Cu. In case of a simple metal with a single Fermi surface pocket, the resistivity reads $e^2 n\rho = p_F/(\tau v_F)$, [26] where e is the electronic charge, τ is the lifetime, n is the carrier concentration, and p_F and v_F are the Fermi momentum and velocity, respectively. Representing the lifetime τ (or inverse scattering rate) of quasiparticles in the form [25, 27]

$$\frac{\hbar}{\tau} \simeq a_1 + \frac{k_B T}{a_2}, \tag{22.2}$$

we obtain [14]

$$a_2 \frac{e^2 n\hbar}{p_F k_B} \frac{\partial \rho}{\partial T} = \frac{1}{v_F}, \tag{22.3}$$

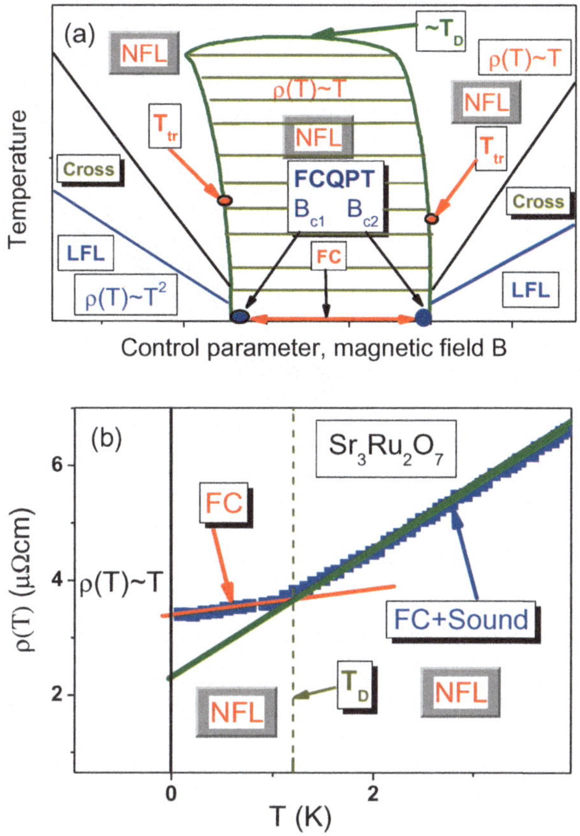

Fig. 22.3 Flat band formed by FC in the HF metal. Panel **a** Schematic phase diagram of the metal $Sr_3Ru_2O_7$ based on experimental observations (see [15, 20–22] and references therein). The topological FCQPTs located at magnetic fields B within the critical magnetic fields $B_{c1} \leq B \leq B_{c2}$ are indicated by the arrows. The ordered nematic phase [15, 20–22] bounded by the thick curve and demarcated by the horizontal lines emerges to remove the entropy excess S_0 that emerges at $T < T_D$. The nematic phase sets in at $T \simeq T_D$. Two arrows label the tricritical points T^1_{tr} and T^2_{tr} at which the lines of second-order phase transitions change to the first order. The NFL behavior with the resistivity $\rho(T) \propto T$ induced by both FC and transverse sound is labeled by NFL. The LFL behavior with $\rho(T) \propto T^2$ is marked by LFL and shown in Fig. 22.4 by the label, while the crossover regions are shown by "Cross." Panel **b** The linear temperature dependence of the resistivity, $\rho(T) \propto T$ at the NFL region. T_D is the Debye temperature, marking the transition from quasi-classic region to the nematic state, shown in panel **a** by horizontal lines. The data are extracted from [15]

where a_1 and a_2 are T-independent parameters. Note that experimental facts corroborate eqrefvfsps22 in case of both strongly correlated metals (HF metals and HTSC) and ordinary metals, provided that these demonstrate the linear T-dependence of their resistivity [15], see Fig. 22.2.

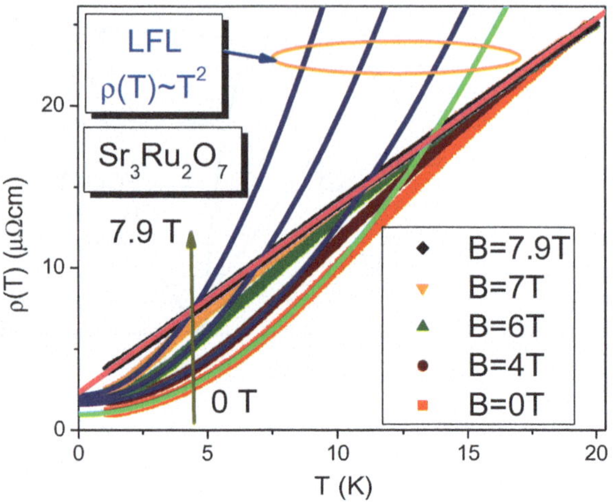

Fig. 22.4 The resistivity of the HF metal $Sr_3Ru_2O_7$ under the application of magnetic fields [15] shown in the right bottom corner. The solid curves are fits of $\rho \propto T^2$ to the low-temperature data. At elevated temperatures, the resistivity becomes $\rho \propto T$, while at growing magnetic fields $B \to B_{c1}$ the LFL behavior vanishes at the critical field $B_{c1} = 7.9$ T, see Fig. 22.3a

22.3 Planckian Limit and Quasi-classical Physics

The analysis of experimental data for various compounds with the linear dependence of $\rho(T)$ shows that the coefficient a_2 is always close to unity, $0.7 \leq a_2 \leq 2.7$, in spite of huge distinction in the absolute value of ρ, T, and Fermi velocities v_F, varying by two orders of magnitude [15]. As a result, it follows from (22.2) that the T-linear scattering rate is of universal form, $1/(\tau T) \sim k_B/\hbar$, regardless of different systems displaying the T-linear dependence with parameter entering (22.3), $a_2 \simeq 1$, [12, 14, 15]. Indeed, this dependence is demonstrated by ordinary metals at temperatures higher than the Debye one, $T \geq T_D$, with an electron-phonon mechanism and by strongly correlated metals which are assumed to be fundamentally different from the ordinary ones, in which the linear dependence at their quantum criticality and temperatures of a few Kelvin is assumed to come from excitations of electronic origin rather than from phonons [15]. We note that in some of the cuprates the scattering rate has a momentum and doping dependence omitted in (22.3) [28–30]. As it is seen from Fig. 22.2, this scaling relation spans two orders of magnitude in v_F, demonstrating the robustness of the observed empirical law [15]. This behavior is explained within the framework of the FC theory, since both for conventional metals and strongly correlated ones the scattering rate is defined by phonons [14]. In case of conventional metals at $T > T_D$, it is well known that phonons make the main contribution to the linear dependence of the resistivity, see, e.g., [26]. On the other hand, it has been shown that the quasi-classical physics describes the T-linear dependence of the resistivity of strongly correlated metals at $T > T_D$, since flat bands, forming the quantum criticality, generate transverse zero-sound mode with the Debye temperature T_D located within the quantum criticality area, see Chap. 17. Therefore, the T-linear dependence is formed by electron-phonon scattering in both ordinary metals and strongly correlated ones. As a result, it is electron-phonon scattering that leads to the near material independence of the lifetime τ that is expressed as

$$\tau T \sim \frac{\hbar}{k_B}. \quad (22.4)$$

Now we turn to the twisted bilayer graphene exhibiting the universal scattering rate [16] and having flat band [4]. It is shown that under the application of pressure, graphene produces increasing correlated behavior, identified by the presence of flat bands at twist angles that increase with growing pressure. Such a behavior signals that it is the correlation that induces the flat bands, and is in accordance with the scenario of FC [1, 6]. We can qualitatively explain this observation as the twisted bilayer graphene can be considered as a quasicrystal, that is, long-range ordered and yet non-periodic. In that case, both the flat band and the corresponding NFL behavior emerge [12, 31]. In fact, without interlayer coupling, a monolayer twisted graphene is not a quasicrystal, for it remains a periodic system, while it is the coupling that makes the quasicrystal, and the application of pressure strengthens the coupling. To support the scenario, we predict that under the application of magnetic field the observed T-linear dependence of the resistivity is changed to the T^2 dependence, since the system transits from the NFL state to LFL one as it is seen from phase diagram 22.1 and the universality of the scattering rate is violated. In the same way, the asymmetrical tunneling conductivity vanishes in the LFL state, as it was predicted within the framework of the FC theory [6, 32–34]. As a result, we can reliably conclude that it is the transverse sound mode that forms the universal scattering rate rather than the Planckian limit does.

The next example is HF metal $Sr_3Ru_2O_7$ having universal scattering rate related to the linear temperature resistivity [15], see Fig. 22.2. This metal represents a useful example because of numerous experimental measurements taken on it, see, e.g., [15, 20–22]. This HF metal is tuned to its quantum-critical line under the application of magnetic field $B_{c1} \leq B \leq B_{c2}$, as it is shown in Fig. 22.3a. The ordered nematic phase highlighted by the horizontal lines is to remove the entropy excess $S(T \to 0) \to S_0$ existing in the absence of an ordered phase, and therefore an ordered phase captures the FC state at $T \to 0$ [6, 12, 14]. At relatively low temperatures, $S(T) \simeq S_0$ and becomes larger than that at LFL state. The second-order phase transition converts to first one at the tricritical points $T^1_{\rm tr}$ and $T^2_{\rm tr}$, as it is shown in Fig. 22.3a. Within the nematic phase at $T \leq T_D$ and at $T_D \leq T$ the system exhibits the NFL behavior with $\rho(T) \propto T$, as it is shown in Fig. 22.3a, b. We note that at $T_D \geq T$ $\rho(T) \propto T$, but the Planckian limit does not take place, for $d\rho(T \geq T_D)/dT > d\rho(T \leq T_D)/dT$, see Fig. 22.3b. Thus, despite the $\rho(T) \propto T$, the universal scattering rate does not occur. This shows that the Planckian limit, quantum-critical behavior, and the T-linear resistivity are not directly related.

Figure 22.4 shows the temperature dependence $\rho(T)$ of $Sr_3Ru_2O_7$ in magnetic fields. As magnetic field $B \to B_{c1}$, the temperature range with the LFL behavior characterized by $\rho(T) \propto T^2$ shrinks, and at $B = B_{c1} \simeq 7.9$ T, the resistivity $\rho(T) \propto T$ over the whole measurement range. At $B \geq B_{c2}$, the magnetoresistivity exhibits a small negative magnetoresistance and the LFL behavior at low temperatures [15], see Fig. 22.3a. We note that such a Planckian metal does exhibit the Planckian limit in its LFL state, and even it does not demonstrate the limit at $T < T_D$, as we have seen above. All these experimental observations are in accordance with general behavior of strongly correlated Fermi systems and can be explained within the FC theory framework [6, 12].

Thus, the physical picture outlined by (22.3) is strongly supported by measurements of the resistivity on $Sr_3Ru_2O_7$ for wide range of temperatures. At $T \geq 100$ K, the resistivity becomes again the T-linear at all applied magnetic fields, as it does at low temperatures and under the application of magnetic fields $B_{c1} \geq B \geq B_{c2}$, see Fig. 22.3a. In the latter case, the coefficient A is lower than that seen at high temperatures [15]. This is because the coefficient A is the composition of two contributions coming from the transverse zero-sound and the FC states, see Fig. 22.3b. If we subtract the FC contribution, A becomes approximately the same at $T \geq T_D$ and at $T \geq 100$ K [14]. Thus, similar strongly correlated compounds exhibit the same behavior of the resistivity at both quantum-critical regime and high-temperature one, allowing us to expect that the same physics governs the T-linear resistivity of the strongly correlated Fermi systems and of conventional metals.

We note that another mechanism can support the T-linear dependence even at $T < T_D$, which lifts the constancy of τ regardless of the presence of T-linear dependence of the resistivity [14, 25]. The mechanism comes from flat bands that are formed by the FC state and contributes both to the linear dependence of the resistivity and to the *residual* resistivity ρ_0, see (22.1). Note that these observations are in good agreement with experimental data [14, 25]. The important point here is that under the application of magnetic field the system in question transits from NFL behavior to LFL one and both flat bands and the FC state are destroyed [6, 12], see the $T - B$ phase diagrams in Figs. 22.1 and 22.3a. Therefore, resistivity $\rho(T) \propto T^2$, magnetoresistance becomes negative, while the residual resistivity ρ_0 decreases abruptly [12, 14, 25]. Such a behavior is in accordance with experimental data, see, e.g., the case of the HF metals $CeCoIn_5$ [19] and $Sr_3Ru_2O_7$ [15] that also demonstrate the universal scattering rate, see Fig. 22.2.

22.4 Universal Scaling Relation

Another experimental result [35] providing insight into the NFL behavior of strongly correlated Fermi systems is the universal scaling relation, which can also be explained using the flat band concept. Reference [35] presents results on the temperature dependence $d\rho/dT$ of the resistivity ρ for a large number of HTSC substances for $T > T_c$. Among them are LSCO and the well-known HF compound $CeCoIn_5$, see Table I of [35]. Remarkable behavior was discovered for all substances considered, and $d\rho/dT$ shows a linear dependence on the penetration depth λ^2. The superconductors considered belong to the London type, for which $\lambda \gg \xi_0$, where ξ_0 is the zero-temperature coherence length (see, e.g., [36]).

It has been shown in [35] that the scaling relation

$$\frac{d\rho}{dT} \propto \frac{k_B}{\hbar}\lambda^2 \tag{22.5}$$

remains valid over several orders of magnitude in λ, signifying its robustness. At the phase transition point $T = T_c$, relation (22.5) yields the well-known Holmes law [35] (see also [37] for its theoretical derivation):

$$\sigma T_c \propto \lambda^{-2}, \tag{22.6}$$

in which $\sigma = \rho^{-1}$ is the normal state *dc* conductivity. It has been shown by Kogan [37] that Holms law applies even for the oversimplified model of an isotropic BCS superconductor. Within the same model of a simple metal, one can express the resistivity ρ in terms of microscopic substance parameters: $e^2 n\rho \simeq p_F/(\tau v_F)$, where τ is the quasiparticle lifetime, n is the carrier density, and v_F is the Fermi velocity. Taking into account that $p_F/v_F = M^*$, we arrive at the equation

$$\rho = \frac{M^*}{ne^2\tau}. \tag{22.7}$$

We note that (22.7) formally agrees with the well-known Drude formula. It has been shown in [36] that good agreement with experimental results [38] is achieved when the effective mass and the superfluid density are attributed to the carriers in the FC state only, i.e., $M^* \equiv M_{FC}$ and $n \equiv n_{FC}$. Keeping this in mind and utilizing the relation $1/\tau = k_B T/\hbar$ (see Chap. 17), we obtain

$$\rho = \frac{M_{FC}}{e^2 n_{FC}}\frac{k_B T}{\hbar} \equiv 4\pi\lambda^2\frac{k_B T}{\hbar}, \tag{22.8}$$

i.e., $d\rho/dT$ is indeed given by the expression (22.5). Equation (22.8) demonstrates that fermion condensation can explain all the above experimentally observed universal scaling relations. It is important to note that the FC approach presented here is insensitive to and transcends the microscopic, non-universal features of the substances under study. This is attributed to the fact that the FC state is protected by its topological structure and therefore represents a new class of Fermi liquids. In particular, consideration of the specific crystalline structure of a compound, its anisotropy, its defect composition, etc. do not change qualitatively our predictions. This strongly suggests that the FC approach provides a viable theoretical framework for explaining universal scaling relations similar to those discovered in the experiments of Božović et al. [38] and Hu et al. [35]. In other words, the fermion condensation of charge carriers in the considered strongly correlated HTSCs, engendered by a quantum phase transition, is indeed the primary physical mechanism responsible for their observable universal scaling properties. This mechanism can be extended to a broad set of substances with very different microscopic characteristics, as discussed in detail in [6, 12].

22.5 Summary

In this chapter, we have explained experimental observations that the scattering rate $1/\tau$ of charge carriers collected on high T_c superconductors, graphene, heavy fermion, and conventional metals exhibits the universal behavior generated by the quasi-classical properties of the above strongly correlated materials, see Chap. 17 for details. In its turn, this universal behavior reveals that HF compounds represent the new state of matter, being formed by the unique topological FCQPT, see Chap. 4 and Sect. 4.1.3. While the Planckian limit may occur accidentally, it is highly improbable that it would be realized in conventional metals, which, obviously, cannot be recognized as Planckian metals with quantum criticality at high or low temperatures. Finally, the fact that we observe the same universal behavior of the scattering rate in microscopically different strongly correlated compounds like HTSC, HF, and conventional metals suggests that some general theory is needed to explain the above body of materials and their behavior in the uniform manner. We conclude that the FC theory is the reliable candidate, explaining the universal scattering rate and a great number of the other experimental observations.

References

1. V.A. Khodel, V.R. Shaginyan, JETP Lett. **51**, 553 (1990)
2. G.E. Volovik, JETP Lett. **53**, 222 (1991)
3. P. Nozieres, J. Phys. (Fr.) **2**, 443 (1992)
4. Y. Cao, V. Fatemi, S. Fang, K. Watanabe, T. Taniguchi, E. Kaxiras, P. Jarillo-Herrero, Nature **556**, 43 (2018)
5. V.A. Khodel, J.W. Clark, V.R. Shaginyan, Solid St. Commun. **96**, 353 (1995)
6. V.R. Shaginyan, M.Ya. Amusia, A.Z. Msezane, K.G. Popov, Phys. Rep. **492**, 31 (2010)
7. V.R. Shaginyan, M.Ya. Amusia, A.Z. Msezane, V.A. Stephanovich, G.S. Japaridze, S.A. Artamonov, JETP Lett. **110**, 290 (2019)
8. MYu. Melnikov, A.A. Shashkin, V.T. Dolgopolov, S.-H. Huang, C.W. Liu, S.V. Kravchenko, Sci. Rep. **7**, 14539 (2017)
9. Y. Liu, H.P. Nair, J.P. Ruf, D.G. Schlom, K.M. Shen, Phys. Rev. B **98**, 041110(R) (2018)
10. A.A. Shashkin, S.V. Kravchenko, Appl. Sci. **9**, 1169 (2019)
11. V.A. Khodel, V.R. Shaginyan, V.V. Khodel, New approach in the microscopic Fermi systems theory. Phys. Rep. **249**, 1 (1994)
12. M.Ya. Amusia, K.G. Popov, V.R. Shaginyan, V.A. Stephanovich, *Theory of Heavy-Fermion Compounds*, vol. 182, Springer Series in Solid-State Sciences (Springer, Dordrecht, 2015), pp. 1–359
13. G.E. Volovik, JETP Lett. **110**, 352 (2019)
14. V.R. Shaginyan, K.G. Popov, V.A. Khodel, Phys. Rev. B **88**, 115103 (2013)
15. J.A.N. Bruin, H. Sakai, R.S. Perry, A.P. Mackenzie, Science **339**, 880 (2013)
16. A. Legros, S. Benhabib, W. Tabis, F. Laliberté, M. Dion, M. Lizaire, B. Vignolle, D. Vignolles, H. Raffy, Z.Z. Li, P. Auban-Senzier, N. Doiron-Leyraud, P. Fournier, D. Colson, L. Taillefer, C. Proust, Nat. Phys. **15**, 142 (2019)
17. V.A. Khodel, J.W. Clark, V.R. Shaginyan, M.V. Zverev, JETP Lett. **92**, 532 (2010)
18. V.R. Shaginyan, A.Z. Msezane, K.G. Popov, G.S. Japaridze, V.A. Khodel, Europhys. Lett. **106**, 37001 (2014)

19. J. Paglione, M.A. Tanatar, D.G. Hawthorn, E. Boaknin, R.W. Hill, F. Ronning, M. Sutherland, L. Taillefer, C. Petrovic, P.C. Canfield, Phys. Rev. Lett. **91**, 246405 (2003)
20. A.W. Rost, S.A. Grigera, J.A.N. Bruin, R.S. Perry, D. Tian, S. Raghu, S.A. Kivelson, A.P. Mackenzie, Proc. Natl. Acad. Sci. **108**, 16549 (2011)
21. A.W. Rost, R.S. Perry, J.-F. Mercure, A.P. Mackenzie, S.A. Grigera, Science **325**, 1360 (2009)
22. A.P. Mackenzie, J.A.N. Bruin, R.A. Borzi, A.W. Rost, S.A. Grigera, Phys. C **481**, 207 (2012)
23. V.R. Shaginyan, V.A. Stephanovich, A.Z. Msezane, P. Schuck, J.W. Clark, M.Ya. Amusia, G.S. Japaridze, K.G. Popov, E.V. Kirichenko, J. Low Temp. Phys. **189**, 410 (2017)
24. N.R. Lee-Hone, J.S. Dodge, D.M. Broun, Phys. Rev. B **96**, 024501 (2017)
25. V.R. Shaginyan, A.Z. Msezane, K.G. Popov, J.W. Clark, M.V. Zverev, V.A. Khodel, Phys. Rev. B **86**, 085147 (2012)
26. A.A. Abrikosov, L.P. Gor'kov, I.E. Dzyaloshinski, *Methods of Quantum Field Theory in Statistical Physics* (Prentice-Hall, London, 1963)
27. P. Aynajian, E. Neto, A. Gyenis, R.E. Baumbach, J.D. Thompson, Z. Fisk, E.D. Bauer, A. Yazdani, Nature **486**, 201 (2012)
28. D.C. Peets, D.G. Hawthorn, K.M. Shen, Y.-J. Kim, D.S. Ellis, H. Zhang, S. Komiya, Y. Ando, G.A. Sawatzky, R. Liang, D.A. Bonn, W.N. Hardy, Phys. Rev. Lett. **103**, 087402 (2009)
29. M.M.J. French, J.G. Analytis, A. Carrington, L. Balicas, N.E. Hussey, New J. Phys. **11**, 055057 (2009)
30. J.W. Alldredge, J. Lee, K. McElroy, M. Wang, K. Fujita, Y. Kohsaka, C. Taylor, H. Eisaki, S. Uchida, P.J. Hirschfeld, J.C. Davis, Nat. Phys. **4**, 319 (2008)
31. V.R. Shaginyan, A.Z. Msezane, K.G. Popov, G.S. Japaridze, V.A. Khodel, Phys. Rev. B **87**, 245122 (2013)
32. V.R. Shaginyan, A.Z. Msezane, G.S. Japaridze, V.A. Stephanovich, Y.S. Leevik, JETP Lett. **108**, 335 (2018)
33. V.R. Shaginyan, A.Z. Msezane, V.A. Stephanovich, G.S. Japaridze, E.V. Kirichenko, Phys. Sci. **94**, 065801 (2019). arXiv:1806.00438
34. V.R. Shaginyan, JETP Lett. **81**, 222 (2005)
35. T. Hu, Yi. Liu, H. Xiao, G. Mu, Yi-f. Yang, Sci. Rep. **7**, 9469 (2017)
36. V.R. Shaginyan, V.A. Stephanovich, A.Z. Msezane, G.S. Japaridze, K.G. Popov, Phys. Chem. Chem. Phys. **19**, 21964 (2017)
37. V.G. Kogan, Phys. Rev. B **87**, 220507(R) (2013)
38. J.I. Božović, X. He, J. Wu, A.T. Bollinger, Nature **536**, 309 (2016)

Chapter 23
Forming High-T_c Superconductors by the Topological FCQPT

Abstract In this chapter, we continue to consider the specific influence of the topological FCQPT on high-T_c superconductors, see also Chaps. 5 and 6. We show that the topological FCQPT, generating flat bands and altering Fermi surface topology, is a primary reason for the exotic behavior of the overdoped high-temperature superconductors represented by $La_{2-x}Sr_xCuO_4$, whose superconductivity features differ from what is described by the classical Bardeen-Cooper-Schrieffer theory. We demonstrate that (1) at temperature $T = 0$, the superfluid density n_s turns out to be considerably smaller than the total electron density; (2) the critical temperature T_c is controlled by n_s rather than by doping, and is a linear function of the n_s; 3) at $T > T_c$ the resistivity $\rho(T)$ varies linearly with temperature, $\rho(T) \propto \alpha T$, where α diminishes with $T_c \to 0$, while in the normal overdoped (nonsuperconducting) region with $T_c = 0$, the resistivity becomes $\rho(T) \propto T^2$. The theoretical results presented are in good agreement with experimental observations, closing the colossal gap between these empirical findings and Bardeen-Cooper-Schrieffer-like theories. (The following chapter contains text from [1] that has been reprinted with the permission of the PCCP Owner Societies.)

23.1 Introduction

Overdoped copper oxides are realized as simple HTSC, whose strongly correlated physics can be understood within the frameworks of the conventional Bardeen-Cooper-Schrieffer theory (BCS), while experimental studies of overdoped high-T_c superconductors (HTSC) $La_{2-x}Sr_xCuO_4$ discovered strong deviations of their physical properties from those predicted by BCS theory [1–6]. These deviations were surprisingly similar for numerous HTSC samples [1–3, 7, 8]. The measurements of the absolute values of the magnetic penetration depth λ and the phase stiffness $\rho_s = A/\lambda^2$ were carried out on thousands of perfect two-dimensional (2D) samples of $La_{2-x}Sr_xCuO_4$ (LSCO) as a function of the doping x and temperature T. Here $A = d/4k_B e^2$ where d is the film thickness, k_B is Boltzmann constant, and e is the electron charge. It has been observed that the dependence of zero-temperature

M. Amusia and V. Shaginyan, *Strongly Correlated Fermi Systems*,
Springer Tracts in Modern Physics 283,
https://doi.org/10.1007/978-3-030-50359-8_23

superfluid density, (the density of superconductive electrons) $n_s = 4\rho_s k_B m^*$ (m^* is the electron effective mass), is proportional to the critical temperature T_c over a wide doping range. This dependence coincides with previous measurements and is incompatible with the standard BCS description. Moreover, n_s turns out to be considerably smaller than the BCS density n_{el} of superconductive electrons [2, 3, 7, 8], which is approximately equal to the total electron density [9]. These observations representing the intrinsic LSCO properties provide unique opportunities for checking and expanding our understanding of the physical mechanisms responsible for high-T_c superconductivity. We note that knowing the responsible mechanism can open avenue for chemical preparation of high-T_c materials with T_c as high as room temperature [10–13].

Here we show that the physical mechanism, responsible for non-BCS behavior of overdoped LSCO, stems from the topological fermion-condensation quantum phase transition (FCQPT) accompanied by the fermion-condensation (FC) phenomenon generating flat bands [1, 10–15]. We note that flat bands and extended saddle point singularity play important role in the theory of HTSC, see, e.g., [13–16]. In order to make our analysis of overdoped LSCO obvious, we use the model of homogeneous heavy-electron liquid [11, 15]. The main experimental facts of [2, 3] represent clear qualitative deviations from those predicted by the classical BCS theory. Therefore, as a first step, we can confine ourself to obtaining transparent analytical results describing quantitatively experimental facts. Our analysis shows that despite drastic microscopic diversity of strongly correlated Fermi systems, they exhibit similar behavior close to FC quantum phase transition point. This is actually related to the altering of Fermi surface topology during FCQPT. We emphasize that the quantum physics of all seemingly different strongly correlated Fermi systems (and overdoped HTSC among them) is universal and emerges regardless of their underlying microscopic details like the symmetries of their crystal lattices. In fact, we deal with momenta transfers that are small compared to those of the order of the reciprocal lattice length (Brillouin zone boundaries), whose contributions have no effect on the topological properties of the systems under consideration [11–13, 15]. Note that despite the highly anisotropic electronic band dispersion in overdoped cuprate HTSC and hence their Fermi surface, our theory still applies for this case. The point here is that after FCQPT the Fermi surface, regardless of initial anisotropy, changes its *topological class*, thus generating all aforementioned salient experimentally observed features, inherent in the fermion-condensation state. In other words, any initially (highly) anisotropic Fermi surface is still *homotopic* to simply spherical one as they can be reduced to each other by continuous deformation [17]. In the superconducting state, to the first approximation, different regions with the maximal absolute value of the d-wave superconducting order parameter are disconnected. Therefore, the order parameter can be either even or odd with respect to $\pi/2$ rotation in the ab-plane [16]. Thus, as a first step, we also neglect the d-wave symmetry of the superconducting order parameter and use the s-wave one.

23.2 Fermion Condensation as Two-Component System

An important problem for the condensed matter theory is the explanation of the NFL behavior observed in HTSC beyond critical point where the low-temperature density of states $N(T \to 0)$ diverges which can generate flat bands, see, e.g., [1, 2, 13–15, 18–20]. In a homogeneous matter, such a divergence is associated with the onset of a topological transition at $x = x_c$ signaled by the emergence of an inflection point at $p = p_F$ (see Chap. 7 for details) [14, 15, 21]

$$\begin{aligned} \varepsilon - \mu &\simeq -(p_F - p)^2, \ p < p_F, \\ \varepsilon - \mu &\simeq \ \ (p - p_F)^2, \ p > p_F, \end{aligned} \tag{23.1}$$

at which the electron effective mass diverges as $m^*(T \to 0) \propto T^{-1/2}$, where ε is the single-electron energy spectrum, p is a momentum, p_F is Fermi momentum, and μ is the chemical potential. Accordingly, at $x \to x_c$ the density of states diverges

$$N(T \to 0) \propto |\varepsilon - \mu|^{-1/2}. \tag{23.2}$$

As a result, both FC state and the corresponding flat bands emerge beyond the topological FCQPT [1, 11, 14, 15, 19], while the critical temperature turns out to be $T_c \propto \sqrt{x - x_c}$ [16]. These results are consistent with the experimental data [2].

At $T = 0$, the onset of FC in homogeneous matter is attributed to a nontrivial solution $n_0(p)$ of the variational equation [10]

$$\frac{\delta E[n(p)]}{\delta n(p)} - \mu = 0, \ p \in [p_i, p_f], \tag{23.3}$$

where E is a ground state energy functional (its variation gives a single-electron spectrum ε) and p_i, p_f stand for initial and final momenta, where the solution of (23.3) exists, see [10, 11, 15] for details. To be more specific, (23.3) describes a flat band pinned to the Fermi surface and related to FC.

To explain emergent superconductivity at $x \to x_c$, we retain the consequences of flattening of single-particle excitation spectra $\varepsilon(\mathbf{p})$ (i.e., flat bands appearance) in strongly correlated Fermi systems, see [11, 13, 15] for reviews. At $T = 0$, the ground state of a system with a flat band is degenerate, and the occupation numbers $n_0(\mathbf{p})$ of single-particle states belonging to the flat band are continuous functions of momentum $\mathbf{p}$, in contrast to standard LFL "step" from 0 to 1 at $p = p_F$, as it is seen from Fig. 23.1. Thus, at $T = 0$, the superconducting order parameter $\kappa(p) = \sqrt{n(p)(1 - n(p))} \neq 0$ in the region occupied by FC [11, 14, 15, 22, 23]. This property is in a contrast to standard LFL picture, where at $T = 0$ and $p = p_F$ the order parameter $\kappa(p)$ is necessarily zero, see Fig. 23.1. Due to the fundamental difference between the FC single-particle spectrum and that of the remainder of the Fermi liquid, a system having FC is, in fact, a two-component system, separated from

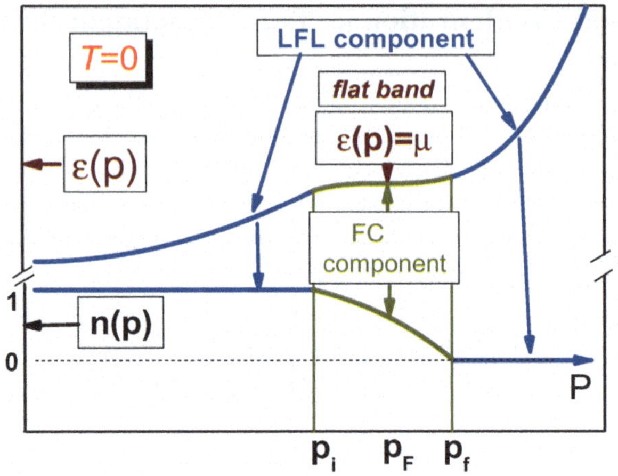

Fig. 23.1 Schematic plot of two-component electron liquid at $T = 0$ with FC. Blue solid line (marked "LFL") shows $n(p)$ for the system without FC, which has ordinary step function shape. Due to presence of FC, the system is separated into two components. The first one is a normal Fermi liquid with the quasiparticle distribution function $n_0(p < p_i) = 1$ and $n_0(p > p_f) = 0$. The second one is FC with $0 < n_0(p_i < p < p_f) < 1$ and the single-particle spectrum $\varepsilon(p_i < p < p_f) = \mu$. The Fermi momentum p_F satisfies the condition $p_i < p_F < p_f$

ordinary Fermi liquid by the topological phase transition. The range L of momentum space adjacent to μ where FC resides is given by $L \simeq p_f - p_i$, see Fig. 23.1.

23.3 Superfluid Density in the Presence of Fermion Condensation

To analyze the above emergent superconductivity quantitatively, it is convenient to use the formalism of Gor'kov equations for Green's functions of a superconductor [15], see Chap. 6. For the 2D case of interest, the solutions of Gor'kov equations [24] determine the Green's functions $F^+(\mathbf{p}, \omega)$ and $G(\mathbf{p}, \omega)$ of a superconductor:

$$F^+(\mathbf{p}, \omega) = \frac{-g_0 \Xi^*}{(\omega - E(\mathbf{p}) + i\,0)(\omega + E(\mathbf{p}) - i\,0)};$$
$$G(\mathbf{p}, \omega) = \frac{u^2(\mathbf{p})}{\omega - E(\mathbf{p}) + i\,0} + \frac{v^2(\mathbf{p})}{\omega + E(\mathbf{p}) - i\,0}. \tag{23.4}$$

Here the single-particle spectrum $\varepsilon(\mathbf{p})$ is determined by (23.3), and

$$E(\mathbf{p}) = \sqrt{\xi^2(\mathbf{p}) + \Delta^2(\mathbf{p})}; \quad \frac{\Delta(\mathbf{p})}{E(\mathbf{p})} = 2\kappa(\mathbf{p}), \tag{23.5}$$

with $\xi(\mathbf{p}) = \varepsilon(\mathbf{p}) - \mu$. The gap Δ and the function Ξ are given by

$$\Delta = g_0|\Xi|, \quad i\Xi = \int F^+(\mathbf{p},\omega)\frac{d\omega d\mathbf{p}}{(2\pi)^3}. \tag{23.6}$$

Here g_0 is the superconducting coupling constant. The function $F^+(\mathbf{p},\omega)$ has the meaning of the wave function of Cooper pairs and Ξ is the wave function of the motion of these pairs as a whole. Taking (23.5) and (23.6) into account, we can rewrite (23.4) as

$$\begin{aligned} F^+(\mathbf{p},\omega) &= -\frac{\kappa(\mathbf{p})}{\omega - E(\mathbf{p}) + i0} + \frac{\kappa(\mathbf{p})}{\omega + E(\mathbf{p}) - i0}, \\ G(\mathbf{p},\omega) &= \frac{u^2(\mathbf{p})}{\omega - E(\mathbf{p}) + i0} + \frac{v^2(\mathbf{p})}{\omega + E(\mathbf{p}) - i0}. \end{aligned} \tag{23.7}$$

In the case $g_0 \to 0$, the gap $\Delta \to 0$, but Ξ and $\kappa(\mathbf{p})$ remain finite if the spectrum becomes flat, $E(\mathbf{p}) = 0$, and in the interval $p_i \le p \le p_f$ (23.7) become [11, 15, 25]

$$\begin{aligned} F^+(\mathbf{p},\omega) &= -\kappa(\mathbf{p})\left[\frac{1}{\omega + i0} - \frac{1}{\omega - i0}\right], \\ G(\mathbf{p},\omega) &= \frac{u^2(\mathbf{p})}{\omega + i0} + \frac{v^2(\mathbf{p})}{\omega - i0}. \end{aligned} \tag{23.8}$$

The parameters $v(\mathbf{p})$ and $u(\mathbf{p})$ are the coefficients of corresponding Bogolubov transformation [24, 26], $u^2(\mathbf{p}) = 1 - n(\mathbf{p})$, $v^2(\mathbf{p}) = n(\mathbf{p})$. They are determined by the condition that the spectrum should be flat: $\varepsilon(\mathbf{p}) = \mu$. It follows from (23.5) and (23.6) that

$$i\Xi = \int F^+(\mathbf{p},\omega)\frac{d\omega d\mathbf{p}}{(2\pi)^3} = i\int \kappa(\mathbf{p})\frac{d\mathbf{p}}{(2\pi)^2} \simeq n_{FC}, \tag{23.9}$$

where n_{FC} is the density of superconducting electrons, forming the FC component, see Fig. 23.1.

We construct the functions $F^+(\mathbf{p},\omega)$ and $G(\mathbf{p},\omega)$ in the case where the constant g_0 is finite but small, such that $v(\mathbf{p})$ and $\kappa(\mathbf{p})$ can be found from the FC solutions of (23.3). Then Ξ, Δ, and $E(\mathbf{p})$ are given by (23.9), (23.6), and (23.5), respectively. Substituting the functions constructed in this manner into (23.7), we obtain $F^+(\mathbf{p},\omega)$ and $G(\mathbf{p},\omega)$. We note that (23.6) and (23.9) imply that the gap Δ is a linear function of both g_0 and n_{FC}. Since $T_c \sim \Delta$, we conclude that $T_c \propto n_{FC} \propto \rho_s$. Note that since we consider the overdoped HTSC case and FCQPT takes place at $x = x_c$, $n_{FC} \propto p_F(p_f - p_i) \propto x_c - x$ with $(p_f - p_i)/p_F \ll 1$ [15, 22, 23], therefore

$$n_{FC} = n_s \ll n_{el}. \tag{23.10}$$

Increasing g_0 causes Δ to become finite, leading to a finite value of the effective mass m^*_{FC} in the FC state [15]:

$$m^*_{FC} \simeq p_F \frac{p_f - p_i}{2\Delta}. \tag{23.11}$$

An important fact is to be noted here. Namely, it have been shown in [11, 15] that in the FC formalism the BCS relations remain valid if we use the spectrum given be (23.11), see Sect. 6.1.3. Thus, we can use the standard BCS approximation with the momentum independence of superconducting coupling constant g_0 in the region $|\varepsilon(\mathbf{p}) - \mu| \leq \omega_D$ so that the interaction is supposed to be zero outside this region. Here ω_D is some characteristic energy, proportional to the Debye temperature. In the frame of these reasonable assumptions, the superconducting gap depends only on temperature and is determined by the equation [11, 14, 15]

$$\frac{1}{g_0} = N_{FC} \int_0^{E_0/2} \frac{d\xi}{f(\xi, \Delta)} \tanh \frac{f(\xi, \Delta)}{2T} + \\ + N_L \int_{E_0/2}^{\omega_D} \frac{d\xi}{f(\xi, \Delta)} \tanh \frac{f(\xi, \Delta)}{2T}, \tag{23.12}$$

where $f(\xi, \Delta) = \sqrt{\xi^2 + \Delta^2(T)}$ and $E_0 = \varepsilon(p_f) - \varepsilon(p_i) \approx 2\Delta(T = 0)$ is a characteristic energy scale. Also, $N_{FC} = (p_f - p_F)p_F/(2\pi\Delta(T = 0))$ and $N_L = m^*_L/(2\pi)$ (m^*_L is the effective mass of electron of the LFL component, see Fig. 23.1) are the densities of states of FC and non-FC electrons, respectively. In the opposite case $T = 0$, as usual, $\tanh(f/(2T)) = 1$ and the remaining integrals can be evaluated exactly. This leads to the following equation relating $\Delta(T = 0)$ to superconducting coupling constant g_0

$$\frac{\delta}{\beta} = B - \delta \ln \delta, \tag{23.13}$$

where $\beta = g_0 m^*_L/(2\pi)$ is dimensionless coupling constant, $\delta = \Delta(T = 0)/(2\omega_D)$, and $B = (E_F/\omega_D)((p_f - p_F)/p_F) \ln(1 + \sqrt{2})$. It is seen that parameter B depends on the width of FC interval so that at $B = 0$ ($p_f = p_F$) system is out of FC and hence is in a pure BCS state. In this case, the solution of (23.13) has standard BCS form $\delta_{BCS} = \exp(-1/\beta)$, while at small β and $B \neq 0$ we obtain the linear relation between coupling constant and gap $\delta = B\beta$ that not only differs drastically from the BCS result, but provides much higher T_c, which is directly proportional to $\Delta(T = 0)$ (in the FC case $T_c \approx \Delta(T = 0)/2$) [11]. In the case of $\beta \sim 0.3$, Fig. 23.2 portrays the solutions of (23.13) for $B = 0.4$ and even smaller $\beta < 0.15$. It is seen that already linear regime provides much higher T_c than BSC case, while nonlinear one comprising the complete numerical solution of (23.13) yields even higher T_c. This means that FC approach is well capable to explain the high-T_c superconductivity. Inset to Fig. 23.2 reports the dependence (23.12) in dimensionless units. This dependence is not peculiar to FC approach as it is qualitatively similar to BCS case. In this case,

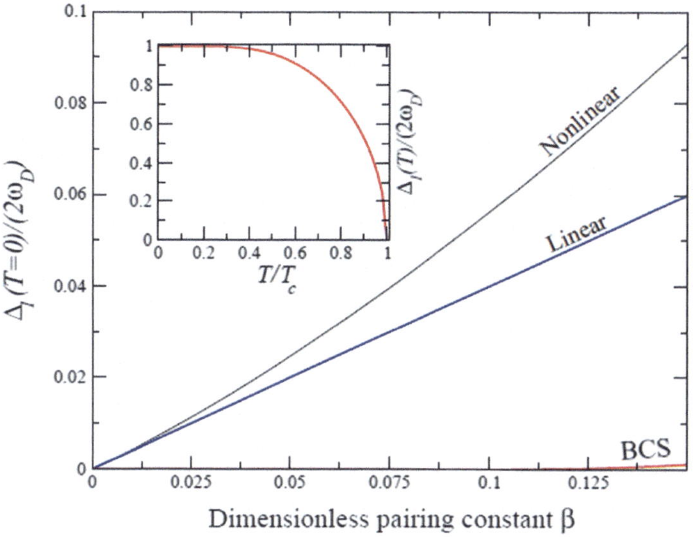

Fig. 23.2 The solution of equation (23.13) in the form $\delta(\beta)$ at $B = 0.4$. Curve marked "Nonlinear" is a direct numerical solution of transcendental equation (23.13). Curve marked "Linear" is a linear dependence $\delta = B\beta$, while that marked "BSC" is a BCS dependence resulting from (23.13) at $B = 0$. It is seen that the FC theory permits to obtain much higher T_c (proportional to superconducting gap $\Delta(T = 0)$) than the BCS approach. Inset reports the temperature dependence of superconducting gap in FC approach at $B = 0.4$

the variation of "FC-parameter" B (and even putting $B = 0$) does not change the situation qualitatively.

Now we analyze the superfluid density n_s for finite g_0. As seen from (23.9) and (23.8), n_s emerges when $x \simeq x_c$ and occupies the region $p_i \le p \le p_f$, so that we denote $n_s = n_{FC} \propto x_c - x$, where n_{FC} is the electron density in FC phase. As a result, we have that in latter phase $n_s \ll n_{el} = n_{FC} + n_L$, with n_{el} and n_L being, respectively, the total density of electrons and that out of FC phase. Note that the result $n_s \sim n_{el}$ does not only follow from BCS theory of superconductivity, but is much deeper and is pertinent to almost any superfluid system, being the result of the Leggett theorem. The short statement of latter theorem is that at $T = 0$ in any superfluid liquid $n_s \sim n_{el}$, here n_{el} denotes the number density of the liquid particles. For this theorem to be true, however, the system should be T-invariant, where T relates to time reversal. Since FC state, being highly topologically nontrivial [11, 27, 28], violates primarily the time-reversal symmetry (actually it also violates the CP invariance, where C is charge conjugation and P is translation invariance, see [11, 15, 27] for more details), the inequality $n_s \ll n_{el}$ is inherent in it, as it is seen from (23.9) and (23.10). This implies that the main contribution to the superconductivity comes from the FC state. We conclude that in the FC case the emerging two-component system violates the BCS condition that $n_s \simeq n_{el}$.

23.4 Penetration Depth, Fermion Condensation, and Uemura's Law

Now we find out if our superconductor belongs to the London type. For that, we write down London's electrodynamics Eqs: $\nabla \times \mathbf{j}_s = -(n_s e^2/m^*)\mathbf{B} \equiv -(n_{FC}e^2/m^*_{FC})\mathbf{B}$ and $\nabla \times \mathbf{B} = 4\pi \mathbf{j}_s$, where $\mathbf{j}_s$ is the superconducting current. These equations imply that the penetration depth is given by relation

$$\lambda^2 = \frac{m^*_{FC}}{4\pi e^2 n_{FC}}. \tag{23.14}$$

Comparing the penetration depth (23.14) with the coherence length $\xi_0 \sim p_F/(m^*_{FC}\Delta)$, we conclude that $\lambda \gg \xi_0$ as the FC quasiparticle effective mass is huge [1, 11]. Thus, the superconductors are indeed of the London type.

It turns out that in FC phase, the penetration depth is a function not only of temperature but also of doping degree x. Then, it follows from Ginzburg-Landau theory that the density of superconducting electrons $n_s \sim T_c - T$. On the other hand, as it has been discussed in [2], the pressure enhances n_s, i.e., the density x of charge carriers is important. Also, it has been shown (see, e.g., [11, 15]) that in superconducting phase with FC it is $T_c \simeq 2\Delta(T = 0)$. This permits to use the relation (23.14) to plot the penetration depth as a function of temperature and doping in the form

$$\frac{\lambda}{\lambda_0} = \frac{1}{\sqrt{1 - y - \tau}}, \tag{23.15}$$

where $y = (x_c - x)/x_c$, $\tau = xT/(2\Delta(T = 0)x_c)$, and λ_0 combines all proportionality coefficients entering the problem. The dependence (23.15) is depicted in Fig. 23.3.

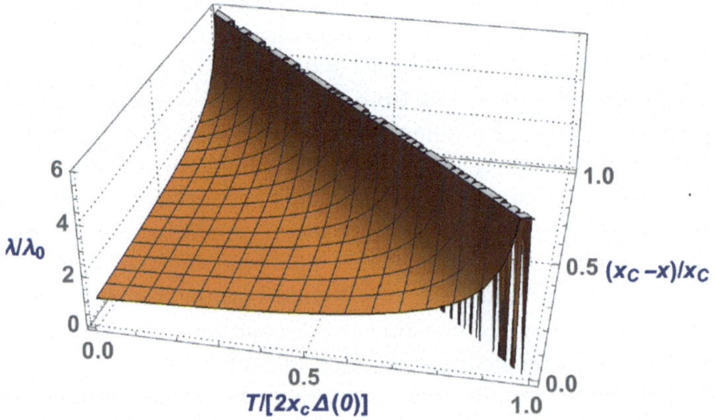

Fig. 23.3 The dependence of dimensionless penetration depth λ/λ_0 (23.15) on temperature and doping

Very good qualitative agreement with experimental data (Fig. 2a from [2]) is seen. Namely, doping-dependent penetration depth λ becomes infinite at the superconducting phase transition temperature. At zero temperature, the divergence of λ occurs at $x \simeq x_c$, corresponding to FC phase emergence, i.e., at $T = 0$ both superconductivity and FC phase arise. In that case, one has that $T_c \to 0$, and $T_c/2\Delta(T \to 0) \simeq 1$. At the same time, at higher temperatures, λ diverges in the region $x < x_c$, i.e., deeply inside the FC phase. This shows the "traces" of FC at finite temperatures. This demonstrates, in turn, that our approach, based on a concept of topological FC quantum phase transition, describes all essential and surprising features of overdoped HTSC. The main input of our model is that in two-component system with FC, occupying a small fraction of the Fermi sphere, $n_s \simeq n_{FC}$ is much less than the total density of the electrons. The latter also allows to verify the validity of the well-known Uemura's law [1, 7] in our case. Indeed, since $T_c \propto n_s/m^* \equiv n_{FC}/m^*_{FC}$, we get from (23.11) and (23.14)

$$\frac{\rho_s}{A} = \lambda^{-2} \simeq \frac{n_s}{m^*} \simeq \frac{n_{FC}}{m^*_{FC}} \simeq 2\Delta \simeq T_c. \tag{23.16}$$

Taking into account that $n_{FC} \propto x_c - x$, we see that (23.16) reproduces the main results of this chapter, being in good agreement with experimental data [1–3]. It is seen that the dependence of T_c on ρ_s is linear, representing the observed scaling law, while T_c is primary controlled by n_s [2]. Note that the results for underdoped HTSC [7] are similar to those for overdoped HTSC, thus being suggestive for underdoped versus overdoped symmetry [2]. As a result, we observe good agreement with Uemura's law in overdoped LSCO as well [2].

At the doping levels $x > x_c$, where FCQPT does not yet occur, the system is in LFL phase with resistivity $\rho \propto T^2$, which is "more metallic" than that exhibited in the FC phase [2, 18, 19, 29, 30]. In the latter phase, the superconductivity appears since FC strongly facilitates the superconducting state. In the normal phase, $T > T_c$, FC causes the linear T dependence of resistivity, $\rho(T) \propto T$ [19, 22, 23, 30], which is in good qualitative agreement with the experimental data on LSCO and $La_{2-x}Ce_xCuO_4$ [2, 18]. In the transition region $x \simeq x_c$, one observes $\rho(T) \propto T^{\alpha}$ with $\alpha \sim 1.0$–2.0 [18, 19, 30].

23.5 Concluding Remarks

In this chapter, we have shown that the main physical mechanism, responsible for the unusual properties of the overdoped $La_{2-x}Sr_xCuO_4$, is the topological quantum phase transition with the emergence of the fermion condensation. This observation can open avenue for chemical preparation of high-T_c materials with T_c up to room temperatures. We have concluded our study of exemplifications of the new state of matter reached by fermion condensation with an exploration of high-T_c superconductors as potential hosts of fermion condensates. In fact, we have shown that the underlying physical mechanism responsible for the unusual properties of the overdoped compound $La_{2-x}Sr_xCuO_4$ (LSCO) observed may very well involve the topological FCQPT that induces fermion condensation [1–3]. Since the topological FC state violates time-reversal symmetry, the Leggett theorem no longer applies. Instead, we have demonstrated explicitly that the superfluid number density n_s turns out to be small compared to the total number density of electrons. We have also shown that the critical temperature T_c is a linear function of n_s, while $n_s(T) \propto T_c - T$. Pairing with such unusual properties is as a shadow of the fermion condensation—a situation foretold by an exactly solvable model [23] long before the experimental observations have been obtained by Božović et al. [2] that demonstrate that both the gap and the order parameter exist only in the region occupied by fermion condensate. Thus, the experimental observations [2] can be viewed as a direct experimental manifestation of FC. Additionally, we have demonstrated that at $T > T_c$ the resistivity $\rho(T)$ varies linearly with temperature, while for $x > x_c$ it exhibits metallic behavior, $\rho(T) \propto T^2$. Thus, the superconductivity formalism adapted to the presence of a fermion condensate captures all the essential physics of overdoped LSCO and successfully explains its most puzzling experimental features, thereby allowing us to close the colossal gap existing between the experiments and Bardeen-Cooper-Schrieffer-like theories. Indeed, these findings are applicable not only to LSCO but also for any overdoped high-temperature superconductor. Thus, the fermion-condensation theory is able to explain both the universal properties of HF compounds that manifest the new state of matter and the specific properties of these compounds that are not exhibited by conventional metals, see Chap. 4 and Sect. 4.1.3.

References

1. V.R. Shaginyan, V.A. Stephanovich, A.Z. Msezane, G.S. Japaridze, K.G. Popov, Phys. Chem. Chem. Phys. **19**, 21964 (2017). arXiv:1702.05804
2. J.I. Božović, X. He, J. Wu, A.T. Bollinger, Nature **536**, 309 (2016)
3. J. Zaanen, Nature **536**, 282 (2016)
4. T. Hu, Y. Liu, H. Xiao, G. Mu, Y.F. Yang, Sci. Rep. **7**, 9469 (2017)
5. V.G. Kogan, Phys. Rev. B **87**, 220507(R) (2013)
6. V. Shaginyan, M. Amusia, A. Msezane, V. Stephanovich, G. Japaridze, S.A. Artamonov, JETP Lett. **110**, 290 (2019)
7. Y.J. Uemura, A. Keren, L.P. Le, G.M. Luke, W.D. Wu, Y. Kubo, T. Manako, Y. Shimakawa, M. Subramanian, J.L. Cobb, J.T. Markert, Nature **364**, 605 (1993)
8. C. Bernhard, C. Niedermayer, U. Binninger, A. Hofer, C. Wenger, J.L. Tallon, G.V.M. Williams, E.J. Ansaldo, J.I. Budnick, C.E. Stronach, D.R. Noakes, M.A. Blankson-Mills, Phys. Rev. B **52**, 10488 (1995)
9. J. Bardeen, L.N. Cooper, J.R. Schrieffer, Phys. Rev. **108**, 1175 (1957)
10. V.A. Khodel, V.R. Shaginyan, JETP Lett. **51**, 553 (1990)
11. M.Y. Amusia, K.G. Popov, V.R. Shaginyan, W.A. Stephanowich, *Theory of Heavy-Fermion Compounds*. Springer Series in Solid-State Sciences, vol. 182 (Springer, Berlin, 2015)
12. G.E. Volovik, JETP Lett. **53**, 222 (1991)
13. G.E. Volovik, in *Quantum Analogues: From Phase Transitions to Black Holes and Cosmology*, ed. by W.G. Unruh, R. Schutzhold, Springer Lecture Notes in Physics, vol. 718 (Springer, Orlando, 2007), p. 31
14. V.A. Khodel, V.R. Shaginyan, V.V. Khodel, Phys. Rep. **249**, 1 (1994)

15. V.R. Shaginyan, M.Y. Amusia, A.Z. Msezane, K.G. Popov, Phys. Rep. **492**, 31 (2010). arXiv:1006.2658
16. A.A. Abrikosov, Int. J. Mod. Phys. B **13**, 3405 (1999)
17. G. Volovik, V. Mineev, JETP Lett. **45**, 1186 (1977)
18. K. Jin, N.P. Butch, K. Kirshenbaum, J. Paglione, R.L. Greene, Nature **476**, 73 (2011)
19. V.A. Khodel, J.W. Clark, K.G. Popov, V.R. Shaginyan, JETP Lett. **101**, 413 (2015)
20. I.M. Lifshitz, Sov. Phys. JETP 1130 (1960)
21. V.R. Shaginyan, A.Z. Msezane, K.G. Popov, J.W. Clark, V.A. Khodel, M.V. Zverev, Phys. Rev. B **93**, 205126 (2016). arXiv:1601.01182
22. V.A. Khodel, V.R. Shaginyan, P. Schuk, JETP Lett. **63**, 752 (1996)
23. J. Dukelsky, V. Khodel, P. Schuck, V. Shaginyan, Z. Phys. **102**, 245 (1997)
24. A.A. Abrikosov, L.P. Gorkov, I.E. Dzyaloshinski, *Methods of Quantum Field Theory in Statistical Physics* (Dover, New York, 1975)
25. V.R. Shaginyan, A.Z. Msezane, V.A. Stephanovich, E.V. Kirichenko, Europhys. Lett. **76**, 898 (2006)
26. E.M. Lifshitz, L. Pitaevskii, *Statistical Physics. Part 2* (Butterworth-Heinemann, Oxford, 2002)
27. V.R. Shaginyan, G.S. Japaridze, M.Y. Amusia, A.Z. Msezane, K.G. Popov, Europhys. Lett. **94**, 69001 (2011)
28. V.R. Shaginyan, K.G. Popov, V.A. Stephanovich, E.V. Kirichenko, J. Alloy. Compd. **442**, 29 (2007)
29. V.R. Shaginyan, JETP Lett. **77**, 99 (2003)
30. V.R. Shaginyan, A.Z. Msezane, K.G. Popov, J.W. Clark, M.V. Zverev, V.A. Khodel, Phys. Rev. B **86**, 085147 (2012)

Chapter 24
Conclusions

Abstract This chapter concludes our book, serves as a road map for material presented, and discusses what else can and should be done. Main emphasis is put on the fact that HF compounds feature the universal behavior in a wide range of the variation of magnetic fields, temperatures, number density, etc., which permits us to suggest that they form the new state of matter. Another emphasis in the book has been put on the ubiquity of the fermion-condensation phenomenon. This phenomenon is ubiquitous as it is capable to describe equally well an impressive diversity of HF compounds represented by heavy-fermion metals; quantum liquids including ^{3}He films; insulating compounds possessing one-, two-, and three-dimensional quantum spin-liquid states; and quasicrystals and beyond, for instance, the Universe itself.

The aim of this book is to demonstrate that a big variety of substances at rather different external conditions demonstrate impressive and unexpected similarity that is determined by a universal phase transition that permits us call this a new state of matter. Since our childhood we know gas, liquids, and solid-state states of matter. Later we learned that states of matter are separated from each other by phase transitions. It is essential that gases, liquids, and solids demonstrate remarkable similarity for each state of matter, while they are constructed from a whole variety of different constituents. Next state of matter became plasma, a system of electrons and nuclei that are formed from atoms of gases, liquids, and solids at temperatures high enough to ionize atoms. There are strong indications that the fifth state of matter exists, the so-called quark-gluon plasma that is created at so high temperatures, at which all nuclei decompose into same constituents, namely, quarks and gluons. In this book, we demonstrate that in very different substances, named HF compounds, under very different external conditions, a universal quantum phase transition known as the topological fermion-condensation quantum phase transition (FCQPT) takes place, determining the macroscopic properties and behavior of numerous substances, the number of which increases as an avalanche. It is quite natural to consider all these substances located near this phase transition as a new state of matter, since their behavior, formed by the phase transition, acquire important similarities making them universal. The idea of such a phase transition, inducing flat bands, started long

M. Amusia and V. Shaginyan, *Strongly Correlated Fermi Systems*,
Springer Tracts in Modern Physics 283,
https://doi.org/10.1007/978-3-030-50359-8_24

ago, in 1990 [1], at first as a curious mathematical option, and now it is a rapidly expanding, vibrant field with uncountable applications.

The flat band studies attract interest of both experimentalists and theorists. From experimental side, one of their most important features that they can contain, being partially filled, an exponentially large number of states. As we have shown in our book, this macroscopic degeneracy leads to novel phenomenon of the fermion condensation, which gives new impetus to the entire strongly correlated physics. Strange as it may seem, such "non-fermionic" substances, as geometrically frustrated magnets, can also be regarded as a kind of strongly correlated fermion systems. Our preceding discussion has shown convincingly that the adopted formalism can describe all experimental details of above quantum frustrated magnets. This is regardless of their microscopic details like kagome of triangular lattice and/or impurities and defects ensemble in each specific substance. Moreover, we have shown that, after scaling procedure, there is one-to-one correspondence between the experimental results obtained in the geometrically frustrated magnets and those for HF compounds. This demonstrates the similarity of the underlying physics of these very different kinds of solids that form the new state of matter. This unifying physical mechanism is indeed fermion-condensation phenomenon.

Combining analytical considerations with arguments based entirely on experimental grounds, we have shown that the data collected on very different strongly correlated many-fermion systems demonstrates a remarkable commonality among them, as expressed in universal scaling behavior of their thermodynamic, transport, and relaxation properties, independently of the great diversity in their individual microstructure and microdynamics, see Chap. 7. The systems considered range from heavy-fermion metals, to quantum liquids including ^{3}He films, to insulating compounds possessing one-, two-, and three-dimensional quantum spin-liquid states, to quasicrystals and beyond. The universal behavior exhibited by this class of systems, generically known as HF systems or compounds, being analogous to that commonality expressed in gaseous, liquid, and solid states of matter, leads us to consider such HF systems as manifestations of the new state of matter arising from the presence of flat bands in their excitation spectra. Such flat bands arise from the formation of the fermion condensate due to FCQPT that transforms the Fermi surface into the Fermi volume, and makes a flat band, see Chaps. 3, 4 and Sect. 4.1.3. Thus, we conclude that the universal features and properties, including the scaling behavior, of the new state of matter are formed with the unique topological FCQPT, see Chap. 7.

We have illustrated our study of the new state of matter reached by the fermion condensation with an exploration of high-T_c superconductors as hosts of the topological FCQPT, see Chap. 23. In fact, we have shown that the underlying physical mechanism responsible for the unusual properties of the overdoped compound $La_{2-x}Sr_xCuO_4$ (LSCO) [2, 3] involves the topological FCQPT that induces fermion condensation. Since the topological FC state violates time-reversal symmetry, the Leggett theorem no longer applies. Instead, we have demonstrated explicitly that the superfluid number density n_s turns out to be small compared to the total number density of electrons. We have also shown that the critical temperature T_c is a linear function of n_s, while $n_s(T) \propto T_c - T$. Additionally, we have demonstrated that at $T > T_c$ the resistivity

$\rho(T)$ varies linearly with temperature, while for $x > x_c$ it exhibits metallic behavior, $\rho(T) \propto T^2$. Such a behavior of high-T_c superconductors has been predicted in 1997 [4]. Thus, development of a superconductivity formalism adapted to the presence of the fermion condensate captures all the essential physics of overdoped LSCO and successfully explains its most puzzling experimental features, thereby allowing us to close the colossal gap existing between the experiments and Bardeen-Cooper-Schrieffer-like theories. Indeed, these findings are applicable not only to LSCO but also for any overdoped high-temperature superconductor, see Chap. 23.

It turns out that the FC occurs in many compounds, generating the non-Fermi liquid behavior by forming flat bands. In this monograph, we have shown both analytically (within the framework of the fermion-condensation theory) and using arguments based entirely on the experimental grounds that the data collected on extremely diverse HF compounds, such as HF metals, compounds with quantum spin liquid, two-dimensional ^{3}He, and quasicrystals, exhibit the universal scaling behavior, see, e.g., Chaps. 8 and 21. This means that different materials with strongly correlated fermions can unexpectedly have a similar behavior despite their microscopic diversity. Thus, the physics of quantum criticality of different HF compounds is universal and emerges regardless of the underlying microscopic details of the compounds. This identical behavior, taking place at relatively low temperatures and induced by the universal quantum-critical physics, allows us to interpret it as the main characteristic of the new state of matter. Our analysis of strongly correlated systems is in the context of salient experimental results, and our calculations of the non-Fermi liquid physical properties are in good agreement with a broad variety of experimental data. Moreover, the fermion condensate can be considered as a defining cause for the non-Fermi liquid behavior observed in various HF metals, liquids, insulators with quantum spin liquids, and quasicrystals. As we have seen, a large variety of the HF compounds exhibit the universal scaling behavior at their quantum criticality. Thus, whichever mechanism drives the system to FCQPT, the system demonstrates the universal behavior. There are lots of such mechanisms or tuning parameters like pressure, number density, magnetic field, chemical doping, frustration, etc.

We have described the effect of FCQPT on the properties of various Fermi systems and presented substantial evidence in favor of the existence of such a transition. We have demonstrated that FCQPT supporting the extended quasiparticle paradigm forms strongly correlated Fermi systems with their unique non-Fermi liquid behavior. Vast body of experimental facts gathered in studies of various materials, such as high-T_c superconductors, HF metals, and correlated 2D Fermi liquids like thin films of ^{3}He, can be explained by the fermion-condensation theory. Description in terms of quasiparticles guarantees that the Kadowski-Woods relation is preserved and that after magnetic field is applied, the Landau Fermi liquid behavior is restored, see, e.g., Chap. 1.

We have found that the differential conductivity between a metal point contact and a HF compound or a high-T_c superconductor can be highly asymmetric as a function of the applied voltage. This asymmetry is observed when a strongly correlated metal is in its normal or superconducting state. We have shown that the application of magnetic field restoring the Landau Fermi liquid behavior suppresses the asymmetry, as it has been predicted for the first time in 2005 [5–7]. We conclude, therefore, that both the time invariance and the charge symmetry are macroscopically broken in the absence of applied magnetic fields, while the application of magnetic fields restores both the Landau Fermi liquid state and the symmetries, see Chaps. 18 and 19. The above features determine the universal behavior of strongly correlated Fermi systems forming the new state of matter.

We have shown how the fermion condensation paves the road for quasi-classical physics in HF compounds. This means simply that systems with FC admit partly the quasi-classical description of their thermodynamic and transport properties. This description permits to gain more insights in the low-temperature non-Fermi liquid physics. The quasi-classical physics starts to be applicable near the topological FCQPT, at which FC generates flat bands and quantum criticality, and makes the density of electron states in strongly correlated metals divergent. Due to the formation of flat bands HF compounds exhibit the classical properties of conventional metal ones like copper, silver, aluminum, etc. In that case, HF compounds demonstrate the quasi-classical behavior at low temperatures rather than elemental metals exhibit a quantum criticality. We have explained experimental observations that the scattering rate of charge carriers collected on high-T_c superconductors, graphene, heavy fermion, and conventional metals exhibits the universal behavior generated by the quasi-classical properties of strongly correlated metals, see Chaps. 17 and 22. Similar universal behavior of the scattering rate in microscopically different strongly correlated compounds like HTSC, HF, and conventional metals suggests that some general theory is able to explain the behavior of these compounds in the uniform manner. We demonstrate that the FC theory is a quite reliable candidate, explaining the universal behavior of the scattering rate.

The FC state forms the topologically protected new state of matter. In the case of Bose system, the equation $\delta E/\delta n(p) = \mu$ describes the ground state. In the case of Fermi systems such an equation, generally speaking, is incorrect. Thus, it is the FC state, taking place behind FCQPT, that makes this equation applicable for Fermi systems. As a result, Fermi quasiparticles can behave as Bose one, occupying the same energy level $\varepsilon = \mu$. This state is viewed as the state possessing the supersymmetry (SUSY) that interchanges bosons and fermions eliminating the difference between them. We have seen that SUSY emerges naturally in condensed matter systems known as HF compounds. The FC state accompanied by SUSY violates the time invariance symmetry, while emerging SUSY violates the baryon symmetry of the Universe, see Chaps. 4 and 18. Thus, restoring one important symmetry, the FC state violates the other. In addition to the already known materials whose properties not only provide information on the existence of FC but also almost shout aloud for such a condensate, there are other materials of impressive interest which could serve as possible objects for studying the phase transition in question. Among such objects are neutron stars, atomic clusters and fullerenes, ultracold gases in traps, nuclei, and quark plasma. Another possible area of research is related to the structure of the nucleon, in which the entire "sea" of non-valence quarks may be in FC state. The combination of quarks and gluons that hold them together is especially interesting because gluons, quite possibly, can be in the gluon-condensate phase, which could be qualitatively similar to the pion condensate proposed by Migdal long ago [8]. We believe that FC can be observed in traps, where there is a possibility to control the emergence of a quantum phase transition accompanied by the formation of FCQPT by altering the particle number density.

Finally, our general discussion shows that the FC theory, supporting the extended quasiparticle paradigm, develops unexpectedly simple, yet good qualitative as well as quantitative description of the non-Fermi liquid behavior of various strongly correlated Fermi systems. Moreover, the FC phenomenon can be considered as the universal reason for both the non-Fermi liquid behavior and the quantum criticality observed in various heavy-fermion compounds, such as HF metals, liquids like 2D ^{3}He, quantum spin liquids, strongly correlated insulators, quasicrystals, and other Fermi systems. This observation permits us to conclude that the topological FCQPT forms flat bands, new quasiparticles, new state of Fermi liquids, and new state of matter in heavy-fermion compounds, and presents an almost unlimited area of research [1, 9–14].

References

1. V.A. Khodel, V.R. Shaginyan, JETP Lett. **51**, 553 (1990)
2. J.I. Bo̊zović, X. He, J. Wu, A.T. Bollinger, Nature **536**, 309 (2016)
3. J. Zaanen, Nature **536**, 282 (2016)
4. J. Dukelsky, V. Khodel, P. Schuck, V. Shaginyan, Z. Phys. **102**, 245 (1997)
5. V.R. Shaginyan, JETP Lett. **81**, 222 (2005)
6. V.R. Shaginyan, K.G. Popov, Phys. Lett. A **361**, 406 (2007)
7. V.R. Shaginyan, K.G. Popov, V.A. Stephanovich, E.V. Kirichenko, J. Alloy. Compd. **442**, 29 (2007)
8. A.B. Migdal, *Theory of Finite Fermi Systems and Applications to Atomic Nuclei* (Wiley, New York, 1967)
9. G.E. Volovik, JETP Lett. **53**, 222 (1991)
10. V.R. Shaginyan, M.Y. Amusia, A.Z. Msezane, K.G. Popov, Phys. Rep. **492**, 31 (2010). arXiv:1006.2658
11. M.Y. Amusia, V.R. Shaginyan, Contrib. Plasma Phys. **53**, 721 (2013)
12. M.Y. Amusia, K.G. Popov, V.R. Shaginyan, W.A. Stephanowich, *Theory of Heavy-Fermion Compounds*. Springer Series in Solid-State Sciences, vol. 182 (Springer, Berlin, 2015)
13. V.R. Shaginyan, V.A. Stephanovich, A.Z. Msezane, P. Schuck, J.W. Clark, M.Y. Amusia, G.S. Japaridze, K.G. Popov, E.V. Kirichenko, J. Low Temp. Phys. **189**, 410 (2017)
14. V.R. Shaginyan, A.Z. Msezane, M.Y. Amusia, SciTech Europa Q. **33**, 56 (2019)

Index

M. Amusia and V. Shaginyan, *Strongly Correlated Fermi Systems*, Springer Tracts in Modern Physics 283,
https://doi.org/10.1007/978-3-030-50359-8

G

H

The manufacturer's authorised representative in the EU is Springer
Nature Customer Service Centre GmbH, Europaplatz 3, 69115 Heidelberg,
Germany. If you have any concerns regarding our products, please
contact ProductSafety@springernature.com

Printed and bound by CPI Group (UK) Ltd, Croydon, CR0 4YY
30/09/2026
02225646-0016